DEVELOPMENTAL AND CELL BIOLOGY SERIES
EDITORS
P. W. BARLOW   P. B. GREEN   C. C. WYLIE

THE CONSEQUENCES OF
CHROMOSOME IMBALANCE
Principles, mechanisms, and models

THE CONSEQUENCES OF CHROMOSOME IMBALANCE

PRINCIPLES, MECHANISMS, AND MODELS

CHARLES J. EPSTEIN

Departments of Pediatrics and of Biochemistry and Biophysics
University of California, San Francisco

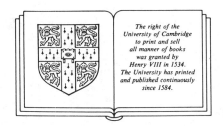

The right of the
University of Cambridge
to print and sell
all manner of books
was granted by
Henry VIII in 1534.
The University has printed
and published continuously
since 1584.

CAMBRIDGE UNIVERSITY PRESS

CAMBRIDGE

LONDON   NEW YORK   NEW ROCHELLE
MELBOURNE   SYDNEY

Published by the Press Syndicate of the University of Cambridge
The Pitt Building, Trumpington Street, Cambridge CB2 1RP
32 East 57th Street, New York, NY 10022, USA
10 Stamford Road, Oakleigh, Melbourne 3166, Australia

First published 1986

Printed in the United States of America

Library of Congress Cataloging in Publication Data
Epstein, Charles J.
The consequences of chromosome imbalance.
(Developmental and cell biology monographs ; 18)
Bibliography: p.
1. Aneuploidy.   2. Human chromosome abnormalities.
3. Down syndrome – Genetic aspects.
I. Title.   II. Series.   [DNLM: 1. Aneuploidy.
2. Chromosome Abnormalites.   3. Down Syndrome –
familial & genetic.   4. Gonadal Dysgenesis – familial &
genetic.   5. Neoplasm – familial & genetic.   QS 677 E64c]
RB155.E77   1986   616′.042   85–12832
ISBN 0 521 25464 7

British Cataloging-in-Publication applied for

To Lois, David, Jonathan, Paul, and Joanna
A New Year's resolution finally fulfilled

Contents

PART III: THE THEORETICAL MECHANISMS AND
            ISSUES: THE PRIMARY AND SECONDARY
            EFFECTS OF ANEUPLOIDY

PART V: THREE MAJOR CLINICAL PROBLEMS OF HUMAN ANEUPLOIDY

Preface

Gene dosage effects have always held a fascination for me that I cannot readily explain. Perhaps it stems in part from the fact that I entered scientific research at a time when the concepts of genetic regulation – induction, repression, feedback control, adaptive enzymes – were so much in the ascendency, and yet there were clear examples in humans and other mammals of situations in which these concepts did not seem to apply. Two papers had a great influence on my thinking at the time, one by Augustinsson and Olssen (1961), concerned, strangely enough, with enzyme activities in pigs, and the other by Allison and Blumberg (1958), dealing with dominantly and recessively inherited disorders in humans.

Despite my intellectual fascination with gene dosage effects and related matters, I did not, aside from a brief letter to the editor (Epstein, 1964), approach the problem seriously until 1965, when, as a "side project," I looked at the relationship between nuclear DNA content and cell volume in polyploid mammalian liver cells (Epstein, 1967; Epstein and Gatens, 1967). I was immediately struck by the beauty and simplicity of the relationship: over a large range of DNA content, cell volume is quite exactly proportional to ploidy. In 1967, I moved from the National Institutes of Health to the University of California, San Francisco, and in the course of this move switched my principal research focus from the genetic control of three-dimensional protein structure to the genetic control of very early mammalian embryonic development. Once again the opportunity to work on gene dosage effects presented itself, this time in the area of X-chromosome expression during oogenesis and preimplantation embryonic development. Although it had not been planned, the fact that this work, the results of which are presented in detail in Chapter 13, also turned out to be relevant to the understanding of a clinical problem (45,X and gonadal dysgenesis) had special significance for me, for reasons to be discussed next.

While working in the laboratory on early embryonic development, I was also very involved in the clinical aspects of medical genetics. This involvement brought me, and still continues to bring me, into continual contact with individuals with a wide variety of genetic problems and other birth defects. Prominent among these individuals were, of course, those with chromosome abnormalities, especially trisomy 21 (Down syndrome), and I often wondered about how the presence of an extra chromosome could lead to the devastating outcomes that we encounter clinically. The same type of thinking also applied to monsomy X (45,X) and its effects. While I thought it would be nice to combine these clinical concerns with my more fundamental research interests, and made a tentative move in this direction with studies on the in vitro rates of proliferation of trisomy 21 fibroblasts (Schneider and Epstein, 1972), it took an unexpected set of coincidences to point the way for work on autosomal aneuploidy. My wife, Lois Barth Epstein, had been among the first to recognize the importance of interferon and was working on problems relating to interferon production and action. Imagine then our surprise and delight when one of the first two genes mapped to human chromosome 21 by Tan, Tischfield, and Ruddle (1973) turned out to be the gene responsible for the cellular response to interferon – what we now know to be the gene (*IFRC*) for the interferon-α/β receptor. Given our interests and a ready access to cell strains and individuals with trisomy 21, an entirely new line of research immediately suggested itself. The results of this work are cited in several places, especially Chapter 12, in this volume. Among the many things of importance that have come out of this work – perhaps the most significant, at least in conceptual terms – is the realization that gene dosage effects are just the beginning of the story. To understand the consequences of aneuploidy, it is necessary to begin with gene dosage effects and to then explore their effects on development and function. Ultimately, it is the latter which really matter.

There is yet one more influence to be mentioned, and that is of Alfred Gropp of Lübeck, whose untimely death in 1983 was a severe loss to all scientists interested in aneuploidy. I met Alfred Gropp in 1974 at a conference in Travemunde, at which time I learned for the first time of his ability to breed mice with any type of whole-chromosome trisomy or monosomy at will. Although my first interest in this system was again concerned with the narrow issue of gene dosage effects, it rapidly broadened to encompass the general area of animal (mouse) models for human chromosomal disorders. And, with the continuing interest and collaboration of Alfred Gropp and the encouragement and assistance from the administrators of the National Institute of Child Health and Human Development, it focused once again on the problem of human trisomy 21, this time in the form of the development of an animal model for this condition. All of this will be dealt with in Chapters 10 to 12.

Having spent over twenty years thinking about and working on problems relating to gene dosage effects and, more generally, to the effects of aneuploidy, and having a sabbatical leave available to me, I felt that this would be a good time to take a look at where the field now stands and where it is going. The writing of this book represents my attempt to do so. I have regarded this task as an intellectual journey, one which has taken me into many areas of clinical genetics and basic biology, some of which I have never glimpsed before. Since it would have been pleasurable to have continued the journey, particularly because of the rapid expansion of relevant literature, it was difficult to know when to stop. However, so that the project could be brought to a conclusion, the survey of the literature was arbitrarily concluded on September 1, 1984, although several additional references were added during the editing of the manuscript.

My overall goals in this volume have been two. The first has been to present a point of view about the effects of aneuploidy – a way of thinking about the problem. In so doing, I have attempted to represent fairly those with whom I might not agree and to identify my own biases for what they are. The second goal has been to bring a sense of coherence to a large mass of clinical and experimental data along with many theoretical considerations. The way in which I have tried to do so reflects, of course, my basic point of view. Nevertheless, the facts still stand for themselves, and their validity is independent of any theoretical construct.

During the writing of this book, there have been many intellectual discoveries. Perhaps the most gratifying of these have been the books and articles that have presented concepts that could be fully appreciated only with hindsight, many years later. Boveri's (1914) theory of the chromosomal basis of malignancy is frequently cited as representing such a view ahead of its time, but two other articles were particularly striking to me – one in which Malcolm Ferguson-Smith (1965) anticipated much of current thinking on X-chromosome aneuploidy and one in which David Comings (1973) did the same for malignancy associated with constitutional aneuploidy. Such discoveries only increase my admiration for the power of the human intellect.

I mentioned earlier some of the individuals who have influenced my thinking along the way, and I would now like to mention a few more: Kurt Benirschke, who introduced me, while I was still a medical student, to the newly emerging field of human cytogenetics; Christian B. Anfinsen, in whose laboratory at the National Institutes of Health I was first exposed to both the intellectual and experimental tools for looking at the relationships between genetics and development; and Arno G. Motulsky, from whom I learned, as a postdoctoral fellow, that even complex problems in clinical genetics can be approached rationally and scientifically.

Throughout my time in research, I have enjoyed the stimulation, collaboration, and assistance of numerous colleagues, fellows, students, and research associates. For much of the work from my own laboratory that is cited throughout this volume, I would like to acknowledge my special debt to my collaborators, David R. Cox, Lois B. Epstein, Terry Magnuson, and Jon Weil, who have devoted much time and effort to the investigation of various problems of aneuploidy and have profoundly influenced my thinking in this area. In addition, I would like to recognize the many research associates who have also acted more in the capacity of collaborators than as technicians in the pursuit of our studies on aneuploidy, in particular Sandra Smith, Joan

Dimpfl, Barbara Hofmeister, Lillian Kwok, Estrella Lamela, Nancy McManus, Bruce Travis, Georgianne Tucker, Della Yee, and Teodosia Zamora. Our work on aneuploidy has been generously supported by the National Institute of Child Health and Human Development, the National Institute of General Medical Sciences, the March of Dimes – Birth Defects Foundation, the American Cancer Society, and the Haas and the Walter Genetic Research Funds.

I wish to thank Jon Weil and David R. Cox for reading and making helpful comments on large sections of the manuscript, Wendy Ovaitt for patiently reducing my execrable scrawl into a finished manuscript on the word processor, Susan Quan for her artistic rendering of the figures, and Norene Parkin for doing so much to facilitate all of our work.

This volume was largely researched and written while I was a Henry J. Kaiser Senior Fellow at the Center for Advanced Study in the Behavioral Sciences, Stanford, California (Gardner Lindzey, director). It is only with the peace and support for sustained intellectual effort that the Center provided that this book could have been completed. And, it is only with the love and encouragement of my wife, Lois, and my children, David, Jonathan, Paul, and Joanna, that the whole project could have been undertaken at all.

Tiburon, June, 1985

Glossary

anal atresia	failure of the anal opening to form
aniridia	absence of the iris of the eye
anophthalmia	absence of the eye
antihelix	the ridge of the external ear that lies anterior to the rim (helix) and behind the concha
arrhinencephaly	absence of the olfactory region of the brain
axial triradial	the point in the palm at which three parallel sets of dermal ridges meet
bicornuate uterus	a uterus which is partially divided into two segments or horns
biparietal diameter	the distance between the parietal bones of the skull (the width of the skull)
blepharoptosis	drooping of the upper eyelid
brachycephaly	a short (in the anteroposterior diameter) head
brachydactyly	short fingers
brachymesophalangia	shortness of the middle phalanges (bones) of the fingers
Brushfield spots	white spots in the iris of the eye (found commonly in Down syndrome)
camptodactyly	flexion contractures of the fingers
caudal hypoplasia	incomplete development of the sacral region
cebocephaly	narrow forehead with flat rudimentary nose and abnormal forebrain
clinodactyly	incurving of the tip of the finger resulting from a wedge-shaped middle phalanx
coloboma	a gap in the iris and/or retina of the eye
concha	the hollow of the external ear into which the ear canal opens
Cornelia de Lange syndrome	a sporadic syndrome characterized by mental retardation, shortness of stature, microcephaly and brachycephaly, bushy

xvii

	eyebrows which meet in the midline, small nose and jaw, and increased hair over the body
craniostenosis	deformity of the skull resulting from craniosynostosis
craniosynostosis	premature closure of the cranial sutures (fibrous joints between the flat bones of the skull)
cri-du-chat	cat cry: a syndrome resulting from del(5p) with a characteristic catlike cry in infancy
cryptorchidism	failure of the testes to enter the scrotum
cubitus valgus	lateral angulation of the forearms at the elbows
cyclopia	a single orbital cavity with or without an eye and with either absence of the nose or a tubular nose above the orbit
digital arches	ridges on the finger tips with an archlike pattern
dolichomesophalangia	long, narrow middle phalanges
dolichocephaly	a long, narrow head
ductus arteriosus	a fetal blood vessel between the aorta and the pulmonary artery that normally closes after birth
edema	fluid within tissues
emphysema	overdistension of the lung tissue
endocardial cushion defect	a defect in the tissue that separates the canal between the atria and ventricles of the heart into right and left and contributes to the formation of the tricuspid and mitral valves
enophthalmos	backward displacement (sunkenness) of the eyes
epicanthal fold	fold of skin over the nasal end of the palpebral tissues
equinovarus	a form of club foot in which the foot points downward and inward (medially)
exencephaly	protrusion of the brain through a defect in the skull
exostoses	bony or cartilaginous growths at the ends of the long bones
falciform folds of retina	curved (sickle-shaped) folds in the retina
fontanelles	the skin-covered soft areas in the infant skull at which bone has not yet formed

foramen ovale	an opening between the atria of the heart that is normally closed after birth
fovea	a small pit at the central point of the back of the retina
gonadal dysgenesis	streaklike ovaries devoid of ova found in 45,X (Turner syndrome)
helix	the prominent rim of the external ear
hemangioma	a benign tumor or collection of dilated blood vessels
holoprosencephaly	failure of cleavage of the forebrain during development, with defective formation of the face in the midline
hydatidiform degeneration/ mole	degeneration and proliferation of the epithelium of the chorionic villi of the placenta to form cysts resembling bunches of grapes
hydronephrosis	distension of the collecting system of the kidney resulting from obstruction of the ureter
hydroureter	distension of the ureter resulting from obstruction
hyperphagia	increased food consumption (usually resulting from increased appetite)
hypertelorism	increased distance between the eyes (as measured between the pupils)
hypotelorism	decreased distance between the eyes
hypotonia	decreased muscular tone
Klinefelter syndrome	the syndrome of testicular atrophy and somatic changes resulting from a 47,XXY chromosome constitution
kyphoscoliosis	a backward (hunchback) and lateral curvature of the spine
laryngomalacia	softness of the cartilage of the larynx
limb reduction	absence of bones in the extremities
lobule	the fleshy part of the ear
lordosis	a forward curvature of the spine (hollowback, swayback)
macula	a yellow depression on the back of the retina in the region that is particularly sensitive to color vision
Marfan syndrome	a dominantly inherited connective tissue disorder characterized by long extremities, dislocated lenses in the eyes, and aortic aneurysm

maxillary hypoplasia	underdevelopment of the maxillary (upper jaw) region of the face
Meckel's diverticulum	an occasional appendage or outpouching of the ileum which is derived from the yolk stalk
metopic suture	the fibrous joint between the right and left halves of the frontal bone
microcephaly	small head
micrognathia	small lower jaw
microgyria	abnormally small malformed convolutions of the brain
micropenis	small penis
microphthalmia	small eyes
microstomia	small mouth
Miller-Dieker syndrome	a syndrome characterized by incomplete brain development, often with a smooth surface, microcephaly, severe mental retardation, and facial anomalies
natal teeth	teeth present at birth
neurofibromatosis	a dominantly inherited disorder with multiple soft-tissue tumors of neural origin, often associated with mental retardation
oblique palpebral fissures	lateral upslanting of the eye slits
occiput	back of the skull
orbital	pertaining to the eye sockets
palpebral fissures	the eye slits (space between the eyelids)
pectus excavatum	a depressed sternum (funnel chest)
philtrum	the vertical groove between the nose and the upper lip
polydactyly	extra fingers or toes
postaxial	on the side of the hand or foot opposite the thumb or great toe
Prader-Willi syndrome	a syndrome characterized by mental retardation, obesity with hyperphagia, hypogonadism (small penis, cryptorchidism), small hands and feet, and a characteristic facies
preauricular sinuses/ tags/pits	defects present anterior to the external ear (toward the face)
proptosis	protrusion of the eyes
pterygium	a web of skin on either side of the neck or at a joint
pyloric stenosis	a narrowing of the outflow of the stomach

radio-ulnar synostosis	a bony fusion of the radius and ulna at the elbow
retinoblastoma	a tumor of the eye
retromicrognathia	a small and receding lower jaw
rocker-bottom feet	feet with prominent curvature of the soles resembling the bottom of a rocking chair
Rubenstein-Taybi syndrome	a sporadic syndrome characterized by mental retardation, short stature, small head, broad thumbs, and a beaked nose
scoliosis	lateral curvature of the spine
simian crease	a single transverse crease crossing the palm
strabismus	squint
syndactyly	fusion of the fingers or toes, involving either soft tissue or bone
synostosis	fusion of adjacent bones, as in the skull
thromboembolism	obstruction of a blood vessel by a clot carried from elsewhere in the circulation
thrombocytopenia	a deficiency of blood platelets
trigonocephaly	a triangular-shaped skull with a sharp ridge over the metopic suture, often associated with arrhinencephaly
triphalangeal thumbs	thumbs with three, rather than two bones (phalanges)
truncus arteriosus	a single blood vessel from the heart receiving blood from both ventricles (a combined aorta and pulmonary artery)
turricephaly	a pointed, tower-shaped skull
uvula	the soft part of the palate
Wilms tumor	a malignant embryonal tumor of the kidney that affects young children

Part I

Introduction

1

The problem of aneuploidy

The human impact of aneuploidy

This book is about a major human problem, one which affects human existence from conception to death. Although sheer numbers cannot tell the whole story, the impact of aneuploidy, in quantitative terms, can be assessed in a variety of ways (Table 1.1) (see also discussions of Golbus, 1981, and of Bond and Chandley, 1983). An exact figure for the frequency of aneuploidy at conception is not available, but it has been estimated to range from as low as 8–10% (Alberman and Creasy, 1977; Kajii, Ohama, and Mikamo, 1978) to as high as 50% (Boué, Boué, and Lazar, 1975; Schlesselman, 1979). The most germane data presently available are those of Martin (quoted in Bond and Chandley, 1983), based on the cytogenetic status of human spermatozoa used for the in vitro fertilization of hamster ova (Rudak, Jacobs, and Yanagimachi, 1978; Martin *et al.*, 1982). In these studies, 8.6% of the human spermatozoa had chromosome anomalies, 5.6% of which were frank aneuploidy. If it is assumed that the frequency of such abnormalities in human oocytes is of a similar or perhaps even greater degree, then the incidence of aneuploidy at conception might be as high as 15–30%. That such a figure is not unreasonable is suggested by the observation that 33% of conceptions prospectively identified by detection of human chorionic gonadotropin in the urine do not survive beyond implantation (Miller *et al.*, 1980).

It is likely that a specific estimate for the frequency of chromosomal abnormality will eventually come from work on human in vitro fertilization, but reliable data are not yet available (Edwards, 1983). However, whatever the figure is, it seems clear that a significant number of aneuploid embryos die in the periimplantation period, including virtually all embryos with autosomal monosomy (Epstein and Travis, 1979), so that by three to five weeks of gestation (after ovulation), 9.3% of therapeutically aborted fetuses are chromosomally abnormal (Yamamoto and Watanabe, 1979). This figure decreases dramatically during pregnancy as more and more such abnormal fetuses are spontaneously aborted or miscarried. Chromosome abnormality thus constitutes the principal known cause of first- and second-trimester abortions, reaching proportions as high as 50–60% between 8 and 20 weeks of gestation (after the last menstrual period) (Warburton *et al.*, 1980). Of the aborted fe-

3

Table 1.1. *Frequency of chromosome abnormalities at various stages of development*

Stage	Nature of data	Derivation of data	Frequency	References
At conception	Chromosome abnormalities in spermatozoa[b]	Observation	8.6%[c]	Martin, in Bond and Chandley (1983, p. 6)
During gestation[a]				
3–4 wks	Chromosome abnormalities in induced abortions	Observation	9.3	Yamamoto and Watanabe (1979)
3–10 wks		Observation	5.0[d]	Yamamoto et al. (1981)
5 wks	Living fetuses with chromosomal abnormalities	Estimation	4.7	Hook (1981)
8 wks			3.8	
12 wks			2.1	
16 wks			0.8	
28 wks			0.33	
Birth	Chromosomally abnormal newborns	Observation	0.37[e]	Higurashi et al. (1979)
			0.31[f]	Bond and Chandley (1983)
			0.27[g]	Hook (1981)
Childhood and adult life	Institutionalized individuals with moderate to severe mental retardation	Observation	10.9[h]	Jacobs et al. (1978)
			15.3	Sutherland et al. (1976)
			32.6	Gripenberg et al. (1980)

[a]Timed from estimated time of conception.
[b]As judged from in vitro fertilization of hamster ova by human spermatozoa.
[c]For all chromosome aberrations; for aneuploidy only, 5.2%.
[d]Adjusted for maternal age.
[e]All chromosome abnormalities.
[f]Aneuploidy only; mosaics excluded.
[g]Clinically significant aneuploidy.
[h]These figures represent the fraction of mentally retarded individuals who have chromosome abnormalities.

tuses with chromosome abnormalities, half have autosomal trisomy, nearly a quarter are polyploid, and a fifth are 45,X (data summarized in Bond and Chandley, 1983).

By 28 weeks of gestation, about 0.33% of fetuses are chromosomally abnormal, and this proportion is about the same at birth. Of the newborns with chromosome abnormalities, nearly three-fifths have sex chromosome anomalies and the remainder have autosomal aneuploidy, overwhelmingly trisomy 21 (data summarized in Bond and Chandley, 1983). The prevalence of cytogenetically abnormal individuals during childhood and the adult years has not been established. Although sex chromosome aneuploidy does not have a high mortality rate associated with it, autosomal trisomies other than trisomy 21 do, and trisomy 21 itself has a mortality of about 50% from birth to age 30 to 40 (Thase, 1982). (The last figure is difficult to estimate because of the decreasing mortality rate, especially in the early years, of individuals with trisomy 21 as medical care and social attention improve.) Therefore, the incidence of all forms of aneuploidy can be roughly estimated as being about 0.25% at age 40, and of autosomal aneuploidy about 0.06%. Although these proportions may seem low, their impact assumes much greater significance when considered in terms of the total number of individuals in the population.

Another way to evaluate the impact of aneuploidy on the human population is to consider its contribution to the causation of mental retardation. Although the estimate arrived at is, of course, influenced by the choice of individuals to be studied, it is clear that a significant proportion of individuals with moderate to severe mental retardation do have chromosome abnormalities (Table 1.1). And, regardless of one's opinions about the nature of the association, it is also clear that there is an increased frequency of sex chromosome aneuploidy (XYY, XXY) among individuals hospitalized or imprisoned for certain forms of behavioral deviance (Hamerton, 1976).

Finally, we may consider the economic burden of aneuploidy. In this regard, trisomy 21 has the greatest impact because of both its frequency and compatibility with long survival. On the average, the expected length of life of a newborn with Down syndrome is close to 36 years (Jones, 1979), resulting in a loss of life calculated to be 53.6 years per 1000 livebirths. In addition, considerable costs are incurred during the period that the affected individuals are alive. A recent analysis placed these costs at approximately $196,000 (in 1980 U. S. dollars), of which $27,000 was for medical costs, $55,000 for education, and $114,000 for residential care (Sadovnick and Baird, 1981). If we add to these figures the costs for long-term medical attention to individuals with some of the sex chromosome abnormalities and the costs for shorter-term management of infants with the more severe autosomal abnormalities, the heavy costs to both individuals and society can be appreciated.

This brief survey indicates that, however it is estimated, the impact of aneuploidy on the affairs of mankind is significant. It probably constitutes a major cause of early postconceptual pregnancy loss and is certainly an important cause of spontaneous abortion. It is a major contributor to infant and childhood morbidity, including both mental retardation and physical abnormalities, and also to morbidity in the adult population. For virtually all forms of aneuploidy, the social, financial, and psychological impact of the condition on the families of affected individuals, on the individual themselves (if they are aware of their condition), and on society at large is profound.

While the discussion so far has been concerned only with constitutional aneuploidy – chromosome imbalance present at birth – it must also be pointed out that acquired aneuploidy is a prominent feature of malignancy, appearing sooner or later in the evolution of virtually every tumor (Sandberg, 1980). Therefore, if we broaden our consideration of the impact of aneuploidy to include malignancy also, it is clear that chromosome imbalance does indeed affect humans from the point of conception to the time of death.

Causation

The "problem" of aneuploidy can be divided into two essentially unrelated parts: causation and effects. The first, that of causation, is concerned with why and how aneuploidy occurs and has received most of the attention in nonclinical discussions of aneuploidy [see, for example, the monograph by Bond and Chandley (1983) entitled simply *Aneuploidy*, and the monograph by de la Cruz and Gerald (1981)]. That this should be the case is understandable, since the premise that prevention is preferable to cure, especially when no cure is in sight, would dictate considerable attention to trying to understand how aneuploidy arises and how it might be prevented. Nevertheless, the prevention of the birth of aneuploid individuals by means other than universal prenatal diagnosis and selective abortion is still only a dream. Therefore, if any practical justification is required for attention to the second part of the aneuploidy problem, that of effects, it is that for the foreseeable future we shall have to be concerned with aneuploid individuals and their problems. To be able to ameliorate their situation in more than a symptomatic manner will require an understanding of the mechanisms by which their difficulties arise. Only by a knowledge of such mechanisms will it be possible to develop strategies that are based on more than empirical and often groundless assumptions.

Approach to understanding the mechanisms of the deleterious effects of aneuploidy

This book will be concerned with issues of mechanism: how does aneuploidy interfere with normal development and function? My intent is not in cataloging the phenotypic abnormalities resulting from chromosome imbalance but, rather, in understanding why and how a particular state of chromosome imbalance produces a specific set of abnormalities. A trivial answer to this question is that the genomes, which we regard as balanced, of humans and other species have evolved so as to optimize the probability of normal development. Therefore, it is quite understandable that any disturbances of its genetic balance should adversely affect the functions which a genome controls. I would not argue against the validity of such a notion, but would only point out that it is more a truism than a statement of cause and effect. Further, while it might say something about *why* imbalance is deleterious, it still does not address the *how*, and this is the matter of principal concern here.

My approach to the analysis of the effects of aneuploidy has a quite straightforward conceptual basis, which is that it will ultimately be possible to deduce the phenotype from the genotype. While not denying a role for environmental and stochastic factors in the final determination of the phenotype, my fundamental premise is that it should be possible to reduce the complex phenotypic effects of an aneuploid state to separable elements which can be attributed to the imbalance of a specific genetic locus or sets of loci. Therefore, the connection between imbalance of a particular chromosome or chromosome segment and the consequent developmental and functional abnormalities will ultimately be explicable in terms of the conventional principles of gene expression, embryology and development, cell biology, and metabolism. Of course, there are many principles which still have not been worked out and can be appreciated only in the most general terms. Nevertheless, it will be these principles and not some group of mysterious effects of chromosome imbalance that will finally provide the sought-after explanations.

This approach to understanding the effects of aneuploidy, which is essentially a reductionist one, is heavily influenced by work on the genetic determination of the three-dimensional (tertiary) structure of protein molecules in which I participated over twenty years ago (Epstein, Goldberger, and Anfinsen, 1963). The essence of the latter, for which Anfinsen received the Nobel prize, was the demonstration that the folding and final three-dimensional structure of a protein molecule is determined solely by the sequence of amino acids which constitutes its primary structure, taking into account, of course, the molecular environment (the milieu) in which the folding occurs. While the precise mechanisms of folding and the rules for translating sequence into three-dimensional structure (and thereby permitting predictions to be made about the effects of amino acid substitutions on structure) are still being worked out, the general validity of the overall principle, which was severely attacked when first enunciated, is firmly accepted today. If we substitute "genetic locus" for "amino acid" and "phenotype" for "tertiary structure," the basis of my conceptual approach to the problem of understanding the effects of aneuploidy becomes clear. However, lest it be argued that this view of the mechanisms of the effects of aneuploidy suffers from the same disabilities as the reductionist or biological determinist approach to intelligence and behavior so roundly excoriated by Lewontin, Rose, and Kamin (1984), I would only echo these authors' own sentiments: "We are in no doubt that, were the processes of development sufficiently well understood, and given a sufficient amount of detailed information about the genotype of an organism, we could predict the phenotype in any given environment." This is just what I am striving to achieve with regard to understanding the consequences of chromosome imbalance.

Earlier in this chapter, I suggested a practical reason for being interested in the mechanistic basis of the effects of aneuploidy. I would now like to suggest two more reasons. The first is that it is an interesting problem in its own right, one which is concerned with a prevalent biological phenomenon. How can one look at a person with a chromosome abnormality syndrome and not wonder how the genetic imbalance produces the phenotype associated with it? The second reason is that the study of aneuploidy may ultimately tell us much about the processes of normal development and function. Just as the study of the inborn errors of metabolism has provided crucial insight into the normal aspects of intermediary metabolism, the study of genetic abnormalities that affect development will provide similar insights into the normal processes of organogenesis and morphogenesis (Epstein, 1978). Furthermore, any mechanisms proposed for processes based on the quantitative aspects of gene expression will become testable in the light of the quantitative perturbations introduced by the aneuploid state. This will occur in much the same way that quantitative differences introduced by heterozygosity for mutant alleles test metabolic relationships and that both heterozygosity and homozygosity for qualitatively aberrant loci test the relationships between the loci and their proposed mechanisms of action (Epstein, 1977). Aneuploidy thus becomes one more class of experiments of nature with which to probe normal development and function.

Organization of the book

The remainder of this book is divided into five parts. In deciding on the organization of the initial part of this volume, I had two choices. The first was to start with a discussion of the critical issues and then to proceed to an analysis of the clinical data relevant to these issues. The other was to do it the other way round, and for the reasons advanced at the beginning of the next chapter, this is what I elected to do. Therefore, this book will begin (in Part II) with an exploration of several aspects of human aneuploid phenotypes, the aim being to determine what general principles about the phenotypic effects of aneuploidy might be inferred from the clinical evidence. This analysis will be restricted to autosomal aneuploidy in order to avoid the complications which might be introduced by the role of X-chromosome inactivation in the determination of the phenotype. The latter will be considered later in the book.

Following the examination of clinical phenotypes, I shall use both this information and other relevant material as the basis for a broad theoretical consideration of the mechanisms which may be involved in generation of the phenotypic effects of aneuploidy (Part III). In addition to the human clinical data, material will be drawn from reports of work involving a variety of different biological approaches carried out on humans and several other organisms ranging from bacteria to mice. The next part (Part IV) of this volume will be concerned with experimental approaches to studying the effects of aneuploidy. It will explore the in vitro and in vivo systems and methods that are now being used and may in the future be developed to investigate the mechanisms whereby aneuploidy produces its effects. Particular attention will be devoted to the development of model systems for studying human aneuploidy.

The next to last part (Part V) will consist of detailed discussions of three situations which might be considered as prototypic problems in aneuploidy. The first, trisomy 21 (Down syndrome), is of course the most frequent of the human autosomal aneuploidies compatible with viability and among the most important causes of moderate to severe mental retardation. The second is X-chromosome monosomy (Turner syndrome, gonadal dysgenesis), the best-studied example of X-chromosome aneuploidy. It has been included to provide a basis for discussing certain aspects of the control of X-chromosome expression which are relevant to an understanding of the effects of X-chromosome aneuploidy. The third situation is not an aneuploid phenotype or syndrome per se but is the intriguing association of aneuploidy and malignancy. The prevalence and specificity (or at least nonrandomness) of this relationship suggest that aneuploidy plays an important role in the establishment or maintenance of the malignant state, and that once again an understanding of the mechanisms by which it produces its effects may be relevant to the prevention or control of these effects.

This volume will conclude (Part VI) with a recapitulation and synthesis of the principal facts and inferences derived from the analyses of the clinical and experimental data and from consideration of the theoretical implications of aneuploidy.

A few definitions are in order at this point. The term "aneuploidy" is used to refer to any state of chromosome imbalance other than polyploidy, irrespective of whether changes in chromosome number are involved. "Trisomy" and "monosomy" are used to refer, respectively, to the addition and loss of a chromosome which, in turn, changes the total number of chromosomes. The addition or loss of a chromosome segment is termed, respectively, a "duplication" or "deletion," although it is sometimes difficult to avoid using the terms "partial trisomy" or "partial monosomy." Finally, in referring to the number of copies of a locus present when there is a trisomy or duplication, the term "triplex" is used. With a monosomy or deletion, the term is "uniplex."

To assist in following the cytogenetic descriptions, the standard human and mouse karyotypes and human cytogenetic nomenclature are presented in the Appendix. A Glossary of clinical terms is also included to facilitate the understanding of clinical syndromes by readers unfamiliar with medical terminology.

Part II

Clinical observations

2

The differentiability and variability of phenotypes

From a purely intellectual point of view, it would have been most satisfying to have first generated a broad theoretical approach to understanding the effects of aneuploidy and then to have matched the theory against clinical and experimental observations. These observations would thus become a test of the predictive qualities of the theory. However, biological hypotheses and theories are rarely formulated in a vacuum, without a background of observations, and it would be misleading to suggest that such is the case for the issues under discussion here. Therefore, I have chosen to begin the body of this work with a survey of certain aspects of clinical cytogenetics. This survey is not intended to be comprehensive or exhaustive, either in considering and discussing all clinical syndromes associated with human chromosome abnormalities, in analyzing every aspect of those syndromes that are considered, or in considering and listing every reported case. Rather, the intent is to determine what general principles, if any, can be deduced from an examination of a wide range of representative clinical data and thereby to learn what specific issues should be addressed by a theoretical formulation of the effects of aneuploidy. Full descriptions of the phenotypes of individual clinical syndromes are contained in Grouchy and Turleau (1984) and Schinzel (1983).

In virtually all instances, the descriptions of syndromes are, unfortunately, couched almost entirely in morphological terms. Only occasionally is there reference to biochemical or cytological changes such as the increased number of granulocyte nuclear projections and elevated concentration of fetal hemoglobin associated with trisomy 13 or dup (13q). I use the term "unfortunately" since the assessment and reporting of morphological abnormalities are often quite subjective, thereby making the data sometimes difficult to interpret. Moreover, even when subjectivity is not a problem, the complexity and interactions of various morphogenetic processes make simple and straightforward interpretations difficult. There are obviously many steps between observable aberrant morphological features and the genetic imbalances which cause them, and opportunities for obscuring the relationship between the two exist at every step. It is often unclear as to just what went wrong during development to cause the features which we recognize as unusual or abnormal. Nevertheless, it is necessary for us to work with what we have, and for this reason the following discussion deals principally with morphological fea-

Table 2.1. *"Discriminating" phenotypic features of chromosomal syndromes*

Chromosome imbalance	Phenotypic features
dup(1q21→qter)	Thymic aplasia
dup(1q25→qter)	Absent gallbladder
dup(3p21.31→pter)	Hypoplastic penis
del(4p15.32)	Marked delay in ossification of carpus, tarsus, pelvis
del(5p15)	Laryngomalacia, premature graying of hair
trisomy 8	Absent patella
del(11p13)	Aniridia-Wilms tumor
dup(11q23→qter)	Hypoplastic penis
dup(13q11→q13)	Increased polymorphonuclear nuclear projections
dup(13q14)	Persistence of fetal hemoglobin
del(13q14)	Retinoblastoma
dup(13q14)	Holoprosencephaly
del(13q31→q32)	Hypoplasia or aplasia of thumb
del(13q31→q32)	Synostosis of 4th and 5th metacarpals
del(13q14→qter)	Anal stenosis
dup(13q31→q34)	Polydactyly
del(13q)	Orbital hypotelorism
del(18q21→qter)	Orbital hypotelorism
del(18q21→qter)	Hypoplasia of 1st metacarpal
dup(21q22.3)	Orbital hypotelorism
dup(22pter→q23)	Anal stenosis
del(Xp)	Hypoplasia of 4th and 5th metacarpals
del(Xp11)	Ovarian aplasia
del(Xp27→qter)	Ovarian hypoplasia
del(Yp)	Testicular aplasia

From Yunis and Lewandowski (1983).

tures. Of course, other nonmorphological characteristics, such as mental retardation and growth retardation, also suffer from the same difficulties.

Differentiability

Even the briefest exposure to clinical cytogenetics makes it clear that the clinical syndromes resulting from chromosome abnormalities are composed of phenotypic features which are rarely unique. Although there are certainly phenotypic changes or abnormalities that are relatively characteristic for certain aneuploid states and as such serve to point to the possibility of their presence, the chromosome abnormality syndromes usually consist of a collection of phenotypic features which tend to be found over and over. Tables illustrating both points have recently been published and are reproduced here as Tables 2.1 and 2.2. The features listed in Table 2.1 (from Yunis and Lewandowski, 1983) have been referred to as "discriminating phenotypes" in that, while not necessarily entirely unique (for example, hypotelorism and anal stenosis), they do significantly limit the range of chromosomal abnormalities which need to be considered. In view of how many different syndromes there are, the list is amazingly brief, containing by the end of 1982 only 25 entries. While this list is small, the number of possible so-called nonspecific features is not. In approaching the problem syndrome delineation by computerized methods based on the concepts of numerical taxonomy, Preus and Aymé (1983) developed a list of 178 phenotypic descriptors, the vast majority of which refer to morphological characters. Even if a considerably shorter list of features is considered, as in Table 2.2 (Lewandowski and Yunis, 1977), it is clear that these features appear repeatedly in different syndromes.

However, even though the features found in the chromosomal syndromes may themselves be nonspecific or undiscriminating, the syndromes are not. The ability of good clinical dysmorphologists to diagnose chromosomal syndromes entirely on the basis of their clinical features is well known. Thus it is not surprising that comparison of the features of the various syndromes listed in Table 2.2 shows that no two of these are identical. While this type of comparative analysis might be regarded as quite informal, the formal analysis by Preus and Aymé (1983) leads to the same conclusion. These investigators used statistical methods to compare the features of four different syndromes: those caused by dup(4p), dup(9p), del(4p), and del(9p). Using the technique of cluster analysis, they showed that the patients with each of these syndromes formed disinct and separable phenotypic groups. Each of the four groups was statistically "equidistant" from each of the others. These groups were so tight that outliers could be suspected and shown to be cytogenetically different from other members of the group (Fig. 2.1). Furthermore, even an apparently single cluster could, by further analysis, be subdivided into two different clusters, and it was therefore possible to discriminate phenotypically patients with dup(9pter→q12) from those with dup(9pter→q22 or 32) (Preus and Aymé, 1983; Preus *et al.*, 1984) (Fig. 2.2). On the basis of their analyses, Preus and Aymé (1983) concluded that "although there are few if any physical features which are exclusive to a particular chromosomal defect, the *pattern* of defects is distinct." As will be expanded on later in this volume, I regard this principle as critical to an understanding of the effects of aneuploidy.

Variability

Although the individual chromosomal syndromes are distinct from one another when groups of patients are compared, there is nonetheless still considerable variability in the expression of the phenotypic features when different individuals with the same cytogenetic lesion are compared with one another. Thus, even in a clinical syndrome as well defined as that associated with

Table 2.2. *Nonspecific phenotypic features of chromosome abnormality syndromes*

	dup 1q	dup 2q	dup 3p	del 4p	dup 4p	dup 4q	del 5p	dup 5p	dup 7q	dup 8	dup 9	dup 9p	del 9p	dup 9q	dup 10p	dup 10q	dup 11p	dup 11q
Mental retardation	+	+	+	+	+	+	+	+	+	+	+	+	+	+	+	+	+	+
Growth retardation	+		+	+	+	+	+	+	+		+	+	+	+	+	+	+	+
Low birth weight				+	+	+	+		+								+	+
Microcephaly		+	+	+	+	+	+				+	+		+		+		
Downward slanting eyes		+		+	+		+				+		+	+		+	+	
Upward slanting eyes						+	+	+					+					
Hypertelorism		+	+	+	+		+	+	+			+				+	+	+
Microphthalmia				+										+		+		
Small palpebral fissures		+		+														
Strabismus		+		+		+	+		+	+			+		+		+	+
Broad nasal bridge				+	+	+	+		+	+	+				+		+	+
Cleft upper lip				+	+										+		+	+
Cleft palate				+		+			+						+			
Micrognathia	+	+		+	+	+	+		+	+	+		+		+	+		+
Low-set ears			+	+	+	+	+	+	+	+		+		+	+	+	+	+
Short neck			+		+					+								
Clinodactyly					+													
Camptodactyly				+			+			+	+		+		+	+		+
Congenital heart disease	+	+	+	+	+					+	+			+	+	+	+	
Cryptorchidism		+		+			+			+	+		+		+		+	
Elevated axial triradius				+											+	+		
Transverse palmar crease				+			+	+	+			+				+		

cont.

Table 2.2 (cont.)

	del 11q	dup 12p	del 12p	dup 13	dup^a 13q^p	dup^a 13q^d	del 13q	dup 14q	dup 15q	dup 18	del 18p	del 18q	dup 20p	dup 21	del 21q	dup 22	del 22
Mental retardation	+	+	+	+	+	+	+	+	+	+	+	+	+	+	+	+	+
Growth retardation	+	+	+	+	+	+	+	+	+	+	+	+		+	+	+	+
Low birth weight	+	+		+							+	+		+	+		
Microcephaly	+		+	+		+	+	+			+	+				+	
Downward slanting eyes			+						+						+	+	
Upward slanting eyes																+	
Hypertelorism	+				+		+	+			+		+	+			
Microphthalmia		+		+			+						+				
Small palpebral fissures							+	+									
Strabismus					+		+	+	+	+	+	+	+	+			
Broad nasal bridge		+		+		+	+								+		
Cleft upper lip				+	+							+					
Cleft palate		+		+				+				+					
Micrognathia	+	+	+	+	+	+		+	+	+	+				+	+	
Low-set ears	+	+		+				+	+	+	+				+	+	
Short neck				+				+	+	+						+	+
Clinodactyly			+						+	+				+			+
Camptodactyly										+				+			+
Congenital heart disease				+			+	+		+		+		+			
Cryptorchidism				+			+	+	+	+		+	+	+	+	+	+
Elevated axial triradius	+	+		+			+			+		+		+	+		
Transverse palmar crease	+	+		+	+		+			+		+		+		+	+

^{a}"p" refers to partial trisomy for the proximal one-third, and "d" to partial trisomy for the distal two-thirds of the long arm of chromosome 13.

Reprinted with permission from Lewandowski and Yunis (1977).

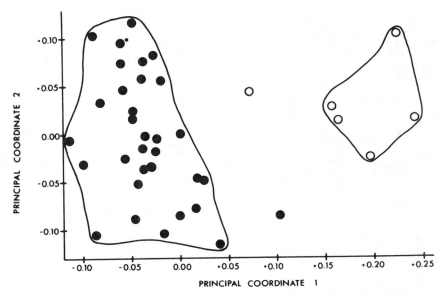

Fig. 2.1. Cluster (principal coordinate) analysis of the phenotypes of patients reported to have dup(9p) (closed circles) and del(9pter→p22) (open circles) based on the analyses of 178 phenotypic characters. The two outliers are considered to be cytogenetically different from the others, with the del(9p) outlier having a shorter deletion and the dup(9p) outlier actually having dup(6p). Figure reprinted by permission from Preus and Aymé (1983), copyright Munksgaard International Publishers Ltd., Copenhagen.

trisomy 13, no single phenotypic feature (other than mental retardation) is present all of the time. Even the cardinal features of polydactyly, microphthalmia, and scalp defects are present only 78%, 88%, and 75% of the time, respectively (see Table 3.9), and a whole host of other abnormalities occur with much lower frequencies. Similar results are obtained when any chromosomal syndrome is examined, and other examples are provided by Table 2.3 and several tables in Chapter 3.

The same is true even within families. If randomly selected series of cases of duplications reported in the literature are compared with regard to the concordance for the positive (abnormal) features which they exhibit, the reported concordances range from 33.3 to 100% (Table 2.3). The mean concordance for all sets of sibs is 71% and is just about the same (74%) if only those cases with fifteen or more reported phenotypic features are compared. Because of the great variability in the reporting of cases and the differences in the precise breakpoints and in the other chromosomes involved among the different cases with the same putative duplication, it is difficult to generate appropriate con-

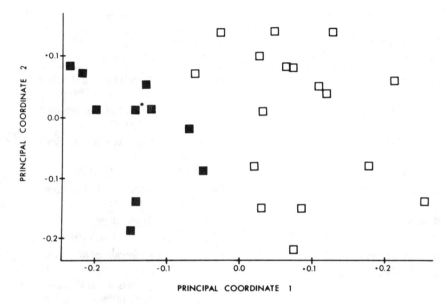

Fig. 2.2. Principal coordinate analysis of the phenotypes of patients with dup (9pter→q22 or 32) (closed squares) and dup(9pter→q12) (open squares), based on the analysis of 178 phenotypic characters (with matches of normal states being excluded from analysis). Figure reprinted by permission from Preus and Aymé (1983), copyright Munksgaard International Publishers Ltd., Copenhagen.

cordance figures for unrelated cases with which to compare the sib data. Subjectively, the affected sibs may appear to resemble each other more closely than they resemble unrelated individuals. For example, Francke (1977), in a discussion of dup(20p) (see Table 2.3), states that "close resemblance between affected members of the same sibship . . . contrasts with the great variability between patients from different families. This could be due to slight differences in the extent of the duplicated segments of 20p as well as to the associated chromosomal deletions involving different regions." Nevertheless, it is quite clear even in the most closely related cases (with the exception of identical twins) that the clinical phenotypes will differ, on the average, with regard to 30% of the features. And, there may be differences even in identical twins. Elder, Ferguson, and Lockhart (1984) reported twins with del (16)(q12.2→q13) in whom one had congenital heart disease and an ectopic anus while the other did not; thirteen other features were present in both.

The sources of this variability in phenotypic expression presumably stem from three sources – genetic, stochastic, and extrinsic – with the last term subsuming all nongenetic influences except for those randomly generated developmental events which are referred to as "stochastic." These matters will

Table 2.3. *Concordance for phenotypic features in siblings with same chromosome abnormality*

Chromosome imbalance	Number of sibs	Number of features	Concordance for positive features[a]	References
dup(2p21→pter)	2	17	100%	Armendares and Salamanca-Gómez (1978)
dup(2p22→pter)	2	17	76	Stoll *et al.*, in Francke (1978)
dup(3p21→pter)	3	12–13	73.5[b]	Rethoré *et al.*, in Say *et al.* (1976)
dup(3p24→pter)	2	9	55.5	Say *et al.* (1976)
dup(3q25→qter)	2	25	56.0	Annerén and Gustavson (1984)
dup(4p12→pter)	2	20	90.9	Reynolds *et al.* (1983)
dup(5q34→qter)	2	12	83.3	Curry *et al.* (1979)
dup(5q31→qter)	2	12	33.3	Rodewald *et al.* (1980)
dup(5q33→qter)	2	8	62.5	Bartsch-Sandhofl and Liersch, in Rodewald *et al.* (1980)
dup(5q31→qter)	2	8	37.5	Osztovics and Kiss, in Rodewald *et al.* (1980)
dup(8q21.2→q22)	3	5–7	85.7[b]	Bowen *et al.* (1983)
dup(9pter→q13)	2	21	71.4	Blank *et al.* (1975)
dup(10p13→pter)	2	23	56.5	Stengel-Rutkowski *et al.* (1977)
dup(10p11→pter)	2	22	50	Cantu *et al.*, in Stengel-Rutkowski *et al.* (1977)
dup(10p11→pter)	2	19	57.9	Schleiermacher *et al.*, in Stengel-Rutkowski *et al.* (1977)
dup(10q24→qter)	2	11	81.2	Fraisse *et al.*, in Klep-de Pater *et al.* (1979)
dup(12p12.1→pter)	2	19	89.5	Hansteen, Schirmer and Hestetun (1978)
dup(15q23→qter)	2	17	76.5	Schnatterly *et al.* (1984)
dup(16p11→pter)	2	9	66.7	Leschot *et al.* (1979)
dup(17q23→qter)	2	19	84.2	Berberich *et al.* (1978)
dup(18q21→qter)	2	8	50	Steele *et al.* (1974)
dup(19q13.3→qter)	2	27	74.1	Schmid (1979)
dup(20p11.2→pter)	3	12	88.9[b]	Centerwall and Francke (1977)
dup(20p11.1→pter)	2	8	87.5	Taylor *et al.*, in Centerwall and Francke (1977)

Table 2.3. *(cont.)*

Chromosome imbalance	Number of sibs	Number of features	Concordance for positive features[a]	References
dup(22q13→qter)	2	19	73.7	Schinzel (1981a)
Mean		15.2	70.5 ± 17.4(SD)	(all)
			73.6 ± 15.2(SD)	(≥15 features compared)

[a]Mental, developmental, and growth retardation not included in comparisons.
[b]Mean of concordances of the three sibs compared two at a time.

be considered later in this volume (Chapters 9 and 12). However, with regard to the present discussion, the important point is that while the individual chromosome abnormality syndromes are, in their overall dimensions, specific and distinct from one another, considerable noise is introduced into their analysis by the significant degree of variability of even those features that best characterize the syndromes clinically.

3

The partitioning of phenotypes

The analysis to this point has permitted us to conclude that characteristic, albeit somewhat variable, syndromes are associated with different chromosome abnormalities. Therefore, clinical data will now be used to determine whether and, if so, to what extent it is possible to recognize the effects of imbalance of specific chromosomal regions in the generation of aneuploid phenotypes. In exploring this issue, I shall make use of two types of information. The first is derived from the examination of double aneuploid states in which one individual is simultaneously aneuploid for two different whole chromosomes or for segments of two different chromosomes. The second is obtained from an analysis of the phenotypic effects of imbalances involving overlapping segments of different lengths from a single chromosome. This analysis will ultimately bring us to a consideration of the issue of phenotypic mapping, i.e., to whether it is possible to assign specific aspects of the phenotype to circumscribed chromosomal regions. In reading the following discussion, the reader must keep in mind the problem of syndrome variability, especially when small numbers of cases are being considered.

Double aneuploidy

Whole chromosomes

Although not qualitatively different from double aneuploidy of chromosome segments, double aneuploidy involving whole chromosomes is easier to examine because the phenotypes of the constituent single aneuploidies are so well characterized. Despite statements in the literature that double aneuploidy is not uncommon, the number of documented cases in which both types of chromosome imbalance are present in the same cells, rather than as mosaics (such as $47, +A/47, +B$ or $48, +A, +B/47, +A/47, +B/46$) or other similar variations which would confuse the situation, and have recognizable phenotypes at the time the patient was studied (not XXY at birth or XXX at any time) turns out to be extremely small. Furthermore, most of the reports of double aneuploidy are from the prebanding period of clinical cytogenetics, so that it is sometimes impossible to identify unequivocably both of the chromosomes present in a karyotype. A list of the best examples of double whole-chromosome aneuploidy that I could find in the literature, given all of these considerations, is presented in Table 3.1.

This list is easily divisible into two parts. The first contains examples of combined sex chromosome and autosomal aneuploidy. In each instance, the full phenotypes of both the autosomal (trisomy 21) and the sex chromosome (47,XXY or 45,X) aneuploidies appear to be expressed. However, when the features of the two are opposed, as are stature in Down syndrome and Klinefelter syndrome, the effect of the former (growth impairment) prevails. Although the 46,X, +21 case involves mosaicism with a 45,X cell line, it is included in the table since the Down syndrome phenotype was clearly expressed and the mosaicism did not involve the sex chromosome constitution.

The difficulties introduced by lack of precise chromosome identification are illustrated by the two entries of 48, +D, +G double aneuploidy (48, +13, +?21; 48, +?D, +21). In one, with phenotypic features of trisomy 13, the identity of the G group chromosome is uncertain. In the other, with features of trisomy 21, it is the D group chromosome that is in doubt. In the putative 48, +18, +21 case of Gagnon *et al.* (1961), the specific identification of the two chromosomes is based as much on the phenotype, which appears to have features of both trisomy syndromes, as on the appearance of the unbanded chromosomes themselves, and the identity of the G-group chromosome is still somewhat in question. This leaves, then, one case of double autosomal aneuploidy documented with chromosome banding: 48, +18, +21 (Grosse and Schwanitz, 1977). In this case, the overall phenotype is that of Down syndrome (with typical facies, short broad hands, simian creases, fifth-finger clinodactyly, deep plantar furrow, hypotonia, and endocardial cushion defect), and the effect of the trisomy 18 appears to be to accentuate the growth and developmental retardation and the microcephaly and to counteract the brachycephaly. This case seems to contradict the suggestion that the phenotype of a double trisomy usually resembles that of the primary trisomy which results in the severest malformations (Hamerton, 1971, p. 316).

Chromosome segments

Examples of double aneuploidy involving chromosome segments are more plentiful that those involving whole chromosomes and are almost always documented by chromosome banding. However, in several instances the phenotype associated with one of the segmental aneuploidies is unknown or the phenotypes of the two aneuploidies do not have clearly distinguishable features, and such cases could not be further considered. Furthermore, since these segmental aneuploidies often result from the meiotic consequences of a balanced translocation or of an inversion, there is a risk of circularity in the reasoning. In many instances, the phenotypes of the individual segmental duplication or deletion syndromes have been derived from the analysis of a

Table 3.1. *Double aneuploidy: whole chromosome trisomies and monosomies*

Karyotype	Age	Phenotype	References
48, XXY, + 21	45 yr	Typical features of Down syndrome and Klinefelter syndrome; height 147 cm	Ford *et al.* (1959*a*); Harnden, Miller, and Penrose (1960)
	10 mo	Typical features of Down syndrome Atrophy of interstitial cells of testes and absence of germinal epithelium	Lanman *et al.* (1960)
	6 yr	Typical features of Down syndrome; height < 3rd percentile; upper segment > lower segment Infantile testes with small tubules devoid of spermatogonia	Hecht *et al.* (1969)
	15 yr	Typical features of Down syndrome and Klinefelter syndrome; height 155 cm	Efinski *et al.* (1974)
46,X, + 21/45,X (86:14)	8 yr	Down syndrome features: brachycephaly, epicanthic folds, upward slant of palpebral fissures, short neck, clinodactyly, 3 ulnar loops and 7 whorls Turner syndrome features: low hairline, slight pterygium, pectus excavatum, wide-spaced nipples, increased carrying angle of forearms Height < 3rd percentile	Townes *et al.* (1975)
48, + 13, + ?21	newborn	Features of trisomy 13: microphthalmia, polydactyly, low-set ears, umbilical hernia, ventricular septal defect No "typical" features of trisomy 21	Gustavson *et al.* (1962)

Table 3.1. *(cont.)*

Karyotype	Age	Phenotype	References
48, + ?D, + 21	4 yr	Typical features of Down syndrome Other features: hypoplastic mandible, proptosis, high-arched palate, mental retardation only "mild"	Becker, Burke, and Albert (1963)
48, + 18, + 21	newborn	Features of trisomy 18: small chin, low-set ears, shield chest with wide-spaced nipples, maxillary hypoplasia, flexed wrist Features of trisomy 21: low nasal bridge, simian creases, flat occiput, small nose Features of both trisomies: congenital heart disease (ASD, VSD) brachydactyly Other features: hypertelorism, hypertrichosis, Meckel's diverticulum, imperforate anus	Gagnon *et al.* (1961)
	13 mo	Typical features of Down syndrome Other features: severe growth (<3rd percentile) and developmental retardation, microcephaly (<3rd percentile), occiput not flat	Grosse and Schwanitz (1977)

number of double aneuploid situations in which imbalance of the segment of interest was the common denominator. Therefore, when possible, I have tried to ascertain the phenotype of the segmental imbalance from descriptions of cases of pure segmental aneuploidy rather than from a collection of such heterogeneous double aneuploid cases. Representative examples of double segmental aneuploidy and the results of their analysis are presented in Table 3.2.

Table 3.2. *Double aneuploidy: duplications and/or deletions*

Duplicated segment	Deleted segment	Phenotype[a]	Number of cases	References
1q32→qter		Some features present	3	Schinzel (1984)
	7q34→qter	No specific features present (cebocephaly or cyclopia with holoprosencephaly, found in all 3 cases, not a common feature of either segmental aneuploidy)		
2p24→pter		Many features present	5	Larson et al. (1982b)
	10q26→qter	Prominant nasal bridge, short philtrum		
2q31→qter		Many features present	1	Wisniewski, Chan, and Higgins (1978)
	18p11→pter	Many features present (cleft palate unexplained)		
5p13→pter		Some features present	2	Liberfarb, Atkins, and Holmes (1980)
	9p22→pter	Some features present		
5q33→qter		Many features present (including high-pitched cry and small jaw)	3	Beemer et al. (1984)
	5p15→pter	Many features present with considerable overlap		
7q32→qter		No specific features present	1	Schinzel and Tönz (1979)
	5p15→pter	Typical cri-du-chat, but more severe		
8p12→pter		Some features present	1	Rethoré et al. (1984)
	5p14	Many features present		
8p22→pter		Some features present	1	Stengel-Rutkowski et al. (1984)
	4p15→pter	Many features present		
9q32→qter		Many features present	1	Mattei et al. (1983)
	9p22→pter	Many features present (prominent chin unexplained)		
10q22→qter		Many features present	2	Juberg et al. (1984)
	12p13→pter	Some features present		
10q24→qter		Typical features present	1	Sills, Buckton, and Raeburn (1976)
	2q31→qter	Microphthalmia		

Table 3.2. *(cont.)*

Duplicated segment	Deleted segment	Phenotype[a]	Number of cases	References
10q24→qter		Many features present which overlap considerably (Large fontanelles, spread metopic sutures, natal teeth, reduction defects unexplained)	2	Pauli *et al.* (1982)
	4p16→pter			
12p11→pter		Many features present	1	Carlin and Norman (1978)
	4p16→pter	Many (but not all) features present		
13p11→qter		Many features present (except microphthalmia and cleft lip)	1	Leisti, Kaback, and Rimoin (1974)
	5pter→?q11	Many features present		
18q11→pter		No specific features present	1	Teyssier and Bajolle (1980)
	18q12→qter	Some features present		
15q15→qter		Some features present	2	Schwanitz *et al.* (1982)
	14q24→qter	No specific features present (hypoplasia of clavicles unexplained)		
18q11→qter		Many features present	1	Bass, Sparkes, and Miller (1979)
	18p11→pter	Many features present (carp-shaped mouth unexplained)		

Duplicated segment	Duplicated segment	Phenotype[a]	Number of cases	References
11q23→qter		Many features present	31	Schinzel *et al.* (1981a)
	22pter→q11 or 12	Many features present (more frequent clefting)		
13q22→qter		Many features present (not polydactyly)	1	Kim *et al.* (1977)
	22pter→q13	Many features present		

[a] Features of segmental aneuploids found in double aneuploids.

Except for two sets of double duplications, both involving chromosome 22, all of the other cases are combined duplications and deficiencies resulting from the segregation of balanced translocations. The effects of double aneuploidy fall into three general categories. The first is one in which the phenotype derives completely or almost completely from that of one of the segmental aneuploidies. An example of this is the case dup(7q32→qter), del(5p15→pter) (Schinzel and Tönz, 1979) which presented as a case of cri-du-chat syndrome. Some unusual features (clenched fingers, defective ossification of the calvarium) cannot be directly attributed to the dup(7q), and conversely, several features of dup(7q31,32→qter), such as receding mandible, cleft palate, and prominent occiput, were not present (although the 7q32 breakpoint is somewhat in doubt). Serving as another example are the cases of dup(2p24→pter), del(10q26→qter), in which the dup(2p) principally determines the phenotype except for the prominent nasal bridge and short philtrum, which are attributable to the del(10q) (Larson *et al.*, 1982*b*). The dup(1q32→qter), del(7q34→qter) described by Schinzel (1984) is also an example, with cebocephaly or cyclopia being a characteristic feature of the combined aneuploidy but not of the individual segmental aneuploidies.

The second category of double segmental aneuploidies is one characterized by the presence of only a few or some features of each of the contributing imbalances, possibly with features not characteristic of either. The two cases of dup(5p13→pter), del(9p22→pter) (Liberfarb, Atkins, and Holmes, 1980) are of this type and are described in Table 3.3 In these patients, several features were apparently attributable to dup(5p) and del(9p). However, many of the features of these two syndromes were not present, and the patients had many abnormalities not directly ascribable to either. In addition, as is frequently the situation, certain common features (for example, upward slanting palpebral fissures, flat or depressed nasal bridge, and ear anomalites) are present in both segmental aneuploids

The third and most interesting group of double segmental aneuploids includes those cases in which the phenotype is a reasonable composite of the individual segmental aneuploid phenotypes, what Carlin and Norman (1978) term a "phenotypic hybrid." Several examples are listed in Table 3.2, and two illustrative cases are summarized in Tables 3.4 and 3.5 In the patient with dup(12p11→pter), del(4p16→ter), many features of both dup(12p) and del(4p) (the Wolf-Hirschhorn syndrome) are present. Although the features of del(4p) appear to predominate, particularly because dup(12p) has so many features in common with it, features characteristic of dup(12p) are also represented – turricephaly, short nose, broad philtrum, and prominent everted lower lip. Moreover, certain aspects of del(4p) are not present, but it must of course be kept in mind that only one individual with this particular double aneuploidy is being considered.

A particularly impressive example of a composite phenotype in double seg-

Table 3.3. *Example of double segmental aneuploidy with some features of each segmental aneuploidy represented*

| | Patients[a] dup(5p13→pter), del(9p22→pter) | | Cases from the literature | |
	Case 1	Case 2	dup(5p13→pter) 4 cases[b]	del(9p22→pter) 6 cases[c]
Prominent forehead	+	+		
Epicanthal folds	+			5/6
Upslanting palpebral fissures	+		3/4	5/6
Broad, flat nasal bridge	+	+	2/4	Depressed in 6/6
Ear anomalies	+	?	3/4	6/6
Cleft palate	+			High arched palate in 6/6
Micrognathia	+	+		6/6
Diaphragmatic hernia	+			
Intestinal obstructions	+			
Kidney abnormality	+			
Heart defect	+			Murmur in 4/6
Long, thin fingers	+	+		6/6
Club foot	+	+		
Hypotonia	+			
Seizures	+			
Psychomotor retardation	+		3/3	
Craniosynostosis	−	+		Trigonencephaly in 6/6
Macrocephaly			4/4	
Hypertelorism	−		2/4	
Anteverted nostrils				6/6
Long philtrum				6/6
Low hairline				4/5
Short neck				6/6
Webbed neck				4/6
Wide-spaced nipples			6/6	

[a]From Liberfarb, Atkins, and Holmes (1980).
[b]From Khodr *et al.* (1982).
[c]From Alfi *et al.* (1976).

mental aneuploidy is the case of dup(13p11→qter), del(5pter→?q11) described by Leisti, Kaback, and Rimoin (1974). This patient, who died at two months of age, had several of the striking features of both the trisomy 13 and cri-du-chat syndromes (Table 3.5): the postaxial polydactyly, hemangioma, hypertonia, overriding fingers, and bicornuate uterus of trisomy 13, and the cat cry,

Table 3.4. *Example of double segmental aneuploidy with a composite phenotype*

	Patient dup(12p11→pter), del(4p16→pter)	From survey of literature	
		dup(12p)	del(4p)
Severe growth retardation	+	−	+
Severe mental retardation	+	mild to moderate	+
Microcephaly at birth	+	−	+
Turricephaly	+	+	Dolichocephaly
High, prominent forehead	+	+	+
Deep forehead wrinkles	+	−	+
Flat face	+	+	±
Glabellar hemangioma	+	−	+
Short nose	+	+	−
Wide nasal bridge	+	+	+
"Greek warrior helmet" appearance	+	−	+
Hypertelorism	+	+	+
Epicanthal folds	+	+	+
Eye abnormalities	Coloboma	−	+
Downslanting palpebral fissures	+	−	±
Philtrum	Broad with strongly marked pillars	Broad	Deep, narrow with strongly marked pillars
Downturned, fish-shaped mouth	+	+	+
Prominent everted lower lip	+	+	−
Retrognathia	+	−	+
Preauricular tag or pit	+	−	+
Low-set ears	+	±	+
Protuberant antihelix	+	+	−
Deep concha	+	+	−
Short neck	+	+	−
Elongated trunk	+	−	+
Sacral or coccygeal dimple	+	−	+
Slender limbs	+	−	+
Proximally placed thumbs	+	+	+
Hypoplasia of ridges	+	−	+
Hypotonia	+	−	+
Median scalp defects	−	−	+
Cleft lip/palate	−	−	+
Seizures	−	−	+
Increased number of arches	−	−	+

From Carlin and Norman (1978).

Table 3.5. *Example of double segmental aneuploidy with a composite phenotype*

	Patient dup(13p11→qter), del(5pter→?q11)	From survey of literature	
		dup(13q)	del(5p)
Cat cry	+	−	+
Hypertonia	+	+	Hypotonia
Microcephaly	+	+	+
Sloping forehead	+	+	
Hemangioma	+	+	
Round face	+		+
Wide-set eyes	+		+
Small chin	+		+
High-arched palate	+	+ often cleft	
Low dysmorphic ears	+	+	low
Postaxial polydactyly	+	+	
Bilateral simian creases	+	+	+
Overriding fingers	+	+	
Bicornuate uterus	+	+	
Patent ductus arteriosus	+	+	
Microphthalmia	−	+	

From Leisti, Kaback, and Rimoin (1974).

round face with wide-set eyes, and small chin of del(5p). Again, not all features of the individual segmental aneuploidies were present in this single case, but, on the other hand, there were few abnormalities that could not be attributed to one of the two segmental aneuploidies. That the latter is not always the case is illustrated by the patients described by Pauli *et al.* (1982) with dup(10q24→qter), del(4p16→pter) (Table 3.2). Included among the abnormalities displayed by these two individuals were large fontanelles with spread metopic sutures, natal teeth, limb reduction defects, and anophthalmia (in one), none of which is characteristic of either dup(10q) or del(4p).

Also included in the composite phenotype category are the double aneuploidies comprised of two duplications: the relatively common dup (11q23→qter), dup(22pter→q11 or 12) (Schinzel *et al.*, 1981*a*) and a single case of dup(13q22→qter), dup(22pter→q13) (Kim *et al.*, 1977). The features of the former are summarized in Table 3.6, and the contributions of the two individual duplications to the phenotype are apparent. However, certain features (cleft palate and/or uvula, micropenis, and cryptorchidism) appear to occur with much higher frequency in the double aneuploid state than would be expected from the single duplications, suggesting that some type of fre-

Table 3.6. *Example of double segmental aneuploidy with a composite phenotype resulting from two duplications*

	dup(11q23→qter), dup(22pter→q11,12) 31 cases	dup(11q23→qter) 7 cases	dup(22pter→ 22q11,12) 7 cases
Severe developmental retardation	17/18	1/3	1/4
Microcephaly	12/15	2/4	1/1
Downslanting palpebral fissures	8/23	0/5	5/5
Short septum, flat tip of nose	16/17	4/5	1/5
Prominent and/or long upper lip	18/20	5/5	4/5
Short mandible (especially at birth)	18/18	5/5	0/4
Cleft palate and/or uvula	22/28	1/7	1/7
Preauricular skin tags/sinuses	12/28	1/7	4/6
Hip dysplasia and/or dislocations	11/18	3/7	1/5
Anal atresia	5/26	0/7	2/7
Congenital heart defect	15/26	4/17	3/7
Micropenis	12/12	2/4	1/2
Cryptorchidism	11/13	0/2	1/3
Hypoplasia of the diaphragm	4/?	?	0/6

From Schinzel *et al.* (1981*a*).

quency reinforcement has taken place. This may also be what is happening in the dup(1q32→qter), del (7q34→qter) cases described earlier (Table 3.2).

Curiously, while there are these examples of increased frequencies of certain abnormalities, there seem to be very few, if any, cases in the many double aneuploids examined of a synergistic effect in terms of the severity of the abnormality. Unlike the situation of traits transmitted in a dominant manner, in which homozygous expression is frequently much more severe than is expression in the heterozygote (Allison & Blumberg, 1958; Pauli, 1983), the combination of two segmental aneuploidies that exhibit certain features in common does not generally lead to an exaggerated expression of those features [see for example the phenotype of dup(12p), del(4p) summarized in Table 3.4]. This suggests either that the final outcome of interfering with a developmental process is the same, irrespective of the point at which the process is affected, or that the effects already represent the maximum perturbations possible in these characteristics. With regard to the latter, consideration of the effects of tetrasomy is of interset, since this state could, for the purposes of the present discussion, be regarded as a double aneuploids of the same chromosome or chromosome segment. It is of course possible that there are instances in which the combined effects of two segmental aneuploidies are so severe as to lead to death prenatally. These would not be recognized or reported.

Combined segmental aneuploidy of single chromosomes

While not double aneuploidy in a formal sense, it is possible to consider aneuploidy of a whole chromosome as the combination of imbalances affecting two or more of its segments. Thus, trisomy for a metacentric chromosome can be regarded as formally analogous to double aneuploidy involving a duplication of each of its arms, or the deletion of part of a chromosome as analogous to double aneuploidy of two smaller contiguous deletions which add up to the larger one. The information from cases such as these can therefore add to our picture of how compartmentalized the aneuploid phenotype actually is. Because of wide variations in the styles of reporting individual cases, data permitting this type of analysis are not easy to obtain. Nevertheless, some interesting examples can be found, and four will be examined here.

The first example is concerned with deletions of 7q, and Gibson, Ellis, and Forsyth (1982) have reviewed several cases of interstitial and terminal deletions. These findings are summarized in Table 3.7. Although the two smaller deletions, 7q32→q34 and 7q35→qter, may not be exactly contiguous and the numbers of cases are quite small, it is apparent, nevertheless, that many of the phenotypic features of the larger deletion can be related to features manifested by the smaller ones. However, it is also obvious that the larger deletion sometimes causes abnormalities, such as cleft lip/palate, micrognathia, and genital anomalies, not described for the smaller ones [micrognathia was described in one case of del(7q35→qter) by Young *et al.* (1984)]. Since none of these apparently new features seems to occur even in a majority of cases, this may be a matter only of the numbers of cases examined. On the other hand, it may constitute a real example of new phenotypic features resulting from the interaction of two segmental chromosomal imbalances, neither of which would produce the feature alone.

A similar comparison can be carried out between the features of putative trisomy 22 (pure or mosaic) and of duplications for the proximal and distal segments of 22q. Selected data taken from the report of Schinzel (1981*b*) are presented in Table 3.8. Again, they are based on a small number of cases, and problems of chromosome identification were present. Nevertheless, even taking this into account, it appears that full trisomy 22 does not produce fea-

Table 3.7. *Segmental deletions of 7q*

	del(7q32→qter) 15 cases	del(7q32→q34) 3 cases	del(7q35→qter) 3 cases
Developmental and mental retardation	14[a]	3	3
Microcephaly	12		3
Prominent forehead/frontal bossing	6		1
Abnormal skull shape	5		1
Abnormal ears	12	2	
Bulbous nasal tip	10	3	2
Large mouth	4–6	3	
Cleft lip ± palate	4		
Micrognathia	5		1
Hypertelorism	5	3	
Ocular anomalies	5	1	1
Shield chest/wide-spaced nipples	5–6		2
Congenital heart disease	2		
Genital abnormalities	9		
Simian creases	6		

From Gibson, Ellis, and Forsyth (1982), with case of Young *et al.* (1984) of del(7q35 → qter).
[a]Number of cases reported as positive. In most instances, negative findings were not recorded.

tures that can be ascribed to duplication of either the "proximal" or "distal" segments of 22q. This conclusion is supported by the brief report of Bendel *et al.* (1982) of two sibs with exactly complementary duplications, 22pter→q13 and 22q13→qter, resulting from a balanced maternal 7;22 translocation. Although they do share certain features in common, such as low-set ears, rocker-bottom feet, and congenital heart disease, additional features of trisomy 22 present with the proximal duplication were microcephaly, cleft lip and palate, and renal malformations; hypertelorism and preauricular sinuses were present with the distal one (Shokeir, 1978; Cantu *et al.*, 1981; Schinzel, 1981*b*). Furthermore, whereas colobomas, anal atresia, and micrognathia were not present, microphthalmia was. Clearly, more cases of trisomy 22 and of segmental duplications of chromosome 22 need to be examined.

Numbers of cases and identification of breakpoints are also somewhat of a problem with the most intensively studied segmental aneuploidies: those involving chromosomes 13 and 18. Several authors (Schinzel, Hayashi, and Schmid, 1976; Niebuhr, 1977; Schwanitz *et al.*, 1978; Rivas *et al.*, 1984; Rogers, 1984) have compared the "proximal" and "distal" segmental dupli-

Table 3.8. *Segmental duplications of chromosome 22*

	Trisomy 22 (mixed pure and mosaic) 5 cases	dup(22pter→~q11) ("proximal")[a]	dup(22~q12→qter) ("distal")
Severe growth retardation	4–5/5	−[b]	+[b]
Hydrocephalus	1/5	−	+
Downslanting palpebral fissures	3/5	+	−
Ocular coloboma	1/5	+	−
Preauricular malformation	4/5	+	−
Congenital heart defect	3/5	+	+
Anomaly of mesentery	2/4	+	?
Renal malformations	3/4	+	+
Anal atresia	2/5	+	−
Genital hypoplasia or malformation	3/5	(+)	+
Distal limb hypoplasia	2–3/4	−	+

[a]See also Table 3.3.
[b]Reported to be present. The subdivision into "proximal" and "distal" duplications is not exact, and there may be some overlap in the region q11-q12.
From Schinzel (1981*b*).

cations of chromosome 13 with one another and with trisomy 13. All comparisons suffer from the fact that the separation into "proximal" and "distal" is somewhat arbitrary with regard to the dividing point, so that there is bound to be overlap at the margin. Therefore, the results obtained may differ, depending on which cases are included in the analysis. This is well illustrated by the data in Table 3.9. Even with these caveats in mind, it is clear that while many features of trisomy 13 are attributable to duplication of the distal segment of chromosome 13, at least some are attributable to the region included in the proximal duplications: the increased numbers of granulocyte nuclear projections, small mouth, receding chin, and scalp defects. As in earlier examples, some features, such as microphthalmia, coloboma, cleft lip/palate, and congenital heart disease, are considerably more severe and/or more frequent in the full trisomy than in the segmental duplications, as is the overall severity of the abnormalities.

In discussing their data (summarized in Table 3.9), Schinzel, Hayashi, and Schmid (1976) commented that several features of trisomy 13, including microphthalmia, cleft lip and palate, hemangiomas, and elevated fetal hemoglobin concentration, were present, although at lower frequencies, in both types

Table 3.9. Segmental duplications of chromosome 13

	Hodes et al. (1978) Trisomy 13 19 cases	Niebuhr (1977) Dup(prox 1/3–1/2) 7 cases	Niebuhr (1977) Dup(dist 1/3–2/3) 17 cases	Schinzel, Hayashi, E Schmid (1976) Dup(prox) 6 cases	Schinzel, Hayashi, E Schmid (1976) Dup(dist) 18 cases	Schwanitz et al. (1978) Dup(13p→q14) 7 cases	Schwanitz et al. (1978) Dup(13q14→qter) 6 cases
Severe retardation	100%(100%)[a]	71%	100%	75%	67%	50%	50%
Microcephaly	86(63)	60	69	17	0	25	
Scalp defects	75(47)			0	50		33
Trigonocephaly	+[b]						
Microphthalmia	88(79)	14	24	17	22	17	33
Coloboma	67(31)	0	12				
Hypertelorism	75	75	46				
Epicanthal folds	83(32)	25	58			25	33
Long, curved eyelashes	+			20	100		67
Stubby nose	+			0	100	25	50
Long philtrum	+			0	100		
Arrhinencephaly	71[c]		100(2/2)				17
Cleft lip and/or palate	63(63)-cleft palate 50(47)-cleft lip	29	29	33	11	50	33
High-arched palate	72(26)			80	0	50	67
Small mouth	+			100	0		17
Receding chin	66(10)			0	89	75	
Malformed ears	87(74)	33	60	20	47		100
Hernias (inguinal or umbilical)	83(26) umbilical			0			50
Polydactyly	78(58)	0	65	0	78		67

Simian crease	64[c]	40	55	17			17
Congenital heart disease	94(84)	33	14		25	25	
Hemangioma	88(37)	16	71	40	69		50
Elevated hemoglobin F	>50[c]	60	50			25	17
Increased number of nuclear projections in granulocytes	≥83[d]	83	20	80	0	75	17

[a] The first number is the percentage of cases in which the feature is specifically mentioned; the number in parenthesis is the percent of total cases (n=19).
[b] Present according to Schinzel, Hayashi, and Schmid (1976).
[c] Quoted in Niebuhr (1977).
[d] From Walzer et al. (1966).

of chromosome 13 duplications. They interpret this to mean that "these features cannot be causally related to the trisomic state of a defined segment of the chromosome, but must be related to a more unspecific type of genetic imbalance." In view of the relatively unusual character of at least some of these abnormalities, I am reluctant to accept their conclusion, their statement that "no common segments seemed to be involved in the cases of the two types of partial trisomy 13" notwithstanding. In looking at the cases included in their analysis, overlaps in the region q13-q14 do appear to be present, and some of these features may well be related to imbalance in this region. Support for this inference is obtained by examining the phenotypes of two pairs of sibs with complementary proximal and distal duplications: the pair described by Noel, Quack, and Rethoré (1976) with the breakpoint at q12, and that of Schinzel, Schmid, and Mürset (1974) with the breakpoint at q14, both summarized by Schütten, Schütten, and Mikkelsen (1978). In the former, increased fetal hemoglobin and cleft lip and palate were associated with distal duplication, whereas in the latter, increased fetal hemoglobin was associated with the proximal duplication, and capillary hemangiomas with the distal one. This seems to place at least the region responsible for increased fetal hemoglobin in the vicinity of 13q14, and cleft lip and palate and hemangiomas in or distal to q12, a conclusion arrived at by Wenger and Steele (1981) in their analysis of duplications of 13q (see below for further discussion of this point). Rogers (1984) places the region responsible for hemangiomas in bands 13q32→qter.

A more limited analysis is also possible for duplications of chromosome 18. Again, although the consequences of full trisomy are more severe, many of the major features are, in the main, attributable to duplications of the proximal or distal end of the chromosome. If the chromosome is divided at 18q11, prominent occiput, congenital heart disease and other visceral anomalies, faunesque ears, and excess digital arches are associated with duplication of the distal segment, and rocker-bottom feet, microstomia, and flexion deformities (overlapping) of the fingers with the proximal duplication (see San Martin *et al.*, 1981, for summary of relevant cases). On the other hand, if 18q21.2 is taken as the dividing point, all features are represented in the proximal segment, suggesting that the major contribution to the phenotype of duplication of the chromosome distal to 18q11 is attributable to 18q11→q12 (San Martin *et al.*, 1981). Turleau *et al.* (1980) place the distal margin at the end of band 18q11.

Inferences

At this point, it is worth considering what can be inferred from this survey of the effects of double aneuploidy, whether involving two different chromosomes or two segments of the same chromosome. It seems clear that there is a wide gamut of possibilities, ranging from situations in which the input from neither individual duplication nor deletion is recognizable, through ones in which the phenotype of only one of the two constituent aneuploidies is expressed, to a phenotype which is a true composite of the phenotypes of the two. In general, it appears that the more flagrant the abnormalities are, the more distinct the features of the two aneuploidies are from one another, and the more these features involve different organ systems and other aspects of morphogenesis, the more likely it will be that a composite phenotype will develop. This situation is consistent with and supportive of the basic notion of the specificity of the effects of aneuploidy, which has already been articulated in this volume.

On the other hand, it must be pointed out that even when these criteria for a composite phenotype are met, the composite phenotype seldom represents the mere sum of the constituent phenotypes. Even in the mixed sex chromosome/autosome double aneuploids, in which both sets of features are readily discernible and in the main fully expressed, the stature is not the mean of that expected from an additive combination of 47,XXY and 47,+21; the effects of the latter on stature appear to prevail. Moreover, in the double autosomal segmental aneuploids, as has already been illustrated, not all features of the individual aneuploidies are represented, and, conversely, features not characteristic of either are expressed. To this extent, then, I would concur in part with the observation of Rehder and Friedrich (1979) that the developmental consequences of two different chromosomal defects may mask the clinical expression of each of them rather than lead to a simple addition of the respective phenotypic characteristics. That such masking might occur or new features should appear is not unexpected, since there is no reason to expect that any developmental event is determined solely by a locus or loci present on a single chromosome or chromosome segment. It is reasonable that an effect caused by imbalance of one region may cancel out the effect resulting from imbalance of another. However, what is most impressive to me is that this expected genetic interaction does not lead to a more frequent obliteration of the constituent phenotypic features and creation of new features. From this I would infer that, at least for those characters that remain unaltered in a double aneuploid (particularly autosomal/autosomal) phenotype, the relevant loci have reasonably strong determinative effects on their development. Conversely, there are instances in which the composite phenotype is a more abnormal one, either in terms of the frequency of the malformations or severity of the overall phenotype with regard to such things as degree of microcephaly or impairment of survival.

Segmental aneuploidies of increasing length

Another approach to elucidating the relationship between aneuploidy and its phenotypic consequences is to examine the effects of unbalancing regions of increasing length from the same chromosome. This permits us to observe

Table 3.10. *Duplications of increasing length involving 5q*

	5q31,33→qter 9 cases	5q34→qter 4 cases
Growth retardation	9/9(severe)	4/4(moderate)
Developmental/mental retardation	9/9(severe)	4/4(mild)
Microcephaly	9/9	1/2(mild)
Low-set dysplastic ears	8/9	1/4
Mild obesity	0/9	3/4
Short, receding forehead	1/7	4/4
Brachydactyly	4/9	2/4
Bulging forehead	2/2	0/4
Carp mouth	5/9	0/4
Microstomia	6.9	0/4

From Rodewald *et al.* (1980).

whether and how phenotypic features of aneuploidy for a particular chromosome segment are altered as that segment becomes included in a larger unbalanced region. It is also one basis for phenotypic mapping, the assignment of individual phenotypic features to circumscribed chromosomal regions. Examples of the approach to phenotypic mapping have already been given in the discussion of proximal and distal duplications of chromosomes 13 and 18. The attempt to localize phenotypic features has both its supporters and detractors, and their arguments, as well as other aspects of the issue, will be considered after some of the available information is reviewed.

I shall begin this part of the analysis by examining two questions: (1) are the phenotypic features of a short aneuploid segment included among those manifested by aneuploidy for a larger segment, and (2) can specific features be identified with aneuploidy of specific chromosome segments? Several series of cases appropriate for consideration have been published, and a few will be considered here. The relevant data are summarized in Tables 3.10 to 3.13. In the terminal duplications of 5q, the longer duplication (5q31,33→qter) has more severe growth retardation, developmental retardation, and microcephaly than the shorter (5q34→qter) duplication, and has abnormalities of the mouth and ears that are not characteristic of the latter (Table 3.10). The contour of the forehead shifts from short and receding to bulging, and the mild obesity is no longer apparent. A phenotypic progression from smaller to larger duplications is apparent in the cases of dup(7q) summarized by Yunis, Ramirez, & Uribe (1980) (Table 3.11). In each instance, the fea-

Table 3.11. *Duplications of increasing length involving 7q*

	q21→qter 2 cases	q31→qter 3 cases	q32→qter 4 cases	q33→qter 1 case
Developmental retardation	1/1	2/2	4/4	1
Low-set ears	1/2	3/3	2/4	1
Small palpebral fissures	1/2	2/2	2/4	0
Skeletal anomalies	1/2	3/3	3/4	0
Retromicrognathia	1/2	3/3	1/4	0
Wide-open fontanelles	1/2	2/2	0/2	0
Hypertelorism	1/2	3/3	0/3	0
Fuzzy hair	1/2	1/2	0/3	0
Ear anomalies	1/2	2/3	0/4	0
Small nose	1/2	2/2	0/3	0
Cleft palate	1/2	3/3	0/4	0

From Yunis, Ramirez, and Uribe (1980).

tures of the smaller duplication are included in the latter, and, up to a point, new abnormalities become manifest as the duplication lengthens. Thus, going from dup(7q32→qter) to dup(7q31→qter) is associated with the appearance of hypertelorism, fuzzy hair, ear anomalies, small nose, cleft palate, and possibly retromicrognathia. The further increase in length of the aneuploid segment to dup(7q21→qter) is not associated with significant changes in the phenotype.

Similar situations obtain for several other sets of duplications. Thus, with duplications of 9q ranging from 9q34→qter to 9q32→qter to 9q31→qter (data summarized by Alleridce *et al.*, 1983), the first transition results in loss of scoliosis but addition of severe congenital heart anomalies and renal abnormalities; the second does not produce any significant changes. A similar progression of features has been noted in duplications of chromosome 9 increasing in size from 9pter→9q21 to 9pter→9q32 (Wilson, Raj, and Baker, 1985). In the 10q25→qter, 10q24→qter, 10q22→qter series, all of the major abnormalities associated with the first duplication persist throughout, but several new abnormalities appear in the second, and possibly an increase in the frequency of one feature (cleft palate) appears in the third (data from Klep-de Pater *et al.*, 1979, and Miró *et al.*, 1980) (Table 3.12). The appearance of new features, with the inclusion of existing ones, also characterizes duplications in the regions 1q23→qter (Michels *et al.*, 1984), 5p13→pter (Khodr *et al.*, 1982), 8q21→qter (Fineman *et al.*, 1979; Bowen *et al.*, 1983), and 14q22→qter (Atkin

Table 3.12. *Duplications of increasing length involving 10q*

	10q22→qter 3 cases	10q24→qter 14 cases	10q25→qter 6 cases
Growth retardation	2/2	11/12	5/6
Psychomotor retardation	1/1	10/10	5/6
Hypotonia		7/8	5/5
High forehead	1/1	9/10	6/6
Flat face		8/8	5/5
Fine, arched eyebrows		9/9	5/5
Downslanting palpebral fissures	2/2	4/7	6/6
Narrow palpebral fissures	1/1	13/13	5/6
Epicanthal folds		5/6	5/5
Hypertelorism	2/2	9/11	5/6
Flat/broad nasal bridge	2/2	7/8	6/6
Short nose	2/2	6/6	3/4
Bow-shaped mouth or prominent lips	1/1	8/8	5/6
Microretrognathia	1/1	10/11	4/4
Low-set ears	3/3	13/13	5/6
Short neck	1/1	8/10	4/5
Kyphoscoliosis	1/1	5/8	4/5
Microcephaly		10/12	1/6
Weight at birth <3%ile	1/1	5/7	0/4
Congenital heart disease		7/9	0/4
Cleft palate	3/3	2/6	0/5

From Klep-de Pater *et al.* (1979) and Miró *et al.* (1980).

and Patil, 1983; Markkanen, Somer, and Nordström, 1984), the one exception being the disappearance of obesity, which characterizes the smallest
dup(5p).

However, lest it be concluded that the situation I have been describing is
applicable only to duplications, a series of deletions of 11q will also be considered (Table 3.13). Once again there is a progression of appearance of phenotypic abnormalities, with about half being recorded in the smallest deletion
and others appearing as the deletion increases in length (Niikawa *et al.*,
1981). And, as before, there is little change once the deletion reaches a certain size, in this case in going from del(q23→qter) to del(q22→qter). A similar
picture occurs for deletions of 4q in the region 4q31→qter (Tomkins *et al.*,
1982). Also, the abnormalities associated with the DiGeorge syndrome and
found with del(22pter→q11) are included in longer deletions or monosomy of
chromosome 22 (see below), and individuals with very small deletions of 5p,

Table 3.13. *Deletions of increasing length involving 11q*

	q22→qter 3 cases	q23→qter 17 cases	q24→qter 3 cases	q25→qter 2 cases as r(11)
Growth retardation	2/2	10/13	2/3	2/2
Psychomotor retardation	2/2	12/12	3/3	2/2
High, narrow palate	2/2	9/9	2/3	2/2
Hypertelorism	2/3	10/13	3/3	1/2
Short nose	2/2	16/16	2/3	1/2
Low nasal bridge	2/2	11/12	3/3	1/2
Microretrognathia	3/3	15/15	0/3	1/2
Low-set, malformed ears	3/3	14/14	2/3	1/2
Cardiac defect	3/3	6/14	1/3	1/2
Thrombocytopenia	1/2	6/11	2/3	1/2
Trigonocephaly	2/2	13/17	3/3	0/2
Epicanthal folds	2/3	10/15	2/3	0/2
Blepharoptosis	1/2	7/12	2/3	0/2
Digital anomaly	2/2	11/13	2/3	0/2
Coloboma of iris	1/1	2/5	0/3	0/2
Carp mouth with thin upper lip	2/2	14/16	1/3	0/2
Deafness	?	3/20	0/3	0/2

From Niikawa *et al.* (1981), with one case of del(11q24 → qter) from O'Hare, Grace, and Edmunds (1984).

e.g., del(p15.1→pter), are said to have a superficial resemblance to classic cri-du-chat syndrome patients (Kushnick, Rao, and Lamb, 1984).

Phenotypic mapping

Having looked at these data, we can now answer the two questions posed in the previous section. With regard to the first, the phenotypic features of short aneuploid segments *are* generally included among the features manifested by aneuploidy for longer segments which include them. Masking of features does sometimes occur, but at least for regions of the size considered here (segments roughly equal in length to chromosome 22, or less), this turns out to be relatively infrequent. Concerning the second question, it *is* possible to identify certain features with aneuploidy for specific chromosome segments. This identification or mapping is generally done by subtraction. If a feature not present in a smaller segment is present in a larger segment that includes the first one, then the feature is provisionally mapped to the region that con-stitutes the difference between these two segments. I have used the term "provisionally," since one cannot really map a phenotypic feature by such a one-way subtraction process. While it is possible that aneuploidy for the region in question does by itself lead to the expression of the feature, it is also possible that the feature results from combined effects of aneuploidy for loci present in both the smaller segment and in the region of difference between the smaller and the longer segment. To prove that aneuploidy for the region does in fact result in the appearance of the feature, it is necessary to show that aneuploidy for this region alone also produces the feature or that this region is the only one in common when the feature appears in association with aneuploidy for overlapping segments that are not contained within one another (Fig. 3.1).

At this point, it is necessary to make clear what is meant by the term "mapping." The term does not mean that there is a specific locus or loci for the feature under consideration. There are no "loci" for cleft lip, polydactyly, or shield chest. Neither does it imply that the aneuploid region contains a locus or loci specifically involved in the development of the tissue or organ which demonstrates the phenotypic features. What is to be understood is that imbalance affecting the region leads directly or indirectly to altered development of the tissue or organ and thereby results in the phenotypic feature. For example, an aneuploidy-caused generalized metabolic effect to which a particular tissue is especially vulnerable may give rise to a phenotypic character, even though the metabolic process affected is not unique to the tissue expressing the feature. It is of course possible that in some instances the locus or loci of concern will be directly involved in the affected developmental or physiological process, but it is likely that the involvement will usually be less direct. This does not mean, however, that it will be less specific. The manner by which the ultimate phenotypic effect is produced does not alter the specific relationship between the chromosome imbalance and the particular features it causes.

The two chromosomes that have received the greatest attention with regard to phenotypic mapping are 13 and 18. (Chromosome 21 has of course also been the subject of considerable interest and will be discussed in Chapter 12.) The reasons for the attention to these chromosomes are quite straightforward. Most importantly, the full trisomies are compatible with viability, thereby making possible complete delineation of the phenotype. Then, many cases of duplications and deletions affecting regions of these chromosomes and again compatible with viability have been detected. And, finally, with particular regard to trisomy and duplications of chromosome 13, certain very striking or unique phenotypic features are present.

The first serious attempt at phenotypic mapping involved chromosome 18. In 1965, Grouchy (1965) published a "tentative map" of the phenotypic features of chromosome 18 deletions. More recent maps of the effects of trisomy and duplications have been published by Turleau and Grouchy (1977) and Turleau *et al.* (1980). The outcome of the attempts to map the duplications of

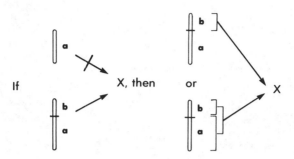

To prove that aneuploidy for region b
is responsible for feature X, it must
be shown that

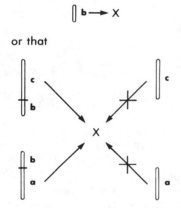

Fig 3.1. Method of phenotypic mapping. Top: While the appearance of a phenotypic feature, X, can be attributed to aneuploidy for region b when aneuploidy of a does not produce the feature and aneuploidy (of the same type) for a + b does, it is also possible that aneuploidy for both segments a and b is required. Bottom: For region b to be unequivocally implicated, it must be shown that either aneuploidy for b alone produces X, or that aneuploidy for a + b and for b + c each produce X while aneuploidy for either a or c alone does not.

chromosome 18 have already been discussed with regard to the effects of duplications of the proximal and distal segments of the chromosome and will not be further considered here.

Several pictorial maps of chromosome 13 have been published (Noel, Quack, and Rethoré, 1976; Niebuhr, 1977; Nichols *et al.*, 1979; Wenger and Steele, 1981; Yunis and Lewandowski, 1983; Rogers, 1984; Tharapel, Wilroy, and Lewandowski, 1984), and features of two of the most recent ones are reproduced in Fig. 3.2. The results of mapping of duplications of chro-

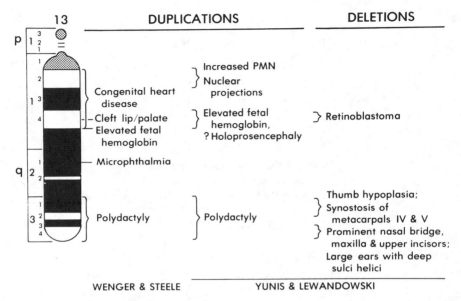

Fig 3.2. Phenotypic maps of chromosome 13. Redrawn from Wenger and Steele (1981) and Yunis and Lewandowski (1983).

mosome 13 have already been anticipated in the discussion of proximal and distal duplications of chromosome 13, in which the mapping of the nuclear projections and elevated fetal hemoglobin is described. The mapping of the latter is still under debate, and the case of Pangalos and Couturier (1981) with dup(q21.3→qter) and a fetal hemoglobin proportion of 39% at 3.5 years of age suggests that it might be somewhat more distal than originally proposed. Although Pangalos and Couturier (1981) suggest that such inconsistencies are not surprising "because the HbF persistence is probably not related to structural genes, but rather to one or more regulatory genes," I would prefer to attribute them to problems with the precise identification of the limits of the duplication. The mapping of the polydactyly to dup(13q22→qter) is based on its observation in cases with just this region duplicated (see Schütten, Schütten, and Mikkelsen, 1978).

The mapping of the phenotypic features associated with deletions has been considerably more problematic. Many of the deletions used for analysis have actually been in the form of rings, with the consequent problems associated with instability of the rings. Nevertheless, reasonably consistent maps of the regions responsible for particular features have been developed. Furthermore, a few highly informative cases of terminal or interstitial deletions have been reported, which by themselves permit localization of several features and a testing of the conclusions obtained from considering the r(13) cases. By this means, it has been possible, for example, to localize the region responsible

Table 3.14. *Overlapping deletions of 13q*

	del(13q22→q32) 1 case[a]	del(13q31.2→q32.3) 1 case[b]	del(13q32→qter) 5 cases (no rings)[c]
Eye abnormalities (coloboma, microphthalmia)	Coloboma	Congenital falciform retinal folds	3/4
Hypoplastic/absent thumbs	Hypoplasia of 5th digits, mild hypoplasia of other digits, soft tissue syndactyly	–	2/2
Malformed toes	+	–	2/3

[a]Nichols *et al*. (1979).
[b]Juberg and Mowrey (1984).
[c]Najafzadeh, Littman, and Dumars (1983).

for eye defects (not retinoblastoma) and for anomalies of the toes and possibly the digits tentatively to 13q32 (Table 3.14). The localization of the region (13q14), which when deleted results in retinoblastoma, does not involve data from cases of r(13) and is on quite solid ground. Several sets of overlapping deletions have been examined, and the regions of overlap converge in the region 13q14 (Fig. 3.3), probably band 13q14.11 (Ward *et al.*, 1984). Similar fine mapping has also been possible for the causal relationship between deletions of 11p13 and the aniridia-Wilms tumor association (Fig. 3.4). Other examples of phenotypic mapping have already been presented in Table 2.1.

Critiques

In 1971, in a discussion of the efforts of Grouchy (1965) to map the phenotypic effects of del(18), Hamerton (1971) expressed a negative view toward the mapping of phenotypic characters and termed the approach "an oversimplification . . . of little value in locating and mapping specific gene loci." His objections were several. They included the technical problems of chromosome identification and estimation of the amount of the chromosome actually deleted, which, while real at the time, have been considerably lessened by the development of the chromosome banding techniques. However, he also had a more sweeping theoretical reservation:

Loss or gain of a chromosome segment leads to loss or gain of many gene loci, some of which may be structural genes and others regulating genes. It may thus affect many facets of cellular development and lead to the production of a whole range of

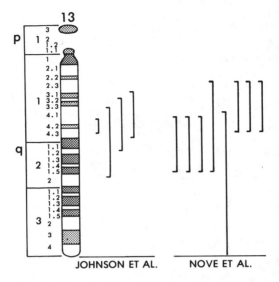

Fig 3.3. Mapping of the region, 13q14, which when deleted is associated with the development of retinoblastoma. Left, summary of four cases analyzed by high-resolution banding, redrawn from Johnson *et al*. (1982); right, summary of cases analyzed by conventional banding, redrawn from Nove *et al*. (1979).

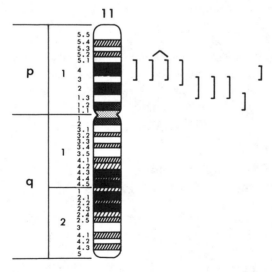

Fig 3.4. Mapping of the region, 11p13, which when deleted is responsible for the development of the aniridia-Wilms tumor association. Redrawn from Francke *et al*. (1979). Narahara *et al*. (1984) have further refined the mapping to 11p1305→p1306.

congenital malformations and developmental retardation. Such generalized features are therefore of little use in chromosome mapping . . . (and) the construction of generalized maps on the basis of gross chromosome "deletions" has little value and may even be misleading.

Other authors, in discussing the possibility of defining the phenotypes associated with specific chromosome abnormalities, have had similar things to say:

Attempts at mapping the proximal segment of chromosome 13 by several investigators . . . have proven only a variability in phenotypic expression. (Gilgenkrantz *et al.*, 1981)

Based on our review of published cases and on the findings in our patients, we find it impossible to assign certain phenotypic characteristics to different portions of the chromosome. (Liberfarb, Atkins, and Holmes, 1980, in reference to mapping 5p)

Even though segmental aneuploidy could explain the varying phenotypes present in trisomy 20p patients, it should be emphasized that differing phenotypes are often observed even when the segmental aneuploidy is relatively consistent, as in the 18p− syndrome. (Funderburk, Sparkes, and Sparkes, 1983)

However, at the other extreme, we have the statement of Riccardi *et al.* (1979), made with reference to the mapping of the region of 13q responsible for retinoblastoma, that "phenotype-karyotype correlations are the keynote of modern clinical investigative cytogenetics." Moreover, Yunis and Lewandowski (1983), long-time proponents of the concept of phenotypic mapping, assert that:

The constellation of congenital abnormalities in the classical chromosome syndromes is usually distinctive, suggesting that a direct relationship exists between certain phenotypic anomalies and specific chromosome defects. Furthermore, since the advent of the metaphase banding techniques, classical chromosome syndromes have been divided into subtypes and many new syndromes resulting from partial duplication or deletion have been established. These new entities also have distinctive and reproducible phenotypes. Although none of the chromosome syndromes have truly pathognomonic features, there are congenital anomalies which appear to be associated with only one or a few chromosome defects. . . . There is now a need to define phenotypes of the band-subband level presently mapped to 2–3 bands. . . . As studies continue to be performed in the future at more detailed levels, phenotypic changes will be mapped to very small subbands corresponding to a DNA segment containing a relatively small number of genes. This will eventually allow for the construction of a more refined phenotypic map for man. As in Drosophila and the mouse, it is anticipated that some sites may be found to be specific for a given phenotypic effect [e.g., aniridia and del(11p13)], others may have a more general or pleiotropic effect [e.g., holoprosencephaly and dup(13)(q11q21)], while in still others, multiple sites may produce a similar phenotype [eg., orbital hypotelorism and del 13q, del(18)(q21qter), and dup(21) (q22.3)]

Where does the truth lie? If we leave aside questions of mechanism, which are implicit in the statement of Hamerton (1971), the truth appears to lie with all of the commentators. The data reviewed in this chapter clearly attest to the fact that, as Gilgenkrantz *et al.* (1981) and Funderburk, Sparkes, and Sparkes (1983) point out, there may be considerable variability in the phenotypic expression of particular chromosome abnormalities. The "noise" introduced by this variability may, in some circumstances, be so great as to obscure the underlying specific pattern of phenotypic features, particularly if the number of cases being analyzed is small, the breakpoints are not identical, and other segmental aneuploidies are added in. On the other hand, when the "signal" can be separated from the "noise," as appears to be more frequently the case than not, then the evidence affirms the assertions of Yunis & Lewandowski (1983) with regard to the reproducibility and subdivisibility of phenotypes. It also affirms that imbalance of different chromosomal regions, while occasionally producing unique abnormalities, may also lead to some of the same phenotypic alterations. Therefore, while in most instances it is not possible to infer that a specific phenotypic feature indicates the presence of aneuploidy for a specific chromosomal segments, the reverse can be asserted: aneuploidy for a particular region will give rise to a defined set of phenotypic features. As stated by Francke (1978), while "quantitative disturbances of genetic material . . . produce malformations that are by themselves not specific for the particular chromosome regions involved, . . . the pattern of malformations may be quite characteristic." We have to be careful not to allow semantic imprecision to obscure the true nature of things.

However, it is also true that as the length of aneuploid segments increased, or when the functionally equivalent process of double aneuploidy occurs, the phenotypic features determined by the smaller segments may undergo modification, either masking or enhancement (with regard to frequency or severity), and new features not characteristic of imbalance of the smaller regions may appear. This does not negate the concept of the relationship between imbalance of specific regions and the expression of defined phenotypic features. It merely reaffirms that many aspects of morphogenesis are influenced by the products of a large number of genetic loci, so that the aberrations caused by the imbalance of one or a few loci are subject to modification when a greater number of loci are simultaneously imbalanced.

Critical regions

At this point, I would like to comment briefly on the concept of critical regions and the complementary issue of the relationship between the length of an aneuploid segment and the severity of its manifestations. Several authors have attempted to define the smallest chromosomal region which, when unbalanced, will give rise to the phenotypic features generally associated with imbalance of larger chromosomal segments which contain that segment. Thus, there are references stating that deletion of band 4p16 is the critical deletion involved in the Wolf-Hirschhorn syndrome (Rivas *et al.*, 1979; Wilson *et al.*, 1981), that band 5p15.2 is the critical segment for the cri-du-chat syndrome (Niebuhr, 1978), that band 5q15 is the critical segment for a del(5q) syndrome (Rivera *et al.*, 1985), that bands 6q26 and/or 6q27 are the critical segment for 6q trisomy (Taysi *et al.* 1983), that deletion of 11q24.1→qter is responsible for the expression of the 11q− phenotype (O'Hare, Grace, and Edmunds, 1984), that 13q31→qter is the critical zone for trisomy 13 features (Rivas *et al.*, 1984), that 16q12.2→q13 is the critical region for the del(16q) syndrome (Elder, Ferguson, and Lockhart, 1984), that deletion of band 18q21.3 is responsible for the typical del(18q) syndrome (Wilson *et al.*, 1979), and that duplications of 18q11 (Turleau *et al.*, 1980; Mücke *et al.*, 1982) or 18q12.3→q21.3 (Cohn *et al.*, 1984) are crucial for the phenotype of trisomy 18. [Hernandez *et al.* (1979*a*) use the term in a somewhat different manner and refer to 18q12.1→q12.2 as the "critical" zone which when duplicated causes the severe stigmata (visceral anomalies, early death) of the complete trisomy 18 syndrome. In this application of the term, which should not be confused with its usage by the other authors, they are using it only to describe the mapping of that region of 18q which causes the most life-threatening (and hence "critical") manifestations of the trisomy 18 phenotype.]

Other authors refer to the lack of relationship between the length of the aneuploid segment and the clinical manifestation or severity of the syndrome. For example, Mulcahy, Pemberton, and Sprague (1979) in their discussion of dup(3q) assert that simple correlations between the length of the trisomic segment and the severity of clinical expression are not readily apparent. Fryns *et al.* (1983*c*) state that "the difference in the amount of chromosomal material involved in the 6p trisomy in these individuals does not appear to affect the striking clinical findings which distinguish the syndrome." Preus *et al.* (1985), using numerical taxonomy, conclude that patients with a minute deletion of 4p exhibit all of the features of a larger del(4p) syndrome and suggest that only a "relatively few genes" are involved. And, Pihko, Therman, and Uchida (1981) conclude that "it is clear that there is no correlation between the length of the trisomic segment and the number and/or severity of symptoms. The obvious conclusion is that most, if not all, symptoms characteristic of partial trisomy for 11q are caused by trisomy for the segment distal to 11q23.2." In the last reference, the authors are, of course, once again defining a critical region.

Are these authors really implying that a whole phenotype is entirely attributable to imbalance of only a small chromosomal region and that imbalance of the surrounding region is irrelevant? I think not. What they are doing is calling attention to the fact that there is a lack of uniformity in the relationship between different regions of an unbalanced chromosome and the generation of phenotypic features − particularly those relatively few features that might be considered as quite characteristic of aneuploidy of the region (for example, the "cat cry" in the cri-du-chat syndrome). Thus, while some regions appear to have little influence on the generation of features, for whatever reason, others seem to have a great impact. Several examples of this, which have already been presented in Tables 3.10 to 3.13, document the fact that increasing the length of a duplication or deletion may produce few or possibly even no additional features, particularly when relatively small chromosomal segments are considered. However, even though certain chromosomal segments do not have significant impacts on the phenotype, this does not mean that they may not have more subtle effects. Thus, while the typical phenotype of the cri-du-chat syndrome may reside in deletion of band 5p15.2, there still appears to be a relationship between the amount of 5p deleted and the severity of the mental retardation (Carlin and Neadle, 1978). Therefore, I believe that while the concept of a critical region may be useful in generating a shorthand expression for identifying the region or regions of a chromosome that seem to have the greatest effect on the observable phenotype, it is important not to fall into the trap of believing that it is imbalance of only this region which has effects on development and function.

Small-deletion syndromes

Having argued that the concept of critical regions must be applied with caution, it is now necessary to take the opposite tack and point out that certain quite distinctive syndromes may result from very small (in gross chromosomal terms) amounts of aneuploidy. The principal case in point is the Prader-Willi syndrome, with its characteristic neonatal feeding difficulties followed by hyperphagia and obesity, hypotonia, mental retardation, growth retardation, and dysmorphic features (narrow bifrontal diameter, almond-shaped eyes, fish-shaped mouth, and small hands and feet) (Holm, 1981). Because of the highly reproducible nature of the disorder and rarity of familial cases, it originally seemed reasonable to attribute it to a de novo mutation, although the combination of mental retardation, growth impairment, and dysmorphic features did, of course, raise the possibility that a chromosomal abnormality might somehow be involved. The first suggestion that this was actually the case came from the observation of translocations involving chromosome 15 in affected individuals (Hawkey and Smithies, 1976), although it is interesting to note that earlier observations of a similar type were ignored (Schneider and Zellweger, 1968; Dunn, 1968).

Following the report of Hawkey and Smithies (1976), there has been a rapidly increasing number of reports describing chromosome 15 abnormalities in the Prader-Willi syndrome, the common denominator being a deletion of region 15q11→q12 or q13. In four recently reported series, with a total of 113 clinically typical cases, 59% had a del(15q) and another 3% had some other chromosome 15 anomaly (Ledbetter *et al.*, 1982; Butler and Palmer, 1983; Mattei, Mattei, and Giraud, 1983; Cassidy, Thuline, and Holm, 1984). The proportion of cases with abnormal chromosomes 15 ranged from 53% (Ledbetter *et al.*, 1982; 40 cases total) to 100% (Cassidy, Thuline, and Holm, 1984; 12 cases total). The "nondeletion" chromosome 15 anomalies included apparantly balanced translocations and an isodicentric chromosome 15 in addition to two normal chromosomes 15, and inv(15) and duplicated 15(pter→q1.3) have also been reported (Kousseff, 1982; Winsor and Welch, 1983). In a pair of familial cases, the affected individuals had del(15pter→q13) inherited from balanced rcp(14;15)(q11.2;q13) parents (Hasegawa *et al.*, 1983). Curiously, when the parental origin of a de novo deleted chromosome 15 could be traced, it was almost invariably (11 out of 13 cases) the paternally derived chromosome 15 which was abnormal (Butler and Palmer, 1983).

The exact significance of the chromosome 15 abnormalities in the Prader-Willi syndrome is still being debated, both because of the failure to find abnormalities in all cases and because of the occasional occurrence of nondeletion abnormalities that are difficult to relate to a deletion [for example, the idic(15)]. However, at the present time, it seems most reasonable to attribute many cases of the syndrome to a deletion of chromosome 15 in the region q11→q12 or q13, and Mattei, Souiah, and Mattei (1984) have gone so far as to attribute all cases to such a lesion, "even if no anomaly is detectable." How many loci must actually be deleted to produce the syndrome is unclear, and it is possible that deletions going beyond q12 or q13 may result in loss of some features of the syndrome, as well as the addition of others (Schwartz *et al.*, 1985*b*). The fact that the syndrome can occur when the chromosome 15 lesion is an apparently balanced inversion or translocation suggests that the number may be quite few, perhaps even one (Fraccaro *et al.*, 1983), although it is also quite possible that even such abnormalities may structurally or functionally (by a position effect) remove a larger segment of chromosome 15.

The DiGeorge syndrome, attributed to developmental abnormalities in the region of the third and fourth pharyngeal arches, is characterized by thymic and parathyroid aplasia, abnormalities of the great vessels (interrupted aortic arch or truncus arteriosus), and a variety of other major and minor anomalies in regions other than the neck or thorax (Conley *et al.*, 1979; Carey, 1980). Despite the latter, the largely localized nature of the consistent abnormalities and occasional familial occurrence led to the conclusion that the condition might be monogenic in inheritance, possibly even autosomal recessive, although a few potential precedents associated with chromosomal abnormalities were cited (Conley *et al.*, 1979). One of these abnormalities was a monosomy 22, in which thymic aplasia, along with multiple other malformations, was reported (Rosenthal, Bocian, and Krmpotic, 1972). A case with del (22pter→q11) and thymic hypoplasia, transposition of the great vessels with pulmonary artery atresia, and numerous other anomalies was reported by

Back *et al.* (1980), who also listed congenital heart disease as a feature of other reported cases of del(22). Nevertheless, it still came as a surprise when reports by de la Chapelle *et al.* (1981) and Kelley *et al.* (1982) of a specific association between del(22pter→q11) and the DiGeorge syndrome first appeared. The four cases reported by de la Chapelle *et al.* (1981) were in the same family and resulted from the unbalanced transmission of the translocation t(20;22)(q11;q11). The three unrelated cases of Kelley *et al.* (1982) were also the result of unbalanced translocations, either inherited (in two) or de novo (in one). They also mentioned four more possible cases (later reported by Greenberg *et al.*, 1984) of a "partial" DiGeorge syndrome or anomalad, with different combinations of congenital heart disease (truncus arteriosus) and thymic and parathyroid aplasia, associated with an unbalanced translocation, t(4;22)(q35.5;q11.2). Several other cases of DiGeorge syndrome studied by high-resolution banding techniques were found to be chromosomally normal (Greenberg *et al.*, 1984). Once again it is not clear whether all cases of DiGeorge syndrome will ultimately be shown to have del(22) [Schinzel (1983) believes they will not] and, for those that do, whether one or several genes are involved in the etiology of the syndrome. de la Chappelle *et al.* (1981) suggests that only one gene, in the region 22q11, is involved.

The Miller-Dieker syndrome has been attributed to del(17p13.3) (Elder *et al.*, 1985). Other conditions associated with small chromosomal deletions, such as retinoblastoma and the aniridia-Wilms tumor association, have already been mentioned in this chapter and will be discussed again in Chapter 14 with regard to the relationship of the deletions to malignancy. However, one other condition requires mention: the Langer-Giedion syndrome (trichorhino-phalangeal syndrome, type II) with characteristic facial abnormalities (including a bulbous, pear-shaped nose), sparse scalp hair, multiple exostoses, and brachydactyly with cone-shaped epiphyses (Hall *et al.*, 1974). Originally believed to result from new mutations of a single gene, recent reports have implicated a deletion of 8(q22→q24) in the etiology of the syndrome (Pfeiffer, 1980*a*; Turleau *et al.*, 1982; Zabel and Baumann, 1982; Bühler, Bühler, and Christen, 1983; Fryns *et al.*, 1983*a*; Zaletajev and Marincheva, 1983; Fukushima, Kuroki, and Izawa, 1983; Wilson, Wyandt, and Shah, 1983; Schwartz *et al.*, 1985*a*). There has been considerable debate about the identity of the exact segment responsible for the syndrome, with 8q22 (Zaletajev and Marincheva, 1983), 8q23 (Turleau *et al.*, 1982; Fryns *et al.*, 1983*a*), and 8q24.3 (Bühler, Bühler, and Christen, 1983) and more recently 8q24.1 (Bühler and Malik, 1984) being implicated. The differences are probably due to difficulties in interpretation.

Not all typical cases of the Langer-Giedion syndrome examined with high-resolution banding have had a demonstrable deletion (Turleau *et al.*, 1982; Gorlin *et al.*, 1982), so once again the question arises as to how much of the chromosome needs to be deleted to produce the syndrome. Turleau *et al.*

(1982) suggest that the region may be very small, possibly involving the mutation of only a single gene, but Fryns *et al.* (1983*a*) argue that the deletion may just be too small to see. In either case, it seems likely that the typical syndrome results from a small lesion and that the deletion of a larger segment may result in the appearance of additional abnormalities (Turleau *et al.*, 1982; Bühler, Bühler, and Christen, 1983).

There are several inferences that can be drawn from these examples of small-deletion syndromes. The first is that there are likely to be more, and conditions with mental retardation, growth retardation, and multiple malformations, such as the Rubinstein-Taybi and the Cornelia de Lange syndromes (Smith, 1982), may be good candidates. The second inference is that small deletions can have quite distinct and highly reproducible phenotypic effects, once again testifying to the specificity of the relationship between the aneuploid segment and the resulting abnormalities. Conversely, specific phenotypic effects can be mapped to small chromosomal regions. However, it is not clear whether one or several loci must be deleted for the characteristic phenotypes to occur. This matter will be resolved only when it is possible to examine those chromosomes without cytogenetically detectable deletions by molecular methods capable of identifying the exact genetic lesions that are present. Nevertheless, however many loci are involved, the experience with these syndromes indicates that there is really no qualitative distinction between aneuploidy involving very small chromosome segments and that affecting longer segments. In the final analysis, it is likely, from the mechanistic point of view, that there will be a continuum between the effects of imbalance of just a single gene and of imbalance of a whole chromosome. Finally, the mechanisms operating to produce the phenotypes of the small-deletion syndromes discussed above should not be confused with those involved in the appearance of malignancy in conditions such as retinoblastoma, which may also be associated with specific small deletions. In the latter, as will be discussed in Chapter 14, the superimposition, on a cell-by-cell basis, of a second deletion or mutation may be required.

Tetrasomy

Earlier in this chapter, mention was made of tetrasomy as a special case of double aneuploidy in which the two aneuploid segments happen to be the same. The reason for being interested in the effects of tetrasomy is for what they may tell us about the relationship between gene dosage and the development of the aneuploid phenotype, the specific question being whether tetrasomy results in new, more frequent, and/or more severe abnormalities than does trisomy for the same region. Autosomal tetrasomy has been observed for chromosome regions 9p, 12p, 14pter→q12.1, 15pter→?q14 [inv dup(15)], 18p [i(18p)], and 22(pter→q11), and for all of chromosome 21. The data on the abnormalities associated with these tetrasomies and with the most closely corresponding trisomies (duplications) are shown in Tables 3.15 to 3.18. The results are quite mixed and range from little or no specific phenotypic differences to clearly distinguishable syndromes.

Tetrasomy for 9p in either mosaic or pure form is quite similar to trisomy 9p, although possibly more severe (Balestrazzi *et al.*, 1983; Shapiro, Hansen, and Littlefield, 1985), and the same also holds for tetrasomy and trisomy 9(pter→q21,22). Possible additional major effects of the second extra dose of the aneuploid segment are hydrocephalus and perhaps urogenital defects (Table 3.15). The same situation of relatively little difference probably also holds for tetrasomy 21. The features of the patient of Jabs, Stamberg, and Leonard (1982) were clinically indistinguishable from typical Down syndrome, although he did have congenital leukemia (see Chapter 12). The patient of Hunter *et al.* (1982) was tetrasomic in all fibroblasts [47,XY, + t(21;21)] but in only 1/30 lymphocytes. Again, his features were those of typical Down syndrome, but he also had in addition certain nontypical features, including dolicocephaly (with a flat occiput); right-sided parietal hair whorl; long, square facies; low, fleshy ears with prominent lobules; excessive facial skin; a high frequency of fingertip whorls; and an exaggerated upslant to the palpebral fissures. The last of these features is of particular interest since it could be interpreted as a more extreme expression of a dysmorphic feature already associated with trisomy 21. Two other cases of mosaic tetrasomy 21 have been reported, one with completely tetrasomic fibroblasts and only diploid lymphocytes (Fryns *et al.*, 1982), and the other with two-thirds tetrasomic fibroblasts and only 1% tetrasomic lymphocytes (Kwee *et al.*, 1984). Neither patient had the facies of Down syndrome, but both had many features similar to the preceding case, including hypertelorism, epicanthal folds, flat nasal bridge, large downturned mouth, small penis, short hands and stubby fingers, and short feet. All of the mosaic cases were hypotonic and severely retarded. Overall, the phenotypes were much more abnormal than that of Down syndrome. It should be noted, however, that Hall (1985) has questioned the diagnosis of tetrasomy 21 in the three cases just described and has suggested that they really represent mosaic tetrasomy 12p. A molecular genetic analysis would be useful to decide the issue definitively.

One case thought to represent mosaic hexasomy 21 has been reported in an infant who also did not have the phenotype of Down syndrome except for upslanting palpebral fissures, short neck with redundant skin folds, hypotonia, and simian lines (Ketupanya *et al.*, 1984). Microphthalmia, hypertelorism, cleft lip and palate, narrow chest with hypoplastic nipples, triphalangeal thumbs, camptodactyly, and anteriorly placed anus were also present.

In comparing the tetrasomic state of chromosome 15 with the corresponding trisomy, one is impeded both by the indecision regarding the breakpoints

Table 3.15. *Features of tetrasomy and trisomy (duplication) of 9p*

	9p			9pter→q21,22	
	Tetrasomy		Trisomy from literature[c]	Tetrasomy (pure) 2 cases[d]	Trisomy from literature[e]
	Pure 2 cases[a]	Mosaic 3 cases[b]			
Microcephaly	1	2	+	1	61%
Hydrocephalus		2			
Open sutures/fontanelles	2	1	+	1	
Enophthalmos	1	1	+	Microph-thalmia(1)	65
Hypertelorism	2	2	+	2	76
Downslanting fissures			±	1	61
Epicanthal folds	1	2			10–50[f]
Bulbous or beaked nose	1	2	+	2	92
Low-set ears	1	1	−	2	67
Dysmorphic ears	1	2	+	2	83
High-arched or cleft palate	2	2	+	2	60(high) 19(cleft)
Micrognathia	1	2		1	10–50[f]
Short neck	2	1	+		68
Congenital heart disease	0	3	±	2	25
Urogenital defect	1			2	
Prominent nasal bridge	1	1			
Large mouth/ downslanting corners	1	1	+	1	95(down-slanting
Wide-spaced nipples	1	1	+		44
Low posterior hairline	2				
Scanty or poor quality hair	1	1			

[a]Garcia-Cruz *et al.* (1982); Moedjono, Crandall, and Sparkes (1980).
[b]Cuoco *et al.* (1982), 100% tetrasomy 9p in lymphocytes, 0 in fibroblasts; Orye *et al.* (1975), 69% in lymphocytes, 63% in fibroblasts; Rutten *et al.* (1974), 86% in lymphocytes. See also Balestrazzi *et al.* (1983).
[c]Rethoré (1977).
[d]Abe *et al.* (1979), tetrasomy 9(pter→q21.01); Wisniewski, Politis, and Higgins (1978), tetrasomy 9(pter→q22). See also Shapiro, Hansen, and Littlefield (1985).
[e]Young *et al.* (1982).
[f]Centerwall and Beatty-DeSana (1975).

associated with the extra inv dup(15) chromosome, which generates the tetrasomy, and by the paucity of cases of the presumably corresponding trisomy. However, examination of the available information summarized in Table 3.16 (see also Gilmore *et al.*, 1984) does not provide strong evidence for any unique abnormalities, although there is a suggestion of more severe neurolog-

Table 3.16. *Features of inv dup(15) and dup(15pter→q14,15)*

	inv dup(15)		dup(15pter→q14,15)	
	15 cases[a]	8 cases[b]	pter→q14[c]	pter→q15[d]
Mental retardation	15/15	8/8	+	+
Abnormal speech	12/14			
Behavioral disturbances	11/13	7/8		
Short stature	4/15	2/8	+	+
Hypotonia	9/15	4/8	−	+
Seizures	11/14	4/8	−	chorea
Downslanting fissures	4/14	3/8	+	
Enophthalmos	4/14		+	
Epicanthal folds	3/11		+	
Strabismus	9/15		−	
Abnormal ears	9/15	6/8	+	+
High-arched palate	5/15		−	cleft
Scoliosis, lordosis, or kyphosis	5/12		−	
Syndactyly toes 2 and 3	5/15		−	−
Low nasal bridge		3/8	−	−
Micrognathia		3/8	−	+
Flat occiput		2/8		
Low hairline		3/8		
Microcephaly			+	+

[a]Wisniewski *et al.* (1979).
[b]Maraschio *et al.* (1981).
[c]Castel *et al.* (1976).
[d]Mankinen, Holt, and Sears (1976).

ical problems (abnormal speech and behavioral disturbances, hypotonia, seizures, instability of the spine). Data on tetrasomy 12p, observed in mosaic form, are insufficient at this point for detailed analysis (Buyse and Korf, 1983).

By contrast, the phenotypic effects of tetrasomy 18p and tetrasomy 22(pter→q11) differ significantly from the equivalent trisomies. While trisomy 18p produces no specific dysmorphic features and is associated with some degree of mental retardation in only some cases, tetrasomy 18p results in a whole host of neurological and physical abnormalities (Table 3.17). The existence of a specific tetrasomy 18p syndrome has been suggested (Batista, Vianna-Morgante, and Richiere-Costa, 1983). An infant with putative tetrasomy 14(pter→q12.1) had bilateral ectrodactyly of the feet and tibial aplasia (Johnston *et al.*, 1985), lesions not reported for dup(14)(pter→q11, q13, or q21) (Schinzel, 1983).

Table 3.17. *Features of tetrasomy and trisomy (duplication) of 18p*

	Tetrasomy 18p [i(18p)] 11 cases[a]	Trisomy (duplication) 18p 6 cases[b]
Psychomotor retardation	11	3
Hypertonia	8	
Microcephaly	9	
Prominent occiput	4	
Flat occiput	3	
Prominent forehead	4	
Low hairline	3	
Facial asymmetry	3	1
Epicanthal folds	1	
Pinched nose	5	
Flat nasal bridge	3	
Small/triangular mouth	7	
High-arched or cleft palate	8	
Micrognathia	3	
Low-set ears	7	
Malformed ears	3	1 (small)
Short neck	3	
Scoliosis	5	
Cryptorchidism/small testes	3	
Simian crease	6	
Finger abnormalities	7	1

[a]Batista, Vianna-Morgante, and Richiere-Costa (1983).
[b]Gardner *et al.* (1978); Jacobsen and Mikkelsen (1968), 4 cases; Taylor *et al.* (1975).

Although contrary opinion has been expressed (Guanti, 1981; Rosenfeld, Verma, and Jhaveri, 1984), the cat-eye syndrome is now believed to result from tetrasomy of 22(pter→q11) (Schinzel *et al.*, 1981*b*). Several features of this syndrome are similar to those associated with trisomy for the same region, either in pure form (Tables 3.8 and 3.18) or as part of the unbalanced rcp(11;22) syndrome (Table 3.6) (Schinzel *et al.*, 1981*a,b*). However, a few features are quite characteristic of the tetrasomic state – especially the ocular coloboma, which gives the syndrome its name, microphthalmia, and the high frequency of anal atresia.

Taken together, the data on the several tetrasomies suggest that when the trisomic state produces a significant set of abnormalities, then the corresponding tetrasomy affects the overall phenotype relatively little, although the patients may be somewhat more severely affected [as perhaps in inv dup(15) and tetrasomy 21] or may display new features [as in tetrasomy 21 and in the cat-eye syndrome]. However, when the trisomy has little or no phenotypic

Table 3.18. *Features of tetrasomy ("cat-eye syndrome") and trisomy (duplication) of 22 (pter→q11)*

	Tetrasomy 33 cases	Trisomy 5 cases
Mild/moderate mental retardation	20/24	2/4
Downslanting palpebral fissures	21/32	4/8
Coloboma of iris	21/34	0/5
Microphthalmia	7/34	0/5
Preauricular malformations	26/34	4/5
Reduced or absent auricles	3/34	0/5
Cleft palate/uvula	3/34	0/5
Anal atresia	22/34	1/5
Congenital heart defects	16/34	1/5
Renal malformations	14	2

From Schinzel *et al.* (1981*b*).

effect, as in trisomy 18p, the additional imbalance introduced by the tetra-somy may result in significant phenotypic abnormalities.

In this context, some findings in the mouse are of interest. Two situations have been described in which progeny carrying two copies of small translo-cation chromosomes in addition to the normal chromosome complement have been produced. In one, involving the translocation, *T(1;13)70H*, duplication (trisomy) of the small regions of chromosomes 1 and 13 contained in the small 1^{13} translocation chromosome is compatible with survival and fertility, although with some developmental abnormalities (see Table 10.4). However, the presence of two translocation chromosomes, producing a tetrasomic state, causes fetal demise (van der Hoeven and de Boer, 1984). On the other hand, tetrasomy for the regions carried by the 5^{12} translocation of *T(5;12)31H* was compatible with birth of a small, infertile male with a shortened head and testes that were two-thirds normal in size. Another such animal, examined at 12.5 days of gestation, was only slightly retarded but had a "twisted spine" (Beechey and Speed, 1981). Thus, in the mouse, tetrasomies of small regions can be compatible with viability, but the phenotypic effects seem to be more severe than in the corresponding trisomies. In the one whole-chromosome mouse tetrasomy studied, tetrasomy 16, the fetuses died earlier and were more severely retarded in development than littermates with trisomy 16 (De-brot and Epstein, 1985).

Although the interpretation is complicated by uncertainties about the com-plexities of the inactivation of X chromosomes in excess of one (see Chapter 13), a similar situation seems to hold for X-chromosome polysomy. While

47,XXX has no distinctive phenotypic features and only mild mental retardation associated with it (Robinson *et al.*, 1983), mental retardation is a consistent feature of tetrasomy X (48,XXXX) and is severe in pentasomy X (49,XXXXX) (Fryns *et al.*, 1983*b*). In addition, pentasomy X produces numerous malformations, including short stature, radio-ulnar synostosis, incomplete development of secondary sexual characteristics, hypertelorism, epicanthal folds, upslanted palpebral fissures, broad flat nose, short neck, skeletal anomalies, congenital heart disease, simian creases, and abnormal dentition (Toussi *et al.*, 1980; Fragoso *et al.*, 1982; Fryns *et al.*, 1983*a*), and the resemblance to Down syndrome has been commented on several times. Therefore, in both the autosomal and X-chromosome situations, while the effects may sometimes be minimal, the overall picture is one of increasing abnormality as the number of times a specific chromosome or segment is represented in the genome is increased.

Part III

Theoretical mechanisms and issues:
the primary and secondary effects of aneuploidy

4

Gene dosage effects

Based on the analysis of the available clinical information, it can be concluded that there is a definite relationship between the identity and therefore the genetic structure of an unbalanced chromosomal region and the phenotypic consequences resulting from its imbalance. Not all regions affect the phenotype equally, and interactions between different unbalanced regions clearly exist. The existence of a quantitative relationship between the degree of genetic imbalance of a specific region and the resulting phenotype is indicated by the results of comparing the phenotypes of tetrasomies with those of their corresponding trisomies. Overall, the clinical data, while pointing to the possible complexity of interactions between and among the products of several different simultaneously unbalanced genes and of the balance of the genome, support the basic premise advanced in the first chapter – that it should be possible to relate the phenotypic effects of an aneuploid state to the cumulative effects of the imbalance of specific loci and sets of loci and ultimately to discover how the loci and the phenotypic effects are connected. Therefore, in this part of the book, I shall use both theoretical and experimental considerations to suggest how imbalance of genes can actually lead to phenotypic alterations.

Central to an understanding of the mechanisms by which aneuploidy produces its consequences is the concept of gene dosage effects. This concept holds that the synthesis, and hence the concentration, of a primary gene product is directly proportional to the number of genes coding for its synthesis. In an aneuploid state, the number of genes, and hence the gene dosage (Junien, Huerre, and Rethoré, 1983, 1984), is of course determined by the number of chromosomes or chromosome segments carrying those genes. A primary gene product is considered to be a protein, or possibly an RNA, for which the structural gene is present on the chromosome in question. The gene dosage effect concept means that in the triplex state, produced by a duplication or trisomy, each of the three loci on the three homologous chromosomes or chromosome segments is operative and is expressing equally. Conversely, in the uniplex state resulting from deletion or monosomy, the one locus present is still expressing at the same rate as it did in the diploid state. The latter situation is not really different from that which obtains in heterozygotes for a large number of enzyme and other protein deficiency states in which the residual activity or concentration is half of normal (Harris, 1980).

65

It can, of course, be argued that the existence of gene dosage effects in the triplex state does not formally prove that all three genes are operative. However, evidence for this is provided by instances in which each chromosome is differentially marked, as for example by carrying three different *HLA* or phosphoglucomutase–1 (*PGM1*) alleles. Several individuals with duplication of 6(p21 or 22→pter) have been reported to express three different *HLA* alleles (McGillivray, Dill, and Lowry, 1978; Pearson *et al.*, 1979; Berger *et al.*, 1979; C. Morton *et al.*, 1980, 1982), and one boy with a duplication of 1(p22.1→p31.4) expressed three different *PGM1* alleles. The products of these alleles, 2 + , 2 − , and 1 + , were demonstrated in red cells and white cells in what appeared to be equal amounts (Cousineau *et al.*, 1981). A similar result was obtained by Farber (1973) in studies of heteroploid mouse cell lines.

By implication, the concept of gene dosage effects further holds that control mechanisms to regulate the final concentration of the gene product at some fixed level do not exist. This does not negate the existence of regulatory or control systems, but does mean that they are not operative for those gene products for which dosage effects are demonstrable. Such control systems, if they did exist, could act either to alter the rate of product synthesis (either at the level of transcription or translation), the rate of degradation, or the access of the product to its appropriate place in the cell. Regulatory interference at any point in this chain could serve to eliminate or quantitatively alter a dosage effect.

In looking for gene dosage effects, it is necessary to compare like with like, in terms of tissues, sites of origin, and method of handling, and to avoid the use of heteroploid materials with chromosome abnormalities other than those for which the dosage effects are being assessed. Once such heteroploidy occurs, there is no way that the aneuploid and control (diploid) materials can be properly matched. Unless these precautions are observed, true dosage effects may be obscured, and nonexistent effects may be demonstrated.

Dosage effects in man and mouse

Evidence in support of the gene dosage effect concept has been gradually accumulating and has been summarized in earlier literature (Krone and Wolf, 1977; Epstein *et al.*, 1977; Feaster, Kwok, and Epstein, 1977; Junien *et al.*, 1980*b*; Ferguson-Smith and Aitken, 1982). The evidence presently available for man and mouse is summarized in Tables 4.1 and 4.2, respectively. The human data are derived from two sources: direct tests for gene dosage effects when the chromosomal locations of the relevant genes are known, and dosage studies carried out to assist in the localization of genes to particular chromosome segments. The latter data have been regarded as admissible only when the genes in question have already been mapped to the chromosomes by other,

Table 4.1. *Gene dosage effects in human cells and tissues*

Chromosome	Gene symbol	Gene product	Tissue[a]	Ratio of aneuploid to diploid (number of subjects)		References
				Triplex	Uniplex	
1q	AT3	Antithrombin III	Plasma		0.60(2)	Winter et al. (1982)
1q	FH	Fumarate hydratase	Fibro	1.53(1)		Braunger et al. (1977)
			RBC/Fibro	1.62(2)		Despoisses et al. (1984)
1q	GUK1	Guanylate kinase-1	RBC	1.72(1)		Dallapiccola et al. (1980b)
2p	ACP1	Acid phosphatase-1	RBC	[0.91 of expected][b]	(1)	Magenis et al. (1975)
					0.60(1)	Ferguson-Smith et al. (1973)
					0.51(1)	Junien et al. (1979c)
				[1.03 of expected]	(4)	Larsen et al. (1982a)
7q	GUSB	β-Glucuronidase	Fibro	1.43(1)		Danesino et al. (1981)
			Fibro		0.35(2)	Ward et al. (1983)
?7q	HAF	Hageman factor	Plasma		0.58	Grouchy and Turleau (1974)
					0.71	Higginson et al. (1976)
8p	GSR	Glutathione reductase	RBC	1.77(3)		George and Francke (1976a)
				1.54(4)[c]		de la Chapelle, Vuopio, and Icén (1976)
				1.63(3)		Sinet et al. (1977)
				14.4(3)		Mattei et al. (1979)
			Lymphoblast (heteroploid)	1.33(6)	0.53(3)	Soos et al. (1981)
9p	AK3	Adenylate kinase-3 (mitochondrial)	RBC	1.39(1)		Jensen et al. (1982)
			Fibro	1.66(1)		Steinbach and Benz (1983)
9p	GALT	Galactose-1-phosphate uridyl/transferase	RBC	1.5(1)		Weber, Muller, and Sparkes (1975)

67

Table 4.1. (cont.)

Chromosome	Gene symbol	Gene product	Tissue[a]	Ratio of aneuploid to diploid (number of subjects)		References
				Triplex	Uniplex	
				1.75(2)		Aitken and Ferguson-Smith (1979)
				1.63(2)		Sparkes et al. (1980a)
				1.66(1)		Eydous et al. (1981)
				1.91(2)		Zadeh et al. (1981)
					0.65(1)	Bricarelli et al. (1981)
			Fibro	1.52(3)		Shih et al. (1982)
				1.40(1)		Steinbach and Benz (1983)
				Tetrasomy		
				2.03(1)		Moedjono, Crandall, and Sparkes (1980)
				2.54(1)		Eydoux et al. (1981)
				2.03(1)		Garcia-Cruz et al. (1982)
				2.09(1)		Balestrazzi et al. (1983)
9q	AK1	Adenylate kinase-1 (soluble)	RBC	1.44(1)		Ferguson-Smith et al. (1976)
				1.77(1)		Mattei et al. (1980)
				1.66(1)		Mattei et al. (1983)
10p	HK1	Hexokinase-1	RBC	1.74(1)		Junien et al. (1979d)
			RBC	2.41(1)		Dallapiccola et al. (1981)
			Fibro	1.49(1)		Schwartz et al. (1984)
			RBC	2.66(1)		Snyder et al. (1984)
			RBC	1.53(1)		Magnani et al. (1983)
10p	PFKF	Phosphofructokinase, fibroblast	Fibro	1.53(1)		Schwartz et al. (1984)

Location	Symbol	Name	Tissue			Reference
10q	PGAMA	Phosphoglycerate mutase A	RBC		1.51(1)	Junien et al. (1982a)
10q	GOT1	Glutamate oxalacetate transaminase-1, soluble	RBC		1.82(1)	Sparkes, Bass, and Sparkes (1978)
					1.53(2)	Aitken and Ferguson-Smith (1978)
			Fibro		1.64(2)	Spritz et al. (1979)
			RBC		1.37(1)	Junien et al. (1979d)
			RBC		1.69(1)	Junien et al. (1982a)
			RBC		1.96(1)]	Tomkins, Gitelman, and Roberts (1983)
			Fibro		1.37(1)]	
11p	CAT	Catalase	RBC	0.28(1)	2.04(1)	Junien et al. (1980c)
			WBC	0.51(1)		
			RBC	0.45(1)		Gilgenkrantz et al. (1982)
			WBC	0.47(1)		
			RBC	0.61(2)		Niikawa et al. (1982)
			RBC	0.47(1)		Narahara et al. (1984)
11p	LDHA	Lactate dehydrogenase A	WBC		1.53(1)]	Rethoré et al. (1980)
			Fibro		1.88(1)]	
11q	UPS	Uroporphyrinogen I synthase	RBC		1.48(3)	de Verneuil et al. (1982)
12p	ENO2	Enolase-2	RBC		1.62(1)	Dallapiccola et al. (1980a)
12p	GPD1	Glycerol-3-phosphate dehydrogenase	RBC		1.42(2)	Rethoré et al. (1976)
			RBC		2.25(1)	Tenconi et al. (1978)
			RBC		1.72(2)]	Serville et al. (1978)
			WBC		1.96(2)]	Suerinck et al. (1978)
			RBC		1.83(1)	
			Lymphoblast		2.31(1)	Junien et al. (1979b)
			RBC		1.44(1)	Dallapiccola et al. (1980a)

69

Table 4.1. (cont.)

Chromosome	Gene symbol	Gene product	Tissue[a]	Ratio of aneuploid to diploid (number of subjects)		References
				Triplex	Uniplex	
12p	*LDHB*	Lactate dehydrogenase B	RBC		0.64(1)	Malpuech et al. (1975)
					0.40(1)	Tenconi et al. (1975)
					0.40(1)	Rethoré et al. (1975)
				1.53(2)		Suerinck et al. (1978)
				1.64(2)		Junien et al. (1979a)
				1.58(3)	0.56(3)	Dallapiccola et al. (1980a)
				1.57(1)		
12p	*TPI1*	Triose phosphate isomerase-1	RBC	1.87(2)		Rethoré et al. (1977)
			RBC	1.42(1)		Tenconi et al. (1978)
			RBC	1.41(2)]		Serville et al. (1978)
			WBC	2.44(2)]		
			RBC	1.16(1)		Suerinck et al. (1978)
			Lymphoblast	1.67(1)		Junien et al. (1979b)
			RBC	1.29(1)		Dallapiccola et al. (1980a)
13q	*ESD*	Esterase D	RBC	1.53(1)		Sparkes, Sparkes, and Crist (1979)
			RBC		0.6(1)	Norum, van Dyke, and Weiss (1979)
			RBC		0.52(3)	Sparkes et al. (1980b)
			Fibro		0.43(1)	
			RBC	1.49(8)	0.53(3)	Strong et al. (1981)
			RBC	1.49(1)	0.47(2)	Rivera et al. (1981)
			Fibro		0.65(2)	
			RBC		0.63(2)	Hoo et al. (1982)
			RBC	1.73(1)]		Junien et al. (1982b)
			Lymphoblast	1.34(1)		
			AFC	1.53(1)		
			Fibroblasts	1.56(1)]		

Location	Symbol	Enzyme	Source	Value	Value	Reference
14q	α1AT	α1-antitrypsin	RBC		0.54(1)	Benedict et al. (1983b)[a]
			Fibro		0.61(1)	
			Lymphoblast		0.58(1)	
			RBC	1.46(1)	0.57(1)	Dryja et al. (1983)
			RBC		0.54(1)	Ward et al. (1984)
14q	NP	Nucleoside phosphorylase	Serum	1.56		Sklower et al. (1982)
			Serum	1.65(1)		Cox, D.W. et al. (1983)
			RBC	1.65(3)		George and Francke (1976b)
				1.54(1)		Frecker et al. (1978)
				2.04(1)		Dallapiccola et al. (1979)
			RBC	1.68(2)		Junien et al. (1980d)
			WBC	1.35(1)		
			RBC	1.65(1)		Abeliovich, Yagupsky, and Bashan (1982)
15q	PKM2	Pyruvate kinase-3	Fibro	1.57(6)		Junien et al. (1980b)
16q	APRT	Adenine phosphoribosyl-transferase	Fibro	1.69(9)		Marimo and Giannelli (1975)
17q	GAA	Acid α-glucosidase	?	1.72(1)		Rethoré et al. (1982)
			AFC	1.55(1)		Sandison, Broadhead, and Bain (1982)
18q	PEPA	Peptidase A	RBC	2.36(7)	0.41(5)	Danesino et al. (1978)
					0.47(1)	Tenconi et al. (1978)
				1.82(2)	0.46(3)	Junien et al. (1980a)
20p	ITPA	Inosine triphosphatase	RBC	1.79(1)		Funderburk, Sparkes, and Sparkes (1983)
20q	ADA	Adenosine deaminase	RBC	2.03(1)	0.49(1)	Rudd et al. (1979)
			RBC		0.55(1)	
			Fibro		0.51(1)	
			Lymphoblast			Philip et al. (1980)

Table 4.1. (cont.)

Chromosome	Gene symbol	Gene product	Tissue[a]	Ratio of aneuploid to diploid (number of subjects)		References
				Triplex	Uniplex	
21q	CBS	Cystathionine-β-synthase	Fibro	1.66(8)		Chadefaux et al. (1985)
21q	IFRC	Interferon receptor (α)	Fibro	1.57(1)	0.64(1)	Epstein et al. (1982a)
21q	PFKL	Phosphofructokinase liver	RBC	1.46(21)		Baikie et al. (1965)
				1.48(2)		Bartels and Kruse (1968)
				1.42(10)		Benson, Linacre, and Taylor (1968)
				1.29(10)		Sparkes and Baughan (1969)
				1.29(28)		Conway and Layzer (1970)
				1.72(15)		Pantelakis et al. (1970)
				1.46(4)		Layzer and Epstein (1972)
				1.47(7)		Vora and Francke (1981)
				1.48(24)		Frischer et al. (1981)
21q	PRGS	Phosphoribosyl-glycinamide synthetase	Fibro	1.64(4)		Bartley and Epstein (1980)
				1.54	0.47	Scoggin et al. (1980)
				~2 (1)		Bradley et al. (1982)
				1.51(2)	0.54(1)	Chadefaux et al. (1984)
21q	SOD1	Superoxide dismutase-1	RBC	1.41(10)		Sinet et al. (1974)
				1.50(11)		Sichitiu et al. (1974)
				1.40(33)		Frants et al. (1975)
				1.38(6)		Priscu and Sichitiu (1975)
			Platelets	1.56(11)		Sinet et al. (1975)
			RBC	1.45(28)		Crosti et al. (1976)
				1.56(14)		Sinet et al. (1976)
				1.41(8)		Gilles et al. (1976)

			Tissue			Reference
			Granulo	1.38(6)		Feaster, Kwok, and Epstein (1977)
			Lymph	1.40(5)		
			Fibro	1.81(5)		
			RBC	1.49(76)	0.60(2)	Garber et al. (1979) Del Villano and Tischfield (1979)
			RBC	1.54(3)		
			RBC	1.52(15)		Baret et al. (1981)
			PMN	1.67(14)		
			Lymph	1.72(11)		
			Platelets	1.11(13)		
			RBC	1.65(15)		Crosti et al. (1981)
			RBC	1.44(24)		Frischer et al. (1981)
			Lymphoblast	1.59(5)		Jeziorowska et al. (1982)
			RBC	1.52(15)		Brooksbank and Balazs (1984)
			Brain	1.60(5)		
			RBC (fetal)	1.96(4)		Lang, Nakagawa, and Nitowsky (1983)
			RBC		0.55(1)	Yoshimitsu et al. (1983)
			WBC		0.46(1)	
			Lymph		0.51(1)	
			Fibro	1.67(29)	0.72(1)	Wisniewski et al. (1983)
			T lymph	1.55(30)		
			B lymph	1.69(30)		
			Mac	1.46(11)		Baeteman et al. (1983)
			RBC	1.27(14)		Ohno et al. (1984)
			RBC	1.49(9)		Björkstén, Marklund, and Hägglöf (1984)
			WBC	1.49(9)		
			Lymph	1.32(12)		
22q	ARSA	Arylsulfatase A	?	~2 (1)		Fryns, Jaeken, and van den Berghe (1979)

73

Table 4.1. (*cont.*)

Chromo-some	Gene symbol	Gene product	Tissue[a]	Ratio of aneuploid to diploid (number of subjects)		References
				Triplex	Uniplex	
			Leuk	1.44(1)		Cantu et al. (1981)
			Leuk	1.69(1)		Hamers, Klep-de Pater, and de France (1983)
22q	IDA	α-L-iduronidase	Fibro	2.13(2)	0.15(3)	Schuchman et al. (1984)

[a]Abbreviations used: AFC, amniotic fluid cells; Fibro, fibroblasts; Leuk, leukemic cells; Lymph, lymphocytes; Lymphoblast, lymphoblastoid lines; Mac, macrophages; PMN, polymorphonuclear leukocytes; RBC, erythrocytes; WBC, leukocytes.
[b]Expectation calculated from subunit composition.
[c]Ts8 of bone marrow with leukemia.
[d]Deletion not visible at 550 bands.

non-dosage-related methods. Because of secondary effects that may alter concentrations of a variety of proteins, direct mapping by gene dosage alone is a procedure fraught with peril – one "which was discarded early in the history of human cytogenetics" (Ferguson-Smith and Aitken, 1982) when trisomy 21 red cells were found to have elevated levels of the X-linked enzyme, glucose–6-phosphate dehydrogenase (Mellman *et al.*, 1964), as well as of several other enzymes which later turned out to be coded for by other chromosomes (Hsia, Nadler, and Shih, 1968). In this regard, the red cell appears to be the principal but not the only culprit, since elevated levels of glutathione peroxidase–1, a chromosome 3-coded enzyme, have been reported in trisomy 21 lymphoblastoid cells and fibroblasts (Sinet, Lejeune, and Jerome, 1979; Frischer *et al.*, 1981).

Of the 37 human gene products listed in Table 4.1, about half were examined in red cells, and half (not necessarily the remaining ones) in fibroblasts. [A recent list of aneuploid human cell lines has been published by Aronson *et al.* (1983).] Analyses were also carried out in lymphoblastoid lines, leukocytes, and platelets. What is perhaps most striking about the tissues examined is that only one gene product has been analyzed in a solid tissue. Superoxide dismutase–1 activity was measured in the brains from five individuals with trisomy 21, and a trisomy:diploid (Ts/2n) ratio of 1.60 was obtained (Brooksbank and Balazs, 1984). Furthermore, only two serum or plasma constituents (perhaps three if the mapping of Hageman factor is validated) have been looked at, antithrombin III and α_1-antitrypsin.

Virtually all gene products studied have been enzymes or, like the two just mentioned, enzymelike products. The only exception to date is the interferon-α cell surface receptor, coded for by the *IFRC* locus on chromosome 21. No quantitative information has been obtained on other cell surface constitutents or on a large variety of nonenzymatic structural proteins, membrane components, receptors, and the like. Therefore, it is still too early to make any specific statements about whether and to what extent gene dosage effects apply to all cellular constituents. However, it would seem likely, if any departure from strict gene dosage effects is to be observed, that it would be with those gene products that must be processed and organized into a complex cellular constituent, such as a membrane or organelle. Gene dosage effects at the time of synthesis could exist, but the final concentration of the product could be altered during postsynthetic events. Many more products of this type need to be examined.

The remarkable feature about the human gene dosage effect data in Table 4.1 is how close to theoretical expectations the data actually come. In the triplex (trisomic, Ts) state, the Ts:2n ratio should be 1.5, while in the uniplex (monosomic, Ms) state, the Ms:2n ratio should be 0.5. The numbers from the individual experiments, which are sometimes not as well controlled as they should be, only infrequently deviate very far from these expectations. However, one unexplained exception is α-L-iduronidase which gives mean Ts:2n and Ms:2n values of 2.13 and 0.15, respectively (Schuchman *et al.*, 1984). Nevertheless, if all of the entries in Table 4.1 are weighted equally, the mean for all (129 in total) of the triplex values is 1.61 ± 0.25(SD), while that for the uniplex data (45 entries) is 0.52 ± 0.11(SD). In the case of the galactose–1-phosphate uridyltransferase (*GALT*) locus in tetrasomy 9p, the mean of the four values in Table 4.1 is 2.17, quite close to the expected value of 2.0.

The data obtained in the mouse (Table 4.2), although much more limited in scope, are similar to those in man. Seven loci, all for enzymes, have been looked at, and again the gene dosage effects are quite precise. The overall means for Ts:2n (15 entries) and Ms:2n are 1.55 ± 0.10(SD) and 0.50 (two estimates only), respectively. In one situation, it was possible to assay phosphoglycerate mutase–1A (*Pgam–1*) in four different tissues, including brain, liver, and heart, in both adult and fetal animals, and in every instance the value was very close to 1.5 (between 1.46 and 1.67) (Fundele *et al.*, 1981).

Very limited work on gene dosage effects at the level of mRNA concentration has been carried out and, as would be expected, is consistent with the protein data. Thus, in human fibroblasts trisomic and monosomic for chromosome 21, the SOD–1 mRNA concentration ratios (to the diploid amount) were 1.7–3.0 and 0.7, respectively (L. Sherman *et al.*, 1983). That the ratios were not closer to 1.5 and 0.5 is probably attributable to the poor matching of the cells used in the comparisons. Similarly, in a series of human epithelial/Burkitt lymphoma hybrid cells, there was a direct relationship between the number of Epstein-Barr virus (EBV) genome copies and the amount of EBV mRNA (Tanaka, Nonoyama, and Glaser, 1977). These findings were extended to the protein level by Shapiro *et al.* (1979), who observed a strict direct quantitative proportionality between the number of EBV genomes per cell and the amount of EBV nuclear antigen. And, again in aneuploidy for chromosome 21, the synthesis of total chromosome 21-specific mRNA was found to be gene dosage dependent, in the ratio of 0.6:2:3 for monosomy, disomy (diploid), and trisomy 21, respectively (Kurnit, 1979).

Taken together, the human and mouse protein data provide very strong support for the existence of quantitatively precise gene dosage effects for genetic imbalances affecting over forty different loci. Since, as has already been noted, virtually all of these loci code for enzymes, the generality of the gene dosage effect concept for loci coding for other types of cellular constituents remains to be established. Nevertheless, in a discussion of the mechanisms responsible for the abnormalities associated with aneuploidy, gene dosage effects at the level of the primary gene product seem to constitute a reasonable point of departure. Contrary to the prediction of Krone and Wolf (1972), gene dosage effects do not appear to "be the exception rather than the rule."

Table 4.2. *Gene dosage effects in mouse cells and tissues*

Chromosome	Gene symbol	Gene product	Tissue	Ratio of aneuploid to diploid (number of subjects)		References
				Triplex	Uniplex	
1	*Idh-1*	Isocitrate dehydrogenase-1	Embryo	1.53(6)		Epstein *et al.* (1977)
7	*c*	Tyrosine hydroxylase		~1.5		Bulfield *et al.* (1974)
7	*Mod-2*	Malic enzyme, mitochondrial	Heart mitochondria	1.44(35)		Eicher and Coleman (1977)
			Heart mitochondria		0.46(50)	Bernstein, Russell, and Cain (1978)
			Kidney mitochondria		0.54(10)	
16	*Sod-1*	Superoxide dismutase-1	Fetal cells	1.35		Cox, Tucker, and Epstein (1980)
17	*Glo-1*	Glyoxylase-1	Fetal cells	1.44		Sawicki and Epstein (unpubl.)
19	*Got-1*	Glutamate-oxaloacetate transaminase-1	Heart	1.58(4)		Fundele *et al.* (1981)
			Liver	1.67(8)		
			RBC	1.59(4)		
19	*Pgam-1*	Phosphoglycerate mutase-1A	Liver	1.48(2)		Fundele *et al.* (1981)
			Brain	1.46(2)		
			RBC	1.58(3)		
			Heart	1.62(2)		
			Fetal liver	1.64(10)		
			Fetal brain	1.67(1)		
			Fetal RBC	1.62(5)		
			Fetal heart	1.50(2)		

Dosage effects in other organisms

Mammals are not, of course, the only organisms in which gene dosage effects have been demonstrated. Similar effects have been observed throughout phylogeny, from bacteria to higher plants and insects. Jacob and Monod (1961), in their pioneering work on transcriptional regulation in *Escherichia coli*, constructed bacteria differing in the dosage of structural genes in the galactose operon. Even under inducing conditions, bacteria diploid for the genes for galactosidase and galactoside acetyltransferase produced between 1.6 and 2.6 times as much of these enzymes as did haploid organisms. Thus, even the classical form of enzyme regulation present in bacteria did not act to nullify the underlying relationship between gene number and product synthesis. Similar results were obtained by Luk and Mark (1982).

Trisomics have been recognized and characterized in a large number of plant species (Khush, 1973; Khush *et al.*, 1984), and dosage effects have been looked at in several because of their potential usefulness for gene mapping (for review, see Birchler, 1983*a*). Strict dosage effects of the type found in mammals have been found in trisomies of the tomato (*Lycopersicon esculentum*) (Fobes, 1980), Jimsonweed (*Datura stramonium*) (Carlson, 1972), and in the appropriate aneuploid and allelic series in maize (Birchler, 1981; Hedman and Boyer, 1982) (Table 4.3); qualitative (not quantitated) effects have been found in wheat (Hart and Langston, 1977) and maize (Nielsen and Scandalios, 1974). In those situations in which quantitation was performed, the trisomic:diploid ratios were very close to the expected value of 1.5.

Numerous dosage studies have been carried out in *Drosophila*, and representative examples are listed in Table 4.4. Because of the existence of regulatory elements which may produce dosage compensation (see below), it has frequently been necessary to look at dosage effects produced by segmental aneuploids with small duplications or deletions rather than at the effects of whole-arm or large-segment trisomies or monosomies. Nevertheless, according to O'Brien and MacIntyre (1978), "the assumption that gene dosage is rate limiting with regard to enzyme activity has been supported in every case of a structural gene (indicated by other criteria) thus far examined." Once again, the values for the trisomic (duplication):control ratios are quite close to 1.5. The consistency of these findings has made the analysis of gene dosage effects, especially in segmental aneuploids, one of the major tools for genetic mapping in *Drosophila* (Stewart and Merriam, 1975). Of particular note in the *Drosophila* data are the findings for those proteins that are located within membranes, for which very few data exist in mammals: α-glycerophosphate oxidase in the mitochondrial membranes, and rhodopsin in the rhabdomere membranes of photoreceptor cells.

Table 4.3. *Gene dosage effects in plants*

Organism	Chromo-some	Gene product	Ratio of trisomic to diploid	References
Tomato	2	Peroxidase-2	1.64[b]	Fobes (1980)
	6	Acid phosphatase-1	1.57	
	10	Peroxidase-4	1.51	
Datura stramonium	2	Isocitrate dehydrogenase	1.63[b]	Carlson (1972)
		Hexokinase	1.56	
	3	6-Phosphogluconate dehydrogenase	1.71	
		Lactate dehydrogenase	1.64	
	5	Malate dehydrogenase	1.53	
	9	Glucose-6-phosphate dehydrogenase	1.58	
		Glyceraldehyde-3-phosphate dehydrogenase	1.52	
	10	Alcohol dehydrogenase	1.60	
	11	Glutamate dehydrogenase	1.74	
Maize	1L 0.71–0.90	Alcohol dehydrogenase	1.48	Birchler (1981)
	—[a]	Starch branching enzyme IIb	1.52[c]	Hedman and Boyer (1982)

[a]Unknown.
[b]All tomato and *Datura* data corrected for mean activity of enzymes not coded for by unbalanced chromosome.
[c]Data from kernels with three and two active alleles at *Ae* (amylose-extender) locus, not trisomy and disomy.

Dosage compensation

It is certainly well known that, in both mammals and *Drosophila*, mechanisms exist for adjusting the synthesis of gene products of the X chromosome so that both males and females will have equivalent quantities of these products. These mechanisms are not of concern to us at this point, although it might be noted that dosage compensation of certain *Drosophila* loci on the X chromosome could be overcome by using small segmental aneuploids (Table 4.4). However, it is also the case that dosage compensation appears to exist for autosomal products as well. At the outset of this discussion, it must be stated that there are no convincing examples of autosomal dosage compensation in the human or the mouse. This does not mean that none does or could exist, but only that of the forty or so loci listed in Tables 4.1 and 4.2, there was no suggestion of compensation. By contrast, in both *Drosophila* and

Table 4.4. *Gene dosage effects in* Drosophila *segmental aneuploids*

Chromosome	Unbalanced region	Gene product	Ratio of duplication (trisomic) to control	References
X	1A-7B	Fumarase	1.47	Pipkin, Chakrabartty, and Bremner (1977)
	2C1-3C4	6-Phosphogluconate dehydrogenase	1.54	Stewart and Merriam (1975)
	16F-20	Glucose-6-phosphate dehydrogenase	1.45	Stewart and Merriam (1975)
2L	22D-24A	Phosphoglycerate kinase	1.31	Devlin, Holm, and Grigliatti (1982)
	25F-26B	α-Glycerol-3-phosphate dehydrogenase	1.42	Rawls and Lucchesi (1974)
	30F-31EF	Malate dehydrogenase, cytoplasmic	1.29	Devlin, Holm, and Grigliatti (1982)
	34B-35D	Alcohol dehydrogenase	1.79	Devlin, Holm, and Grigliatti (1982)
	36EF-37D	DOPA decarboxylase	1.41	Devlin, Holm, and Grigliatti (1982)
2R	50C-52E	α-Glycerophosphate oxidase	1.46	O'Brien and Gethmann (1973)
	55B-56C	Trehalase	1.41	Oliver, Huber, and Williamson (1978)
	T(Y;2)C	Kynurenine hydroxylase	1.38	Sullivan, Kitos, and Sullivan (1973)
3L	66B-67C	Isocitrate dehydrogenase	1.53	Rawls and Lucchesi (1974)
3R	*Dp(3;f)ry +*	Xanthine dehydrogenase	1.39	Grell (1962)[a]
	87B-E	Acetylcholinesterase	1.55	Hall and Kankel (1976)
	91B-93F	Sorbitol dehydrogenase, soluble	1.57	Bischoff (1976)
	92A-92B	Rhodopsin	1.23	Scavarda, O'Tousa, and Pak (1983)

[a]Ratio for monosomic (deficiency) using *Df(3R)*, 0.48.

plants, there are clear examples of dosage compensation which, because of the ease of genetic manipulation in these species, could be well analyzed. A few will be discussed here.

Virtually all of the *Drosophila* data in Table 4.4 were obtained by using segmental aneuploids. When the unbalanced segments were small enough, there was no evidence for compensation. However, when the results obtained with these aneuploids are compared with those obtained with whole-arm trisomics, some striking differences are found. Devlin, Holm, and Grigliatti (1982) prepared larvae trisomic for chromosome arm 2L and quantitated the activity of five enzymes coded for by genes on this arm. In two, the expected gene dosage effects were observed. If appropriate corrections are made for the differences in activities in the parental stocks, the observed trisomic:diploid ratios for alcohol dehydrogenase and DOPA decarboxylase were 1.78 and 1.65, respectively. However, when similar comparisons were made for three other enzymes, the results were quite different, and the following ratios were obtained: phosphoglycerate kinase, 1.17; α-glycerol–3-phosphate dehydrogenase, 0.99; cytoplasmic malate dehydrogenase, 1.05. These values should be compared with the values of 1.31, 1.42, and 1.29, respectively, obtained with the appropriate small segmental aneuploids (Table 4.4). Therefore, it has been inferred that there is a regulatory locus or loci on 2L which, in the trisomic state, are able to compensate for the increased number of structural genes in the distal part of the chromosome arm. Regions on 2L at which such loci may reside are suggested by the results of Rawls and Lucchesi (1974) in their work on segmental aneuploids. That the dosage compensation for the enzymes on 2L occurs at the level of transcription is supported by the finding that trisomics with three different electrophoretic alleles of α-glycerol–3-phosphate dehydrogenase do synthesize all three forms of the enzyme, but each in one-third less an amount than would have been expected (Devlin, Holmes, and Grigliatti, 1982).

Quite similar results were obtained when dosage studies were carried out on alcohol dehydrogenase (ADH) in maize. When plants with one to four copies of the distal 80% of chromosome arm 1L were prepared, the activities of alcohol dehydrogenase, the structural locus for which is known to be on 1L, were in the ratio of 1.6:2.0:2.3:2.4 instead of the expected 1:2:3:4 (Birchler, 1979). However, when segmental aneuploids of the region 1L 0.72–0.90 were examined, the ratios (corrected for the differences in activity of the different *Adh* alleles used in the cross) were 2.0:3.2:4.2 for two, three, and four copies, respectively (the trisomic figure is the mean of two different determinations giving relative values of 2.8 and 3.6) (Birchler, 1981). Once again, a locus or loci outside of the structural locus were affecting the activity of the enzyme, and the region 1L 0.20–0.72 was implicated as a site for this locus or loci. When measured directly, three copies of this region reduced the ADH activity produced by two copies of the *Adh* locus by about a quarter (23%), and comparison of trisomy for the two regions combined (1L 0.20–0.90) with control disomics showed that complete compensation had occurred (Birchler, 1981). In theory, a reduction of one-third would just compensate for an increase by 50% ($2/3 \times 3/2 = 1$). The concentrations of several other enzymes and proteins mapped to 1L were also compensated to some degree.

Dosage compensation has also been reported for esterase–1 in the tomato, with a trisomic:disomic ratio of 0.98 (Fobes, 1980). Evidence for the participation of all three loci in the trisomic was obtained from electrophoretic analyses of the products of different *Est–1* alleles.

Dosage compensation, possibly of a different type, has been reported in yeast (Osley and Hereford, 1981). Duplications in a haploid organism of the region carrying the genes for histone 2B and for another protein designated as protein 1 result in respective 2.1- and 2.0-fold increases in the transcription rates of these two loci. However, when the steady-state concentrations of the mRNAs of these two proteins were examined, the ratios were 1.2 and 2.2, indicating that compensation for the concentration of the histone H2B mRNA was occurring. By direct measurement, this was shown to be a posttranscriptional twofold increase in the rate of H2B mRNA turnover. Similarly, both posttranscriptional and translational dosage compensation have been shown for the ribosomal protein, L3, in yeast (Pearson, Fried, and Warner, 1982), and posttranscriptional or translational compensation for the ribosomal protein, S20, in bacteria (Parsons and Mackie, 1983). Dosage compensation of the reverse type has been observed in anucleolate mutants of *Xenopus*, with amplified production of oocyte ribosomal RNA occuring at the same level in eggs with one or two copies of the rRNA loci (Brown and David, 1968).

Special situations

It was stated earlier that autosomal dosage compensation is not known to occur in mammals. However, several special situations do deserve mention. While mammalian immunoglobulin loci, being autosomal, occur in pairs (on homologous chromosomes), only one allele at each locus is expressed in any immunoglobulin-producing cell. This suppression of the second set of alleles, referred to as allelic exclusion, is mediated by the genomic rearrangements which occur during ontogeny of the immunoglobulin-producing system in B lymphocytes (Perry *et al.*, 1980; Coleclough *et al.*, 1981). Only one allele at each locus is permitted to reach a genetically active configuration in a diploid cell, and immunoglobulin production is therefore quantitatively independent of the number of genes actually coding for each region of the immunoglobulin heavy and light chains (Taub *et al.*, 1983). However, it is not known what would happen in a lymphocyte progenitor cell that is trisomic for an immunoglobulin locus-carrying chromosome. This is a question that is now approachable experimentally in the mouse (see Chapter 10)

Examination of two-dimensional polyacrylamide electrophoretic gel patterns of proteins extracted from trisomic and normal cells has not in general revealed the number of gene dosage-dependent changes that might be expected. For example, in analysis of cells from four different mouse trisomies involving chromosomes ranging in size from largest to smallest, only a total of 13 changes compatible with strict gene dosage [mean trisomy:diploid ratios of 1.68 ± 0.51(SD)] were found in 1250 polypeptides (Klose and Putz, 1983). The expected number would be of the order of 200–250. Klose and Putz (1983) suggested, therefore, that a stable regulation may exist for the cellular concentration of many proteins, a form of regulation that would be equivalent to dosage compensation. Whether this is the correct explanation for the electrophoretic findings remains to be determined. Similar conclusions were suggested by Klose, Zeindl, and Sperling (1982) on the basis of their failure to find any consistent dosage effects in human trisomy 21 cells. However, Van Keuren, Merril, and Goldman (1983), in a similar analysis of proteins from human trisomy 21 fibroblasts, found two quantitatively altered proteins (superoxide dismutase–1 and an unknown peptide #377) among 400 spots, about a third the number that would be expected in view of the fact that chromosome 21 represents about 1.7% of the human genome on the basis of its R-band content (Martin and Hoehn, 1974).

It might be expected that the concentration of serum albumin would be subject to reasonably tight control, by alterations in either the rate of synthesis or of degradation. The rare families in which a gene for analbuminemia, an autosomal recessive condition, is segregating provide an interesting test of this possibility. In the family reported by Boman *et al.* (1976), the father and mother of the affected individual, who were obligate heterozygotes, had serum albumin concentrations of 3.3 and 2.8 g/dl, respectively, while several other possible heterozygotes had concentrations averaging 3.2 g/dl. The 95% range in the laboratory was given as 3.7–5.2 g/dl, and I therefore presume a mean of about 4.5 g/dl (assuming a reasonably symmetrical distribution). Therefore, both the obligate and the presumed heterozygotes had albumin concentrations about 70% of normal. Similar albumin concentrations, 3.3 and 3.0 g/dl, were also reported for the parents of the original cases reported by Bennhold, Peters, and Roth (1954). It thus appears that the heterozygote for analbuminemia does compensate to some extent for the loss of one albumin gene, but this compensation, however accomplished, does not appear to be complete. Rats heterozygous for analbuminemia appear to compensate completely (Nagase, Shimamane, and Shumiya, 1979).

A special type of dosage compensation mechanism, which has been defined for prokaryotes but for which only a few possible precedents exist in mammals, is what has been termed "autogenous regulation" (Goldberger, 1974). In this form of regulation, a protein specified by a structural gene is itself a regulatory element which in some manner moderates the expression of the very same gene. A possible example of dosage compensation of this type was reported by Guialis *et al.* (1977). They observed that the activity of RNA polymerase II in Chinese hamster ovary (CHO) hybrid cells containing both wild-type and α-amanitin-resistant polymerase was the same when the cells were grown in the presence or absence of α-amanitin. Since the wild-type polymerase is inactivated by the α-amanitin, this means that the activity of the resistant enzyme had to rise proportionally to maintain the total enzyme activity at a constant level. RNA polymerase II is, of course, essential for all RNA synthesis, and these findings were thought to indicate that the enzyme might be regulating its own rate of synthesis. A very similar result was obtained by Somers, Pearson, and Ingles (1975) in cells presumably heterozygous for a mutation that results in the production of an α-amanitin-resistant RNA polymerase II.

An example of supposed dosage compensation in response to an alteration in the number of genes for U1 small nuclear RNA (snRNA) has been reported by Mangin, Ares, and Weiner (1985). When multiple active copies of the gene for human U1 snRNA were introduced into cultured mouse cells and maintained extrachromosomally, the total amount of cellular U1 snRNA (mouse plus human) remained relatively constant, irrespective of how much human U1 snRNA was made. The point in the synthesis of U1 snRNA at which the level of total U1 snRNA is regulated is not known.

Finally, humans heterozygous for a deficiency in the C complementation group (*pccC*) of propionyl coenzyme A carboxylase (PCC) deficiency have PCC activities indistinguishable from normal, as opposed to A complementation group heterozygotes with 52% of normal activity (Wolf and Rosenberg, 1978). The explanation suggested for this difference and apparent lack of gene dosage effect for *pccC* is the following. PCC is a multimeric enzyme composed of two nonidentical subunits. Therefore, if one subunit, the product of *pccC*, is synthesized in sufficient excess to the product of the other locus, *pccA*, a 50% reduction in its synthesis would not be expected to have any measureable effect on PCC concentration. While not really compensation, this is an example of how a gene dosage effect would fail to become manifest at the level of the functional gene product.

Although demonstrations of autosomal dosage compensation per se have so far been restricted almost entirely to organisms other than mammals, the underlying mechanisms may still have relevance to the mammalian situation. The work on dosage compensation in *Drosophila* and maize has pointed to the existence of regulatory loci which modulate the expression of other loci so as to maintain the outputs of the latter at constant levels. Such loci need not be tightly linked to the structural loci, and changes in their gene dosage may affect the transcription of more than one structural locus. Whether such regulatory loci exist in mammals is still a matter for conjecture. However, because of the implications which they may have for understanding the effects of aneuploidy in mammals, they will be considered again in Chapter 6.

5

Metabolic pathways, transport systems, and receptors

Secondary effects of aneuploidy

In earlier discussions of the effects of aneuploidy (Weil and Epstein, 1979; Epstein *et al.*, 1981), the effects were divided into two categories – primary and secondary. The term "primary" was used to refer to the gene dosage effects discussed in the previous chapter, and "secondary" to denote all of the effects consequent to the primary ones (Fig. 5.1). The major reason for drawing this distinction was to focus attention on the fact that the primary gene dosage effects do not in themselves explain the deleterious outcomes of aneuploid states and that it is necessary to consider the functional or secondary consequences of these quantitative alterations on the synthesis of gene products. The significance of an aneuploidy-produced increase or decrease in the synthesis of a gene product must therefore be understood in terms of the function of the product and the alterations in this function which result from a change in the quantity of the gene product. In this context, the term "function" is being used in a very general sense and refers to whatever role the gene product has in the development, structure, or metabolism of the organism.

In addition to the distinction between primary and secondary effects of aneuploidy, a distinction was also made between two categories of secondary effects – direct and indirect (Fig. 5.1). The direct effects were considered to be those effects on functions directly mediated by the gene products of the unbalanced chromosome or region, while the indirect effects were those on functions mediated by products of other chromosomes or regions. This distinction, admittedly somewhat arbitrary, was made to separate those phenomena, principally regulatory in nature, which might affect the expression of genes on many chromosomes from those outcomes which could be traced directly back to imbalance of specific genes. Its usefulness lies in reminding us that both types of possibilities must be considered when we attempt to explain or predict the effects of genetic imbalance at any specific locus or set of loci.

In the following discussion, I shall consider several of the mechanisms by which a gene dosage effect could be translated into an abnormality of development, structure, or metabolism. While principally theoretical in nature, the discussion will include specific examples when relevant information is available. In many respects, the issues to be raised here are quite similar to those

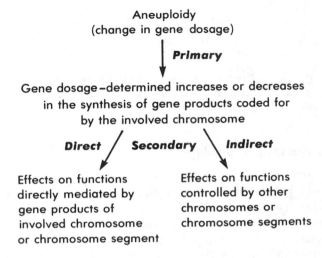

Fig. 5.1. The primary and secondary effects of aneuploidy.

applicable to inherited metabolic disorders in general, especially in the parallels between the effects of monosomy or deletions and of heterozygosity for an enzyme or other protein deficiency state [see Harris (1980) and Vogel (1984) for general discussions of the effects of heterozygosity]. However, it must be kept in mind, when considering the lessons to be learned from examples of heterozygosity, that we are often dealing with qualitative as well as quantitative abnormalities, while in aneuploidy the abnormalities are purely quantitative.

Metabolic pathways and enzyme reactions

Theory

Since so many of the demonstrated gene dosage effects have involved enzymes, I shall begin this analysis of the secondary effects of aneuploidy with a consideration of the effects of altering enzyme concentrations on metabolic pathways and enzyme reactions. The real question here is whether changing the concentration of one component of the pathway, by an amount equal to plus or minus 50%, will affect either the transit (the flux) of metabolites through the pathway or the concentration(s) of the substrates, products, and metabolic intermediates of the pathway, or both. In its simplest form, this question might be thought of as being tantamount to asking whether the enzyme in question is "rate-limiting" for the pathway, so that a change in its concentration is accompanied by an equivalent change in metabolite flux (Fig. 5.2). However, as Kaczer and Burns (1973) point out in their theoretical

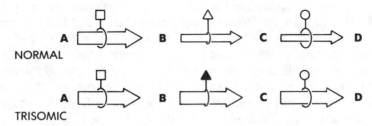

Fig. 5.2. Hypothetical example of a direct secondary effect of trisomy on a metabolic pathway controlled by a "rate-limiting" enzyme. In the pathway shown, reaction B→C is controlled by "rate-limiting" enzyme, BC. In the trisomic state (below), the amount of BC is increased and the rate of transit of metabolites is commensurately raised, thereby resulting in a greater production of metabolite D. Reprinted by permission from Epstein *et al.* (1981).

exposition on the control of flux in metabolic reactions, only one enzyme in a pathway can be of this type and approach full control of the rate of flux. In their view, it is equally likely that none of the enzymes in a pathway will be of major (rate-limiting) importance, a conclusion they consider to be consistent with the fact that the usual 50% reduction in enzyme activity in heterozygotes for enzyme deficiency is rarely accompanied by readily demonstrable metabolic changes (see below). Stated another way, this means, according to them, that "in a system containing a reasonable number of enzymes, almost all enzymes will appear to be 'in excess' in the sense that, choosing any one, its quantity or activity can be reduced (sometimes considerably) without appreciable effect on the flux." Presumably the same conclusion would hold for commensurate increases in enzyme activity.

However, these latter considerations should not be taken to mean that changing the enzyme concentration will have no effect on either flux or metabolite pool sizes, and it is useful to consider briefly the theoretical aspects of this problem. In their treatment of the problem, Kaczer and Burns (1973) develop a general equation relating several parameters and variables for an enzyme pathway of the form

$$X_1 \xrightarrow{E_1} S_1 \xrightarrow{E_2} S_2 \to S_{n-1} \xrightarrow{E_n} X_2$$

where X_1 is the substrate; X_2 the product; $S_1, S_2, \ldots S_{n-1}$ the intermediates in the reaction; and $E_1, E_2 \ldots E_n$ the several enzymes in pathway. The equation, expressed in terms of the pathway flux, F, at the steady state and in the absence of any fully saturated enzymes is

$$F = (X_1 - X_2 / K_1 K_2 \ldots K_n) \left/ \left[\frac{M_1}{V_1} + \frac{M_2}{V_2 K_1} + \frac{M_3}{V_3 K_1 K_2} + \ldots \right] \right. \qquad (5.1)$$

where X_1 and X_2 are the substrate and product concentrations, respectively, and K, M, and V are the respective equilibrium constant, Michaelis constant, and maximal velocity for each of the enzymes in the pathway. A change in the concentration of one enzyme is thus a change in the value of V for that enzyme, and the effect of this change in V will be determined by the values of all of the other parameters (K, M, and V) and variables (X_1, X_2). Since all of the parameters of the reaction are considered to be fixed, the change in V must alter any, some, or all of the variables X_1, X_2, and F, depending on how the system is otherwise constrained. The magnitude of the alteration will be determined by the quantitative relationships among the several parameters.

Similar considerations also apply to the concentrations, S, of the intermediate metabolite pools. At the steady state, these concentrations are determined by equations of the form

$$F = \frac{V_1}{M_1}\left[X_1 - \frac{S_1}{K_1}\right] = \frac{V_2}{M_2}\left[S_1 - \frac{S_2}{K_2}\right] = \frac{V_n}{M_n}\left[S_{n-1} - \frac{X_2}{K_n}\right] \quad (5.2)$$

Therefore, a change in F resulting from an increase or decrease in the concentration (V) of one of the enzymes in the pathway may affect the concentrations of all of the intermediates in the pathway. According to Kaczer and Burns (1973), the issue therefore comes down not to whether a given enzyme is truly rate limiting, but to how great the effect of changing its concentration will be on fluxes and metabolite concentrations. Depending on the parameters of the system, which, in turn, determine where an organism is on the curve of flux versus enzyme concentration (Fig. 5.3), this effect may range from virtually nil to directly proportional to the magnitude of the change in enzyme concentration. For convenience in representing the change in flux to be expected from a change in enzyme concentration, Kaczer and Burns (1973, 1981) define the relationship between flux and enzyme concentration in terms of a sensitivity coefficient, Z, which is a function of the slope of the curve of flux, F, versus enzyme concentration, E, so that:

$$Z = \frac{dF}{dE} \cdot \frac{E}{F} = \frac{dF}{F} \bigg/ \frac{dE}{E} \quad (5.3)$$

This relationship is illustrated in Fig. 5.3. The importance of the sensitivity coefficient is that it can be shown, for any multienzyme metabolic pathway of the type depicted above, that the sum of the sensitivities, ΣZ_i, of all of the enzymes of the pathway must be equal to 1. It is this conclusion which forms the basis for the assertions of Kaczer and Burns (1973) with regard to the possible existence of rate-limiting enzymes cited at the beginning of this section.

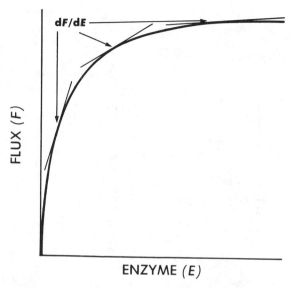

Fig. 5.3. The sensitivity coefficient as a function of the enzyme/flux relationship. The sensitivity coefficient, Z, is defined as

$$Z = \frac{dF}{dE} \cdot \frac{E}{F}$$

where F is the flux and E is the enzyme concentration. Redrawn from Flint *et al.* (1981).

Experimental and clinical evidence

The theoretical treatment just outlined is, of course, a simple one which ignores a variety of situations that complicate real-life metabolic reactions, such as feedback inhibition, branching pathways, and irreversible reactions. However, it does provide us with a conceptual framework with which to consider the potential effects of aneuploidy on enzyme-mediated metabolic reactions. At this point, therefore, it is appropriate to look at the experimental and clinical data which bear on the theoretical points just discussed. Although much of these data are derived from states of enzyme deficiency and excess rather than aneuploidy, they are still applicable to the issues under consideration.

In support of their theoretical analysis, Kaczer and Burns (1981) provide several experimentally derived examples of flux-enzyme concentration relationships, and these are shown in Fig. 5.4. All of the curves are quite similar in form to that shown in Fig. 5.3. It can easily be appreciated that, for most systems, reducing enzyme activity by 50% has a demonstrable but usually quite small effect on product generation (flux). Nevertheless, in the case of the tyrosine hydroxylase mutants, melanin production is significantly reduced

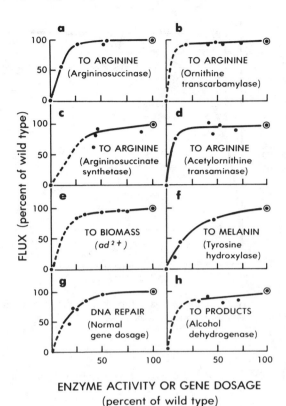

Fig. 5.4. Experimentally determined relationships between flux and enzyme concentration or gene dosage. In all figures, the circled point represents wild-type activity, which is scaled to 100. (a) to (d) *Neurospora* heterokaryons constructed with differing activities of the indicated enzymes in the arginine synthetic pathway. (e) *Saccharomyces* strains with different activities at the phosphoribosylaminoimidazole carboxylase (*ad–2*) locus. (f) Mouse mutants at the albino (*c*) locus. The intermediate points are *c*^ch*c*^ch (chinchilla) and albino heterozygotes (*cc*^ch,*Cc*). Enzyme activities and melanin production were estimated in skin extracts of 5- to 6-day-old mice. (g) DNA repair, estimated by thymidine incorporation, in tissue culture heterokaryons between normal nuclei and nuclei from xeroderma pigmentosum cells. (h) *Drosophila* mutants at the alcohol dehydrogenase (*Adh*) locus. Redrawn from Kacser and Burns (1981).

at the 50% enzyme concentration point. What these curves do not show, however, is what happens to the concentrations of the metabolites in the affected enzyme pathways. If we examine the metabolite pool data in the experiments on the *Neurospora* arginine pathway represented by parts (a)-(d) in Fig. 5.4 (Flint *et al.*, 1981), we observe the changes in pool sizes summarized in Table 5.1. Although the overall flux to arginine is virtually unchanged, there are

Table 5.1. Changes in pool sizes resulting from ~50% reductions in enzyme concentrations in the arginine pathway of Neurospora crassa[a]

	Glutamate→Ornithine		Ornithine carbamoyl- transferase →	Citrulline	Argininosuccinate synthetase →	Arginino- succinate	Arginino- succinase →	Arginine→Protein (soluble)	
Wild type	122	11		3.7		1.3		24	130
48% Ornithine carbamoyltransferase	105	21	[b]	3.4		1.3		8	149
47% Argininosuccinate synthetase	91	8		25	[b]	1.3		22	129
54% Argininosuccinase	127	15		6		6.6	[b]	24	126

Taken from data of Flint et al. (1981).
[a] Pool sizes expressed as micromoles per gram dry weight.
[b] Point in pathway at which enzyme concentration is reduced.

significant increases (from two to seven fold) in the sizes of the intermediate metabolite pools, particularly at the points in the pathway just proximal to the enzymes that are reduced in concentration.

In mammals, the principal data comparable to those just presented for *Neurospora* are derived from the examination of individuals heterozygous for an enzyme deficiency. With few exceptions, the metabolic effects are minimal. For example, in mice heterozygous for histidase deficiency, the mean concentration of histidine in the liver was increased by about 30%, with substantial overlap in the distribution curves for normal and heterozygous individuals (Bulfield and Kacser, 1974). The plasma concentrations of phenylalanine (P) in individuals heterozygous for phenylketonuria were increased by 20% in the fasting state (Ford and Berman, 1977) and 46% in a semifasting state at midday (Griffin and Elsas, 1975), and the tyrosine concentrations (T) were decreased by 5 and 16%, respectively. In lymphocytes, the phenylalanine concentration was increased by 33% (Thalhammer, Lubec, and Königshofer, 1979). While the differences are statistically significant, there was in all instances a sizable overlap between phenylalanine concentrations in the heterozygotes and normals. This overlap could be largely eliminated in the semifasting plasma values by using either P^2/T or a discriminant function of the form $0.0239 P - 0.0170 T$ (Griffin and Elsas, 1975). Heterozygotes for cystathioninuria (γ-cystathionase deficiency) are also sometimes identifiable by increased excretion of cystathionine in the urine (Mudd and Levy, 1978), and heterozygotes for argininosuccinic aciduria (argininosuccinase deficiency) by increased excretion of argininosuccinate.

While relatively few heterozygous conditions are characterized by demonstrable abnormality in the basal or unstressed state, many more can be identified by stressing the system by a tolerance or loading test. This may result in a decreased ability to handle the increased level of substrate, an outcome well known for phenylketonuria (Cunningham *et al.*, 1969; Ford and Berman, 1977), and also recognized for several other inborn errors including galactosemia (Donnell, Bergren, and Ng, 1967), pentosuria, homocystinuria, cystathioninuria, hypersarcosinemia, and the Crigler-Najjar syndrome (Stanbury, Wyngaarden, and Frederickson, 1978, 1983). Another interesting example is provided by heterozygotes for the 21-hydroxylase deficiency type of congenital adrenal hyperplasia. Prepubertal and early pubertal heterozygous children had a 2.5-fold higher basal serum level of 17-hydroxyprogesterone and a 3.4-fold higher value six hours after administration of ACTH (Lorenzen *et al.*, 1979).

Taken together, these results indicate that, for many inborn errors of metabolism, the heterozygous state, with a demonstrated or presumed 50% reduction in enzyme activity, may produce subtle but recognizable alterations in metabolism. These are observable particularly under conditions of metabolic stress, which may actually be closer to real life situations than is the fasting state. But, even if demonstrable, do these changes have any implications for the development and function of the organism? Here the data are very difficult to obtain, but once again phenylketonuria provides some intriguing information. Studies of intelligence quotients (IQ) in phenylketonuria heterozygotes have been reported to reveal a small (5 IQ points) but still statistically significant decrease in total IQ scores when compared either to noncarrier sibs (Ford and Berman, 1979) or to histidinemia heterozygotes (Thalhammer *et al.*, 1977), and Vogel (1984) has pointed to other "peculiarities" including a higher risk of psychoses. How these might occur, if the findings are real, is still unknown. One possibility is, of course, a low level of phenylalanine toxicity, and in this regard the findings in hereditary fructose intolerance are of note. It has been reported that incomplete treatment of this condition (in the homozygous form) may result in growth retardation even in the absence of any symptoms of either complete or chronic fructose intoxication (Mock *et al.*, 1983).

Precedents from the inborn errors of metabolism

From the foregoing discussion, it would appear that, because of the built-in buffering capacity of metabolic pathways, decreases in enzyme activity of the order of 50% have relatively small, if any, effects on fluxes through the pathways. The effects on metabolite pools may be greater, but in most instances, appear to be innocuous for the individual. However, as the several dominantly inherited porphyrias indicate, this does not tell the whole story. Individuals heterozygous for one of several mutations involving heme biosynthesis may be quite severely affected. Through the efforts of a large number of investigators, the situation of the porphyrias, heretofore quite complicated, now appears to be reasonably straightforward [for summaries, see Meyer and Schmid (1978) and Sassa and Kappas (1981)]. The pathway of heme biosynthesis, starting with the condensation of succinyl-CoA and glycine to δ-aminolevulinic acid (ALA) by ALA synthase is mediated by a series of porphyrin biosynthetic enzymes (Fig. 5.5). At each step of the pathway, an enzyme defect has been identified and associated with a specific disease, all but one of which, congenital erythropoietic porphyria, is dominantly inherited. And, as might be expected for conditions with a dominant mode of inheritance, the reductions in enzyme activity are of the order of 50% (Table 5.2). The only exception is ferrochelatase in protoporphyria which, for reasons that are unclear, has a considerably lower enzyme activity although, consistent with a ~50% decrease in enzyme activity, the elevation in protoporphyrin IX concentration in cells is only 2.4 fold.

How does a 50% reduction in the activity of one of the enzymes of the pathway result in disease? Unfortunately, the explanations are still incomplete and tentative (Meyer and Schmid, 1978; Sassa and Kappas, 1981), but in

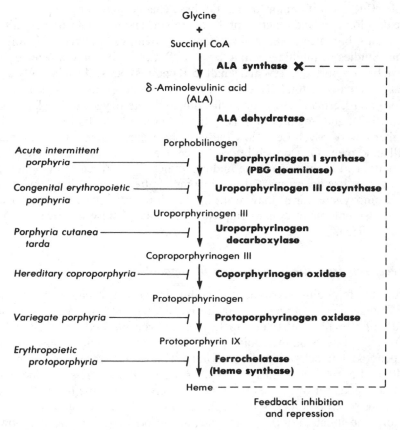

Fig. 5.5. The pathway of heme biosynthesis. Each of the intermediates and catalyzing enzymes (right) is shown as well as the disorders (left) associated with deficiencies of specific enzymatic steps. All disorders, with the exception of congenital erythropoietic porphyria, are dominantly inherited.

general terms they go as follows. Reduced enzyme activity at any of several points in the pathway may result in an impairment of heme biosynthesis. This reduction in heme produces, in turn, an increase in the activity of ALA synthase by a combination of enzyme induction (derepression) and possibly of decreased inhibition of the enzyme by heme (Sassa and Kappas, 1981). As ALA synthase activity increases, the synthesis of ALA is increased and more precursor is thereby fed into the heme biosynthetic pathway, only to accumulate proximal to the point of obstruction. Depending on the properties of the particular precursors (porphobilinogen, aminolevulinic acid, and one or more porphyrin intermediates and metabolic derivatives) which are accumulated, toxic effects may be produced in the nervous system and the skin, the latter because of the ability of porphyrins to induce light sensitivity. What is

Table 5.2. *Enzyme activities in the dominantly inherited porphyrias*

Condition	Enzyme	Tissue	Activity[a]	References
Acute intermittent porphyria	Uroporphyrinogen I synthetase	Liver	37%[b]	Strand *et al.* (1970)
		Red cells	45	Strand *et al.* (1972)
Porphyria cutanea tarda	Uroporphyrinogen III decarboxylase	Liver	47	Felsher *et al.* (1982)
		Red cells	56	Elder *et al.* (1983)
Hereditary coproporphyria	Coproporphyrinogen oxidase	Fibroblasts	53	Elder *et al.* (1976)
		Lymphocytes	48	Nordmann *et al.* (1977)
Variegate porphyria	Protoporphyrinogen oxidase	Fibroblasts	43	Brenner and Bloomer (1980)
Protoporphyria	Ferrochelatase	Fibroblasts	21	Bloomer (1980)
		Liver	13	Bonkowsky *et al.* (1975)
		Lymphocytes	(2.4[c])	Sassa *et al.* (1982)

[a]Relative to normal controls.
[b]Only two subjects, with two determinations on one.
[c]Increase in protoporphyrin IX in mitogen-stimulated lymphocytes.

perhaps most remarkable about several of the porphyrias is that individuals may have a genetic defect, with a demonstrable decrease in enzyme activity, and still not have any clinical manifestations. What appears to be required is some additional environmental factor, endogenous or exogenous, to destabilize the system. Among the recognized agents that may precipitate acute attacks are a large number of drugs and steroids which are capable of directly inducing ALA synthase, thereby exacerbating the situation further. It is possible that these factors, rather than a decreased synthesis of heme, are actually the more important ones in the pathogenesis of the disease (Sassa and Kappas, 1981). But, whatever the cause, it is of interest that administration of exogenous heme can prevent or abort attacks by inhibiting and repressing ALA synthase (Lamon *et al.*, 1978).

Do the considerations of Kacser and Burns (1973, 1981) discussed earlier apply to the porphyrias? In the absence of appropriate measurements, it is difficult to say. Furthermore, the situation is complicated by the fact that secondary changes in this system, particularly the induction of ALA synthase, play an important role in the outcome. Nevertheless, it is possible to look at the porphyrias as situations in which the system attempts to maintain flux at an appropriate level. Because of the reduced activity of one of the enzymes in the pathway, this can be done only at the expense of expanding the pools of certain of the intermediates in the pathway. And, because of the particular properties of these intermediates, it is exactly this increase in their concentrations which leads, in whole or in part, to the development of the abnormalities associated with this group of conditions. Thus, a control mechanism that is presumably advantageous under normal genetic circumstances becomes deleterious when the gene dosage of one of its constituent enzymes is altered.

Another condition worthy of mention, similar to the porphyrias in some respects, is deficiency of the mitochondrial enzyme, ornithine carbamoyltransferase (OCT) (Shih, 1978). Since this is X-linked (Short *et al.* 1973), rather than autosomal, it cannot be safely assumed that the overall enzyme activity in the livers of all heterozygotes is 50% of normal or that assay of enzyme activity in another tissue, such as duodenum, necessarily reflects the activity in the liver. Furthermore, the reduced enzyme activity does not affect every cell. Rather, because of the X-inactivation process, some cells have normal activity while others are totally deficient (Ricciuti, Gelehrter, and Rosenberg, 1976). This cellular mosaicism makes interpretation of the results of activity measurements in liver specimens obtained by needle biopsy quite difficult (see Glasgow, Kraegel, and Schulman, 1978). Nevertheless, it has been shown that all heterozygotes given a protein load have an abnormal response with an elevated excretion of orotic acid in the urine (to as high as 20 times normal) (Batshaw *et al.*, 1980; Haan *et al.*, 1982), and many develop hyperammonemia as well (Batshaw *et al.*, 1980). It has been suggested that this outcome may, in large measure, be the result of the mosaic nature of the heterozygous liver rather than of a simple ~50% reduction in enzyme activity (Glasgow, Kraegel, and Schulman, 1978). As with the porphyrias, the heterozygotes can remain asymptomatic or develop acute attacks. In addition, and reminiscent of the heterozygous state for phenylketonuria, it has been claimed that OCT-deficiency heterozygotes, presumably because of their exposure to repeated asymptomatic episodes of hyperammonemia, have a slightly lower IQ (5.6 ± 1.7 points) than their unaffected sibs (Batshaw *et al.*, 1980).

An interesting case of partial deficiency of plasminogen, the precursor of the fibrinolytic enzyme, plasmin (Fig. 5.6), has been reported in an an individual with recurrent thromboses (Aoki *et al.*, 1978). The person, heterozygous for an inactive structural variant of plasminogen, had 41% of normal plasminogen activity in a casein hydrolysis assay. The minimal amount of activity necessary to maintain effective fibrinolysis in vivo may be very critical, as may the existence of other precipitating factors, since several other

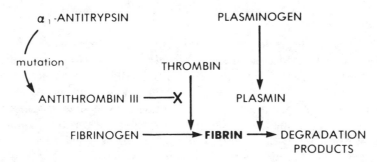

Fig. 5.6. Schematic view of a small part of the fibrin-generating and -degrading system. A decrease (~50%) in antithrombin III, an inhibitor of thrombin, or in plasminogen, a precursor of the fibrin-degrading protease, plasmin, results in a hypercoagulable state and thromboembolism. Conversely an increase (~50%) in antithrombin III activity, resulting from the presence of a mutant form of α_1-antitrypsin, produces a bleeding disorder. Thus, 50% increases or decreases in critical components of a delicately balanced enzyme system can have deleterious effects.

putative heterozygotes with plasminogen activities between 40 and 60% were asymptomatic. Similar findings were also obtained by Hasegawa, Tyler, and Edson (1982) in three unrelated patients and their families.

Although not an enzyme deficiency condition, heterozygosity for α_1-antitrypsin deficiency is another condition in which a decrease in protein concentration may be associated with abnormality. It has been claimed that heterozygotes are more susceptible than normal individuals to the development of chronic pulmonary disease with emphysema, but this relationship has been questioned. Nevertheless, it does appear that there is some evidence that heterozygotes may acquire subtle differences in the mechanical properties of the lungs at an earlier age than nonheterozygotes and that these differences may be accentuated by smoking (Morse, 1978).

Another antienzyme deficiency which produces disease is antithrombin III deficiency (Fig. 5.6), antithrombin III being a thrombin inhibitor which is structurally very closed related to α_1-antitrypsin (Owen *et al.*, 1983). Heterozygotes for this disorder have a mean plasma antithrombin III concentration of 45% of the midpoint of the normal range and a functional activity of about 58% of the midpoint of the normal range (Halal *et al.*, 1983). With few exceptions, all heterozygotes are symptomatic as adults and display a wide variety of thromboembolic phenomena. Thus, as both this disorder and partial deficiency of plasminogen described earlier indicate, the coagulation system is apparently quite delicately balanced and, because of the amplification deriving from the cascade nature of the system, a concentration change of the order of 50% in one of the components is quite capable of causing serious problems.

Increased enzyme activity

All of the examples cited so far involve situations of enzyme or enzyme inhibitor deficiency and as such are potentially applicable to the consideration of monosomy (deletion) but not of trisomy (duplication). From the results presented in Fig. 5.4, it might be expected that an increase in the activity of a single enzyme in a pathway would have little or no effect unless the enzyme was indeed functioning in a rate-limiting manner with a sensitivity coefficient, Z, close to 1.0. The example of antithrombin III deficiency just discussed could provide a precedent for what might be expected if, rather than an antienzyme being deficient, the enzyme to be inhibited were itself increased in amount so that the normal concentration of antienzyme would no longer be sufficient to inhibit it to the appropriate degree. No examples for this are available, but another mutation affecting α_1-antitrypsin does provide a fascinating example of how an increase in an antienzyme can have deleterious effects. α_1-Antitrypsin Pittsburgh (358 Met→Arg) is a mutant form of α_1-antitrypsin in which a single amino acid substitution converts α_1-antitrypsin into a molecule with antithrombin III activity (Fig. 5.6) (Owen *et al.*, 1983). The effect of this is to have about a 50% excess of circulating antithrombin III activity, which in turn produces a coagulation state equivalent to that resulting from full heparinization. The patient with the mutant α_1- antitrypsin had an episodic bleeding disorder which ultimately proved fatal to him.

Despite the cases just cited, very few examples are actually known of genetically determined increases in enzyme activity in the absence of aneuploidy, and of these only one has been investigated to any degree. It is believed that the regulation of the formation of phosphoribosylamine is the single most important factor in the control of de novo purine biosynthesis (Wyngaarden and Kelley, 1978). Phosphoribosylamine concentration is determined by the concentration of the precursor, phosphoribosylpyrophosphate (PRPP), which in turn, is controlled by the enzyme, PRPP synthetase. Several families have now been reported in which the activity of PRPP synthetase, an X-linked enzyme, is increased and with it the synthesis of purines and hence of uric acid. In the fibroblasts of affected male hemizygotes, PRPP synthetase activity may be increased about three fold, and in the cells of female heterozygotes about two fold (Becker *et al.*, 1982; Sperling *et al.*, 1978). This results in parallel increases in PRPP generation and concentration and in purine synthesis (Becker *et al.*, 1982). Thus, in a situation in which an enzyme does appear to function in a rate-determining manner, increased activity is accompanied by increased product generation, in this instance of a product (uric acid) which can have deleterious effects if in too high an amount. An experimental system for examining the relationship of metabolite flux to the activity of a rate-limiting enzyme has been described in *Escherichia coli* by Walsh and Koshland (1985).

Two additional pieces of evidence concerning the effects of increased enzyme activity are to be found in the literature on aneuploidy. In Fig. 5.4(f), a graph is shown relating melanin synthesis by mouse skin to the activity of tyrosine hydroxylase. Male mice carrying the Cattanach translocation, *T(X;7)Ct*, in which a segment of chromosome 7 is inserted into the X chromosome, are triplex for the *c* locus, which controls the synthesis of tyrosine hydroxylase. In these animals, the tyrosine hydroxylase activity of the skin is increased approximately 1.5 fold while the melanin content is increased 1.2 fold (Bulfield *et al.*, 1974) – about what would have been expected by extrapolating the curve in Fig. 5.4(f) to 150% tyrosine hydroxylase activity. In an entirely different vein, red cell glucose metabolism has been investigated in an individual with a duplication of chromosome arm 10p, the site of the locus for hexokinase. This enzyme, the first and presumed rate-limiting step in glycolysis, was increased 53% in red cells, while the activities of all of the other glycolytic and hexose monophosphate shunt enzymes were unchanged (Magnani *et al.*, 1983). Presumably as a result of this increase in hexokinase activity, the utilization of glucose by the trisomic red cells was increased two fold, and the concentrations of several of the glycolytic pathway intermediates were also increased. Conversely, in a patient with an interstitial deletion of 9p13.3→p23, red cell galactose–1-phosphate uridyltransferase activity was decreased by 35% and the conversion of [1–^{14}C]galactose to ^{14}CO$_2$ by 58% (Bricarelli *et al.*, 1981). In a nongenetic case, the stimulation of platelets with arachidonic acid or thrombin resulted in malondialdehyde and thromboxane production in amounts inversely related to the activity of intracellular glutathione peroxidase (Guidi *et al.*, 1984).

Inferences

These, then, are the data concerning the consequences of decreasing or increasing enzyme activities by about 50% on the metabolic reactions in which they are involved. The paucity of recognized dominantly inherited enzyme deficiency states would suggest that these consequences are likely to be minimal at least insofar as their ability to cause disease is concerned. This inference is supported both by theoretical considerations and by a review of the available literature. In most instances in which changes in enzyme activity of the amounts under consideration here do have demonstrable clinical or in vitro metabolic effects, the enzymes seem to be particularly critical in controlling the entry or transit of metabolites through a specific pathway. However, even for these systems, it appears that additional extrinsic stresses are required to bring about the development of symptoms or of detectable metabolic changes. In some cases these stresses might occur naturally, as by the ingestion of food or increases in hormone levels; in others, they are clearly artificial. While there is some suggestion that long-term exposure to these naturally occurring stresses may have untoward effects, particularly in terms of intelligence, more information bearing on this question is certainly required.

Admittedly, the data are sparse. However, from these considerations I would for the moment infer that individual changes in enzyme activity are unlikely, except in rare instances, to have major deleterious effects on function. While this conclusion applies particularly to trisomies (duplications), it is likely to be generally true for monosomies (deletions) as well. What effect these changes in enzyme activity might have on development is unknown, but the absence of visible developmental abnormalities in heterozygotes for a large number of inborn errors of metabolism suggests once again that individual changes in enzyme activity are unlikely to have any significant deleterious developmental effects.

It is, of course, obvious that any real aneuploid state, however small the involved segment, differs from the situation I have just been discussing in that it involves the imbalance of not one but many loci, the actual number being determined by the size and genetic composition of the segment. It is certainly conceivable that the accumulation of many small metabolic disturbances, each in itself relatively or wholly innocuous, might adversely affect development and function. The data on the IQs of phenylketonuria and OCT-deficiency heterozygotes suggest that a quality such as intelligence might be particularly vulnerable to an aggregation of results, especially in a monosomic situation. However, at present we really have no data to support either this conclusion or any conclusion relating to the importance of such results to either developmental abnormality or death. The findings in *Drosophila* regarding the effects of combining individual segmental aneuploidies, none of which by itself produces lethality, into larger segmental aneuploidies that are lethal (Lindsley, Sandler *et al.*, 1972; see Chapter 9) might be regarded as evidence for such an inference. But, other possibilities also need to be considered at this point and will be dealt with in the following sections.

Transport systems

Quite similar to the considerations applicable to enzyme-mediated reactions are those concerned with transport systems. Again, several relevant examples can be found in the literature of the autosomally determined inborn errors of metabolism. X-linked errors are not considered relevant to the present discussion because of the special localized geographical conditions resulting from the mosaic nature of heterozygote tissues which could affect transport capacity (Epstein, 1977). The prototypic active transport defect in humans is cystinuria which, in the homozygous state, produces defective renal reabsorption and intestinal absorption of cystine, lysine, arginine, and ornithine (Segal and Thier, 1983). This disorder is actually a group of three or more disorders which can be distinguished on the basis of differences in intestinal amino acid transport and in the urinary excretion of amino acids by heterozygotes. In types II and III, urinary cystine and lysine excretions are significantly above normal in the heterozygotes, although still much less than in the homozygotes. Unfortunately, the structure of the dibasic amino acid transport system has not been elucidated. Therefore, it is not possible to comment specifically on why the heterozygotes should show manifestations of the defect, except to infer that the system is unable to handle the load presented to it during the time the substrate is present. A situation similar to that in cystinuria seems to obtain in certain families with iminoglycinuria, in which heterozygotes may manifest hyperglycinuria (Scriver, 1983), and in renal glycosuria (Krane, 1978). In the former, the transport maxima, T_m, for the reabsorption of proline and hydroxyproline are, as expected, reduced to about half. By contrast, heterozygotes for glucose-galactose malabsorption do not appear to be clinically symptomatic (Elsas *et al.*, 1970).

Cystinosis is presumed to be a condition resulting from a defect in an active transport system, in this instance one involved in the efflux of cystine from lysosomes (Jonas *et al.*, 1982; Gahl *et al.*, 1982). As a result of the 50% decrease in transport capacity, the rate of cystine efflux from lysosomes (Gahl *et al.*, 1982, 1984) and whole leukocytes (Steinherz *et al.*, 1982) of heterozygotes is also reduced by half. This reduction in the rate of efflux appears to be sufficient to permit, by means unknown, the accumulation of low but still increased (four to seven times normal) concentrations of free cystine in leukocytes and fibroblasts (Schneider *et al.*, 1968; Jonas *et al.*, 1982).

All of the examples of genetically determined defects in transport systems involve deficiencies, and, in the context of the present discussion, are relevant principally to monosomies or deletions. It is of course possible to speculate on what effect trisomies or duplications might have. In the case of absorption or uptake systems, the effect is likely to be negligible or nonexistent if the capacity of the system is already sufficient to meet its needs. However, if the uptake system is limited, increasing its capacity could result in increased intracellular concentrations of the transported metabolites, with whatever consequences that might have. For efflux systems, the results could be the reverse, with reduced intracellular concentrations of the transported metabolites. Once again, the determining factor is the relationship between demand and capacity.

Receptors

Closely related conceptually to enzyme-mediated reactions and transport systems are receptor-mediated phenomena. In the latter, which affect every aspect of metabolism, growth, physiological regulation, neural function, and presumably development, the binding of a ligand to a receptor is often the first step in a metabolic sequence which may involve one or more series of reactions to produce one or more effects. If it is assumed, as present evidence would suggest, that the number of receptors in a cell is gene dosage dependent, then it is possible to calculate the effects of aneuploidy on certain receptor-mediated phenomena (C. J. Epstein *et al.*, 1981).

Theory

The simplest model for ligand-receptor interaction follows the Langmuir absorption isotherm, which can be written in the form

$$S = \frac{L}{K_D + L} \tag{5.4}$$

where S is saturation of receptors, L is ligand concentration, and K_D is the dissociation constant. This equation is derived from the equilibrium relationship

$$\frac{R \cdot L}{RL} = K_D \tag{5.5}$$

where R is the free receptor concentration and RL the concentration of receptor-ligand complexes.

Since, as Equation (5.4) indicates, the degree of saturation is independent of receptor concentration, the relationship of the numbers of receptors occupied in the normal and aneuploid states can be derived from the expression

$$S = \frac{\text{number of occupied receptors}}{\text{total receptors}} = \frac{RL_N}{R_{ON}} = \frac{RL_A}{R_{OA}} \tag{5.6}$$

when RL_N and RL_A are the concentration or number of occupied receptors in or on the normal and aneuploid cells, respectively, and R_{ON} and R_{OA} are the respective total number of receptors. From Equation (5.6), we obtain

$$\frac{RL_A}{RL_N} = \frac{RL_{OA}}{R_{ON}} = \alpha \tag{5.7}$$

where α is the gene dosage factor (1.5 for trisomic cells, 0.5 for monosomic cells).

A plot of the number of sites occupied as a function of ligand concentration, as calculated from Equation (5.4), is shown in Fig. 5.7. As predicted by Equation (5.7), at any specific ligand concentration, trisomic cells will always have 1.5 times as many occupied sites as normal cells, and monosomic cells 0.5 times as many. The consequences of this can be appreciated in several ways. The first is in terms of the relationship between the degree of response and the number of receptors occupied. If the two are directly related to one another, such that the degree of response is proportional to the number

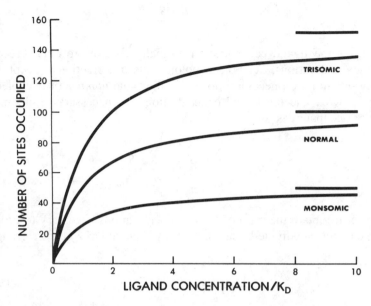

Fig. 5.7. The number of receptor sites occupied, as a function of ligand concentration, on normal and aneuploid cells. The total number of receptors on a normal cell is arbitrarily set at 100, and, assuming strict gene dosage effects, the numbers on trisomic and monosomic cells become 150 and 50, respectively. These totals are indicated by the horizontal lines on the right. As predicted by Equation (5.7), for any ligand concentration there are 1.5 times as many sites occupied on trisomic cells as on normal cells and 0.5 times as many on monosomic cells. The curves are calculated from Equation (5.4). Reprinted by permission from Epstein *et al.* (1981).

of occupied receptors [as, for example, occurs for steroid and thyroid hormones (Baxter and Funder, 1979)], then, for any ligand concentration, the aneuploid cell will have a proportionally greater or lesser response when compared to normal cells. Direct evidence for this has been obtained in the analysis of cells trisomic and monosomic for chromosome 21 and will be considered later (Chapter 12). Furthermore, even with ligand concentrations that lead to (virtually) complete saturation of the receptors, monosomic cells will not be able to obtain the same level of response as normal cells, and trisomic cells will be able to attain levels of response which normal cells never could.

Another way to look at the effect of aneuploidy on receptor-mediated phenomena is to consider the amount of ligand required to produce a specific response, again assuming that response is proportional to the number of receptors occupied. If normal (diploid) cells require a certain fractional receptor occupancy, S_N, to achieve a certain effect, perhaps because of a threshold phenomenon, then Equation (5.4) can be rearranged to give

$$L_N = \frac{K \cdot S_N}{1 - S_N} \qquad (5.8)$$

where L_N is the ligand concentration normally required to obtain receptor occupancy (saturation), S_N. For aneuploid cells, the fractional receptor occupancy or saturation necessary to obtain the same *number* of occupied receptors is S_N/α, and the ligand concentration, L_A, necessary to obtain this degree of occupancy is

$$L_A = \frac{K \cdot \dfrac{S_N}{\alpha}}{1 - \dfrac{S_N}{\alpha}} = \frac{K \cdot S_N}{\alpha - S_N} \qquad (5.9)$$

Therefore, to obtain the ratio of ligand concentrations necessary to produce the same effect, we divide Equation (5.9) by Equation (5.8) and rearrange to obtain

$$\frac{L_A}{L_N} = \frac{1 - S_N}{\alpha - S_N} \qquad (5.10)$$

This relationship is shown graphically in Fig. 5.8 for the monosomic case, and in Fig. 5.9 for the trisomic one. It is quite clear from the former that very substantial increases in ligand concentration will be required if monosomic cells are to achieve a receptor occupancy equivalent to a 40% or greater saturation of diploid cells. And, as has already been pointed out, they will never be able to realize occupancies greater than that equivalent to 50% of diploid cell saturation. Therefore, the higher the receptor occupancy required, the less sensitive monosomic cells will appear to be. The reverse is true for the trisomic cells. The higher the occupancy required, the more sensitive will the trisomic cells appear to be. In fact, sensitivity can be regarded as the reverse of the relationship shown in Equation (5.10) and plotted in Figs. 5.8 and 5.9. For the trisomic case, this has been plotted in Fig. 5.10. It is clear that the sensitivity is heavily dependent on the proportion of receptors which must be occupied to produce the measured effects in normal cells. Trisomic cells will be 1.5 times more sensitive to the ligand only if production of the measured effect requires a small fraction of receptors to be occupied. The differential sensitivity between trisomic and normal cells increases to two fold when 50% of the normal number of receptors must be occupied, to three fold at 75% occupancy, and to four fold at 83.3% occupancy. Therefore, if the normal threshhold is reasonably high, trisomic cells will respond at substantially lower ligand concentrations than would normal diploid cells.

It is possible to speculate briefly on the developmental implications of this result. Consider, for example, the possibility that a developmental event is initiated when the occupancy of some group of receptors by a particular ligand reaches a certain point (the threshold) and that the ligand is gradually

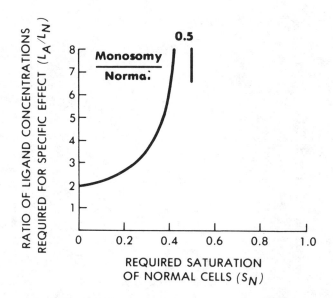

Fig. 5.8. The relative requirements of monosomic and normal cells for a ligand that induces a receptor-mediated function. This curve, calculated from Equation (5.10) with $\alpha = 0.5$, represents the ratios of ligand concentrations necessary to produce the same number of occupied receptors, plotted as a function of the proportion of receptors that must be occupied in normal cells to obtain the measured effect. It approaches a limiting value at a saturation of 0.5 since monosomic cells, with only half as many receptors as normal ones, cannot obtain a receptor occupancy greater than half that of normal cells. Modified from Epstein *et al.* (1981).

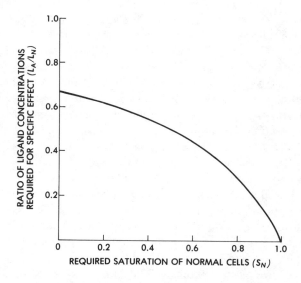

Fig. 5.9. The relative requirements of trisomic and normal cells for a ligand that induces a receptor-mediated function. This curve was calculated and plotted from Equation (5.10), with $\alpha = 1.5$, as indicated in the legend to Fig. 5.8. It approaches a ratio of 0 at $S_N = 1$ since diploid cells cannot in theory be completely saturated at any finite ligand concentrations.

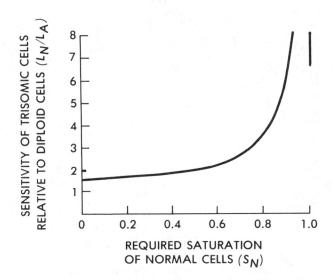

Fig. 5.10. The relative sensitivity of trisomic cells to a ligand that induces a receptor-mediated function. This curve is the reciprocal of that shown in Fig. 5.9.

increasing in concentration as development proceeds. By the principles outlined above, it would be predicted that trisomic cells would be triggered earlier and monosomic cells later, if at all, than normal cells (Fig. 5.11). In either case, the temporal relationships between the developmental event in question and other developmental processes would be disturbed. An aneuploidy-produced change in receptor number is thereby converted to an alteration in timing. Similarly, one can envision some receptor-mediated inhibitory response which normally requires a relatively high receptor occupancy, perhaps one not usually obtained with the ligand concentrations normally present. In this situation, trisomic cells, which are capable of responding at lower ligand concentrations, would be inhibited when the normal cells were not. Conversely, for any reaction requiring occupancy of more than 50% of the number of receptors present on diploid cells, monosomic cells would become totally nonresponsive. It should be noted that similar results would be obtained if aneuploidy affected the concentration of the ligand rather than the receptor.

All of the foregoing calculations have been made on the assumption that response is a linear function of receptor occupancy. Situations may exist in which the response is a function of some higher power of the number of receptors occupied. Suppose, for example, that binding to a receptor results in the formation of two or more products which together are required to produce an observable outcome. What effect would this have on the relative sensitivity of aneuploid and diploid cells? A sample calculation of what would happen in the trisomic situation if effects were proportional to the square or fourth

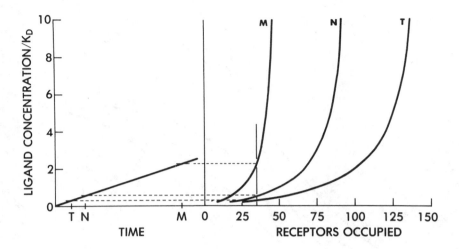

Fig. 5.11. The effect of aneuploidy-produced changes in receptor number on developmental timing. Consider a ligand linearly increasing in concentration as a function of time (left) and a developmental event which requires a certain receptor occupancy (in this example, 35) for initiation (right). The times at which the ligand reaches the concentration required to trigger trisomic (T), normal diploid (N), and monosomic (M) cells are estimated as shown. The graph on the right is the same as in Fig. 5.7 with the axes rotated.

power (n = 2 or 4) of receptor occupancy, *RL*, is shown in Fig. 5.12. As can be seen, the sensitivity of trisomic cells, expressed in terms of the relative concentrations of ligand necessary to produce 50% of the total effect achievable by the diploid cells, increases from 2 fold, for an effect directly proportional to receptor occupancy, to 2.7 and 4.1 fold for effects proportional to RL^2 and RL^4, respectively. In general, at any given ligand concentration and power relationship, n, between occupancy and response, the differential response between aneuploid and diploid cells would be equal to α^n, where α is the gene dosage factor. Thus, however we look at it, there would be a further amplification of the differences between the diploid and aneuploid cells if the response brought about by ligand binding were a function of some power of receptor occupancy higher than n = 1. An example of such apparent amplification, still unexplained with regard to mechanism, has been observed in the response of trisomy 21 cells to interferon and will be discussed in detail in Chapter 12.

In certain receptor-mediated reactions, occupancy of very few receptors is required to trigger a maximal biological response, and there thus appear to be "spare" receptors in the system. This situation usually obtains for polypeptide hormones and catecholamines (Baxter and Funder, 1979). In these instances, the differential sensitivities of aneuploid and diploid cells are reduced to the

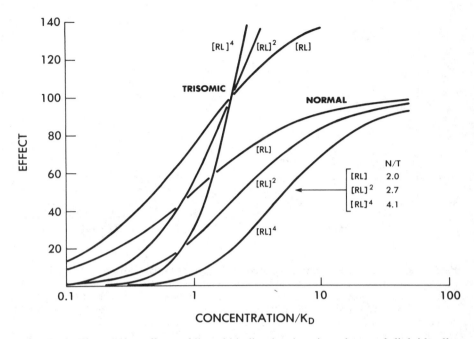

Fig. 5.12. The relative effects of ligand binding in trisomic and normal diploid cells if effect or response is proportional to RL^n, where n is a power higher than one of the number of sites occupied. Response curves are plotted for n = 1, 2, and 4, with the maximal response achievable by the diploid cells being set equal to 100 in each instance. The relative sensitivities for a 50% of total normal response are shown. Note that unlike previous figures in this section, the abscissa is log(ligand concentration/K_D), thereby imparting a sigmoid shape to the curves.

simple gene dosage differences between them. Thus, trisomic cells will give equivalent responses with a ligand concentration two-thirds (1/1.5) as great as required for normal cells, and monosomic cells will require twice (1/0.5) the normal amount. This result, which can be derived from Equation (5.4) for all ligand concentrations, L, much less than K_D, holds even if as few as only one receptor in or on each responding cell need be occupied to obtain the desired response. For a further discussion of receptor-mediated reactions, the reader is referred to Williams and Lefkowitz (1978), Ariëns *et al.* (1979), Baxter and Funder (1979), and Baxter and MacLeod (1980).

Clinical and experimental precedents

Once again we can look to the inborn errors of metabolism for possible examples of genetic abnormalities affecting receptors. To my knowledge there are no instances of genetically produced increases in receptor number, and unfortunately most of the examples of decreases are caused by mutations on the X chromosome, with the usual difficulties in interpreting findings in heterozygotes. The best-studied autosomal receptor disorder in humans is familial hypercholesterolemia, which results from a defect in cell surface low density lipoprotein (LDL) receptors. This disorder can, in a sense, also be considered as a defect in a transport system since the receptor also acts to transport LDL into the cell. Heterozygotes for this condition have approximately 50% the normal number of LDL-binding sites on the surfaces of fibroblasts and lymphocytes (Goldstein *et al.*, 1976; Bilheimer *et al.*, 1978). This reduction in receptor number, which remains proportionally the same whatever the concentration of LDL in the medium, results in a proportional reduction in LDL internalization and degradation. It also leads to a reduction in the suppression of the activity of the enzyme, 3-hydroxy–3-methylglutaryl coenzyme A (HMG-CoA) reductase, a rate-limiting enzyme of cholesterol biosynthesis, by the internal cholesterol released from the degraded LDL (Fig. 5.13). As a result of the lack of normal suppression of HMG-CoA reductase, it is necessary for the level of serum LDL cholesterol to rise to a concentration which will produce sufficient LDL binding to bring about the proper steady state of cholesterol synthesis and degradation. As would be predicted from the fact that, as illustrated in Fig. 5.13, suppression of the reductase occurs at relatively low levels of receptor occupancy, the rise in LDL cholesterol necessary to bring about the new steady state is of the order of two to three fold (see Fig. 5.11). This is what is actually observed in familial hypercholesterolemia heterozygotes: mean heterozygote LDL cholesterol, 249 mg/dl; mean control LDH cholesterol, 113 mg/dl; ratio, 2.2 (Bilheimer *et al.*, 1978). Similar observations have also been made in heterozygous Watanabe heritable hyperlipidemic rabbits, the animal counterpart of human familial hypercholesterolemia, in which the ratio is 2.4 (Goldstein, Kita, and Brown, 1983).

A very interesting demonstration of the relationship between receptor number and biological effect is provided by the experimental studies of Kono and Barham (1971) on insulin binding to fat cells. By treating these cells with trypsin for varying periods of time and then allowing them to recover, cells differing approximately two fold (actually 2.25) in the number of insulin receptors were obtained. The concentration of insulin required to produce a binding of 1.5 μU/ml to the fat cell surface membranes, which yielded maximal stimulation of glucose oxidation, was four times greater in the cells with fewer receptors, and, as the curves in Fig. 5.14 indicate, the same was true for lesser responses. Once again, the biological response behaved as would be expected from a consideration of the mathematics of receptor binding.

A beautiful example of the effect of altering receptor number on a receptor-mediated function was reported by Bourgeois and Newby (1979) for the glucocorticoid receptor system. By cell fusion, they were able to produce tetra-

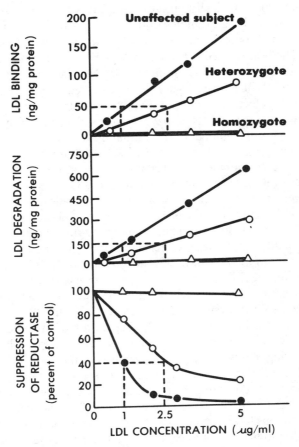

Fig. 5.13. Functions of the low density lipoprotein (LDL) receptor in the cultured fibroblasts of one representative familial hypercholesterolemia homozygote, one familial hypercholesterolemia heterozygote, and one unaffected subject. LDL binding, LDL degradation, and LDL-mediated suppression of HMG-CoA reductase activity were carried out at LDL concentrations below saturation. Redrawn from Goldstein and Brown (1975).

ploid murine thymoma lines with one to four copies of a functional structural gene for the glucocorticoid receptor, and, as expected, the glucocorticoid receptor content of these cells was proportional to the number of receptor genes (in the ratio of 0.9:2.3:3.0:4.0 for one, two, three, and four active genes, respectively). Further, the cytolytic response of the hybrid cells was directly proportional to receptor number, so that the greater the number of receptors, the greater the sensitivity of the cells to the cytolytic effects of dexamethasone (Fig. 5.15). In view of the nature of the response curves, it would be expected

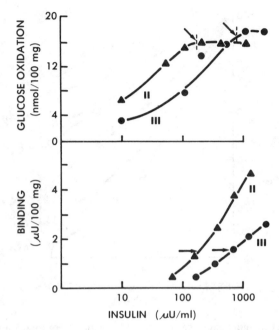

Fig. 5.14. Relationships between insulin receptor concentration, insulin binding, and physiological effect (glucose oxidation). The curves marked II are for cells with 2.25 times as many receptors as the cells depicted in curves III. This difference in receptor number results in a displacement of both the insulin binding and physiological response curves, so that approximately fourfold differences in insulin concentration are required to obtain the same physiological effect. Redrawn from data of Kono and Barham (1971).

that increased levels of glucocorticoid receptors, such as might be found in a trisomic state, would produce even greater sensitivity to the cytolytic and probably other activities of glucocorticoids. In a somewhat different type of experiment, the time required for T lymphocytes to respond to interleukin–2 stimulation of DNA synthesis was found to be inversely related to the number of IL–2 receptors (Cantrell and Smith, 1984).

Filmus *et al.* (1985) have described a very interesting example of a change in the response of a breast cancer cell line to the growth inhibitory effect of epidermal growth factor (EGF) brought about by an increase in the dosage of the EGF receptor gene. Whereas cells with a low number of receptors are resistant to this effect of EGF, mutant cells with high receptor numbers are growth inhibited at low EGF concentrations. These results have been interpreted as being consistent with a threshold of EGF receptor occupancy that must be exceeded if growth inhibition is to occur, a threshold that can be reached only if a sufficiently large number of receptors are present.

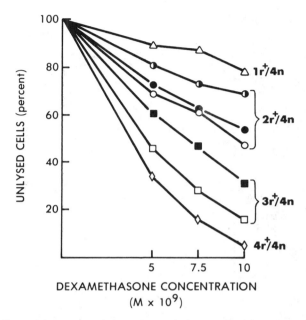

Fig. 5.15. Effect of increasing the numbers of glucocorticoid receptors on the cytolytic response of murine thymoma cell lines to dexamethasone. The numbers of functional glucocorticoid receptor structural genes present in tetraploid hybrid cells are shown on the right side of the curves. Modified from Bourgeois and Newby (1979).

Although not a receptor per se, metallothionein, a copper-binding protein, has interesting properties when present in increased amounts in yeast. Such organisms become progressively more resistant to the toxic effects of copper as gene dosage is increased, and this change has been proposed as a diagnostic test for the induction of aneuploidy (Fogel and Welch, 1982).

I have devoted considerable attention to receptor phenomena because it is my feeling that they will turn out to be very important for understanding the mechanisms of the consequences of aneuploidy. Relatively small changes in receptor number can, under appropriate circumstances, have quite large effects on the outcome of ligand binding. As I have illustrated, these effects can involve changes in sensitivity, in timing, and in the attainment of thresholds. Given the vast number of growth and regulatory processes in which receptor systems are known to be involved, it is difficult to believe that perturbations of the type described will not be of importance. Furthermore, it is quite likely that many as yet still undefined developmental events are controlled by receptor-mediated processes and that a significant number of the developmental abnormalities associated with aneuploidy are the result of the alterations in receptor response.

6

Regulatory systems

Theoretical considerations

Soon after the appearance of the work of Jacob and Monod (1961) on the regulation of gene expression in prokaryotes, attempts were made to explain the effects of aneuploidy on the basis of regulatory disturbance. Thus, Huehns *et al.* (1964) proposed two different regulatory models to explain the continued appearance of embryonic and fetal hemoglobins in the erythrocytes of patients with trisomy 13 (see below). Rohde (1965) generalized this notion to explain, at least in concept, the origin of the congenital anomalies associated with an aneuploid condition. However, his hypothesis invoked quantitative defects in the feedback regulation of induction or repression resulting from the aneuploidy-produced changes in metabolic pathways (secondary to changes in structural gene dosage) rather than a direct effect of aneuploidy on regulatory loci themselves (Rohde and Berman, 1963; Rohde, Hodgman, and Cleland, 1964). The latter was explicitly considered by Yielding (1967), who quoted data of Sadler and Novick (1965) showing that the presence of two copies of the lactose operon repressor locus in *E. coli* produced a 90% reduction in the already low levels of β-galactosidase present in uninduced cells containing only one repressor locus. This effect was eliminated when enzyme synthesis was induced, and a strict gene dosage effect could be observed under inducing conditions when an extra copy of the structural gene locus was also present (Fig. 6.1) (see also Luk and Mark, 1982). Yielding (1967) therefore proposed that "the presence of extra genes for synthesis of repressor (in the absence of a compensatory accelerated removal of repressor) would increase the steadystate concentration of free repressor within the cell," with consequent increased repression of the regulated locus or loci. He further predicted that "extra chromosomes could result in decreased, unchanged, or increased genetic expression in the case of 'regulated' genes depending on the concentrations present of specific inducers" and depending on whether the structural gene and regulatory loci were on the same or on different chromosomes. Gaudin (1973) also invoked a gene dosage-produced change in the number of repressor loci and therefore in the concentrations of repressor molecules to explain the basis of malignancy in aneuploid cells.

Krone and Wolf (1972) considered the effects of aneuploidy-induced

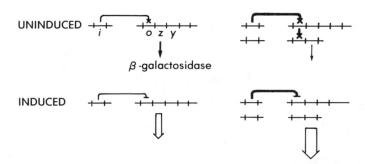

Fig. 6.1. Effect of aneuploidy on regulation of β-galactosidase induction in *E. coli*. Normal (haploid) bacteria, when induced (left), increase the synthesis of β-galactosidase by 25 fold. Bacteria with two copies of the repressor gene, *i*, reduce the uninduced level of enzyme synthesis by ~90%, even if two copies of the structural gene are also present (upper right). However, when induced, the latter shows a gene dosage effect for β-galactosidase synthesis (lower right). A schematic version, based on the Jacob and Monod (1961) regulatory system, of data taken from Sadler and Novick (1965) and presented by Yielding (1967).

changes on eukaryotic regulation as proposed by Britten and Davidson (1969), and Vogel (1973) made a strong argument for regulatory disturbances being the principal causes of the phenotypes associated with aneuploidy. While admitting that many phenotypes were in fact distinguishable, he asserted that there is a "striking overlap" of the features of different autosomal chromosomal syndromes, with most features being independent of the specific chromosome involved and whether the amount of genetic material is increased or decreased. Furthermore, he pointed out that conditions known to be autosomal dominant in inheritance are "hardly ever" observed as a regular part of a chromosome deletion syndrome, presumably because autosomal dominant conditions require the interaction of normal and abnormal products. As has already been described, this is true in many but not all instances. Based on these considerations, Vogel (1973) concluded that the most likely hypothesis is that the majority of symptoms of chromosome imbalance are not caused by too few or too many "classic" structural genes but by disturbances in the regulation of gene activities in space and time. He attempted to show how this might operate by appealing to the Britten and Davidson (1969) model of genetic regulation (see below) and postulating the need for quantitative balance between the amounts of activator RNAs synthesized and the numbers of receptor-inducer units with which they interact.

As illustrated in Fig. 6.1, the Jacob and Monod model of bacterial genetic regulation is one of negative control in which gene expression is governed by the presence or absence of a repressor. The function of an inducer, usually a low-molecular-weight metabolite or metabolite analog, is to prevent the repressor from interfering with gene transcription, which would otherwise occur constitutively. Because of this formulation, early attempts to apply regulatory concepts to eukaryotic gene regulation therefore postulated systems of negative regulation. However, it was soon discovered that positive control systems also exist in bacteria (Englesberg and Wilcox, 1974). In these systems, expression of the gene(s) of interest does not occur in the absence of any specific regulatory product, and a product from a specific regulatory gene – an activator – is required for the initiation of transcription. As will soon become apparent, both negative and positive control systems have been proposed to explain eukaryotic gene regulation.

Models of genetic regulation in eukaryotes

Despite the attractiveness and simplicity of the regulatory systems proposed for prokaryotes, it is not clear to what extent they are applicable to the eukaryotic and, more specifically, the mammalian situation. In fact, reviews of gene regulation in eukaryotes often include statements pointing to the high level of our ignorance: "We must confess complete ignorance of the primary event (or events) controlling molecular differentiation" (Caplan and Ordahl, 1978); "The central issue of this book must remain an unanswered question: how is gene expression controlled?" (Lewin, 1980); "We do not yet understand the control of any single eukaryotic gene with the molecular detail with which we understand the lac operon of *Escherichia coli* or lambda phage genes" (Brown, 1981). Nevertheless, there is no shortage of models for eukaryotic gene regulation, and a large number of potentially relevant facts are, of course, known. Unfortunately, as they now stand, these facts and models do not help us a great deal in understanding how gene regulation may be perturbed by an aneuploid state, and I shall therefore discuss them only briefly before going on to experimental data which, for the moment, may be more relevant.

A major difference between prokaryotic and eukaryotic regulatory systems is the greater number of possible control points or levels of control available to the latter (Brown, 1981; Darnell, 1982; Weisbrod, 1982). These points or levels of control are summarized in Table 6.1. Of specific interest to the present discussion are the points at which regulatory molecules, the concentrations of which could be altered by aneuploidy, might operate, and three potential sites have been implicated. Following the original Jacob and Monod model, the first point of control is at the level of initiation of transcription. Several authors [see, for example, Brown (1981) and Darnell (1982)] feel that this is the most important, particularly for the highly specialized genes which make products characteristic of differentiated cells (Brown, 1981). Even Davidson and Britten (1979), who do not award primacy in regulation to transcriptional control, agree that it does apply to the regulation of what they term

Table 6.1. *Points or levels of control of eukaryotic gene expression*

Chromatin structure
  Packing of DNA – active and inactive conformations
  Nonhistone proteins
DNA structure
  Gene loss
  Gene amplification
  Gene rearrangement
  Methylation
Transcription
  Initiation of mRNA synthesis
  Termination of mRNA synthesis
Posttranscription (processing)
  Capping of mRNA with m^{7}Gp
  Polyadenylation
  Methylation
  Splicing (removal of introns)
  Degradation
Translation
  Stability of mRNA
  Efficiency of translation of mRNA

"superprevalent mRNA," present in many ($\sim 10^4$) copies per cell and clearly specific to particular states of differentiation. What Davidson and Britten (1979) argue for in general is posttranscriptional control at the level of RNA processing. This control, they believe, determines the fraction of any continually produced mRNA that will be successfully processed and transported to the cytoplasm. The third possible point of control, originally proposed by Tomkins *et al.* (1969) to explain certain types of enzyme induction, is at the level of translation. Since it was put forward before the mechanisms of RNA processing had been elucidated, it is quite possible that much of the evidence in support of this notion would actually be applicable to posttranscriptional control as well (Gelehrter, 1976).

Involved in all of these suggested models for control of gene expression is the synthesis and action of regulatory molecules, usually thought to be RNA in nature, which either promote (Davidson and Britten, 1979) or interfere with (Tomkins *et al.*, 1969) mRNA synthesis, processing, and stability. In the former mode, they would be considered as inducers or activators [Davidson and Britten (1979) refer to them as integrating regulatory transcripts]; in the latter, as repressors. Their synthesis could either be constitutive, with their ability to act being determined after synthesis, or could itself be regulated by another molecule. It is also conceivable that, rather than functioning in their

RNA forms, these regulatory molecules could be translated into proteins with the same functions. In all instances, the genes for these regulatory molecules need not be located near the structural genes whose activities they ultimately regulate. Regulation is therefore visualized as operating *trans*, with the product of one regulatory locus controlling a structural gene or genes not linked to it.

The actions of regulatory molecules are presumably influenced in some way by their concentrations. It is possible, therefore, to visualize that a change in the dosage of a regulatory gene would affect the ability of the regulatory molecules which it produces to act. However, without knowledge of the quantitative aspects of the process of regulation, it is not possible to make specific predictions about what these aneuploidy-produced alterations in regulation might be. However, one set of examples, coming out of experiments on somatic cell hybridization, will be discussed later in this chapter.

One reason for being interested in the role of regulatory systems in understanding the effects of aneuploidy is the possibility that imbalance of one regulatory locus could affect the expression of many structural genes present throughout the genome. Even in the original Jacob and Monod (1961) scheme, one repressor is able to control the expression of at least three structural genes which constitute the *lac* operon, and more recent eukaryotic models have often involved large networks of regulatory interactions. In the highly elaborated model of Davidson and Britten (1973, 1979), originally proposed in 1969 (Britten and Davidson), repetitive gene sequences are postulated to be involved in the coordinate control of sets of functionally related structural genes, especially during development and differentiation. These repetitive sequences are presumed to include multiple regulatory genes (referred to as "integrators" or "integrating regulatory transcription units") and, at least for control at the transcriptional level, multiple receptor sites in the DNA adjacent to the regulated structural genes (Fig. 6.2). By invoking regulatory networks such as these, it is easy to visualize how imbalance of one chromosome or chromosome segment, which could contain one or several sets of regulator gene loci, could adversely affect many developmental processes. These notions are still quite hypothetical since, as Davidson and Posakony (1982) themselves point out: "Despite their ubiquity, their quantitative prominence, their apparent developmental regulation and the amount of interest they have aroused, the repetitive sequence transcripts of animal cells remain a phenomenon in search of a physiological meaning."

One final point requires mention before considering the experimental evidence. Genetic regulation may be regarded as operating in two spheres. The first is concerned with development and differentiation and the maintenance of the differentiated state. Sometimes the terms "commitment" and "determination" are used in place of "differentiation." This process, once it occurs, is regarded as being stable and irreversible (Caplan and Ordahl, 1978). The

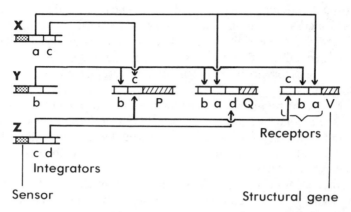

Fig. 6.2 The Britten-Davidson model for genetic regulation, with redundancy of both receptor sites and integrator elements. Each sensor element causes synthesis of activator RNA at each of the integrator genes adjacent to it. By including the same integrator gene sequence under control of more than one sensor, the set of genes responding to its activator RNA may be activated in response to more than one cellular signal. Conversely, the same structural gene may have more than one receptor element adjacent to it and may, therefore, respond to more than one activator RNA. Thus, sensor X controls structural genes P, Q, and V, while structural gene V can be activated by activator molecules from sensors X, Y, and Z. This scheme is based on the original Britten and Davidson (1969) model. In the modified version of the model (Davidson and Britten, 1979), the integrators become integrating regulatory transcription units, the activator RNAs become integrating regulatory transcripts, and the interactions between the latter and the structural genes take place at the level of mRNA processing rather than transcription. Modified from Lewin (1980).

second sphere of regulation is in the control of expression of specific genes characteristic of the determined or differentiated cells, once the differentiated state has been established. Enzyme induction would fit into this latter category. At present, it is not known whether the same or different regulatory mechanisms operate in these two spheres. Tomkins *et al.* (1969) expressed the "prejudice . . . that transcriptional control is most important in the sequential gene activation which occurs during development, and in maintaining the differentiated state of metazoan cells. Posttranscriptional regulation and messenger stabilization might therefore be involved in finer regulation such as enzyme induction." Davidson and Britten (1979) see it the other way around: "The only clear evidence for cell-specific regulation in the transcription of structural genes pertains to the superprevalent mRNA class. . . . Single-copy structural genes giving rise to complex [one to several copies per cell] and moderately prevalent class mRNAs are transcribed continuously, at more or less similar rates. . . . The direct implication would be that both the quantitative and qualitative structure of cytoplasmic mRNA populations are controlled posttranscriptionally." However, even in the Davidson-Britten model, it is necessary to postulate transcriptional control of the regulatory (integrator) genes themselves and possibly some developmental alteration of the genetic structures (sensors) that control them.

Regulatory systems in mammals

Several interesting mammalian and nonmammalian control systems have been the subject of investigation, with the goal being to delineate the regulatory processes which are operative. All of these systems involve the control of enzyme synthesis and provide an insight into the application of the theoretical constructs just discussed. None has yet been implicated or even studied in any aneuploid state.

Temporal genes

Many enzymes in mammals undergo marked changes in their concentrations and tissue distributions during development and growth, and evidence has been accumulating that these changes are under genetic control. This evidence has given support to the concept of temporal genes, genes which control the developmental programs of enzymes and other cellular functions and constituents. Paigen (1980), the leading proponent of this concept, defines temporal genes as "the set of genetic elements determining the program for enzyme concentration during development. Mutations in these elements alter the developmental sequence and relative tissue distribution of an enzyme." It should be noted, however, that although the term "temporal" is used, the mutations do not alter the time in development at which changes occur, but only whether and sometimes to what extent they occur at all. Paigen (1980) believes that the genetic programming encoded in these temporal genes "is at least one appreciable aspect of developmental regulation and quite possibly the major driving force."

Examples of what are considered to be temporal genes have been found in a variety of organisms, including the mouse, *Drosophila*, and maize, and these genes appear to divide into two groups. The first includes the proximate genes or elements that map close to the structural genes which they control. In most instances they operate in a cis manner; that is, they regulate the expression of the structural genes to which they are physically linked and not the expression of the homologous structural genes on the other chromosome in the pair. These proximate genes thus occupy locations similar to those of the operator genes of the Jacob and Monod model and the receptor elements of the Britten and Davidson model. Since each proximate temporal gene controls the expression of the structural gene to which it is linked, it would be expected that additive inheritance should be demonstrable, with the enzyme activity of an organism heterozygous for high and low activity temporal gene alleles having an intermediate enzyme activity. This has been observed in several systems, including esterase E_1 in maize, aldehyde oxidase in *Drosophila*, and H–2 in the mouse (Paigen, 1979).

The second group of temporal genes consists of those, termed "distant," that are located at a distance from the structural genes which they control; these temporal genes act in a trans manner. While such distant loci may actually be on the same chromosome as the structural loci, as for example the *map* gene in *Drosophila* which is 2 centimorgans (cM) away from the amylase locus (*Amy*) which it regulates (Abraham and Doane, 1978), they are usually completely unlinked. Distant loci have been identified for the regulation of murine α-galactosidase and H–2 and of *Drosophila* amylase (Paigen, 1979, 1980). In the murine β-glucuronidase system, the temporal *Gus-t* locus, although proximate in location, is trans acting (Lusis *et al.*, 1983). Thus, most control systems have both proximate and distant (or at least trans acting) loci, with only aldehyde oxidase in *Drosophila* and esterase in maize having proximate loci alone, and only α-galactosidase in the mouse having just a distant site. In every instance, a specific temporal gene, whether proximate or distant and whether cis or trans acting, controls only one structural gene.

Two examples of the operation of trans-acting distant temporal genes are shown in Figs. 6.3 and 6.4. The first illustrates the developmental regulation of murine β-galactosidase activity in the liver and shows that the final enzyme activity attained is independent of the specific structural genes that are present. The second depicts the regulation of the spacial distribution (and hence also the activity) of amylase in the posterior midgut of *Drosophila* by the *map* locus.

Of particular interest with regard to trans-acting temporal loci is the nature of the dosage effects which have been observed. In the *Drosophila map*-amylase system, heterozygotes carrying both the A and C alleles of the *map* locus have an A-type pattern (see Fig. 6.4) but with somewhat reduced enzyme activity (Abraham and Doane, 1978). In the murine α-galactosidase system, heterozygotes for high- and low-activity temporal alleles controlling only the activity of liver enzyme showed additive inheritance: CBA, 5.32; C57BL/6, 9.40; F_1 (heterozygote), 6.89 (expected, 7.36) (Lusis and Paigen, 1975). Similarly, in the murine β-glucuronidase system in which both cis- and trans-acting temporal genes are operative, heterozygotes for the trans-acting locus, *Gus-t*, have an enzyme activity in the adult liver midway between those of the pure heterozygotes (DBA, 56.1; C3H, 5.7; F_1, 33.4 (expected, 30.9) (Lusis *et al.*, 1983). It thus appears that there is a quantitative relationship between the expression of the temporal genes and the activity and presumably synthesis of the structural gene products which they regulate. Paigen (1979) regards this situation as surprising and draws the following inferences from it:

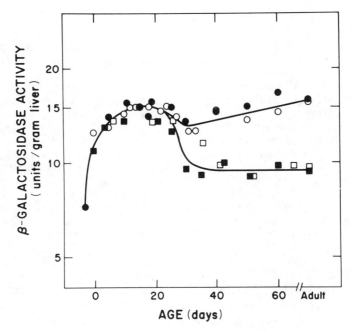

Fig. 6.3. Regulation of the final activity of murine β-galactosidase in the adult liver by a temporal gene. β-Galactosidase specific activity in developing liver is plotted for the two progenitor strains, BALB/c (closed squares) and C57BL/6 (closed circles), together with two of the recombinant inbred strains each of which received the BALB/c form of the structural gene. One recombinant inbred strain (open squares) shows BALB/c type development, but the other (open circles) shows C57BL/6 type development because of the presence of the C57BL/6 allele of the temporal gene. Reprinted from Paigen (1979) by permission.

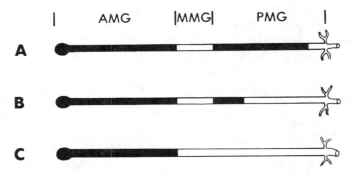

Fig. 6.4. Regulation of the distribution of amylase in the midguts of *Drosophila melanogaster* by the temporal gene, *map*. Three phenotypes have been defined. The C phenotype completely lacks enzyme in the posterior midgut (PMG); the B phenotype lacks enzyme in the terminal region of the posterior midgut; and the A phenotype possesses enzyme throughout the posterior midgut. All flies show enzyme in the anterior region (AMG) and lack it in the central region of the midgut (MMG). Redrawn from Abraham and Doane (1978).

The further implication of additive inheritance [of trans-acting temporal genes] is that if a gene does act in the regulation of enzyme synthesis, the molecular nature of its function in the control mechanism is unlike any of the functions seen in microbial systems, where mutations of unlinked regulatory genes are invariably recessive or dominant. Microbial trans-acting regulatory proteins, either repressors acting negatively or activators acting positively, react with an operatorlike DNA sequence to regulate an adjacent structural gene or group of genes. When different alleles of the genes producing the repressor or activator proteins are combined in the same cell, one allele is almost always dominant over the other. In each case, the reasons for this can be understood as coming either from the kinetic parameters that describe the interaction between the regulatory protein and its operator site of action or from the nature of subunit interaction between the monomers making up the regulatory protein. These structural and kinetic constraints are required to maintain a functional regulatory system. Thus, the observation of additive inheritance for regulatory elements in eukaryotes implies a very different molecular basis for control.

Whether this is really the case must remain an open question. The molecular basis of the eukaryotic temporal gene regulatory systems is still completely undefined, and without knowing what kinetic or subunit requirements are operative, it is difficult to formulate a specific prediction regarding additivity. Furthermore, in those eukaryotic induction systems in which hormone-receptor complexes seem to be the regulatory substances that initiate gene expression, additivity has been demonstrated (Bourgeois and Newby, 1979), suggesting that additivity in trans-acting eukaryotic regulation is certainly not without precedent. However, what is important from the point of view of our concern with aneuploidy is that, whatever the mechanism, additive effects are demonstrable for trans-acting temporal genes. This means that changing the dosage of such genes can affect the synthesis of the structural gene products they regulate, most likely to an extent proportional to the change in their number. This should ultimately become testable experimentally.

One last point with regard to temporal genes deserves mention, and that is whether the developmental modulations of enzyme activity which they regulate are involved in the basic mechanisms of cellular differentiation. Paigen (1979) argues that it is likely that they are, since there is "no reason to suspect that qualitatively different genetic systems are responsible for larger and smaller effects in the regulation of enzyme levels during development. . . . It is unlikely that life has evolved two sets of genetic regulatory mechanisms, one for early modulations of nuclear differentiation and enzyme activity and another for later on." And, he goes on, even if it did, "their equivalent importance in deciding the final phenotype of the organism, and the paucity of knowledge about either, makes it a matter of importance to investigate whichever is amenable to study."

While not a temporal gene in the sense just discussed, the locus involved in the *lin–14* mutations in the nematode, *Caenorhabditis elegans*, is concerned with the timing of certain developmental decisions with regard to cell lineages (Ambros and Horvitz, 1984). It is of considerable interest, therefore,

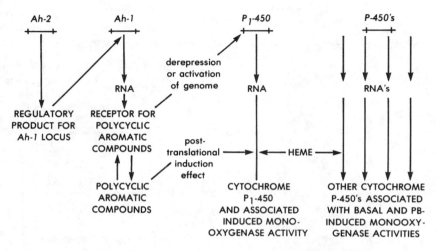

Fig. 6.5. Hypothetical scheme for regulation of the arylhydrocarbon hydroxylase system. A receptor-mediated positive control system in which the product of the regulator locus, *Ah–1*, a polycyclic hydrocarbon receptor, binds to an inducer to form an inducer-receptor complex which directly activates the cytochrome P–450 structural gene locus or loci. Redrawn with modification from Nebert *et al.* (1978).

that the dosage of *lin–14* gene activity may specify the fates of cells at several stages of development, high levels specifying earlier fates and low levels later fates.

The arylhydrocarbon hydroxylase system

A regulatory system with a quite different function (although it might also include temporal elements) is that which controls the induction of the cytochrome P–450-mediated monooxygenases by a variety of polycyclic aromatic hydrocarbons (Nebert and Jensen, 1979; Nebert *et al.*, 1978, 1981, 1982). The genetic dissection of this system in the mouse has led to the identification of two regulatory loci, *Ah–1* and *Ah–2*, which are involved in the regulation of the cytochrome P–450 structural genes. Although the mode of operation of the *Ah–2* locus is not known, it appears to control the activity of *Ah–1*, which is possibly located on mouse chromosome 17 (Legraverend *et al.*, 1984) and is considered to be the structural gene for a cytosolic receptor. This receptor binds polycyclic aromatic hydrocarbons (inducers) and, following transit to the nucleus, activates transcription of the cytochrome P–450 genes. Two hypothetical schemes of how the system operates are shown in Figs. 6.5 and 6.6. In one, the inducer-receptor complex acts directly to activate a series of cytochrome P–450 structural genes. In the other, cast in the Britten-Davidson model, the inducer-receptor complex interacts with a sensor

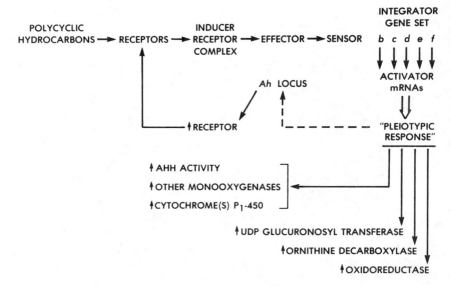

Fig. 6.6 Hypothetical scheme for regulation of the arylhydrocarbon hydroxylase system. A receptor-mediated control system following the Britten-Davidson model. The inducer-receptor complex is visualized as interacting with a sensor gene, thereby activating the synthesis of a set of activator RNAs which, in turn, influence the synthesis of a large number of induced enzymes. Redrawn with modification from Nebert *et al.* (1978).

gene which, in turn, activates the synthesis of a number of regulatory molecules that turn on the synthesis of many gene products. The latter scheme is proposed to explain the pleiotypic response brought about by activation of the *Ah* system, since several other enzymes in addition to cytochrome P–450 monooxygenases are induced (Nebert *et al.*, 1978).

Once again, the principal question of interest here is the effect on induction of the products of the system brought about by quantitative differences in the product of the *Ah–1* locus, the cytosolic receptor. Several alleles of this locus have been defined in inbred strains of mice, as have several patterns of response in the heterozygotes (Nebert *et al.*, 1982). The alleles can be divided into two groups, those resulting in a high rate of induction and those with little or no induction. Heterozygotes with both high- and low-induction alleles of *Ah–1* may have one of three characteristics: induction at or very close to the level of the high responder (as for example, in F_1s from C57BL/6 × DBA/2 crosses), induction at a level intermediate between that of the two pure types (C3H × DBA/2 heterozygotes), or, in a rare instance, induction (or lack of induction) at the level of the low responder. In some instances the response may vary from tissue to tissue. Thus, while the liver of C57BL/6 ×

DBA/2 heterozygotes has about 85% the enzyme activity present in homozygous C57BL/6 animals after induction, the bowel, albeit lower in enzyme activity over all, has only about 57% (Nebert *et al.*, 1978). The basis for these differences has not been fully elucidated, and it is not known to what extent qualitative, as opposed to quantitative, differences in the receptors are involved. However, it has been shown that, at least for the C57BL/6 and DBA/2 situation, the responses are a direct linear function of the number of inducer-receptor complexes present in the nuclei of animals treated with an inducer (Tukey *et al.*, 1982). C57BL/6 livers have a high number of intranuclear receptors, while the DBA/2 livers have only 20–25% as many; C57BL/6 × DBA/2 heterozygotes have an intermediate number.

The presence of intermediate patterns of enzyme induction in heterozygotes and the relationship between inducer-receptor concentration and enzyme induction indicates that aneuploidy-produced changes in the dosage of the gene for the regulator/receptor could potentially affect the sensitivity and response of the system. The types of effects are likely to be those already discussed with regard to aneuploidy and receptors. In fact, the arylhydrocarbon hydroxylase system, as well as the steroid receptor system mentioned earlier, serves to illustrate that there is a significant overlap between the principles applicable to receptors in general and those which may govern the operation of regulatory systems, particularly when receptors are intimately related to the operation of the regulatory systems. Although I have not discussed it in any detail, the steroid hormone-mediated induction system has many similarities to the arylhydrocarbon hydroxylase system [for review, see Gelehrter (1976)].

Lethal albino deletions

I should like to make brief mention of one other locus or region in the mouse that has been the subject of considerable investigation. Although it has not been proven to involve a regulatory system, it has been postulated that such a system is involved. A series of deletions of mouse chromosome 7 covering the region of the albino locus, *c*, have been induced and shown, when homozygous, to be associated with preimplantation, postimplantation, or perinatal lethality (Gluecksohn-Waelsch, 1979). When the livers of animals homozygous for one of the prenatal lethal forms were analyzed, they were found to be deficient in the activities of several enzymes (glucose–6-phosphatase, tyrosine aminotransferase, serine dehydratase, UDP glucuronosyltransferase, and glutamine synthetase), serum proteins (albumin, α-fetoprotein, and transferrin), and hormone receptors (glucocorticoid, insulin, epidermal growth factor, glucagon), and to have severe ultrastructural abnormalities (Gluecksohn-Waelsch, 1979; Goldfeld *et al.*, 1981, 1983; Shaw and Gluecksohn-Waelsch, 1983). The fact that significant quantities of many of these proteins are still present indicates that the mutation has not affected the structural loci (which in the homozygotes would be totally deleted), and this has been proven directly for tyrosine aminotransferase and glucose–6-phosphatase (Cori *et al.*, 1983). It is likely that not all abnormalities are the direct result of the deletion, and that the decreased activities of some products, such as enzymes, are the result of decreases in others, particularly the hormone and growth factor receptors (Shaw and Glucksohn-Waelsch, 1983). Several of the deficient liver enzymes are normally inducible by glucocorticoids or insulin. In view of the widespread nature of the effects of the mutation, which are limited to the liver, it has been postulated that the basic problem is in the regulation of gene expression that specifically characterizes the liver cell, with one or more genes with trans-acting regulatory functions being located within the deleted segment (Glucksohn-Waelsch, 1979; Goldfeld *et al.*, 1983; Shaw and Glucksohn-Waelsch, 1983; Schmid *et al.*, 1985).

Heterozygotes for the albino deletions can be considered as aneuploid for the involved segment, sometimes as long as 5 cM, and it is therefore of interest to know whether they display any of the abnormalities characteristic of the homozygotes. According to Glucksohn-Waelsch (1979), they do not. The activities of the five affected liver enzymes listed above were identical in homozygous normal and heterozygous animals, as was the response to glucocorticoid induction. Therefore, the affected regulatory system, if that is indeed where the lesion is, is one which is not demonstrably gene dosage sensitive and thus not likely to be affected by an aneuploid state.

Expression of differentiated functions in mammalian cell hybrids

With this theoretical introduction, we can now proceed to a consideration of experimental and clinical data relevant to the issue of regulatory perturbation in aneuploidy. A line of investigation that has been particularly rewarding in providing information about regulatory systems in mammalian somatic cells has made use of the formation of heterokaryons (cells with two or more unfused nuclei) and cell hybrids by cell fusion techniques. Several reviews of work in this area have appeared and may be consulted for a more detailed discussion than is presented here (Ephrussi, 1972; Davis and Adelberg, 1973; Davidson, 1974; Ringertz and Savage, 1976; Weiss, 1977). The central focus of this work has been to determine the effects of combining two cell types on the expression of differentiated functions and thereby to deduce what mechanisms are operating to maintain the expression of these functions.

Observations

In her brief review of the experimental evidence, Weiss (1977) divided the phenomena observed – she refers to them as interactions – into four categories: extinction, gene dosage effects, reexpression, and activation.

The phenomenon of "extinction" is the disappearance of differentiated, facultative (Ringertz and Savage, 1976), or, as Ephrussi (1972) termed them, "luxury functions" expressed by one or both parent cell types when hybrids are made between dissimilar cell types. It does not occur when both parent cells are expressing the same function, for example, hemoglobin synthesis in mouse erythroleukemia cells and human bone marrow cells (Deisseroth *et al.*, 1975). Several types of differentiated cells, including melanomas (melanin synthesis), erythroleukemias (hemoglobin synthesis), gliomas and neuroblastomas (a variety of nervous system specific functions), and hepatomas (synthesis of liver-specific proteins and enzymes), have been examined, either in combination with one another or with cells, such as fibroblasts, lymphoblasts, or leukocytes, which do not express these functions. The usual outcome, irrespective of the cell types used, has been the rapid disappearance of the differentiated function(s) (Davidson, 1974; Weiss, 1977). In an even more general demonstration of extinction, Bravo *et al.* (1982) showed that more than one-third of mouse polypeptides discernible by two-dimensional polyacrylamide gel electrophoresis were not expressed in a tumorigenic transformed Chinese hamster lung fibroblast × diploid mouse embryo fibroblast hybrid, even though at least one copy of each mouse chromosome was identified in the hybrid cell line.

Despite the frequency of its occurrence, extinction is not the universal outcome after hybrid formation, and one of the three other phenomena listed by Weiss (1977) may occur, either immediately or with time. One, which she refers to as a "gene dosage effect" (an unfortunate use of the term since all of the phenomena are probably gene dosage effects), is the failure of extinction to occur if the effective ploidy or overall gene dosage of the cell expressing the differentiated function(s) of interest is double that of the other cell type used in the hybrid. (Since many of the cells used in such hybrids are aneuploid, the expressions "1s" and "2s" are used to refer, respectively, to near-diploid and near-tetraploid chromosome complements.) In hybrids of this type, whether inter- or intraspecific, functions of the 2s parent, which would have been extinguished in a 1s × 1s hybrid, are retained. Several examples of this phenomenon are listed in Table 6.2.

A contrary example of extinction in 2s × 1s hybrids has been presented by Hyman, Cunningham, and Stallings (1981). When a 2s line of Abelson leukemia virus-induced lymphoma that did not express the Thy–1 antigen was fused with a 1s line that did, expression of the Thy–1 antigen was extinguished in the resulting hybrid.

Reexpression of previously extinguished functions has been observed in hybrids in which chromosomes contributed by the nonexpressing parental cell type are lost as the hybrid is carried in culture. Numerous instances of this phenomenon have been reported, again both in inter- and intraspecific hybrids. For example, reexpression of aldolase B activity, alanine aminotransferase activity and inducibility, and tyrosine aminotransferase inducibility has

Table 6.2. *Gene dosage-dependent retention of expression of differentiated functions in 2s × 1s cell hybrids*

2s Parent	1s Parent	Differentiated function	References
Rat glial cells	Mouse fibroblasts	Glycerol-3-phosphate dehydrogenase induction	Davidson and Benda (1970)
Syrian hamster melanoma	Mouse fibroblasts	Melanin synthesis	Davidson (1972) Fougère, Ruiz, and Ephrussi (1972)
Mouse neuroblastoma	Mouse fibroblasts	Electrical excitability	Minna *et al.* (1971)
		Acetylcholinesterase activity	Minna *et al.* (1971)
		Acetylcholine sensitivity	Peacock, McMorris, and Nelson (1973)
Mouse erythroleukemia	Human fibroblasts	Mouse globin[a]	Alter *et al.* (1977)
Human sarcoma	Human fibroblasts	Tumorigenicity	Stanbridge *et al.* (1981, 1982)

[a]Analyses carried out in heterokaryons.

been observed in rat hepatoma × rat epithelial cell hybrids as chromosomes (of the order of 40%) are lost (Weiss and Chaplain, 1971; Sparkes and Weiss, 1973; Bertolotti and Weiss, 1974). Similarly, resumption of the production of melanin or of albumin has been observed in rat hepatoma × mouse melanoma hybrids concomitant with chromosome loss (Fougère and Weiss, 1978). In some instances, there were sequential shifts from the expression of one function to the expression of the other, but the two were never expressed at the same time. And, in rat hepatoma × human somatic cell hybrids, a large number of liver-specific enzymes were expressed as many of the human chromosomes disappeared (Kielty, Povey, and Hopkinson, 1982*a,b*).

It is noteworthy that, in all of the hepatoma hybrids in which several liver cell functions were being examined, the reappearances of these different functions occurred independently of one another. On the assumption that this outcome was not the result of differential loss of structural genes in the different hybrid segregants, Bertolotti and Weiss (1974) inferred that "these cases of independent reexpression of differential functions of the hepatoma parent show that there exist, in hybrid cells, multiple elements regulating the expression of the spectrum of enzymes which characterize hepatic differentiation." Unfortunately, because of the difficulties involved in karyotyping the hybrid cells, most experiments of this type do not correlate loss of a chromosome or chromosomes with reappearance of a function. Therefore, the studies of Stanbridge *et al.* (1981, 1982) are interesting in this regard. In the analysis of human intraspecific hybrids between tumorigenic HeLa cells and nontumorigenic fibroblasts, they found that the hybrids, originally nontumorigenic, regained tumorigenicity when one copy each of chromosomes 11 and 14 were lost. What is most remarkable about this observation, granting that tumorigenicity is a difficult "differentiated" function to work with, is that the shift in tumorgenicity occurred when each of the two chromosomes was reduced in number from four to three. This finding suggests that it is the number of chromosomes, not just their presence or absence, that is important in determining the cellular phenotype (see Chapter 14 for a further discussion of potential mechanisms involved in the control of malignancy).

In some respects, Weiss's fourth phenomenon, *activation*, is the most interesting of all. In this situation, not only are the differentiated functions of the differentiated parental cell type retained, but some of the same functions are expressed by the genome of the previously nonexpressing cell type. This phenomenon has of necessity been demonstrated only in interspecific hybrids, since species differences in the differentiated products have been used to establish their genetic origin. Once again, a gene dosage phenomenon appears to be operative in that such activation occurs only when many of the chromosomes of the nonexpressing parent are lost, as in the hybrids studied by Kielty, Povey, and Hopkinson (1982*a,b*), or Pearson *et al.* (1983), or when the chromosome constitutions of the expressing and nonexpressing parental cells are 2s and 1s, respectively. Several examples of 2s × 1s hybrids and heterokaryons are listed in Table 6.3. Of particular note in this list are the mouse myotube × human amniocyte heterokaryons (Blau, Chiu, and Webster, 1983) in which chromosome loss is not an issue. In this case, and presumably also in many of the others listed in which large-scale chromosome loss is not a problem, the activation or reprogramming (Blau, Chiu, and Webster, 1983) of the nonexpressing gene locus or loci must be attributed to a gene dosage effect. The same argument can be made for retention of expression of mouse globin synthesis in the mouse erythroleukemia × human fibroblast heterokaryons (Alter *et al.*, 1977) listed in Table 6.2, as well as for several of the hybrids listed in that table. According to Mével-Ninio (1984), activation and expression are linked in that, in hepatoma × fibroblast hybrids, activation of the fibroblast albumin gene(s) does not occur without reexpression of the hepatoma albumin gene(s) – although the converse can take place.

Table 6.3. *Gene dosage-dependent expression of differentiated functions in 2s × 1s interspecific cell hybrids and heterokaryons*

2s Parent	1s Parent	Differentiated function	References
Rat hepatoma	Mouse fibroblasts (3T3)	Mouse albumin	Peterson and Weiss (1972)
Rat hepatoma	Mouse fibroblasts (L)	Mouse albumin	Mével-Ninio and Weiss (1981)
Rat hepatoma[a]	Mouse lymphoblasts	Mouse albumin	Malawista and Weiss (1974)
Rat hepatoma	Mouse lymphoblasts	Mouse tyrosine aminotransferase, aldolase B	Brown and Weiss (1975)
Mouse hepatoma[b]	Human leukocytes	Human albumin	Darlington, Bernhard, and Ruddle (1974)
Mouse hepatoma[c]	Human skin fibroblasts	Human albumin	Darlington, Rankin, and Schlanger (1982)
	Human lymphoblastoid cells	Human albumin, α_1-antitrypsin	
	Human amniocytes	Human albumin, transferrin, α_1-antitrypsin, ceruloplasmin	
Mouse erythroleukemia	Human fibroblasts	Human α- and/or β-globin	Willing, Niehuis, and Anderson (1979)
Mouse erythroleukemia	Human fibroblasts Human lymphoblasts	Human β- and γ-globin	Zavodny, Roginski, and Skoultchi (1983)
Mouse myotubes[d]	Human amniocytes	Human myosin light chains 1 and 2, creatine kinase M and B subunits	Blau, Chiu, and Webster (1983)

[a]Also get mouse albumin with 1s rat hepatomas. The mouse cell parent contains about half the number of chromosomes present in the 3T3 cells used by Peterson and Weiss (1972).
[b]58 mouse chromosomes (diploid = 40); many human chromosomes lost.
[c]Many human chromosomes lost in hybrids.
[d]Myotubes are multinucleate, nondividing cells. The fusion products are actually heterokaryons with mouse and human nuclei in the ratio of 2.2:1.0.

Mechanisms

Given all of the phenomena exhibited by cell hybrids and heterokaryons, how should the regulation of differentiated functions be conceptualized? Weiss (1980) puts the issue quite succinctly:

Extinction, dosage effects and activation provide insight into the problem of overt expression of differentiated functions, where the final effect is either negative (extinction) or positive (activation). Since these interactions occur when two genomes in different functional states are simply confronted, we are forced to conclude that diffusible substances are involved, although we have as yet no idea of their chemical nature nor of their sites of action. In the case of activation, however, it appears highly probable that the substance is acting at the gene level.

In their heavily theoretical discussion of the use of cell hybrids for the analysis of differentiation, Davis and Adelberg (1973) distinguished among three "alternative, all-inclusive" hypotheses to explain the appearance of a differentiated function: autonomous expression of the structural genes, continuous production of an activator, and discontinued production of a repressor. They then went on to describe many of the experimental pitfalls which make it difficult to discriminate among these possibilities and concluded that, up to 1973, most of the reported experiments were wanting.

In considering the issue further, Davidson (1974) hypothesized that, on the basis of observed extinction (suppression) of differentiated functions in a large variety of hybrids and the failure of this to occur when gene dosage was increased, "mammalian cells are continually producing a large number of specific regulator substances that act to suppress the expression of the majority of the differentiated functions that the cells do not exhibit." That continued expression or reexpression occurred when the gene dosage was altered, either directly or by chromosome loss, was taken to indicate that the putative regulator substances are produced in limited quantities. That different functions could reexpress independently suggested that each differentiated function is controlled by its own specific regulator. To explain activation of previously unexpressed functions, Davidson (1974) postulated the existence of regulatory substances which induce rather than suppress gene expression, with these substances again being produced in limiting amounts. These two postulated mechanisms, suppression and induction, are illustrated in Fig. 6.7.

These prototype mechanisms can be considered as simple representatives of sets of more complicated regulatory systems that might be proposed. However, even in their simplest forms, it may be quite difficult to distinguish between them experimentally (Davis and Adelberg, 1973). Furthermore, there are certain conceptual difficulties which require resolution. Of chief concern to me is the notion of limiting amounts of regulator substances and the interpretation of the gene dosage experiments. If continued expression of a function takes place in a 2s × 1s but not in a 1s × 1s hybrid, it would be inferred

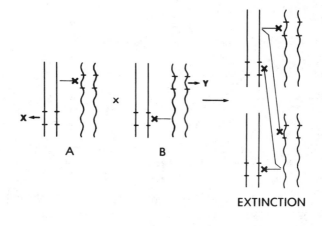

EXTINCTION

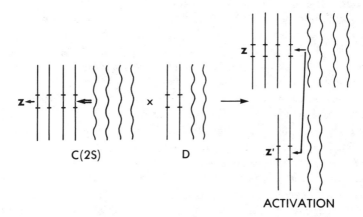

ACTIVATION

Fig. 6.7. Models for actions of regulatory substances in somatic cell hybrids. Top: Extinction of differentiated functions X and Y characteristic of differentiated cell types A and B, respectively, by diffusible suppressor substances in A × B hybrids. Bottom: Activation of previously unexpressed gene for function Z′ in genome of cell type D in C(2s) × D(1s) hybrids. The arrowheads indicate activation, the crosses suppression.

from Davidson's (1974) hypothesis that the 1s genome can no longer produce enough regulator to suppress the function in question (since there are now twice as many sites competing for it). However, this would imply that it is the stoichiometric amount of the regulator substance, not its concentration, which determines its interaction with whatever site(s) it interacts with. While conceivably true, this notion does violence to the principles of receptor-ligand interactions outlined earlier which ought to govern molecular interactions of this type. It can be speculated that the issue is still one of concentration if it is assumed that the volume of a 2s × 1s hybrid is greater than that of a 1s × 1s hybrid, thereby permitting the effective receptor concentration to drop below the necessary threshold. At present, we do not have the experimental data to test this possibility. The same concerns do not apply to experiments demonstrating the expression of previously unexpressed functions by the 1s genome, since in this case the concentration of the putative inducing regulator substance produced by the 2s genome would actually be increased.

At present, there is no real direct evidence for the existence of the postulated suppressing and inducing regulator substances. However, it has been shown that anucleate mouse fibroblast cytoplasm, when fused to rat hepatoma cells, produces a rapid but short-lived suppression or extinction of albumin synthesis which lasts less than 48 hours (Kahn *et al.*, 1981), presumably the time required for its degradation. The same rapid extinction is also observed even in 2s rat hepatoma × 1s mouse fibroblast hybrids, with reexpression of rat albumin synthesis and activation of mouse albumin synthesis occurring several days later (Mével-Ninio and Weiss, 1981). Reexpression of the rat protein precedes activation of the mouse albumin gene by several days, and rat albumin reappears earlier in 2s × 1s than in 1s × 1s rat hepatoma × mouse fibroblast hybrids. The latter observation suggests that previously active genes are more rapidly expressed than previously silent ones, possibly because of differences in the conformation of the chromatin. These results are consistent with the existence of both suppressing and inducing regulator substances, with the former coming into play immediately or soon after fusion and the latter considerably later, possibly because of the time required for the necessary concentration to be achieved. The possibility that de novo methylation of the albumin gene might be involved in the extinction phenomenon has been ruled out by Ott, Sperling, and Weiss (1984).

Further evidence in support of trans-acting positive (inducing) and negative (suppressing) regulators comes from the work of Gopalakrishnan and collaborators in cell × cytoplast (anucleate cytoplasm) fusions to generate "cybrids." When 1s mouse erythroleukemia cells capable of hemoglobin synthesis were fused with mouse neuroblastoma or fibroblast cytoplasms, hemoglobin synthesis was suppressed, even after many cell divisions (Gopalakrishnan, Thompson, and Anderson, 1977). Conversely, when the same cells were fused with rat hepatoma cytoplasts, hemoglobin synthesis was not suppressed. Surprisingly, however, the resulting cybrids were able to produce phenylalanine hydroxylase, a liver-specific enzyme, of the mouse type (Gopalakrishnan and Anderson, 1979). Similar results were obtained by fusing erythroleukemia cells with mitomycin C- or iodoacetamide-treated hepatoma cells to form "pseudocybrids" (Gopalakrishnan and Littlefield, 1983). Furthermore, transfection of erythroleukemia cells with rat hepatoma RNA resulted in the appearance of erythroleukemia cells (1 per 10^7 to 1 per 10^6) capable of synthesizing mouse phenylalanine hydroxylase. It was suggested, therefore, that this RNA or its translation product might be the positive regulator substance responsible for activation of the mouse phenylalanine hydroxylase gene and that it might be possible to clone the DNA segment involved in phenylalanine hydroxylase regulation (Gopalakrishnan, 1984).

Other experiments of interest can be mentioned. A mutant line of rat hepatoma cells with low purine nucleoside phosphorylase activity was isolated. When fused with wild-type cells, the hybrids also displayed low enzyme activity (Hoffee *et al.*, 1983). However, with chromosome loss, segregants appeared which again displayed high nucleoside phosphorylase activity. These results were interpreted as providing evidence for the appearance, as a result of mutation, of a trans-acting regulator (suppressor) of purine nucleoside phosphorylase expression. While probably not the same type of regulator postulated for the extinction phenomenon, its behavior in the hybrid system is certainly quite similar. However, Killary and Fournier (1984) believe that they have identified and mapped a locus on mouse chromosome 11 and human chromosome 17 designated *Tse–1* (tissue-specific extinguisher–1) that is responsible for the extinction of liver functions in hepatoma × fibroblast hybrids. Similarly, human chromosome 1 has been identified as carrying a locus that is capable of suppressing the transformed phenotype in BHK × human fibroblast hybrids (Stoler and Bouck, 1985).

In interpreting the results of the cell hybrid work, several authors (Ephrussi, 1972; Davis and Adelberg, 1973; Chen, Worthy, and Krooth, 1978; Weiss, 1980), as in the earlier discussion of regulatory systems, have again distinguished between cellular determination, which refers to the commitment of the cell to exhibit a specific stable differentiated phenotype (for example, that of a hepatocyte, erythrocyte, or neuron) and expression or modulation or even differentiation, which refers to the expression of particular genes characteristic of the differentiated phenotype. Determination or commitment is considered to be a highly stable, heritable process, since, as the reexpression data indicate, it is not lost when cells of different types are combined in the form of hybrids. Nevertheless, all of the chromosomes in a particular genome cannot be considered as irreversibly locked into expression of a fixed set of differentiated functions, as Caplan and Ordahl (1978) propose, since, as has already been discussed, activation of genes expressing functions characteristic of another differentiated cell type has been demonstrated. However, what the cell hybrid experiments seem to be testing primarily appears to be the control of expression or modulation and, as outlined above, the evidence favors the existence of diffusible or trans-acting regulator substances.

Relevance to aneuploidy

What implications do these observations have for our understanding of the effects of aneuploidy? In this regard, the two most important conclusions to be drawn from the hybrid work are the following. First, consistent with the theoretical models of gene regulation discussed earlier, diffusible trans-acting regulator substances appear to be involved in the expression of differentiated functions. These regulator substances may be either inhibitory (suppressing) or activating (inducing) in nature, and any differentiated function may be governed by one or, more likely, both types of substances. Second, and of greatest significance, the activity of these regulator substances is gene dosage and therefore presumably concentration dependent. Twofold differences in gene dosage can make the difference between extinction and nonextinction, retention and nonretention, and activation and nonactivation of specific differentiated cellular functions. Furthermore, as the experiments of Mével-Ninio and Weiss (1981) indicate, such gene dosage differences may also affect the temporal control of expression of a differentiated function.

Several possible effects of aneuploidy on such regulatory systems can be visualized and are shown schematically in Figs. 6.8 and 6.9. For example, if the activity of some function were maintained by a regulator substance of the inducing type (Fig. 6.8), a decrease in its concentration, such as might result in a monosomy or deletion, could lead to cessation of expression of that function. Conversely, the trisomic state could result in premature expression, excessive expression (if there is any quantitative relationship between expression and regulator concentration), or perhaps, as the 2s × 1s activation data would suggest, even inappropriate expression (expression in the wrong cell type). If the regulator substance were of the suppressing or repressing type, the effects would be just the opposite (Fig. 6.9). In a monosomy or deletion, genes ordinarily suppressed and therefore inactive in a specific cell type might become active because of a drop in concentration of the suppressor substance, and even in the appropriate cell type might become prematurely active or overactive. In the trisomic state, suppression or delayed appearance of differentiated functions could occur. In all instances, the effects would be expected to be highly specific. All or even many functions of a single cell type would not necessarily be involved, while functions characteristic of more than one type could. Such effects could, of course, be deleterious or even disastrous, particularly early in development. And, while the models derived from cell hybrids pertain principally to the expression of differentiated functions in cells that are already determined or differentiated, they can, in general terms, also be considered generally applicable to the processes of differentiation as well.

Regulatory disturbances in aneuploid mammalian cells

Is there any evidence that these hypothetical effects which result from imbalance of regulatory loci actually occur in aneuploid cells or organisms? To date, only one directly relevant report has been published, but the results are most fascinating. Krooth and his collaborators (Chen, Worthy, and Krooth, 1978), after reviewing the cell hybrid data summarized earlier, concluded, as

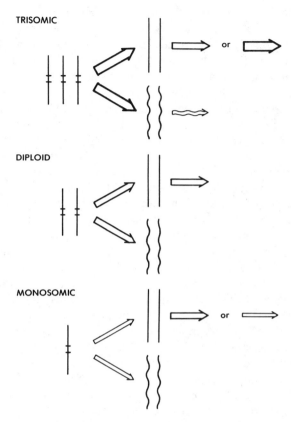

Fig. 6.8. Hypothetical effects of aneuploidy-produced changes in concentrations of regulatory substances in aneuploid cells. Effects of aneuploidy on a system with an activating (inducing) regulatory substance. The straight chromosomes on the left are the sites of production of the regulatory substances for a specific cell type, and the straight and wavy chromosomes on the right are those involved in the expression of the differentiated function in the same (straight) and different (wavy) cell types. The arrowheads between the two sets of chromosomes indicate activation.

have others, that there were several potential problems with the reported studies. These included the use of parental cells which were frequently transformed (malignant) and karyotypically unstable, the karyotypic instability and interspecific nature of many of the hybrids themselves, and the fact that the dosage of only whole-chromosome sets rather than of individual chromosomes could be varied. The last of these problems is, of course, particularly germane to our present concerns. Therefore, to overcome both these and the other difficulties, Chen, Worthy, and Krooth (1978) set about, in a prospective manner, to look for the analog of the activation phenomenon described in hybrids – the appearance of an inappropriate differentiated function

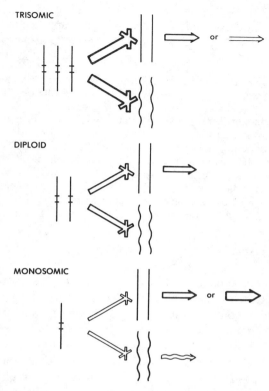

Fig. 6.9 Hypothetical effects of aneuploidy-produced changes in concentrations of regulatory substances in aneuploid cells. Effects of aneuploidy on a system with a suppressing (repressing) regulatory substance, as in legend to Fig. 6.8, with crosses indicating suppression.

in a human cell with trisomy of a single chromosome. And, what is most remarkable, they appear to have found one. Using acetylcholinesterase as a marker of neural cell function, they found three fetal human cell strains derived from chorion and/or amnion with a markedly elevated activity of this enzyme, and each was trisomic for chromosome 2. The mean enzyme activity for trisomy 2 cells was 14.2 ± 0.97(SE) pmol/min/μgDNA; the mean for control strains (not trisomy 2) was 0.50 ± 0.17, and the ratio of trisomy 2:control activities was 28.4. Pseudocholinesterase activity, used as an enzyme activity control, was the same in both the trisomy 2 and control cell strains, and xanthine oxidase and choline acetyltransferase were not detectable in either. Comparison of the acetylcholinesterase activity in trisomy 2 cells with that in human brain indicated that expression in the former was only 1.5% of that in the latter.

To explain the selective increase in acetylcholinesterase activity in the trisomy 2 cells, Chen, Worthy, and Krooth (1978) made use of a model similar to those described here, actually one in which both suppression and activation of gene expression are operating. They then went on to suggest that their results imply that:

The balance between genetic elements favoring gene expression (for a special product) and those opposing it is, at least at some loci, a delicate one. If this were not the case, a mere 1.5-fold change in chromosome dosage would hardly be expected to perturb the mechanism for repression sufficiently to render the cellular concentration of special product easily measurable. The present interpretation of the somatic cell hybridization experiments likewise implies that the balance can be conspicuously perturbed by small changes in gene dosage. . . . Our results lead us to suspect that we have observed such an occurrence [inappropriate expression of a differentiated function] once. It is conceivable, therefore, that the absolute number of loci (for special products) which are inappropriately expressed in the tissues of a trisomic fetus is very great, at least in the case of fetuses which are trisomic for one of the larger chromosomes.

This belief that a small change in the concentration of a regulatory element could have a large effect on an all-or-nothing phenomenon such as the synthesis of a special differentiated protein is not shared by Zuckerkandl (1978). He has visualized the existence of buffering systems to produce a stable form of regulation and to counteract just such small changes in the concentrations of inducers and repressors.

In the discussion of the work just described, the authors (Chen, Worthy, and Krooth, 1978) have reminded us of another possible example or set of examples of inappropriate gene expression – the secretion of hormones by tumors of nonendocrine origin. And, they cited a previous paper by Krooth (1964), written shortly after the appearance of the work of Jacob and Monod on gene regulation in bacteria, in which it was postulated that this phenomenon may be related to chromosome loss (monosomy, nullisomy) or gain (trisomy) by just the type of regulatory mechanisms discussed above. Whether inappropriate peptide hormone secretion has a basis in aneuploidy is still unknown. Odell *et al.* (1977) believe that it is a universal consequence of malignancy and that it represents a less efficient repression of peptide hormone synthesis which is never totally repressed even in normal cells. Thus, while regulatory alterations are undoubtedly involved, they may be related to the malignant state per se and not necessarily to any concomitant karyotypic changes. However, Sandberg (1983c) argues that the removal of the peptide hormone loci from the normal cis-acting suppressors by chromosomal deletion or rearrangement may be the explanation (see Chapter 14).

There are two clinical observations that could conceivably be related to developmental regulatory disturbances of the type under consideration. The first is the appearance, in the erythrocytes of human newborns with trisomy 13, of a variety of embryonic and fetal hemoglobins (Huehns *et al.*, 1964). These include, in addition to increased quantities of hemoglobin F ($\alpha_2\gamma_2$), low quantities of hemoglobin Bart's (γ_4) (~1.5%), hemoglobin Gower–2 ($\alpha_2\epsilon_2$) (~0.5%), and possibly (Hecht, Jones, and Koler, 1967) of hemoglobin Portland 1 ($\zeta_2\gamma_2$). Conversely, hemoglobin A2 ($\alpha_2\delta_2$) is decreased in amount (Walzer *et al.*, 1966; Wilson *et al.*, 1967). These discrepancies between the trisomy 13 and normal infant hemoglobin patterns continue for several months, suggesting a delay in the normal switch(es) between fetal and adult hemoglobin chain synthesis. These delays affect not only the γ to β and δ transition, but also the embryonic to fetal transition from ζ and ϵ and to α and γ, respectively, which occurs very early in development (usually by 10 weeks of gestation) (Gale, Clegg, and Huehns, 1979).

To explain these delays, Huehns *et al.* (1964) proposed two Jacob and Monod-type regulatory models. One model postulated that the linked operator and structural genes for ϵ and γ chains are in the triplex state and that the amounts of repressor required for suppression of ϵ- and γ-chain synthesis are therefore insufficient. This model is clearly untenable, since the loci in question are now known to be on chromosome 11. The other model postulated that the regulator loci are on chromosome 13, so that trisomy 13 results in an excess or increased concentration of the ϵ and γ-chain inducers. However, in view of the fact that the synthesis of multiple hemoglobin chains is affected, it seems more reasonable to postulate that trisomy 13 affects a more general hemoglobin-switching process rather than the synthesis of two chains. Unfortunately, very little is known about the control of hemoglobin switching, although it has been suggested that modulation of the methylation pattern of the DNA in the closely flanking sequences of globin genes may represent a key mechanism for regulatory expression of human globin genes during the embryonic → fetal and fetal → adult hemoglobin switches in human (Mavilio *et al.*, 1983). For the β-globinlike chains, a strong correlation exists between hypomethylation of the flanking sequences and globin gene expression, and although the actual mechanism is in doubt (Kolata, 1984), a methylation inhibitor, 5-azacytidine, can increase hemoglobin F synthesis in vivo (Ley *et al.*, 1982). How trisomy 13 affects the methylation process, if it is indeed causally related to hemoglobin switching, remains to be determined. Despite intense interest and work on the mechanisms controlling hemoglobin switching (for a general review, see Peschle *et al.*, 1983, and Stamatoyannopoulos and Nienhuis, 1983), no model has been put forward which makes it possible even to speculate on the role of chromosome 13 in the process and the way in which trisomy 13 interferes with it.

The second clinical observation relates to the situation of immunoglobin A (IgA) in abnormalities of human chromosome 18; this situation, while still somewhat confused, suggests that a locus or loci regulating IgA synthesis and secretion may be located on chromosome 18. A structural gene abnormality cannot be involved since the IgA heavy chain genes are located on chromosome 14 and the light-chain genes, which serve all immunoglobulins, are on chromosomes 2 (κ) and 22 (λ). Abnormally low levels and even absence of serum and salivary IgA have been reported in several but not all cases of ring 18 and del(18q), and also in cases of del(18p) (Stewart *et al.*, 1970; Kunze, Stephan, and Tolksdorf, 1972; Wilson *et al.*, 1979). The differences, even in the del(18q) cases, are not simply the result of different regions of the chromosome being deleted, since even within a single series of cases with comparable deletions, significant differences in IgA levels are found (Wilson *et al.*, 1979). Moreover, deficiency of IgA is not attributable to an absence of B lymphocytes capable of producing it. Lewkonia, Lin, and Haslam (1980) reported two cases of IgA deficiency with del(18q) in whom the proportion of B lymphocytes positive for IgA fluorescent staining was normal.

At the present time, not enough is known about the regulation of immunoglobulin synthesis to make possible a meaningful explanation of the relationship of IgA deficiency and deletions of chromosome 18. Since structural loci cannot be implicated, regulatory abnormalities have, of course, been suggested (Stewart *et al.*, 1970; Lewkonia, Lin, and Haslam, 1980), with the variability being attributed either to environmental factors (Lewkonia, Lin, and Haslam, 1980) or to possible allelic differences in the regulatory locus (loci) present on the nondeleted chromosome (Stewart *et al.*, 1970). In support of the existence of such regulatory loci, although not relevant to the role of chromosome 18, are the reports of several cases of selective IgA deficiency which are apparently inherited in an autosomal dominant manner (Ammann and Hong, 1971; Levitt and Cooper, 1981). Thus, for both trisomy 13 and deletions of chromosome 18, regulatory gene disturbances, while attractive as hypothetical explanations for observed phenomena, do not presently have any firm evidence to support them.

Search for regulatory disturbances by analysis of cellular proteins

Aside from trisomy 2 and acetylcholinesterase, there are no clear examples in mammals of aneuploidy-related regulatory disturbances of the type expected from consideration of the cell hybrid data. This is perhaps not surprising, since they really have not been searched for in any systematic manner, and many possible instances affecting development probably would not be recognized as such. The closest to what might be considered a general search for such regulatory abnormalities has been the use of electrophoretic techniques to examine a large number of cellular proteins simultaneously. The earliest work of this type was not carried out in mammals, but in plants, and several potential examples of regulatory disturbances associated with aneuploidy were reported in barley (McDaniel and Ramage, 1970), Jimsonweed (*Datura stramonium*) (Smith and Conklin, 1975), and grain sorghum (Suh *et al.*, 1977) trisomics. The work with *Datura* trisomics, which interestingly had distinct leaf phenotypes, involved an analysis of peroxidase isoenzymes, and what was most striking was the occasional profound decrease in the intensity of a specific band associated with a particular trisomy, especially the decrease in peroxidase band 3 intensity from 2.7 to 0.1 units in trisomy for chromosomes 11 and 12. A similar result, termed "suppression" by McDaniel and Ramage (1970), was observed in barley, with protein band 9 missing in plants trisomic for chromosome 6. In grain sorghum, both band suppression and the appearance of new bands were noted (Suh *et al.*, 1977).

To look for major regulatory disturbances in gene expression in mammalian cells, Jon Weil and I (Weil and Epstein, 1979) decided to use two-dimensional (2D) polyacrylamide gel electrophoresis. This technique is capable of resolving 500–1000 polypeptides in these cells. For our studies, we compared the 2D gel patterns of the radioactively labeled proteins of carefully matched pairs of human trisomy 21 and diploid cell strains. The gels were analyzed visually, with the method being capable of detecting twofold or greater differences in spot intensities. Of the 850 polypeptides so compared, only four showed a pattern of variation that could conceivably be related to trisomy 21, and none of these differences was either profound or entirely consistent. Similar studies were also carried out by Klose, Zeindl, and Sperling (1982). They visually examined about 800 Coomassie Blue-stained polypeptides and found four decreases regularly associated with trisomy 21 when trisomic and diploid fibroblasts were compared. However, only one difference was found when either cultured epithelioid amniotic fluid cells or cloned fetal lung fibroblasts were compared.

Because of the difficulties inherent in visual comparisons, Van Keuren, Merril, and Goldman (1983) used an automated scanning system to analyze silver-stained 2D gel patterns of trisomy 21 and diploid control cells. Only two examples of gene dosage effects (superoxide dismutase–1 and an unidentified peptide) were found among the 400 peptides analyzed, and no really unequivocal examples of either marked increases or decreases in specific peptide concentrations.

In addition to the work with human cells, my collaborators and I looked visually at the 2D gel patterns of proteins from cells with mouse trisomy 17 and *Drosophila* larvae trisomic for chromosome arm 2L (Epstein *et al.*, 1981). In neither case were major differences found between aneuploid and diploid patterns. Klose and Putz (1983) also looked at proteins made by cells from four different mouse trisomies (1, 12, 14, and 19), this time using a combination of visual scanning and computerized quantitation. Roughly 1000 Coomassie Blue-stained polypeptides were examined, and two types of differences were found. For each of the trisomies, between 1 and 7 (mean = 4) peptides were found that showed a chromosome-specific increase in concentration. The mean increase in the concentrations of these peptides was 1.68 ± 0.51(SD), as might be expected for a gene dosage effect. In addition, they also detected about 12 peptides that were reduced in concentration and one that was increased in a chromosome-nonspecific manner. The actual decreases or increase were not described in quantitative terms. However, what is most remarkable about this report is the assertion that "most of the variants [i.e., the chromosome-nonspecific ones] observed regularly in a particular trisomy occurred also regularly in several or all other trisomies investigated." This observation is difficult to explain on any theoretical basis. The limited number of changes implies that there should be a high degree of specificity, with aneuploidy for each chromosome having a specific effect. Yet, just the opposite appears to be the case!

Taken in the aggregate, the electrophoretic studies suggest that major changes in polypeptide synthesis that can be attributed to regulatory disturbances are quite rare, if they exist at all, in aneuploid mammals (specifically man and mouse). While slightly more suggestive evidence exists in the plant literature, it is by no means strong. In considering the situation in mammals, certain qualifications must be kept in mind. Most of the human and animal work was carried out with cultured fibroblasts or fibroblast-like cells. This cell type has, despite its synthesis and secretion of collagen and other macromolecules, often been accused of being relatively undifferentiated. The question might therefore be raised as to whether results obtained with fibroblasts are representative of what would be obtained with other functionally differentiated cells, such as the ones (hepatocytes, pigment cells, neural cells, and the like) used for hybridization studies. This objection does not, however, really apply to the work with aneuploid mouse embryos. Samples were obtained from uncultured brain, liver, and mixed skin and connective tissue (Klose and Putz, 1983), and still the results were the same.

Another potential objection to the human studies is that they were carried out only with trisomy 21 cells, with trisomy 21 being the least severe of the human autosomal trisomies. Similarly, the mouse studies of Klose and Putz (1983) were performed with material from four of the trisomies compatible with the longest survival in utero. Whether the results would be different with aneuploid states resulting in earlier lethality or more severe abnormalities is unknown. However, our own studies (unpublished) with mouse trisomy 17 and Klose and Putz's (1983) analysis of head tissue from exencephalic fetuses with trisomy 12 or 14 argue against such an inference. Perhaps the strongest criticism might be that tissues were not investigated early enough during the period of active development and differentiation. However, once again the mouse studies argue against this, since both brain and liver in 10- to 13-day mouse embryos are certainly undergoing active development and differentiation.

Therefore, despite the implication of the cell hybrid results that gene dosage changes could result in alterations in gene expression such that novel products might appear or customary products would be suppressed, strong supportive evidence is not forthcoming from the analysis of either the newly synthesized or steady-state polypeptide composition of aneuploid cells or tissues. Perhaps unidentified polypeptides are not the right thing to be looking at, and more experiments of the type carried out by Chen, Worthy, and Krooth (1978) with specific functional markers of differentiated cell expression are required. It is conceivable that the quantitatively most prominent polypeptides, such as are revealed by the electrophoretic techniques, are primarily structural and "housekeeping" proteins which one would not expect to be affected. According to Klose (1982), only 32–35% of spots in a 2D electrophoretic gel pattern are tissue specific, and these may be the only ones worth considering.

It is somewhat peculiar that the number of changes consistent with gene dosage effects are many fewer than would be expected. For example, as has already been noted, in looking at 1000 peptides, presumably coded for all twenty pairs of chromosomes in the mouse, Klose and Putz (1983) found only about four peptides per trisomy that were increased 1.5 fold. If these numbers are representative of all trisomies, gene dosage effects would exist for only about 80 peptides (even ignoring X-chromosome inactivation in females), an order of magnitude less than the number of polypeptides actually being looked at. What the 2D electrophoretic gel patterns see is amazingly stable; perhaps what they do not see is actually of much greater interest. However, this is only conjecture, and for the present it must be concluded that there is no evidence that aneuploidy (trisomy) produces extensive or even moderate disturbances in the regulation of the synthesis of gene products of already differentiated cells and possibly of differentiating tissues as well. We can only hope that it will be possible to approach the problem in a more straightforward manner once the identities and locations of mammalian regulator genes are known. At the present time, unfortunately, no clear-cut human regulatory loci have been identified (Human Gene Mapping 7, 1984) and only about 15 regulatory loci, nearly all of the temporal type, are known in mouse (Green, 1981).

Regulatory disturbances in aneuploid nonmammalian cells

In the earlier discussion of gene dosage effects and dosage compensation (Chapter 4), mention was made of loci in maize and *Drosophila* that regulate the expression of other loci so as to produce dosage compensation in certain aneuploid states. However, when the dosage of the regulated structural genes is not altered, aneuploidy of the regions carrying these regulatory loci produces a negative inverse effect on structural gene expression (Rawls and Lucchesi, 1974; Birchler, 1979, 1981, 1983*b*; Birchler and Newton, 1981). Several examples of such regulatory effects on enzyme activity are shown in Table 6.4. Similar results have also been obtained for protein bands demonstrated by gel electrophoresis (Birchler and Newton, 1981). In the triplex

Table 6.4. *Negative inverse effects on gene expression in aneuploids*

	Dosage of putative regulator locus				
	1	2	3	4	References
Maize					
Glucose-6-phosphate dehydrogenase	1.56[a]	1.0	0.79	0.64	Birchler (1979)
6-Phosphogluconate dehydrogenase	1.44	1.0	0.64	0.66	Birchler (1979)
Isocitrate dehydrogenase	1.29	1.0	0.89	0.72	Birchler (1979)
Alcohol dehydrogenase	1.25	1.0	0.77		Birchler (1981)
Drosophila					
α-Glycerophosphate dehydrogenase		1.0	0.66		Rawls and Lucchesi (1974)
		1.0	0.73		Birchler (1983*b*)
Alcohol dehydrogenase		1.0	0.70		Birchler (1983*b*)

[a]Numbers express relative enzyme activity; activity in diploid set at 1.0.

situation, the reported relative amounts of regulated gene activity were in the range of 0.66–0.89 (compared to the diploid), while in the uniplex state, the ratios were in the range of 1.25–1.56. If the purpose of such regulatory loci were to produce dosage compensation of unbalanced structural genes, the expected ratios would be, of course, 0.67 and 2.0, respectively. By measurement of mRNA levels, the regulatory effects for one locus, *Drosophila* alcohol dehydrogenase, have been shown to be at the level of transcription or posttranscriptional processing (Birchler, 1983*b*).

Of even greater interest than the existence of regulatory loci capable of affecting dosage compensation is the fact that more than one regulatory locus may affect a particular structural gene, and several genes may be affected by changes of a single regulatory locus or closely linked group of loci (Paigen and Birchler, personal communication). These regulatory loci may be scattered throughout the genome and need not be closely linked or even on the same chromosome with the regulated genes (see also Ayala and McDonald, 1980). Thus, aneuploidy for any region of the genome is potentially capable of altering, in a specific manner, the synthesis of a number of gene products coded for by other regions of the gene.

Clearly comparable mechanisms have not as yet been demonstrated in mammals. The situations closest to those in maize and *Drosophila* are the reductions in polypeptide concentration or synthesis observed in 2D electrophoretic gels of trisomic cells. However, as was discussed in the previous section, such effects are relatively infrequent, even with whole-chromosome trisomies, and have a degree of nonspecificity greater than would be expected. Quantitative data are not available for the work of Klose and Putz (1983), so it is not possible to determine whether the decreases they reported resemble those found in maize and *Drosophila*.

Birchler and Newton (1981) believe that these negative effects of aneuploidy, taken in combination with positive (direct gene dosage) effects, especially in monosomies, provide a way to understand the effects of aneuploidy: "Aneuploidy produces numerous reductions of specific proteins in both partial monosomy and trisomy and allows the interpretation that the reductions *per se* are a major factor contributing to the aneuploidy phenotypes." The negative regulatory perturbations, if they did exist in mammals, could certainly result in significant structural and metabolic alterations by the mechanisms outlined in this part of the book. However, there is at present no compelling evidence that such effects do exist in mammals, and the available evidence argues that, if they do exist at all, they are not a major phenomenon.

7

Assembly of macromolecules, cellular interactions, and pattern formation

Assembly of macromolecules

It is of interest that many of the dominantly inherited disorders for which biochemical abnormalities have been defined involve the synthesis of normal or decreased amounts of structurally abnormal protein molecules (C. J. Epstein, 1977). Included in this group are, for example, the dysfibrinogenemias (Graham *et al.*, 1983), some of the hemoglobinopathies associated with unstable hemoglobins (Weatherall and Clegg, 1981), several of the connective tissue defects (osteogenesis imperfecta, Ehlers-Danlos syndrome) in which abnormal collagen chains are synthesized (Byers *et al.*, 1983; Chu *et al.*, 1983; Prockop *et al.*, 1983; Stolle, Myers, and Pyeritz, 1983; Wenstrup, Hunter, and Byers, 1983; Prockop and Kivirikko, 1984), red cell disorders (elliptocytosis, spherocytosis) with abnormal membranes (Coetzer and Zail, 1981; Tomaselli, John, and Lux, 1981; Wolfe *et al.*, 1982), and, in *Drosophila*, a meiotic mutant with an abnormal β-tubulin (Kemphues *et al.*, 1980). In these disorders, it is easy to visualize how a structurally abnormal subunit interferes with the formation, stability, and/or function of a multisubunit macromolecular complex (C. J. Epstein, 1977). However, in the situation of aneuploidy, we are not dealing with qualitative alterations in subunit structure but only with quantitative changes in rates of synthesis and hence, in concentrations. Can such changes result in abnormality?

It is possible to generate hypothetical examples for how a change in the concentration of one subunit in a multisubunit protein could lead to the formation of abnormal or unusual products. Two such examples, formulated in terms of an increased subunit concentration associated with trisomy, are shown in Fig. 7.1. Another, based on the behavior of certain drug resistance genes in fungi, has been put forward by Lewis and Vakeria (1977). While hypothetical, these examples do have a real basis in fact – the situation that gives rise to the abnormalities associated with heterozygosity for either α- or β-thalassemia.

The genes for α-globin exist in tandem pairs on human chromosome 16, so that an individual normally has four α-globin genes. In α-thalassemia 2, one α-globin gene is deleted or inactive; in α-thalassemia 1, two are deleted. Therefore, in appropriate combinations, it is possible for individuals to have

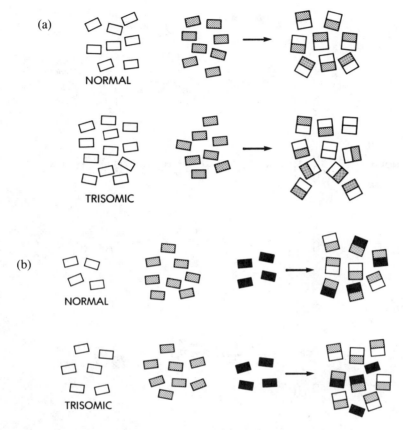

Fig. 7.1. Hypothetical examples of how a change in the concentration of one of the subunits of a multisubunit protein could affect formation of the protein. (a) A multisubunit protein with two types of subunits. In the normal situation (above), the rates of synthesis of the two different subunits are balanced, and only mixed dimers are produced. In the trisomic condition, synthesis of the open subunits is increased, thereby permitting the formation of unusual homodimers containing only one type of subunit. (b) Multisubunit proteins which share a common subunit. In the normal situation (above), the rates of synthesis of all three types of subunits are balanced, and only two types of mixed dimers are formed. In the trisomic state, the synthesis of the open subunit (left) is increased. If the affinity of this subunit is greater than that of the closed subunit (right) for the shared common subunit (center, hatched), then there will be an increase in the number of mixed dimers with the open subunit, a deficiency of mixed dimers with the closed subunit, and excess closed subunits in a free state. Reprinted by permission from Epstein *et al.* (1981).

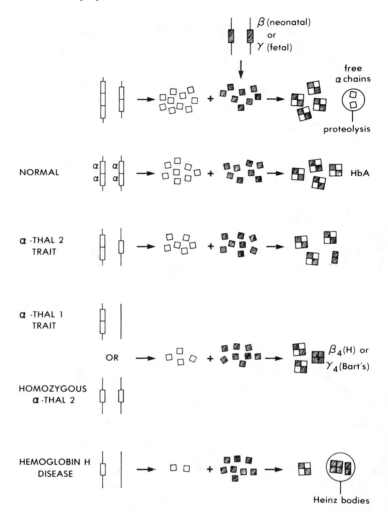

Fig. 7.2. Globin chain synthesis and tetramer formation in individuals with differing numbers of functional α-globin chain genes. In sufficient concentration (as in hemoglobin H disease), hemoglobin β_4 (H) precipitates and forms Heinz bodies. The decreasing number of hemoglobin (Hb) A molecules with decreasing α-chain synthesis is apparent, as is the increasing synthesis of β_4 and, in fetuses, γ_4. Excess β or γ chains always aggregate as tetramers (not in the dimers shown here to maintain numerical balance in the figure).

zero (hydrops fetalis), one (hemoglobin H disease), two (α-thalassemia minor), three (silent carrier), or four (normal) functional α-globin genes (Fig. 7.2) (Ottolenghi *et al.*, 1974; Taylor *et al.*, 1974; Dozy *et al.*, 1979; Orkin *et al.*, 1979). And, since unequal crossing over of the α-globin loci has apparently given rise to chromosomes with three α-globin genes in tandem, individuals can have five and, in theory, even six α-globin genes (Goossens *et al.*, 1980; Higgs *et al.*, 1980). In the α-globin deficiency states, the symptoms are determined by a combination of the decreased oxygen-carrying capacity of the blood and the effects of the unstable atypical hemoglobin molecules formed. Since red cell precursors normally produce only a slight excess of α chains, a decrease in the synthesis of α chains will result in a relatively proportional decrease in the concentration of hemoglobin A ($\alpha_2\beta_2$) and a commensurate decrease in normal oxygen binding and release. Furthermore, depending on their concentration, the excess β chains, and in the case of the fetus, γ chains, aggregate together to form tetramers: β_4 (hemoglobin H) and γ_4 (hemoglobin Bart's). As a result, individuals with three α-globin genes have 1–2% hemoglobin Bart's at birth, whereas those with only two α-globin genes have 5–15% (Weatherall, 1978). This abnormal hemoglobin disappears with development and is not replaced by an equivalent amount of hemoglobin H, although some hemoglobin H may be demonstrable in the latter situation. However, when only one α-globin gene is functional, the red cells will contain 5–30% hemoglobin H and small amounts of hemoglobin Bart's and hemoglobin δ_4. The hemoglobin H, which is unstable, eventually precipitates intracellularly in the form of Heinz bodies. These are removed from the red cells on passage through the spleen, but the ultimate outcome is a significantly shortened red cell survival (Weatherall and Clegg, 1981).

The presence of excess α chains (five genes instead of four) does not cause any clinical abnormalities, and it is presumed that the extra α chains are removed by proteolysis (Higgs *et al.*, 1980). However, the capacity to remove free α chains must be limited, since, in the β-thalassemias resulting from β-chain deficiency, excess α chains can precipitate to form inclusion bodies which may be injurious to the erythrocyte membrane. This occurs to a small extent in β-thalassemia heterozygotes and to a much larger extent in the homozygotes (Weatherall and Clegg, 1981). In the heterozygotes, ineffective erythropoiesis secondary to unbalanced α- and β-chain synthesis and the precipitation in the bone marrow of the small excess of α chains, rather than injury to the membrane, is considered to be the principal cause of anemia (Fig. 7.3). Thus, in the heterozygous states for both the α- and β-thalassemias, a decrease in synthesis of one of the major hemoglobin subunits results in a decreased synthesis of the normal hemoglobin A tetramers and the appearance of either soluble (such as hemoglobins H and Bart's) or precipitated (a-chain inclusion bodies, Heinz bodies) products which may secondarily affect the metabolism of the erythrocytes.

Quantitative deficiencies of structural protein subunits can also have adverse affects. Thus, heterozygotes for deficiency of red cell skeletal membrane protein band 4.1, with a mean proportion in the membrane 48% of normal, have "modest" elliptocytosis and a minimal increase in the osmotic fragility of their erythrocytes (Tchernia, Mohandas, and Shohet, 1981). Similarly, heterozygotes for deficiencies of collagen-chain subunits develop con-

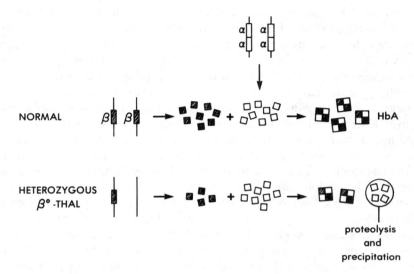

Fig. 7.3. Globin chain synthesis in normal individuals and individuals heterozygous for β^0-thalassemia. Most of the excess α chains in the heterozygote are removed by proteolysis, with the remainder precipitating to form α-chain inclusion bodies.

nective tissue abnormalities. One example of this is type I osteogenesis imperfecta, which is characterized by bone fragility with fractures during childhood, blue sclerae, and frequent hearing loss. The rate of synthesis of $\alpha 1(I)$ procollagen chains, which ordinarily is twice as great as that of $\alpha 2(I)$ procollagen chains, is decreased by half (Barsh, David, and Byers, 1982). As a result, the formation of the $[\alpha 1(I)]_2 \, \alpha 2(I)$ trimers which make up type I collagen is also decreased by half, resulting in a 50% reduction in the synthesis and secretion of type I collagen in vitro and a 50% reduction in type I collagen concentration in skin and presumably in bone. The net effect is a weakness or fragility of the connective tissue and tissues with high collagen I concentrations. Since the structural locus for collagen $\alpha 1(I)$ is on human chromosome 17 (q21→q22), it could be predicted that one of the consequences of deletion of this region would be a mild osteogenesis imperfecta. Situations similar to that of type I osteogenesis may obtain for several of the other connective tissue disorders mentioned earlier in this section, in which deficiency, in addition to or instead of the presence of abnormal subunits, may play a role in the pathogenesis of disease.

At present, I am unaware of any other examples of abnormalities resulting from imbalance in the synthesis of mixed multimer subunits and hence in the composition of these multimers. Nevertheless, in view of the large number of such multisubunit complexes to be found among enzymes and structural proteins of all types, the possible importance of this mechanism cannot be ignored. In fact, if I were to hazard a guess, I think that it is particularly in the structures of cellular constituents such as membranes and the components of the cytoskeleton that aneuploidy-produced defects of the kinds discussed are likely to be found, especially in monosomies and deletions.

Cellular interactions

The various secondary effects of aneuploidy that have been discussed thus far have been principally concerned with intracellular events. However, if we consider the several primary processes involved in development – cell division, migration, cell recognition and adhesion, differentiation, and cell death (Cowan, 1978; Edelman, 1983) – it is clear that some of them involve considerations quite different from the others, considerations of how cells recognize one another and interact. The issue here is not how aneuploidy affects the integrity of cells in terms of their metabolism and structure, but how it affects the ways cells interact with one another. As stated by Gallatin, Weissman, and Butcher (1983), "The involvement of specific cell-surface molecules in the selective association of cells has become a central tenet of modern cell biology." These interactions are clearly paramount in histogenesis and morphogenesis. Therefore, in view of the prominence of morphogenetic abnormalities in aneuploid phenotypes, it would seem that aberrations of cellular interactions could well play an important role. A possible role for changes in cell adhesiveness in the causation of congenital heart disease in Down syndrome has been suggested by Wright *et al.* (1984) and will be discussed in greater detail in Chapter 12. Similarly, alterations in the number of intercellular gap junctions in *Drosophila* wing disks have been implicated in the genesis of abnormalities of wing development: "Alterations in the number or distribution of gap functions may be as disruptive to normal growth and development as their complete absence" (Ryerse and Nagel, 1984).

Several categories of cellular interaction or recognition can exist during embryonic development (Moscona, 1980). These include (1) germ layer-specific cell recognition, which mediates segregation of early embryo cells into three primary germ layers; (2) tissue-specific cell recognition, which enables cells with the same histological identity to group together into a tissue-forming assemblage; and (3) a cell type-specific recognition, which determines the positioning and organization of individual cells within the tissue framework. The last of these, in turn, can be of several types – (a) homotypic interactions between cells of the same type, (b) allotypic interactions between similar but not identical cells, and (c) heterotypic interactions between different but functionally affiliated cell types; moreover, different cell surface domains on the same cell may also be involved in different types of interaction.

Edelman (1983) distinguished three kinds of supramolecular systems mediating cellular interactions: (1) cell-cell adhesion, mediated by cell adhesion molecules (CAM); (2) cell-substrate adhesion, mediated by substrate adhesion molecules (fibronectin, collagen, and laminin); and (3) cell contacts via intercellular junctions, such as gap and tight junctions and desmosomes. Little is known about the regulation of the quantitative aspects of junction formation, and it is therefore difficult to speculate on how aneuploidy might affect it. However, some quantitative information is available for the two other systems of cellular interaction.

The substrate adhesion molecules (SAM) have been implicated in numerous aspects of cell behavior, including adhesion (cell-cell and cell-substrate), alignment, shape, surface architecture, and motility and migration (Yamada, Olden, and Hahn, 1980). These effects may be either positive or negative, depending on the specific event or process being analyzed. One assay for the effects of SAM involves the measurement of cell attachment to collagen-coated plastic dishes in the presence of varying concentrations of SAM. Studies of this type have been carried out with a variety of cell types, including established Chinese hamster ovary (CHO) cells and freshly isolated intestinal epithelial cells. At low concentrations of fibronectin, the number of cells attached is proportional to the concentration of fibronectin present (Yamada, Olden, and Hahn, 1980; Burrill *et al.*, 1981). However, the effect is much less pronounced at higher concentrations and is affected by numerous factors such as the chemical state of the fibronectin and the presence of serum.

A number of cell adhesion molecules, some requiring calcium for activity, have been described. Some have been shown to have distinct immunological and cell type/tissue specificities, and it has been suggested that the latter might be of crucial importance for specific communication between particular cell types in the heterogeneous cell populations of tissues (Yoshida-Noro, Suzuki, and Takeichi, 1984). The nervous system has been a particularly favored subject for the investigation of cell interaction (Gottlieb and Glaser, 1980), and recent quantitative studies of a homotypic cell-adhesion molecule obtained from brain (neural CAM, N-CAM) have provided very interesting information with regard to the potential developmental effects of relatively small changes in concentration. This molecule is believed to play an important role in a variety of developmental events involving neural tissue and muscle (Rutishauser, 1984). Hoffman and Edelman (1983) have developed an assay to measure the rate of aggregation of either synthetic lipid vesicles containing purified N-CAM or of native brain membrane vesicles with endogenous N-CAM. In these assays, the rate of aggregation is considered to be a measure of adhesion. The results of the assays with the synthetic vesicles are shown in Fig. 7.4. When the embryonic N-CAM was treated with neuraminidase to produce a molecular form similar to that present in adult animals, the rate of aggregation was proportional to $N\text{-}CAM^{5.2}$. For untreated N-CAM, the rate was proportional to $N\text{-}CAM^{3.5}$, and the same result was also obtained when neuraminidase-treated embryonic and adult brain vesicles were compared.

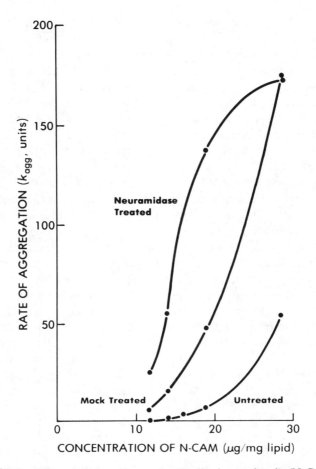

Fig. 7.4. Effect of concentration of neural cell adhesion molecule (N-CAM) on the rate of aggregation of synthetic lipid vesicles containing N-CAM of the embryonic type. Mock treated vesicles were exposed to the conditions used for neuraminidase treatment without neuraminidase being present. From data of Hoffman and Edelman (1983).

According to Edelman (1983), "there are at least three epochs in specific CAM function: the early embryonic related to inductive events and formation of organ rudiments, the embryonic related to detailed histogenesis, and the adult, which may be concerned with surface regulation of cellular metabolic states." With regard to N-CAM and development of the nervous system, the critical factors are the proportions of embryonic- and adult-type N-CAM (the latter, lower in sialic acid, produces a 5- to 7-fold increase in the rate of aggregation when compared to the former in synthetic vesicles) and the concentration of N-CAM in the membrane (Hoffman and Edelman, 1983). If it is assumed that the concentration of N-CAM in a neural cell membrane is directly gene dosage dependent, the potential effect of aneuploidy can readily be appreciated. For early embryonic cells, with only the embryonic type of N-CAM, trisomy would produce an 8.2-fold ($1.5^{5.2}$) increase in adhesivity while monosomy would produce a decrease to 0.027 ($0.5^{5.2}$) of normal. For cells with adult N-CAM, the changes would be 4.1 ($1.5^{3.5}$) and 0.088 ($0.5^{3.5}$), respectively. If the quantitative aspects of whatever the aggregation assay is measuring have any real developmental and functional significance, changes of this degree could certainly produce very major perturbations. As stated by Edelman (1984), "Cell surface modulation of the amount of a CAM during embryogenesis would be expected to have striking effects on binding, movement, and ultimately upon form." Unfortunately, no clear-cut examples of genetic abnormalities affecting cell adhesion molecules have as yet been described in vertebrates (Loomis, 1983).

For the preceding calculation, the assumption was made that the concentration of cell surface molecules is gene dosage dependent. The only directly relevant data based on an aneuploid state is for the interferon-α receptor in trisomy 21 and monosomy 21. This receptor, coded for by the human chromosome 21 *IFRC* locus, shows the expected gene dosage effect in cultured fibroblasts (Epstein *et al.*, 1982*a*) (Table 4.1). However, other data obtained from studies of individuals heterozygous for mutations affecting cell surface constituents, such as the LDL receptor (already discussed in Chapter 5), and for erythrocyte antigens, major histocompatibility antigens, and lymphocyte markers are also consistent with the assumption. Thus, Masouredis, Dupuy, and Elliot (1967) showed that erythrocytes from persons heterozygous for Rh factor D (*Dd*) bound half as much anti-D as did erythrocytes from homozygotes. Similar results were obtained by Caren, Bellavance, and Grumet (1982) for the Duffy antigens (Fy^a, Fy^b), for s, and, with some differences attributed to haplotype differences, for the Rh antigens, D and e. The same was true for the binding to lymphocytes of antibodies directed against several human HLA-A specificities (White, da Costa, and Darg, 1973; Dumble and Morris, 1975; Madsen, 1981; Gladstone, Fueresz, and Pious, 1982), against mouse TL (thymus leukemia) antigen (Boyse, Stockert, and Old, 1968), and against rat major histocompatibility antigens (Fuks and Guttman, 1983). However, that the ability of the membrane to accommodate increased quantities of certain molecules may be limited is indicated by the results of Boyse and Cantor (1978), who showed that there is a reciprocal relationship between the concentrations of H–2D and TL antigens. In the diploid state, the latter results in a commensurate decrease in the former. Furthermore, the expected gene dosage effects were not always found when cells from F_1 hybrids were compared to parental cells with regard to H–2 antigen concentrations (O'Neill and Blanden, 1979*a*,*b*; O'Neill, 1980). It is not clear what effect trisomy for human chromosome 6 or mouse chromosome 17, the chromosomes carrying the major histocompatibility locus, would have on antigen expression, and this is now being investigated.

Pattern formation

Wolpert (1978) has identified three main processes in embryonic development: cytodifferentiation, the development of form, and pattern formation. Cytodifferentiation refers, as has already been discussed, to the acquisition of molecular characteristics which distinguish one cell type from another, and development of form is concerned with changes in shape. Pattern formation is concerned with the spatial organization of cytodifferentiation, with how cells of the same type may come to occupy different positions in the developing organism. Cell-cell interactions, which have just been discussed, are certainly an element in pattern formation and probably in the development of form as well. However, a major interest of Wolpert and his collaborators has been in the role that concentration gradients of morphogens may play in pattern formation, with a particular emphasis on morphogenesis of the limb. In this section, I shall discuss only these concepts, but for a broader view of the problems of pattern formation, the reader is referred to Gaze *et al.* (1981), Fallon and Caplan (1982), and Slack (1983).

The principal goal of a theory of pattern formation is to explain how a cell or group of cells can recognize its position in space so as to develop into the appropriate structure. With the limb, for example, the problem is to instruct a group of mesenchymal cells with regard both to what type of bone or muscle to form and to which digit these structures should generate. The basis of the gradient theories of pattern formation is the assumption that mechanisms can be envisioned by which a morphogen or group of morphogens produced at specified points in space can establish stable concentration gradients. These gradients, in turn, can specify a number of discrete and persistent cell states, even after cessation of morphogen synthesis (Lewis, Slack, and Wolpert, 1977; Meinhardt, 1978).

It is quite instructive to see how such a morphogen gradient might specify the formation of a set of contiguous structures, for example, the digits of a hand or wing. Since the major experimental work has been done with chick embryos, we shall consider, for the moment, a wing. Experimental work with developing chick wings has shown that changes in the shape of a morphogen gradient can affect both the number of digits and their exact identities (Tickle, Summerbell, and Wolpert, 1975). The putative morphogen is believed to be secreted at the posterior end of the limb bud, in the zone of polarizing activity, and then to diffuse anteriorly. On the basis of theoretical considerations, it has been proposed that a specific developmental state is established each time the concentration of a morphogen (in this instance, only one is considered) crosses a particular threshold, irrespective of whether the threshold is crossed

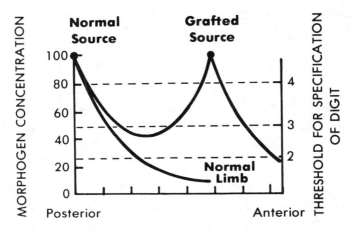

Fig. 7.5. An idealized diagram to illustrate how the source and concentration of a morphogen (the shape of the morphogen gradient) could specify the pattern of digits in the chick wing. In the normal limb it is assumed that the source of a morphogen is kept at a concentration of 100. Since the morphogen is broken down, an exponential gradient is set up. The thresholds for the digits are shown, and in the normal limb the pattern of digits is *2, 3, 4*, in an anteroposterior sequence. In the diagram, the posterior margin of the limb is at the left-hand side. If a new source of morphogen is grafted near the original one, then the pattern of digits will be *2 3 4 4 3 3 4*. Note that a digit is specified each time the threshold is crossed. Redrawn from Wolpert (1978).

from a higher to a lower or a lower to a higher morphogen concentration. This is illustrated in Fig. 7.5. Changing the shape of the morphogen gradient – for example, by grafting another source of morphogen (obtained from the posterior aspect of another limb bud) onto the limb bud – will increase the width of the limb bud and alter the pattern of digits in accordance with how the gradient now intersects with the threshold (Smith and Wolpert, 1981; Wolpert and Hornbruch, 1981). For example, when the new source of morphogen is grafted midway in the limb bud, the resulting digits are increased in number, with the pattern specified by the thresholds crossed (Fig. 7.5). This process is concentration dependent, and the use of different numbers of grafted cells will alter the pattern in the expected manner (Tickle, 1980, 1981). Similarly, the application of different concentrations of retinoic acid soaked in filter paper, which will substitute for grafts, also alters the pattern in a reasonably predictable manner (Tickle *et al.*, 1982). For example, when 2.5-, 5.0-, and 7.5-mg/ml solutions are applied to the anterior margin of the limb bud, the patterns of digits obtained are, respectively, *2 3 4* (unchanged), *2 3 4* (50%) or *2 2 3 4* (40%), and *2 2 3 4* (38%) or *4 3 3 4* and *4 3 4* (50%). It should be noted that, depending on the concentration, extra digits may appear. Similar results were also obtained by Summerbell (1983).

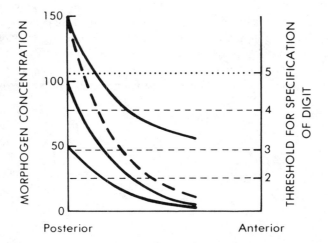

Fig. 7.6. Diagram to illustrate how a change in the concentration of a morphogen could affect the pattern of digits. In this case, it is assumed that the concentration of the morphogen at the source is 150, as would occur in a trisomic state, or 50, as would occur in monosomy. In the trisomic case, either no effect would be produced, a new postaxial digit would appear (if another threshold were postulated), preaxial digits would be deleted (if the concentration of morphogen declined with a curve parallel to the original), or preaxial digits would be deleted and a postaxial digit added (if there were a threshold 5 and the morphogen concentration declined in parallel to the original). In the case of monosomy, postaxial digits would be deleted.

This process of pattern formation is apparently not unique to the chick wing, as grafts taken from mouse limb buds will produce the same effect on chick wing development as do chick grafts themselves (Tickle *et al.*, 1976). However, experiments of a similar nature have not been carried out on mouse limb buds, and there is no specific evidence that the model described here is applicable to mammals. It has been suggested, however, that it is (Tickle, 1980). Nevertheless, because of its predictive value with regard to quantitative effects, it is of interest to consider what effects aneuploidy might have if the synthesis of the morphogen is controlled by the chromosome that is unbalanced. If we use exactly the same model described for the chick wing, the following could occur (Fig. 7.6): for a 1.5-fold increase in morphogen concentration at the source, with the same proportional decline with distance from the source and the same thresholds, there would be no significant effect. On the other hand, if there were another threshold above that specifying digit *4*, then an extra postaxial digit would form. Conversely, if the rate of decline was, for some reason, parallel to that normally present, there then would be a loss of preaxial digit *2* and possibly *3*. If an additional threshold, *5*, were also postulated, then a new postaxial digit would appear. In the case of a 50% decrease in morphogen, deletion of one or two digits would occur.

All of this is, of course, pure conjecture, but the point I am trying to make is, I believe, a significant one. Developmental models such as the one based on a morphogen gradient are potentially capable of relating changes in concentrations of morphogens of the order of plus or minus 50% to significant aberrations of morphogenesis. More experimentation is required to see whether such changes do in fact produce the effects that might be expected.

8

Type and countertype

Following reference to the reciprocal nature of syndromes associated with del(5p) and dup(5p) (Lejeune *et al.*, 1964*b*), the concept of type and counter-type (contre-type) was proposed by Lejeune *et al.* (1964*a*) to describe what they believed to be the opposing physical and physiological effects of trisomy 21 and monosomy 21 (actually complete and mosaic partial monosomy 21). They were impressed by certain striking and opposing phenotypic character-istics in the latter two syndromes – such as the slant of the eyes, the tone of the muscles, the configuration of the nasal bridge (Table 8.1). Reference was made to the existence of *Drosophila* with triplo-IV and haplo-IV, which were said to furnish the classic example of type and countertype. [However, this is probably not the case since haplo-IV flies have numerous somatic abnormal-ities (Bridges, 1921), while the triplo-IV are hard to distinguish from ordinary diploids (Hochman, 1976)].

The monosomy 21 or del(21) syndrome later came to be known as anti-mongolism (Reisman *et al.*, 1966), conjuring up the image that the trisomic and monosomic phenotypes had a large number of complementary or oppo-site phenotypic characters. From antimongolism, the term "antisyndrome" evolved as a synonym for countertype. Following the initial description, the concept, however named, was then extended to cover abnormalities of chro-mosome arms 5p and 18q (Lejeune *et al.*, 1964*b*; Lejeune, 1966), and then 13q (Lejeune *et al.*, 1968), 4p (Rethoré *et al.*, 1974), 9p (Hernandez *et al.*, 1979*b*), 10qter (Turleau *et al.*, 1979), and 12p (Rethoré *et al.*, 1975). Grou-chy and Turleau (1977, 1984) briefly summarized the features that they re-garded as confirming the existence of type-countertype, and these features are listed in Table 8.1. What is perhaps most striking about these lists is that the putative phenotypic characters that exhibit opposite effects are in reality relatively few. This is in contrast to the lengthier lists originally published by Lejeune *et al.* (1964*a*).

The notion of countertype or antisyndrome has not received strong support outside of Europe and has for the most part been discounted by the majority of workers in the field. The reasons for this are, I think, obvious. First, the concept has been rejected on principle as being mechanistically so unlikely as not to warrant serious consideration (Opitz and Gilbert, 1982*b*). This is unfortunate since, as I shall suggest below, it is not without merit. The second

159

Table 8.1. *Counterstate characters regarded as producing type and countertype*

Chromo-some arm	Characters	Monosomy or deletion	Trisomy or duplication
4p	Forehead	Prominent	Flat
	Glabella	Aplastic	Prominent
	Nasal bones	Prominent	Hypoplastic
	Nasal apex	Square	Round
	Chin	Receding	Protruding
	Neck	Long	Short
	Dermatoglyphics	Excess of arches	Excess of whorls
9p	Head	Trigonocephaly	Brachycephaly
	Eyes	Exophthalmos	Enophthalmos
	Palpebral fissures	Upslanting	Downslanting
	Nares	Anteverted	Oriented downward
	Nose	Short	Bulbous
	Upper lip	Long	Short
	Hands	Dolichomesophalangia	Brachymesophalangia
	Dermatoglyphics	Excess of whorls	Excess of arches
18q	Nose	Midface retraction	Thin and well developed
	Chin	Prominent	Retromicrognathia
	Ears	Prominent antihelix	Hypoplastic antihelix
		Deep concha	Faunesque ears
		Helix well-folded	Hypoplastic helix
	Pelvis	Froglike	Narrow
	Hands	Tapered fingers	Short, overlapping fingers
	Dermatoglyphics	Whorls	Arches
21q (distal)	Tone	Hypertonia	Hypotonia
	Cranium	Protuberant occiput	Flat occiput
	Nasal bridge	Protuberant	Aplastic
	Palpebral fissures	Downslanting	Upslanting
	Nares	Broad and horizontal	Small and oriented forward
	Ears	Large; large concha, large auditory canals	Small; small concha, small auditory canals
	Dermatoglyphics	Hypermature ridges	Immature ridges
	Palmar creases	Normal	Single transverse palmar crease

From Grouchy and Turleau (1977, 1984).

reason is that the advocates of the concept were imprecise in their own definition of the term. When one examines the lists of features originally proposed as supporting the concept, it is clear that many supposedly opposed features do not really represent true phenotypic opposites of one another. Thus, in the list of features of the del(5p) and dup(5p) syndromes published by Stoll *et al.* (1975), of the nine features listed, only four at most can be considered opposites: for example, microcephaly is not the opposite of a normal head circumference (macrocephaly would be); hypertelorism is not the opposite of normal eye spacing; the cat cry is not the opposite of a normal cry; and so on. On the other hand, a flat and a prominent nasal bridge and a small and prominent mandible in del(5p) and dup(5p), respectively, could be considered as opposites.

Because of the potential mechanistic implications of the concept of type and countertype, I believe that it is worth reexamination at this time. In doing so, we must separate the existence of individual pairs of opposing characteristics – counterstate characters in the terminology of Preus and Aymé (1983) – from the full type-countertype or syndrome-antisyndrome concept. The latter refers to a whole constellation of phenotypic effects that express in an opposite manner in a trisomy (duplication) – monosomy (deletion) pair. In other words, it is the aggregate of counterstate characters that produces a syndrome and antisyndrome. The problem in defining an antisyndrome is, of course, in knowing how many sets of counterstate characters are required to establish its existence, and, as I have indicated, this is what has made investigators wary of the concept in general. However, whether or not one considers an antisyndrome as having been unequivocally demonstrated, the existence and potential importance of the counterstate characters themselves must be considered on their own merits.

The search for countercharacters

As a first approach to testing the concept of type and countertype, I compared the phenotypes of individuals within the same family with complementary duplications and deletions resulting from the meiotic products of insertional or reciprocal translocations or of pericentric inversions. This was done to ensure that the unbalanced chromosomal segments were the same and to reduce, to the extent possible, the noise introduced by genetic differences. Unfortunately, relatively few families of this type have been described, and the published descriptions are frequently insufficient to permit detailed comparison. Nevertheless, it is worth considering briefly the information forthcoming from those cases that have been reported. These are listed in Table 8.2, along with those features that might be considered as examples of counterstate characters. Although many phenotypic abnormalities were noted in both the duplication and deletion patients, with the latter usually being more severely

Table 8.2. *Possible examples of counterstate characters in familial complementary duplications and deletions*

Chromosome segment unbalanced	Chromosome constitution of balanced carriers	Phenotypic character	State of character in		References
			Duplication	Deletion	
1(q25→q31)	ins(1)(p32;q25q31)	Palate	High, narrow	Cleft (also lip)	Pan *et al.* (1977)
1(q25→q32)	ins(1)(p32;q25q32)	None			Garver, Cioca, and Turack (1976)
3(q22.1→q24)	dir ins(11;3) (q22.1;q22.1q24)	Midface Dermatoglyphics	Slight hypoplasia 10 long whorls	Long 9 loops, 1 arch	Williamson *et al.* (1981)
4(p14→pter)	rcp(4;12)(p14;p13)	Eyes	Deep set	Hypertelorism	Mortimer *et al.* (1978); Mortimer, Chewings, and Gardner (1980)
5(p14→pter)	t(5;13)(p14;qter)	None			Lejeune *et al.* (1964*b*)
5(p14→pter)	inv(5)(p14;qter)	Eyes Jaw Nasal bridge	Hypotelorism Macrognathia Deep	Hypertelorism Micrognathia Flat	Warter, Ruch, and Lehman (1973)
5(p15.1→pter)	t(3;5)(p27;p15.1)	Nasal bridge	Deep	High	Andrle, Erlach, and Rett (1981)
7(q36.1→qter)	t(7;22)(q36.1;pter)	None			Lowry *et al.* (1983)
7(p15→p21)	ins(7)(p1500p2104; q22)	None			Miller *et al.* (1979)
10(p13→pter)	rcp(1;10)(q44;p13)	Palpebral fissures	Upslanting	Downslanting	Slinde and Hansteen (1982)
10(q24.2→25.3)	ins(5;10)(q13 or 14;q24.2q25.3)	None			van de Vooren *et al.* (1984)

11(p11.3→p14.1)	inv ins(11)(q14.5; p14.1p11.3)	Occiput	Prominent	Flat	Hittner, Riccardi, and Francke (1979); Strobel et al. (1980)
11(q23→qter)	t(3;11)(p27;q23)	Lips Dermatoglyphics	Thin Low total ridge count	Thick High pattern intensity	Ridler and McKeown (1979)
13(q12.5→q22.1)	inv ins(12;13) (p11.2;q22.1q12.5)	Occiput Ear pinnae	Flat Narrow	Prominent Large lobules	Riccardi et al. (1979)
13(q21→q22)	inv ins(3;13) (q12;q21q22)	None			Toomey et al. (1978)
13(q13→qter)	t(13;17)(q13;p13)	Digits and toes	Polydactyly (hands and feet)	4 fingers bilaterally; 4 toes left foot, 4-5 syndactyly on right foot	Wilroy et al. (1977)
14(pter→q13)	rcp(5;14)(p15;q13)	Eyes	Hypotelorism	Hypertelorism	Abeliovich, Yagupski, and Bashan (1982); Fried et al. (1977)
18(p)	t(18;G)(p1;p1)	None			Jacobsen and Mikkelson (1968)
18(p)	i(18p)	None			Taylor et al. (1975)
18(p11→pter; q21→qter)[a]	inv 18(p11;q21)	Dermatoglyphics	10 arches; total ridge count 0	7 whorls, 3 ulnar loops; total ridge count 138	Vianna-Morgante et al. (1976)
18(q22→qter)	t(14;18)(p12;q22)	Facies Nose Mouth	Heavy, long Long, prominent Small	Square Midface retraction Large	Turleau and Grouchy (1977)

Table 8.2. (cont.)

Chromosome segment unbalanced	Chromosome constitution of balanced carriers	Phenotypic character	State of character in		References
			Duplication	Deletion	
		Angle of mandible	Effaced	Prominent	
		Dermatoglyphics	10 arches	10 loops	

[a] "Duplication" = dup(q21→qter) + del(p11→pter); "deletion" = the reciprocal dup and del.

164

affected, the number of phenotypic features or characters that might be construed as counterstate characters, even by the most liberal interpretation, is very small. And of these, only a few are at all reasonably convincing: hypotelorism and hypertelorism in the 5(p14→pter) and the 14(pter→q13) comparisons and the dermatoglyphic differences, high versus low total finger ridge counts, in the 11(q23→qter), 18(p11→pter; q21→qter), and possibly 18(q22→qter) cases. These limited numbers of examples, while not negating the concept of type and countertype, do not provide very strong support for it, and one is forced to look at groups of collected cases for further support.

From comparisons of unrelated indivduals, other possible examples of countercharacters not already mentioned include the following: craniostenosis (craniosynostosis) in del(7p21→pter) and widely separated sutures in the corresponding duplication (Moore *et al.*, 1982); high nasal bridge in del(10qter) and flat nasal bridge in dup(10qter) (Klep-de Pater *et al.*, 1979; Mulcahy *et al.*, 1982); upslanting palpebral fissures, thick upper lip, and hypoplastic ear lobules in del(16q21→qter) and downslanting palpebral fissures, thin upper lip, and thick ear lobules in the corresponding duplication (Balestrazzi *et al.*, 1979). Other possibilities are listed in Appendix VII of Grouchy and Turleau (1984). However, the most detailed and objective analysis of large groups of cases has been carried out by Preus and Aymé (1983), using the methods of numerical taxonomy and cluster analysis, which have already been alluded to in the discussion of the differentiability of phenotypes in Chapter 2.

A major difficulty in the analysis of the complex phenotypes associated with chromosomal aneuploidy is the highly subjective nature of both the descriptions of the phenotypes and of their comparative analysis, especially when a large number of cases are being studied. In an attempt to overcome these problems, Preus and her collaborators have been attempting to develop phenotypic descriptors that can be objectively assigned and can be used for subsequent statistical analysis. This approach has been applied to two pairs of duplication and deletion states, those involving chromosome arms 4p and 9p. Based on their analysis of what they term "state-order characters," in which the normal condition is the intermediate character state (Preus, 1980), Preus and Aymé (1983) identified several characters which showed opposite states in these trisomic and monosomic conditions. These so-called counterstate pairs are shown in Table 8.3. Although many pairs were identified in the two sets of comparisons, especially that of duplication and deletion of 4p, only some of the differences are statistically significant at the 5% level (Table 8.3). Moreover, of all possible counterstates, only 10.6% and 5.6% were significantly different for the 4p and the 9p comparisons, respectively (Preus and Aymé, 1983). Whether these proportions are regarded as large or small will ultimately depend on one's beliefs about what ought to be expected.

Despite the presence of those counterstates that could be considered as sta-

Table 8.3. *Counterstate characters in deletion and duplication of 4p and 9p[a]*

Character	Deletion 4p (N = 13)	Duplication 4p (N = 22)	Deletion 9p (N = 5)	Duplication 9p (N = 18)
Cranial shape	Dolicocephaly	Brachycephaly		
Face shape (1)	Round	Long		
Face shape (2)	Square	Triangular		
Mala	Flat	Prominent		
Forehead height	High	Low		
Forehead width			Narrow	Wide
Orbits			*Shallow ridge	+
Palpebral fissures			Up	Down
Orbital ridge	Shallow	Prominent		
Brow slant	Up	Down	Up	*Down
Brow arch	Increased	Linear		
Brow density	Sparse	Heavy		
Medial brow	Absent	Increased		
Nasal size (1)	Small	Large	Small	Large
Nasal size (2)	Long	Short		
Nasal position	Flat	*Prominent		
Nasal root			Flat	Prominent
Nasal bridge	Beaked	Depressed		
Nasal tip	Small	*Full	Small	Full
Nasal wings	Thin	Thick	± Thin	± Thick
Philtrum length	Short	Long	Long	Short
Philtrum depth	Deep	Flat		
Upper lip	Thick	*Thin	Thin	Thick
Chin length	Short	*Long		
Mandible	Retrognathia	*Prognathia	*Retrognathia	
Ear size			Small	Large
Ear shape	Triangle	Reversed triangle		
Helical rim	Thin	*Thick		
Lobe size	Small	*Large		
Torso length	Long	Short		
Nervous tone	Hypotonia	Hypertonia		
Nail size			Large	*Small
Finger patterns	Arches	Whorls	Whorls	*Arches
Palm length	Short	Long		
Digits			Long	Short

Taken from Preus and Aymé (1983).
[a]Characters marked by asterisk (*) are statistically significant at the 5% level by χ^2 test.

tistically acceptable, Preus and Aymé reject the antisyndrome hypothesis for two principal reasons. The first is that the number or proportion of significant differences between complementary duplications and deletions just mentioned is not always greater than is observed between pairs of unrelated syndromes. For example, when two obviously unrelated abnormalities, del(4p) and dup(9p), are compared, 10.8% of the possible counterstates are significantly different. The second reason is that the calculated "phenotypic distances" between the pairs of complementary syndromes are again no greater than between unrelated syndromes: all possible comparisons of duplications and deletions of 4p and 9p give calculated distances between 0.840 and 0.885.

Although based on the intercomparison of just two duplication-deletion pairs, the conclusion arrived at by Preus and Aymé (1983) is consonant with the reservations generally expressed about the antisyndrome concept. It may be just as easy to put together a list of counterstates by pairing any two unrelated chromosome-aneuploidy syndromes as by analyzing a complementary duplication-deletion pair. However, a sample of two pairs, even as well analyzed as the two just described, seems somewhat small to permit us to regard the antisyndrome concept as having been definitively disproved. The possibility that meaningful aggregations of counterstates specifically related to the chromosome abnormalities being compared are lost in the random noise of the comparisons cannot be discounted on so small a sample.

How countercharacters could be generated

Considering the complexity of developmental processes, with the likely involvement of the products of many chromosomes in any morphogenetic event, it is probably unreasonable a priori to expect all or even perhaps a substantial number of phenotypic characters of a duplication state to be replaced by the opposing characters in the corresponding deletion state. Such an event would imply that all of these characters are directly controlled by the chromosome region in question. However, as unreasonable as it may be to expect a whole phenotype to be mirrored in an antiphenotype, it is not unreasonable to believe that individual counterstates may in fact represent direct reflections of reciprocal gene dosage effects rather than mere random juxtapositions. Thus, while the whole antisyndrome concept may have "no biological basis" (Preus and Aymé, 1983), individual counterstates or small sets of such counterstates well may. For this to occur, it would be necessary for a product or set of products of the unbalanced chromosome segment to affect the developmental process producing the character in question in a manner related to the concentration of the gene product. This notion may be most easily considered in terms of the control of the rate of growth of a specific cell type, tissue, or organ, by some substance acting as a morphogen. In the

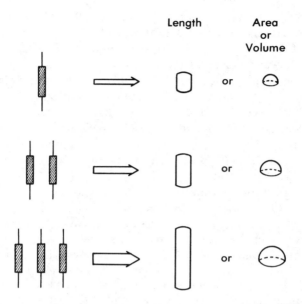

Fig. 8.1. Hypothetical relationship of morphogenetic event to gene dosage in genesis of countercharacters. The production of a morphogen (arrow) is proportional to the number of genes coding for its synthesis, and the developmental response, such as growth in length, area, or volume, is a direct function of the quantity of morphogen. This function need not be a linear one.

simplest situation, the growth of the target is a direct function of the concentration of the morphogen, so that growth is proportional to morphogen concentration and, therefore, to the number of genes coding for synthesis of the morphogen (Fig. 8.1). Such a hypothetical morphogen could, among other things, be a growth factor, a growth factor receptor, or a factor controlling the rate of energy production or of DNA, RNA, or protein synthesis. The number of such morphogens coded for by a single chromosome segment that would be expected to have a directly observable and quantifiable phenotypic effect (as opposed to a more weakly interactive effect) is likely to be small, and therefore, the number of such strong effects is also likely to be small (Fig. 8.2). That the same apparent effect should be produced by imbalance involving different chromosome regions (see Table 8.3) does not negate this concept, since it is not unreasonable to expect that a specific phenotypic character could be strongly affected by more than one morphogen (Fig. 8.2).

It is, of course, possible that the effect of a morphogen may be inversely related to its concentration. In terms of our growth control example, this would mean that the greater the concentration of the morphogen (in the trisomic as opposed to the monosomic state), the lower the growth rate. Actu-

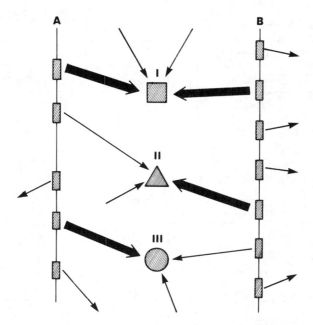

Fig. 8.2. Generation of countercharacters. Chromosomes A and B are each represented as controlling the synthesis of morphogens, some with strong effects (wide
arrows) and others with weak interactive effects (narrow arrows) on a variety of
developmental processes, I, II, and III (represented by the square, triangle, and
circle, respectively). Weak morphogens from other chromosomes or affecting other
processes are also indicated. Those processes controlled by strong morphogens will
give rise to counterstate characters. Thus, imbalance of A will affect processes I and
III, while imbalance of B will affect I and II. Although a morphogen produced by A
is shown as interacting with process II, counterstates of II will not be produced by
imbalance of A (similarly with B and III). However, some alteration in character II
could result from imbalance of A.

ally, this would have to be the case, since otherwise it would be expected that
features of the trisomic state would always be characterized as bigger or
longer, and of the monosomic state as smaller or shorter. While this is often
the situation (see Table 8.3 for examples), it is not invariably so. An example
of proteins that inhibit epithelial cell growth in a dose-dependent manner was
recently published by Holley *et al.* (1980) and by Nilsen-Hamilton and Holley (1983). These proteins are able to antagonize the growth-stimulating effects of epidermal growth factor.

Although these concepts are most directly applied to linear growth phenomena, they may also be applied to two- or even three-dimensional phenomena (such as the slant of the orbits, shape of the ears) which may be the result
of differential growth rates affecting more than one region or tissue. As has

Table 8.4. *Digital patterns in complementary duplications and deletions*

Chromosome segment	Digital patterns in		References
	Duplication	Deletion	
4p	↑ Whorls, ↑ TRC[a]	Arches, ↓ TRC	Reed (1981)
6q21→qter	↑ Arches	Whorls and ulnar loops	Clark *et al.* (1980); Fitch (1978); Bartoshesky, Lewis, and Pashayan (1978)
7p13→pter	↑ Whorls, ↑ TRC	↓ TRC	Moore *et al.* (1982)
9p	↑ Arches, ↓ TRC	↑ Whorls, ↑ TRC	Reed (1981); Young *et al.* (1982); Preus and Aymé (1983)
11q23→qter	↓ TRC	High pattern intensity	Ridler and McKeown (1979)
13q	↑ Arches	↑ Whorls	Reed (1981)
18q	↑ Arches	↑ Whorls	Reed (1981)

[a]TRC, total finger ridge count.

already been discussed in Chapter 7, they may also be applied to other morphogenetic processes, such as intercellular adhesivity, cell migration, and tissue turgor, changes which may produce quantifiable phenotypic effects.

Digital dermal ridge patterns

An example of how the concept of a linear relationship between gene number and growth might produce an unexpected phenotypic effect is provided by the dermatoglyphic changes associated with aneuploidy. In examining the phenotypes of chromosomal duplications and deletions, several examples of extreme and opposite deviations from normal digital dermal ridge patterns (fingerprints) are found, and these have been brought together in Table 8.4. The deviations are not always in the same direction, so a duplication of one chromosome segment may result in an increased number of whorls, while duplication of another results in a high frequency of arches. The question, then, is how such opposing effects might be produced.

The formation of the digital patterns during embryonic and fetal development has been the subject of considerable discussion, although of only limited investigation. The seminal thinking in this area is that of Cummins, and is concisely presented in the monograph by Cummins and Midlo (1961):

Ridge alignment is conditioned by the stresses and tensions incident to the general growth of the part. . . . There are no predeterminations of ridge direction other than those which operate through their control of specific contours. . . . Volar pads in the normal fetus are sites of differential growth, each being responsible for production of one of the local configurations comprised in the morphologic plan of dermatoglyphics. The skin possesses the capacity to form ridges, but the alignments of these ridges are as responsive to stresses in growth as are the alignments of sand to sweeping by wind or wave.

These concepts have been taken up by later authors, who have proposed that the parallel dermal ridges might correspond either to lines of curvature (Smith, 1979) or to lines of stress (Mulvihill and Smith, 1969). Reasoning from either premise, it is concluded that a high volar pad will produce a whorl, a low pad an arch, and an intermediate-sized pad a loop. The direction of a loop depends on which side of the pad is the steeper. The determination of dermal ridge pattern thus translates into an issue of volar pad growth, a phenomenon which is easier to relate to our mechanistic proposals. A theoretical model for the generation of spacially periodic patterns, such as dermal ridges, has been presented by Harris, Stopak, and Warner (1984).

In the balanced diploid state, the concentrations of those products that control the growth of the volar pads are such as to produce mainly intermediate-sized volar pads and consequently digital patterns with loops predominating (64%) (Lin, Crawford, and Oronzi, 1979). If growth of the pads is a direct function of the concentration of some gene product, then the triplex state associated with trisomy or duplication will increase growth and hence lead to an increased tendency to form whorls; conversely, the uniplex state associated with monosomy or a deletion will have the effect of producing a decreased growth rate and will thereby lead to the formation of arches. On the other hand, if the gene product in question is one which inhibits or moderates growth, the opposite effects will be produced – arches in the triplex state and whorls in the uniplex. As the data in Table 8.4 indicate, both types of situations have been observed.

Green and Thomas (1978) have suggested that it is not the growth of underlying tissue (the pads) but the growth and movement of the overlying epidermal cells themselves that may determine pattern formation. In this regard, it is of interest that epidermal growth factor can stimulate epidermal cells to form whorls in vitro (Green and Thomas, 1978) and that the gene for its receptor (*EGFR*) is located at 7p13→pter (Robson and Meera Khan, 1982). Duplication of this region increases the frequency of whorls and the total ridge count; deletion decreases the total ridge count (Moore *et al.*, 1982).

Other aspects of hand morphogenesis, such as the lengths of the fingers and toes (long versus short) and the existence of polydactyly versus syndactyly and/or reduction defects, can also be examined to look for reciprocal phenotypic characters that could be related to the dosage-dependent effects of the gene products of an aneuploid segment. However, once the specific chromosome segments present in reported duplications and deletions are appropriately matched, examination of the data summarized by Reed (1981), as well as of individual case reports, does not provide many convincing examples for reciprocal effects of these kinds. Although wide and narrow palms were described as countercharacters for 12p duplication and deletion, respectively (Rethoré *et al.*, 1975), the only really solid example appears to be the duplication and deletion of 13q31→34. The former produces postaxial polydactyly, and the latter syndactyly (bony) and/or hypoplastic or absent thumbs (Noel, Quack, and Rethoré, 1976; Najafzadeh, Littman, and Dumars, 1983; see also Wilroy *et al.*, 1977).

In the chromosome 13 case, it ought to be possible to propose a hypothetical mechanism relating the growth of the embryonic limb buds to a gene product coded for by the chromosome segment in question, with an increase causing polydactyly and a decrease, oligodactyly. Unfortunately, there is one major difficulty. The opposite of postaxial polydactyly would not be absence of the thumb, which is actually a preaxial structure; it would be absence of the fifth digit. I had hoped that it would have been possible to apply Wolpert's (1978) morphogen gradient theory of pattern formation in the vertebrate limb to the digit anomalies resulting from chromosome 13 abnormalities. The theory has already been discussed in Chapter 7. However, if we construct a morphogen gradient diagram analogous to Fig. 7.6 that could explain the postaxial polydactyly in dup(13q), it turns out that it predicts, as expected, loss of the postaxial digits in del(13q). Therefore, at the moment we have no theoretical basis on which to explain the situation of the digits in abnormalities of chromosome 13.

Finally, while not strictly countercharacters in the sense just described, it is worth mentioning that both duplication and deletion of 11(p11.3→p14.1) result in abnormalities of eye development (Hittner, Riccardi, and Francke, 1979; Strobel *et al.*, 1980). While the latter resulted in the well-known aniridia with associated cataracts and absence of the maculae in a girl, her half brother with the dup(11p) had nystagmus, thin-appearing maculae without distinct foveolar reflexes, encroachment of blood vessels on the fovea, and abnormal rod responses by electroretinography. It would appear, therefore, that some gene product or products of 11p may have a reasonably direct and presumably dose-dependent effect on morphogenesis of the eye.

Do antisyndromes and countercharacters really exist?

Looking over all of the data presented in this section, it can be concluded that true antisyndromes have not been shown to exist. Duplications and deletions of the same segment do not produce syndromes that are mirror images (in a morphogenetic sense) or, better, complete opposites of one another. On the other hand, there are several hints that a limited number of individual phenotypic features of different pairs of duplications and deletions may in fact be real countercharacters. A mechanism by which such countercharacters can be generated, dependent upon the dose-dependent response of tissues to growth-promoting or growth-inhibiting morphogens, has been proposed. The existence of such a mechanism would imply a relatively close and strong relationship between a locus or loci in the aneuploid region and morphogenesis of the affected organ or tissue. This relationship should ultimately be explicable after the identities of the loci in the aneuploid region are known and their functions understood.

9

Nonspecific effects of aneuploidy

In the previous chapters in Part III, I considered a variety of ways by which changes in gene dosage can effect alterations in metabolism, regulation, cellular structure, and pattern formation. In each instance, the attempt was made to demonstrate how a 50% increase or decrease in gene product could produce a characteristic effect based on the function of the product in question. The relationship between the change in gene number and the phenotypic change or changes that ultimately resulted was regarded as a specific one. In some cases the relationship might be a direct one, but in others the actual pathway between genetic imbalance and phenotype might involve several steps and be difficult to trace.

However, not everyone shares this belief in the specificity of aneuploid effects. It is appropriate, therefore, to consider the arguments made for the existence of nonspecific effects, effects which are not wholly a product of the particular loci that are unbalanced but derive from a more general perturbation in genetic structure or balance.

Regulatory disturbance

Mention has already been made of the possible major role of regulatory disturbances, as proposed, for example, by Krone and Wolf (1972) and by Vogel (1973), to explain the effects of aneuploidy. These effects would not need to be nonspecific if we were dealing with the imbalance of specific regulatory loci, no matter how many structural loci they interacted with. However, if the source of the disturbance were in some more general aspect of gene expression and regulation, such as the overall processes of transcription or translation, then the effects would certainly lose specificity. Abnormalities affecting RNA polymerases, splicing and posttranscriptional modification, transfer RNAs, or the components of the protein synthesis system would be of this type. Many of these possibilities were considered by Krone and Wolf (1972, 1978), although no evidence in support of them was provided.

Alterations in the cell cycle and retardation of development

The cell cycle and cellular proliferation

Again starting with the assumption that it does not seem likely that the abnormalities associated with trisomy can be accounted for by the presence of genes on the additional chromosome, Mittwoch (1971, 1973) proposed that aneuploidy produces its effects by altering the rate of cellular proliferation. This idea was further elaborated for sex chromosome imbalance by Barlow (1973). This alteration in rate, which is presumed to have its root in changes in the cell cycle, is attributed to an effect of the total quantity of chromatin, with the length of the mitotic cycle being in some way directly related to the amount of chromatin that must be replicated. Both hetero- and euchromatin are presumed to have an effect, so that the cell cycle is shortened by loss of even a heterochromatic chromosome, such as the inactivated X chromosome in 45,X (Barlow, 1972), and lengthened by the presence of any extra chromosome. Additionally, male cells, with X and Y chromosomes, are postulated to proliferate slightly more rapidly than female cells with two X chromosomes.

In making these suggestions, Mittwoch (1971) was not wholly oblivious to the obvious phenotypic differences among trisomies. She proposed, therefore, that in addition to its general effect on the cell cycle, a particular extra chromosome might also, by virtue of its gene dosage effects, exert a more specific effect on the differential mitotic rate at certain developmental stages. This, then, is an attempt to introduce some degree of specificity into a mechanism which cannot otherwise account for the observed phenomena.

The hypothesis of nonspecific effects on cell proliferation has several difficulties. It is not clear that significant changes in the cell cycle are associated with all aneuploid states or that they necessarily occur in the direction predicted. Thus, in a study of mitogen-stimulated lymphocytes from individuals with del(5p), 45,X, or trisomy 21, the cell cycle was found to be increased about 10% (not statistically significant) in the first (not decreased as predicted), decreased in the second, and normal (not increased) in the third (Frias and Carnevale, 1983). By contrast, Heidemann, Schmalenberger, and Zankl (1983) found that trisomy 21 lymphocytes had a distinctly shorter cell cycle. Furthermore, despite suggestions that the rate of fibroblast proliferation is slower in trisomy 21 (Kaback and Bernstein, 1970; Schneider and Epstein, 1972; Segal and McCoy, 1973; Paton, Silver, and Allison, 1974), no significant differences were observed by Hoehn *et al.* (1980). The last group of investigators also could not find statistically significant differences from normal in the rates of proliferation of trisomy 13 (mean rate increased 16% over control) and trisomy 18 (mean rate decreased 14%). Additionally, in contrast to the earlier data of Barlow (1972), 45,X and 47,XXX cell strains cloned from the same mosaic population had identical proliferative rates.

Significant increases in doubling times were observed with aneuploid cells of all types (including trisomies, 45,X, and triploids) established from abortuses (Boué *et al.*, 1976). However, it is difficult to know how to judge these results. In the case of trisomy 21, this effect was found for spontaneous but not induced abortions, thereby suggesting an adverse effect of death in utero on the subsequent behavior of cells in vitro. Hassold and Sandison (1983) could not find any effect of chromosome constitution on the time required to grow enough cells from an abortus to obtain a successful karyotype. In mouse aneuploids, cultures of fetal trisomy 1, 12, and 19 cells were reported to show normal cell-cycle kinetics (Gropp, 1982; Nielsén, Marcus, and Gropp, 1985), but trisomy 13 cells had a longer cell-cycle time and a decreased rate of growth in vitro (Hongell, 1981; Hongell and Gropp, 1982). In mixed cultures of trisomic and normal cells, trisomy 12 and 19 cells did not appear to have a selective advantage or disadvantage (Nielsén, Marcus, and Gropp, 1985).

Another difficulty with the Mittwoch hypothesis is that it would predict that the degree of growth retardation or general severity of the effects of a trisomic state is in some way proportional to the size of the chromosome affected. This is not uniformly the case for the nineteen mouse autosomal trisomies (see discussion below and also Chapter 10). Furthermore, the hypothesis would not explain the finding that deletions are, in general, more severe in their effects than the corresponding duplications. Also, it is difficult to see why changes in chromatin content of the order of a few percent should produce proportionally much larger alterations in cell-cycle length. In fact, just the opposite is generally the case in the work on lower organisms, which is often cited in support of the altered cell-cycle hypothesis. For example, Rees (1972), in a survey of mitotic cycle times in plants, does show a linear relationship between the length of the mitotic cycle and the amount of nuclear DNA. However, except for one series of changes in DNA content resulting from the presence of small B chromosomes, in which a 5% increase in DNA content produces a 9–10% increase in mitotic cycle time, the observed changes in cycle time are closer to 2–3% for each 5% change in DNA content. Similarly, in vertebrates (frogs), the time required to reach a specific stage of tadpole development – obviously requiring many cell divisions – is again increased only about 3% for a 5% increase in DNA content (Bachmann, Goin, and Goin, 1972).

The quantitative implications of a change in cell cycle have not been fully considered by the proponents of the cell replication hypothesis. For example, a 5% decrease in the rate of cell replication would produce, in 20 cell divisions, a 64% decrease in cell numbers. Conversely, a 5% increase in rate would result in a 165% increase in number. These predictions are clearly in the wrong direction for the growth changes associated with the sex chromosome anomalies and autosomal deletions, and the weight reduction in the autosomal trisomies is rarely of the degree that would be expected. Also, the evidence from the mouse trisomy ↔ 2n chimeras, in which reasonably balanced proportions of diploid and aneuploid cells are frequently observed (Table 10.6), argues against a generalized major growth differential between diploid and trisomic cell populations.

Finally, a major part of the rationale for claiming an effect of chromosome number on the cell cycle was the need to explain the developmental abnormalities associated with the sex chromosome anomalies (Mittwoch, 1971, 1973; Barlow, 1973). According to the strict interpretation of the X-inactivation hypothesis, which held that only one X chromosome is active in any cell (Lyon, 1970), alterations in the number of X chromosomes would not be expected to have any phenotypic effects – but clearly they do (Barlow, 1973). Therefore, the presence of extra heterochromatic X chromosomes or even the absence of one was visualized as affecting development in a nonspecific manner by altering the cell cycle. However, as is discussed in detail in Chapter 13, it is now known that X inactivation is not complete and that certain loci at the end of Xp are expressed by both X chromosomes in normal female cells (Shapiro and Mohandas, 1983). This makes other explanations based on the gene-dosage effect concepts developed in the preceding chapters possible and removes the need to postulate a strong nonspecific effect based on the quantity of chromatin present.

On the basis of these considerations, I would conclude that while changes in the rates of cell proliferation might occur in aneuploid states, these changes are not a nonspecific effect of alterations in chromosome number and do not have a general but nonspecific etiologic role in the genesis of phenotypic abnormalities. I would, however, agree with Mittwoch's (1971) suggestion that changes in the rate of cell proliferation could occur in individual tissues at particular stages of development and that such changes could have specific developmental effects. For the same reasons that I reject a general cell-proliferation disturbance hypothesis, I find it difficult to accept the suggestion (Green and Thomas, 1978) that the inverse relationship between total fingertip ridge count and the number of sex chromosomes is a product of a slight inhibition of cell movements and distortion of cell shape necessary to produce curvature of the ridges brought about by "the bulkier nucleus resulting from the presence of extra chromosomes."

General retardation of development

An extension of the concept of impaired cellular proliferation is that the phenotypic effects of aneuploidy are to a large extent the result of impaired fetal growth and development. This view was first advanced by Hall (1965), who argued that the extra chromosome in a trisomy "acts mainly as a source of diffuse disturbance" which causes abnormalities that frequently represent delayed embryonic development. As examples of this, in trisomy 21 he pointed to the clinodactyly resulting from dysplasia of the middle phalanx of the fifth finger and to the overfolding of the helix of the external ear, both of which, he claimed, are normal findings during fetal development. A similar mechanism was presumed to operate in all trisomies, thereby resulting in marked phenotypic similarities among them. Although qualitatively different effects resulting from imbalance of specific genes could be envisioned, they were considered to be less important than the delayed development, which was thought to be quantitatively related to the amount of chromosome imbalance.

Recent arguments in favor of this developmental delay notion have been advanced by Daniel (1979) and Aziz (1981). Daniel (1979) cites the following types of evidence: In autosomal aneuploidy, there is generalized growth retardation which may be attributable to prolongation of the cell cycle. There is also a great similarity of phenotype among different aneuploid conditions. In his analysis, the following features were found in 20% or more of trisomic and monosomic children: mental and growth retardation, microcephaly, micrognathia, low-set ears, hypertelorism, higharched/cleft palate, epicanthic folds, and congenital heart defects. In addition, factors that affect growth and development in a general way produce phenotypes "reminiscent" of monosomy and trisomy, a specific example being the fetal alcohol syndrome. To these arguments, Aziz (1981) adds several of his own. One is the finding that certain muscular anomalies in trisomy 13 and 18 infants represent delayed development rather than anatomical malformation. Moreover, he quotes Down's original observation that the anomalies of Down syndrome are due to arrested development, and he lists examples of persistent expression of embryonic or fetal proteins (such as the fetal hemoglobins in trisomy 13) in liveborn trisomics. Finally, he refers to the statement of Lewis (1964) that many of the anomalies in trisomy 18 represent arrests in development and to a similar statement by Mottet and Jensen (1965) about trisomy 13.

That aneuploidy is associated with intrauterine growth retardation is unquestioned, and evidence has been summarized in the earlier literature (Brent and Jensh, 1967; Naeye, 1967). However, this retardation is not uniform and is not necessarily the same in all conditions. In fact, Naeye (1967), in his analysis of trisomies 13, 18, and 21, points out that "in some instances abnormal growth patterns in individual organs seem to be as characteristic for an individual trisomic disorder as are body and organ malformations." Barr (1984) reached similar conclusions. Similarly, as has already been discussed and will be discussed in the following section, there is clearly considerable overlap in certain features found in aneuploid states, but this overlap is not so extensive as to obscure the individuality of the different phenotypes. Furthermore, despite the fact that the fetal alcohol syndrome does have some of the common dysmorphic features in Daniel's (1979) list, such as growth and mental retardation, microcephaly, and congenital heart defects, and occasionally cleft palate and micrognathia (Smith, 1982), its phenotype is distinct from those of the defined chromosome abnormality syndromes. It is also distinct from the phenotypes associated with other general teratogens that affect fetal growth – e.g., anticonvulsants (hydantoin, trimethadione), aminopterin, congenital rubella, and maternal phenylketonuria – as they are from one another (Smith, 1982). Even in experimental animals given large doses of ethanol, the abnormalities that result appear to be quite specific (Chernoff, 1977; Webster *et al.*, 1983).

In considering the reference to the comments of Lewis (1964) on the pathology of trisomy 18, we would do well to go back to the original report:

It seems that development up to about the fifth week is overtly normal, and that several anomalies may be traced to a period of origin in the second half of the second month of gestation. . . . Some of these features represent delays in normal processes (e.g., syndactyly, Meckel's diverticulum, interventricular septal defect), whereas others are aberrations (e.g., polydactyly, small jaw, malformed ears, bifid ureters).

The period of teratogenesis is not, however, confined to the sixth to eighth weeks. There is a group of malformations, such as patency of the ductus arteriosus and foramen ovale, equinovarus deformities of the feet, cryptorchidism and inguinal hernias, all of which may be thought of as delays in processes normally occurring during the last trimester or even in postnatal life. There is then a large and heterogeneous residuum, presumably originating shortly after the first group (ninth to fourteenth week) and comprising mostly aberrations rather than arrests. There are somatic defects in all areas of the body, such as prominent occiput, short sternum, ventral midline defects including small external genitals, and peculiar but often diagnostic anomalies of hands and feet.

A similar point of view was expressed by Mottet and Jensen (1965) in their discussion of the pathology of trisomy 13, although it is true that they do stress the retarded development concept:

This chromosomal change is manifested by alteration of structures that involve embryonic developmental events at 5 to 6 weeks gestation. . . . The events involved were predominantly decreased proliferation and defective differentiation of mesodermal structures. Anomalies of the brain and organs of special sense are possible exceptions, although these lesions may also be explained as the result of altered inductive influences by cephalic mesodermal structures. . . . The anomalies associated with trisomy 13–15 seem to be manifestations of retarded developmental events occurring in embryos of 5 to 6 weeks (6.5 to 15 mm) gestation.

Evaluation of these two reports leads to two conclusions. First, not all abnormalities are attributed to retarded development. Second, the term "retarded development" does not, except with reference to features such as cryptorchidism or patent foramen ovale, mean just a slowing down of the rate of development. It refers, rather, to an apparent permanent arrest of development at a particular point in embryogenesis. This arrest could, of course, be the result of a reduced rate of cellular proliferation such that a tissue does not reach the critical size necessary for a morphogenetic event to occur (for example, fusion of the palatal shelves). But, again the retardation of development affects specific tissues at specific times rather than being a general process affecting the whole organism. Furthermore, it is also possible that what is interpreted as retarded or arrested development could have other causes independent of changes in the rate of cellular proliferation or in the progress of a developmental program.

On the basis of these considerations, it is concluded once again that a generalized process such as retardation of development cannot explain the phenotypes associated with aneuploidy. To be sure, as has just been discussed, many abnormalities can be interpreted as being the result of retarded or arrested (incomplete) development at various stages of gestation (Smith, 1982). However, each developmental insult occurs at a specific time and in a specific place, and the problem again becomes one of translating imbalance of particular chromosomes into particular phenotypes.

Disruption of developmental and physiological homeostasis

A strong challenge to the notion of a specificity in the relationship between genetic imbalance in aneuploidy and the resulting phenotype has been made by B. Shapiro (1983*a*). In a heavily documented review of Down syndrome, he argues that aneuploidy results in a generalized genetic imbalance and a consequent altered response to the genetic and environmental factors to which all individuals are exposed. This, in turn, produces *much* of the aneuploid phenotype. It is asserted that

despite the generally held notion to the contrary, clinical findings in DS (Down syndrome) are nonspecific. Unquestionably, phenotypic differences among the autosomal trisomy syndromes occur that usually separate them from one another and from the general population. . . . Nevertheless, the overlap of anomalies among these trisomies and the general population is as dramatic or perhaps more dramatic than are the clinical differences.

Further, he refers to several instances in Down syndrome in which there is an increased variance of metric traits, both developmental (anatomical) and physiological, and points to what he terms a generally increased morbidity (congenital malformations, neoplasia, infection, and mortality) in affected individuals. Based on these considerations, and reference to certain aspects of developmental theory, the following assertions are made:

The multiplicity of signs in DS, the phenotypic parallels with other trisomic conditions and the general population, and the increased variance of metric traits in DS cannot be explained as a direct result of one or some very few loci on chromosome 21. . . .

In a trisomic individual, the presence of a whole extra functioning chromosome or large chromosome segment causes a generalized disruption of evolved genetic imbalance. Because of the obligatory integration of the entire genotype this disruption affects not only products of the trisomic chromosome but of other chromosomes as well. The disruption of evolved balance results in decreased developmental and physiological buffering against genetic and environmental forces. Decreased buffering leads to decreased developmental and physiological homeostasis.

If generally decreased homeostasis exists in cells of these individuals, one can understand the parallel, but amplified prevalence, precosity, severity, multiplicity, and chronicity of abnormalities in DS in comparison with chromosomally balanced individuals and also the many shared findings among autosomal trisomic conditions.

Similar arguments have also been advanced by Barden (1983) and by Opitz and Gilbert (1982*a*,*b*). Arguing from the same basic premises, the latter assert that aneuploidy should have several predictable effects on development. These include (1) multiple minor anomalies which represent, for the most part, "altered means and variances of anthropometric traits"; (2) major malformations which represent mainly "nonspecific developmental noise" and are the result of altered thresholds; (3) growth disturbances; (4) dysplasias ("buffering defects of the fine tuning of histogenesis"); (5) maturational disturbances; (6) CNS abnormalities; (7) gonadal defects; (8) increased right-left asymmetry; (9) atavisms (features expressed only during fetal life); and (10) hyperreactivity to teratogens (Opitz and Gilbert, 1982*a*; Gilbert and Opitz, 1982). Since the midline is believed to be a "weakly buffered developmental field," they further predict that aneuploid individuals will have a greater liability to midline defects than nonaneuploid individuals (Opitz and Gilbert, 1982*b*). Ethnic differences are also predicted by Langenbeck *et al.* (1984).

The concept of developmental buffering, referred to as "canalization" by Waddington (1942), is explained by B. Shapiro (1983*a*) in the following manner:

> In its evolution, man's gene pool has responded to the pressures of selection so that there has evolved a balanced coadapted gene pool. This provides for well coordinated developmental and physiological pathways. . . . One can visualize the buffering systems as funneling the coordinated products of the species' genes toward a standard mean, and environmental and genetic trauma (eg, mutation or chromosomal rearrangements) as tending to diverge the funneled product from that standard. . . . This balanced genome results in developmental and physiological pathways that are buffered against disturbing factors and ordinarily lead to a standard phenotype. The multiplicity of interactions among multiple pathways gives such a system great plasticity as well as stability.

How can I reply to such a proposal which, if taken at face value, clearly runs counter to the concepts on which this monograph is based? In fact, B. Shapiro (1983*a*) states that "the model presented does not propose specific molecular mechanisms for the pathogenesis of abnormalities in trisomic conditions. A model that did so would be inappropriate. . . ." Perhaps the best reply is given by B. Shapiro (1983*a*) himself, and I shall quote from it directly:

> The DS phenotype does, of course, have a gestalt about it that usually permits its clinical detection and its distinction from other aneuploid syndromes. Nevertheless, overwhelming data exist that suggest that most of what is observed in DS is the result of nonspecific and inconsistent effects of the extra chromosome on complex traits. How then can one account for the DS phenotype? Most likely those recognizable portions of human aneuploid states that permit us to distinguish them from one another and from chromosomally balanced states are attributable to gene contents of

the triplicated chromosome. Different doses of specific gene products exist in each cell of affected individuals and evolved dominance relationships of gene products are changed. These alterations will affect gene products throughout the cell. Particular sets of gene products will affect cell mechanisms in subtle ways that differ depending on what sets of loci are triplicated and lead to the differences we observe. Thus the impressions that permit us to distinguish aneuploid states from one another and from chromosomally balanced states must ultimately relate to multiple interactions of extra gene products coded for by the extra chromosomes with each other and with gene products of other chromosomes through lengthy, serpentine, metabolic, physiological, and epigenetic pathways. The anatomical variations in trisomies 13, 18, and 21 described by Pettersen and Bersu (1982) presumably depend on the genic content of the particular chromosome involved.

The genesis of the DS phenotype resolves itself to the basis for i) its distinguishing characteristics and ii) its nonspecific characteristics. The first must be due to 21 chromosome loci and their interaction with complex developmental physiological and pathological pathways and the second to a generalized disruption of evolved developmental and physiological systems. No one or few characters in the first category are fixed or found only in DS, but as a composite they include those features [usually regarded as "cardinal signs" of DS, such as abundant neck skin, hypotonia, flat face, oblique palpebral fissures, simian creases, and the like]. In the second category fall findings such as those described above under Generalized Increased Morbidity. The impact of these latter conditions on the individual with DS seems to be more profound in terms of morbidity and mortality than those conditions that permit us to make a clinical diagnosis of DS.

Similar ideas are also expressed by Opitz and Gilbert (1982a):

Although *individual* minor anomalies are shared by other aneuploidy syndromes and also occur as developmental variants in the normal population . . . it is their *particular* combination which is characteristic of given chromosome defects. . . . Certain *combinations* of primary [major] malformations are more characteristic of one syndrome than of another.

Perhaps Shapiro, Opitz and Gilbert, and I are not so much in disagreement after all. Although I believe that some of the evidence presented by B. Shapiro (1983a) is open to more than one interpretation, the concept that aneuploidy decreases developmental homeostasis is certainly an intriguing one. It has the virtue that certain aspects, such as the predicted increased susceptibility to environmental factors, are now experimentally testable in animal model systems, And, it does have an attractiveness as a possible explanation for some of the apparently nonspecific features of the syndrome. However, where Shapiro and I clearly differ is in the weight to be given to specific gene-dosage effects in the genesis of the features of an aneuploid syndrome. The matter is not one of either-or, but of relative emphasis. I certainly give it considerably more weight than he does. As I have attempted to demonstrate earlier in this volume, I do not believe, as he does (B. Shapiro, 1983a), that "the unlikelihood of a triplicated dose of an enzyme being responsible specifically for a complex trait ought to be apparent" or, as has already been quoted, that "the overlap of anomalies among [the] trisomies and the general population is as dramatic or perhaps more dramatic than are the clinical differences." The fact that the constellations of abnormalities – whether regarded as specific or nonspecific – that constitute the aneuploid syndromes are distinguishable from one another would argue for a significant degree of specificity, even in the face of a loss of developmental homeostasis or buffering. Since the latter is thought to be brought about in Down syndrome by "mutual interactions of gene products of chromosome 21 and other chromosomes," presumably by affecting a variety of regulatory or metabolic interactions, imbalance of one chromosome would not be expected to produce the same effects, even on developmental homeostasis, as imbalance of another. At present, there is no evidence, at the molecular level, that aneuploidy, in general, produces a generalized disruption of genetic balance. Therefore, while not denying a potentially important role for decreased developmental homeostasis in the genesis of certain features of aneuploid syndromes, I believe that even it must ultimately be considered as the consequence of the effects of unbalancing particular genes on particular chromosomes.

Variability in expression

Regardless of my feelings about the weight to be attached to the B. Shapiro (1983*a*) hypothesis, there is no way of disputing the fact that there is considerable variability in the manifestation of any particular state of chromosome imbalance. Examples of this have already been given in Chapter 2. Since this variability has been interpreted as indicating nonspecificity of effects, it is worth considering the possible explanations for it. However, before doing so, it is necessary to make one point, and that is that variability in expression is no stranger in other types of genetic and teratogenic disorders (O'Donnell and Hall, 1979). It has long been recognized in numerous autosomal dominant conditions [neurofibromatosis and Marfan syndrome being two of the prototypes (Pyeritz, Murphy, and McKusick, 1979; Riccardi, Kleiner, and Lubs, 1979], and widely divergent manifestations may occur within the same family. At present, we have no convincing explanations for this variability, although it has been suggested that differences in genetic constitution, even between sibs or a parent and child, may be responsible [for a general discussion see Erickson (1979)]. But, as Vogel and Motulsky (1979) point out, "it is again no more than a label for our ignorance when we invoke the 'genetic background' or the action of all other genes for help. Analysis using the methods of formal genetics has contributed little to a better knowledge, . . . and even biochemistry and molecular biology have not been very useful so far." This criticism may be a little harsh since, in the mouse at least, genetic modifiers of penetrance and expression can be demonstrated (see, for example, Hummel, 1958, 1959; Wallace, 1976).

Not only do autosomal dominant conditions vary in expression within families, but X-linked and autosomal recessive conditions may do so as well. To cite just two examples, male sibs with X-linked Hunter syndrome (iduronate sulfatase deficiency) may have either severe or mild mental retardation (Yatziv, Erickson, and Epstein, 1977). Similarly, sibs with autosomal recessively inherited Farber disease (lysosomal acid ceramidase deficiency) differing widely in clinical manifestations have been reported (Antonarakis *et al.*, 1984), and both severely and mildly retarded individuals with phenylketonuria may be present in the same sibship (Primrose, 1983). The differences were attributed to the action of modifying genes or the influence of variable environmental factors. And, finally, even in a teratogenic situation, such as maternal phenylketonuria, in which exposure during pregnancy ought to be reasonably comparable from sib to sib, sibs may differ significantly in manifestations such as congenital heart disease or microcephaly (Stevenson and Huntley, 1967; Fisch *et al.*, 1969; Lenke and Levy, 1980). Similar findings have also been made in mice exposed to phenytoin in utero. Although the features observed were generally the same, the frequencies of individual characteristic features differed dramatically for each inbred strain tested (Finnell and Chernoff, 1984). Therefore, the existence of variability of expression in a chromosome imbalance state does not mean that the basis of the manifestations which are varying is necessarily any more nonspecific than it is in a Mendelian or teratogenic disorder.

Nongenetic factors

Returning to the chromosomal disorders, three possibilities, not wholly independent of one another, may be considered as causes for phenotypic variability: stochastic factors, extrinsic factors, and genetic differences. Stochastic factors refer to the inherent variability normally present in any developmental process so that, all else being equal, more than one outcome is possible. Perhaps the best evidence for the existence of such factors is the lack of complete concordance for congenital abnormalities in identical (monozygotic) twins. In these conditions, which include club foot, congenital dislocation of the hips, cleft lip and palate, pyloric stenosis, and congenital heart disease and are presumed to be multifactorially determined, the highest concordance rates are no greater than 50% (Vogel and Motulsky, 1979). Many of these conditions are regarded as threshold characters in which the aggregate of a constellation of relatively weak genetic and nongenetic determinants exceeds a certain hypothetical threshold and thereby leads to an abnormality (Fraser, 1976). These nongenetic determinants, generally referred to as environmental, include both extrinsic factors and those which represent the variability present in the biological processes themselves, however tightly controlled. It is this latter group of factors that I refer to as stochastic, realizing of course that it may be quite difficult in practice to separate the two types of nongenetic factors from one another.

The implication of the existence of stochastic factors and the threshold concept is that relatively weak genetic perturbations, such as might exist in aneuploid states, may not always be sufficient to lead consistently to the appearance of a particular abnormality. An example of this in a theoretical model of congenital heart disease in Down syndrome is discussed in Chapter 12. This can also be expressed in terms of the concepts of developmental buffering and canalization discussed earlier in this chapter. It is not necessary to assume that buffering and canalization are always decreased by an aneuploid state and that it is this decrease that leads to the increased but variable appearance of many or most congenital abnormalities (B. Shapiro, 1983*a*). While this mechanism could certainly operate in certain instances, it is also quite possible that these hypothetical homeostatic mechanisms might, if unaltered themselves by aneuploidy, bring about the same outcome in a different manner. By being capable of partially counteracting the effects of the genetic imbalance, they would, while not eliminating them completely, prevent the consistent development of a particular defect or set of defects. In this situation, the variability is introduced not by increasing the spread of the developmental determinants so that they more often but not always exceed some developmental threshold, but by limiting the perturbation of the pathway so that the determinants do not always exceed the threshold. The differences between these two concepts, which produce the same outcome, are shown in Fig. 9.1.

The second group of factors that could influence the appearance of aneuploidy-induced abnormalities consists of the extrinsic environmental factors. These would include variations in the maternal anatomy and metabolism, as well as specific external agents acting as teratogens. Rohde, Hodgman, and Cleland (1964) expressed this concept in the following manner:

It does not seem likely that the phenotypic variability can be assigned to mutant genes in the genome or to microscopically undetected deletions and translocations. It is suggested that trisomy gene expression is partially governed by environmental concentrations of certain hormones, enzymes, or vitamins. These latter substances could be brought into critical shortage during early pregnancy, and in gene trisomies the concentrations sufficient for normal embryos might be made relatively inadequate. . . . Phenotypic variability in the chromosomal syndromes may be governed by the flux of essential metabolites which control embryonic development. (Rohde, 1965)

As already mentioned, it may be difficult to separate these putative extrinsic factors from the intrinsic stochastic factors discussed previously, and the same general considerations would apply to both with regard to the multifactorial congenital malformations and the chromosomal disorders. Unfortunately, although the existence of extrinsic factors is plausible, there is as yet no experimental evidence available to support these suggestions. However, the experimental systems available in the mouse now make it possible to examine the role of extrinsic factors in a critical and controlled manner.

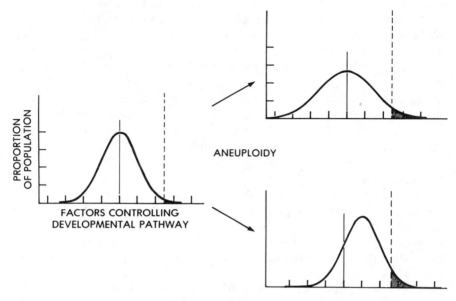

Fig. 9.1. Possible roles of developmental buffering or canalization in the variable expression of congenital abnormalities in aneuploid states. Left: The factors controlling a developmental pathway in normal individuals are represented as having a mean value (vertical solid line) and a spread or variance. Those individuals falling on the extreme right of the distribution exceed a threshold (vertical dashed line) which results in the development of a particular defect. Upper right: According to the model of B. Shapiro (1983a), aneuploidy reduces buffering or canalization and produces an increase in variance, without necessarily changing the mean, so that a greater proportion of the population exceeds the threshold and enters "an alternative developmental channel." Lower right: According to the alternative notion proposed here, aneuploidy may perturb the developmental pathway, displacing the mean to the right, without increasing the variance. Again, a greater proportion of the population exceed the threshold. In this instance, however, developmental buffering serves to prevent the mean from being displaced even farther to the right.

Genetic factors

We come then to the potential role of genetic differences in bringing about variability of expression. Once again, differences of this type can be fitted into the threshold models discussed earlier in this section and become just one more source of the variance attached to developmental processes. Genetic differences can be thought of as being two kinds – differences in the chromosomes or chromosome segments that are unbalanced, or, conversely, differences in the balanced remainder of the genome. Differences of the first type have been suggested by Engel (1980), who pointed out that in some instances two of the three chromosomes in a trisomy might be wholly or partially identical (what he terms "isodisomic"). Such identity of loci could, he suggests, lead to effects, by means unspecified, different from those which might be produced by triplication with three nonidentical chromosomes. Another suggestion implicating differences in the genetic constitution of the chromosomes in the unbalanced set was made by Carothers (1983). He calculated that if two alleles differing in the quantities of gene product produced exist at an unbalanced locus, not only will the mean concentration of product be increased, but so also will be the variance. This idea, put forward to explain the increased variance in metric traits observed by B. Shapiro (1983a), was rejected out of hand (B. Shapiro, 1983b). I would agree that it does not seem to be capable of explaining the variability in the appearance of a wide variety of phenotypic features.

The other and probably much greater source of genetic variability is in the still balanced remainder of the genome. A state of chromosome imbalance does not operate in a vacuum but is superimposed on the overall genetic constitution of the person or animal – a genetic constitution that clearly differs from individual to individual. The great differences in development encompassed within the concept of "normal" are such that it would not be surprising that the genetic perturbations associated with aneuploidy would not always act in a uniform manner. What is perhaps more surprising is that even subtle phenotypic manifestations, such as the facies in Down syndrome, can be detected so readily, even though superimposed on vastly different backgrounds.

However, these assertions do not explain or even prove that genetic differences do affect the expression of aneuploid states. Unfortunately, the human data discussed in Chapter 2, which compare phenotypes within and among different families, are insufficient to permit any definitive conclusion. Likewise, the results of investigations carried out specifically to look for either genetic or environmental influences on mental retardation, a prominent aspect of the Down syndrome phenotype, are still inconclusive. Although several authors have reported a positive correlation between parental education level or IQ and the IQ of their child with Down syndrome (Golden and Pashayan, 1976; Fraser and Sadovnick, 1976; Libb *et al.*, 1983), not all investigators agree (Bennnett, Sells, and Brand, 1979). Furthermore, in view of the demonstrated positive benefits of substituting a home for an institutional environment (Centerwall and Centerwall, 1960; Stedman and Eichorn, 1964), and possibly of early stimulation programs as well (Ludlow and Allen, 1979), it is very difficult in practice to dissociate genetic from environmental factors. Nevertheless, the former may well be present. In this regard, Smith's (1977) comment about children with Down syndrome is of interest. He attributes to Brook *et al.*, in a personal communication, the finding that the genetic background for stature exerts a similar role for Down syndrome children as for their normal siblings. He then goes on to make the following interesting assertion: "In similar fashion it would be anticipated that the smartest set of parents would, on the average, have the smartest Down syndrome children. I have considered this to be true from clinical experience for many years." It turns out, however, that in their published paper, Brook *et al.* (1977) do not find any relationship between the height of individuals with trisomy 21 and their parents, although quite good correlations are found in 45,X and 47,XXY!

Although the human studies on IQ are difficult to interpret, the results of animal work do make a significant contribution to the question of genetic factors, and three relevant examples from the mouse trisomies may be cited. The first is mouse trisomy 18, for which two phenotypes have been described. One, obtained using 3H1 females mated to *Rb3H/Rb6Rma* males (see Chapter 10 for discussion of this nomenclature), is characterized by moderate hypoplasia and death usually by day 12 (Evans, Brown, and Burtenshaw, 1980). The other, from NMRI females mated to *Rb6Rma/Rb10Rma* males, manifests only slight developmental retardation with survival to term, but 75% of the animals have a cleft palate and ventricular septal defect (Gropp, Gropp, and Winking, 1981). The differences are striking, but whether they result from differences in the constitution of the chromosomes 18 (only one of the 18s is the same in both crosses) or of the remainder of the genome cannot be determined, since both were varied simultaneously. This is somewhat less of a problem with the analysis of the role of genetic factors on the manifestations of mouse trisomy 12. In this case, two of the three chromosomes 12 were from the same source. When the same males were bred to females of different outbred and inbred strains, a variety of phenotypes were obtained (Gropp and Grohé, 1981). Thus, with C3H females there is marked developmental retardation, and death occurs by day 12 or 13. With NMRI females, the trisomic fetuses survive to day 16 or 17 and show slight, if any, developmental retardation even though weight may be reduced. Ventricular septal defects occur in somewhat less than half, but the most striking feature is a complete exencephaly coupled with an apparent microcephaly (Putz *et al.*, 1980; Pexieder, Miyabara, and Gropp, 1981). On the other hand, with BALB/c females, 10% of the trisomic fetuses are not exencephalic, survival beyond day 15 is rare, and some fetuses are severely retarded and hypoplastic. When C57BL/6 females are used, the frequency of exencephaly is reduced to 60%, and some of the trisomic progeny are even liveborn. In fact, there seems to be a curious direct relationship between the presence of the malformations and survival.

In the third mouse example, partial trisomy 2 (2H3–2H4) can be produced by mating animals carrying the balanced translocation, *T(2;4)1Sn*, to normal animals. When the latter are C57BR/cd, the trisomic progeny are runted and have exencephaly. However, when the normal parent is DBA/2, one-third of the trisomic progeny have this phenotype while two-thirds, although still runted, are not exencephalic and live for several days (Washburn and Eicher, 1975).

One other example might be mentioned. In our original work with mouse monosomy 19, death of virtually all monosomics occurred about 3.5 days after conception, just after blastulation (Epstein and Travis, 1979; Magnuson, Smith, and Epstein, 1982). However, when the male breeding stock was re-constituted on a new genetic background because of difficulties in maintaining one of the translocation chromosome-bearing strains, the results were different (Table 10.2). Monosomy 19 fetuses with ICR mothers did not die until 5–6 days after conception, although about half had died by 4.5 days of gestation. Breeding the same males to inbred females of several strains revealed that similar results were obtained with BALB/c and DBA/2 females, but the monosomic progeny of C57BL/6 mothers were fully viable until at least 4.5 days of gestation (Magnuson *et al.*, 1985). When the monosomic embryos were examined, it was found that the time of death was directly related to the rate of cellular division. The more rapidly cleaving embryos died earlier, whereas those that took longer to attain the same number of cleavages died later. Thus, genetically determined alterations in cleavage rate resulted in changes in the time of death.

The work with the mouse aneuploids indicates quite clearly that genetic differences can affect both the frequency and severity of specific developmental defects, as well as the overall viability of the aneuploid individuals. Since it is capable of even more stringent genetic control and manipulation than has been used to date, this experimental system should ultimately make it possible to dissect out and hopefully identify these genetic influences and to establish their role in causing variability in phenotypic expression.

Experiments carried out with the stocks used to generate mouse aneuploids have produced results which suggest another way in which genetic factors may lead to variability in the manifestations of an aneuploid state. Using either Robertsonian fusion or reciprocal translocation stocks, it is possible to generate genetically balanced mice in which both copies of a chromosome or chromosome segment are derived from a single parent. In several instances, the phenotype of these animals with so-called "parental disomy" is affected by the parental source of the chromosome, even though the animals are genetically identical (Searle and Beechey, 1978; Cattanach and Kirk, 1985). For example, mice in which both chromosomes 11 are derived from the father (paternal disomy) are about twice as large as are those with maternal disomy, and the responsible genetic region could be localized to the proximal part of chromosome 11. Similarly, a paternal source of both copies of distal chromosome 2 resulted in hyperkinetic mice with short, square bodies and short, flat backs, while maternal partial disomy yielded hypokinetic animals with long, flat-sided bodies and arched backs.

On the basis of these and related observations, Cattanach and Kirk (1985) have postulated that

a differential functioning of maternal and paternal chromosomes or chromosome regions is indicated, which suggests the existence of a form of chromosome imprinting that affects gene activity. The contrasting phenotypes (small and large; hypokinetic and hyperkinetic) depart from the normal in opposite directions and this suggests excess versus shortage of gene activity. This could mean either single or earlier activity of the paternal chromosome regions in the cases cited, and inactivity or later activity of the corresponding maternal regions. Paternal duplication would then give excess/earlier activity and the reciprocal subnormal/later activity.

The idea of contrasting phenotypes is, interestingly, reminiscent of type and countertype.

Although the concept of differential activity of maternally and paternally derived chromosomes has not been examined directly and its applicability to humans is completely unknown, its potential relevance to both mouse and human aneuploidy cannot be ignored. Two of the three unbalanced chromosomes or chromosome segments in a trisomic animal are, of course, derived from the same parent. Therefore, the functional state of this set of chromosomes or chromosome segments would depend on parental origin if regions of differential expression were present on them. Thus, a trisomy or duplication with two maternal and one paternal chromosome or chromosome segment might produce less imbalance (because of the lower activity of the maternally derived chromosomes) in terms of gene activity than would the reciprocal case in which two paternal chromosomes or chromosome segments were present. Similar considerations should also apply to monosomies and deletions. It would therefore be expected that it may be possible to produce animals with the same aneuploidy but different (qualitatively or quantitatively) phenotype. The availability of appropriate mouse stocks and breeding schemes (see Chapter 10) makes it feasible to test this possibility directly.

Determinants of lethality and severity

In an early discussion of the effects of partial trisomy, Patau (1962) commented on the relationship between the size and identity of the aneuploid region and the phenotypic consequences. Lethality was explained by the existence of a relatively small number of "trisomy lethal" genes which were scattered throughout the genome and had a lethal effect when present in triplicate. Chromosomes 13, 18, and 21 were considered to have fewer than the required number of lethal genes, and their respective trisomies were therefore compatible with viability, and the same situation presumably also applied to the viable duplications or partial trisomies. By contrast, Patau (1962) suggested that

whereas it appears likely that the number of genes responsible for trisomy lethality is relatively small, the opposite may be assumed for nonlethal effects. . . . If the number of genes involved in trisomy effects is large, we may expect that the complexity, though not necessarily the clinical severity, of partial trisomy syndromes will be in a very rough way proportional to the length of the additional euchromatic chromosome segment.

In this proposal, Patau thus distinguished between lethality and other pheno-typic manifestations and between the effects of a single or small number of genes and of large numbers of genes on these outcomes, and these ideas have been considered in greater detail in both *Drosophila* and mammalian aneu-ploidy.

Shortly after Patau (1962) made his proposal, Yunis (1965) suggested that lethality in the human trisomies is in fact a function, not of the length of the aneuploid segment per se, but of the amount of early replicating and presum-ably active chromatin (euchromatin). Subsequent investigations and analyses to examine these relationships have been carried out with both *Drosophila* and mammalian aneuploids. Probably the most influential of these studies has been the extensive series of investigations carried out by Lindsley, Sandler, and their collaborators (1972) in *Drosophila*. They used a system which per-mitted them to generate segmental aneuploidies, both duplications and defi-ciencies, covering 85% of the two large autosomes, and then to systematically examine the relationship between lethality (preadult death) and segment size and identity. Their principal conclusions were the following (the nomencla-ture, while different from that used here, should be readily understandable):

The Drosophila genome contains 57 loci, aneuploidy for which leads to a recogniz-able effect on the organism: one of these is triplolethal and haplo-lethal, one is triplo-abnormal and haplo-abnormal, one is hyperploid-sensitive, ten are haplo-abnormal, 41 are Minutes, and three are either haplo-lethals or Minutes.

The viability of terminal hyperploids declines regularly with the amount of tripli-cated material, becoming generally lethal when more than one-half an autosomal arm is present in three doses . . . ; viability reaches zero when 500 bands are in-cluded in the terminally hyperploid regions irrespective of arm.

Because of the paucity of aneuploid-lethal loci, it may be concluded that the delete-rious effects of aneuploidy are mostly the consequence of the additive effects of genes that are slightly sensitive to abnormal dosage . . . and not by the individual effects of a few aneuploid-lethal genes among a large array of dosage insensitive loci. . . . Moreover, except for the single triplolethal locus, the effects of hyper-ploidy are much less pronounced than those of the corresponding hypoploidy.

The conclusion in *Drosophila* is therefore inescapable. The lethal genes proposed by Patau (1962) cannot explain the lethality associated with aneu-ploidy. However, the size of the aneuploid segment, which he thought affected only the complexity of the phenotype, apparently can.

The one locus, designated *Tpl*, identified by Lindsley, Sandler, *et al.* (1972) as being lethal in the triplo- (as well as haplo-) state is intriguing, but unfor-tunately nothing is known about its mode of action. On the basis of mutational studies, it has been suggested that it is not a "typical" structural gene, but either a very small, reiterated or otherwise complex locus or one functionally insensitive to base-pair substitutions (Keppy and Denell, 1979). It was suggested that this locus might be involved in sex chromosome dosage compensation. This does not appear to be the case, although a specific X chromosome locus has been implicated in suppressing the triplo-lethal phenotype of *Tpl* (Roehrdanz and Lucchesi, 1981).

In considering the Lindsley, Sandler, *et al.* (1972) data just cited, it must be borne in mind that the endpoint in their analysis was death prior to adulthood, and the relationship between segment size and outcome was presented in terms of death. What about nonlethal aneuploidy? Unfortunately, very few data are given, and the relationship between segment size and phenotype was apparently considered in only fairly rough terms (Lindsley, Sandler, *et al.*, 1972): "Intermediate levels [of hyperploidy] cause reduced survival, small size, and a variety of morphological anomalies such as rough eyes, abnormal wings and bristle patterns, and a misshapen abdomen. These morphological effects do not seem to be characteristic of any particular region of the chromosome, but are more appropriately thought of as a hyperploid syndrome." However, this conclusion may not be terribly firm, since only a year later it was stated that "almost certainly . . . more intensive examination would discriminate among different hyperploids because enzyme levels for several different enzymes have been shown to vary with gene dosage. . . . The apparent nonspecificity of aneuploid phenotypes in Drosophila may thus be partially a function of less discriminating diagnostic methods [than are used with humans]" (Sandler and Hecht, 1973). Furthermore, as has already been quoted, there were several examples of phenotypic abnormality associated with a small deletion or, rarely, duplication. Therefore, the Lindsley and Sandler findings cannot and should not be taken as evidence for a nonspecific relationship between degree of aneuploidy and phenotypic effects other than lethality.

To what extent are the *Drosophila* results generalizable to mammals? Were the matter only one of genetics, it would be quite tempting to attempt to apply them directly. However, the issue of generalizability is probably more one of metabolism and development than it is of genetics, and here the differences between flies and mammals are quite extensive. Therefore, it is necessary to look directly at the mammalian data.

In their comparison of *Drosophila* and human results, based only on chromosome lengths, Sandler and Hecht (1973) suggested that in humans the probability of lethality is proportional to the extent of duplication or deletion, "but not strictly so." The maximum imbalance considered compatible with viability was 5–6% of the haploid autosomal complement for a duplication and 2–3% for a deletion. Daniel (1979), in a study of aneuploidy associated with translocations, also concluded that a major factor in the survival to term of aneuploid fetuses was the size of the chromosome abnormality, irrespective of the origin of the chromosome region. The limits of the amount of the genome imbalanced were estimated to be 4% and 2%, respectively, for duplications and deletions. When both duplications and deletions are present simultaneously, again as the results of translocations, duplications greater than 2% combined with deletions greater than 1% of the haploid autosomal complement were found to be nearly always lethal (Davis *et al.*, 1985). However, there was considerable variation from chromosome to chromosome, suggesting that factors other than only the lengths of the segments were involved.

Despite the rough correlation between chromosome length and lethality, other authors have used methods for expressing human chromosome content which they believe more accurately reflect the active genetic content of the chromosomes. One such measure was "relative gene content," a function of total DNA content and proportion of dark areas by R-banding (Martin and Hoehn, 1974; Hoehn, 1975). This correlated much better with survival than did chromosome size (DNA content), and the three chromosomes with the lowest relative gene contents were, in order, 21, 18, and 13. However, the correlation was much less satisfactory for other chromosomes and chromosome segments, and the approach generally lacked predictive ability.

Looking at the problem in a somewhat different manner, Korenberg, Therman, and Denniston (1978) attempted to correlate very early human embryonic lethality prior to recognizable and analyzable spontaneous abortion with various parameters of chromosome structure and content. These included length, chiasma densities, and Q-band brightness. They found that those chromosomes (referred to as "hot spot" chromosomes) that had the highest density of mitotic chiasmata (in Bloom syndrome lymphocytes) – chromosomes 3, 6, 11, 12, 17, and 19 – were among those which were *least* recognizable in trisomic abortuses. This relationship, which they considered to be most significant, was not simply a function of chromosome length. However, it is also known that there is a very high correlation between chiasma density and Q-band darkness (Kuhn, 1976). The latter, like R-band darkness, is again considered to represent the regions with high densities of active loci (Korenberg, Therman, and Denniston, 1978). Therefore, the results of Korenberg, Therman, and Denniston (1978), who looked at the most lethal trisomies, and of Martin and Hoehn (1974), who considered the least lethal ones, are generally compatible with each other and with the conclusion that lethality is in some way related to the number of putative active loci unbalanced in an aneuploid state.

The mouse ought to be particularly useful for analyses of the type under consideration here since all of the whole-arm trisomies can be prepared and examined (see Chapter 10), but relatively few studies have as yet been carried out. However, when survival in fetal life is plotted as a function of chromosome size (Fig. 9.2), there is a modest clustering of longer survival for the trisomies involving the smaller chromosomes. Although the linear correlation coefficient is quite low ($r = -0.42$ for the regression of survival on chromosome length, excluding chromosome 19), comparison of the survivals of tri-

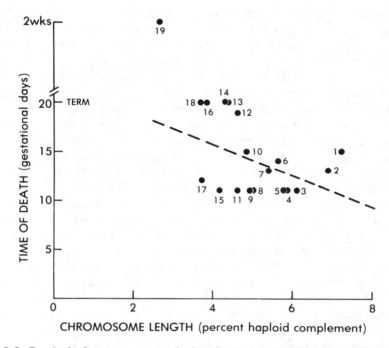

Fig. 9.2. Survival of mouse autosomal trisomics during fetal life. Plotted from data in Epstein *et al.* (1984*a*) and Epstein (1985*a*). Chromosome lengths are taken from Nesbitt and Francke (1973). The number of each chromosome is indicated next to the point representing it. The linear regression is calculated without the point for trisomy 19.

somies for the nine largest chromosomes, as a group, with those of the nine or ten smallest chromosomes does reveal a significant difference ($p < 0.05$) by the two-tailed Wilcoxon rank sum test. However, while trisomies for the smaller chromosomes are generally compatible with longer survival, there are major exceptions – especially trisomy 1, which survives much longer than might be expected, and trisomies 15 and 17, which survive for a much shorter time. While the genetic maps of the first two are not striking, chromosome 17 does carry the major histocompatibility (*H–2*) and *t* complexes which contain many genes apparently involved in development. Interestingly, the major histocompatibility complex in man is located on chromosome 6 in the region of one of the chiasma "hot spots," and it has been suggested (Korenberg, Therman, and Denniston, 1978) that this is a region of particularly high gene activity. The exceptional placements of chromosomes 1 and 17 are not altered by consideration of Giemsa-staining dark and light bands, since light-band content in the mouse is generally proportional to chromosome length (Epstein and Travis, 1979). However, chromosome 17 does appear quite dark after quinacrine staining (Q-banding) (Davisson, 1981), but quantitative studies have not been carried out. Clearly more work with the mouse trisomies is required, but it should ultimately be possible to determine the number of genes on each chromosome that function during development and to correlate this with survival.

Taken together, the *Drosophila*, human, and mouse data are consistent with the conclusion that aneuploidy-induced lethality is a function of the amount of active genome that is imbalanced, although certain regions, probably quite few in number, may play a disproportionately large role. This generalization does not, however, tell us anything about why or how embryonic and fetal death occurs and does not necessarily imply that the causes of death are non-specific. In some of the mouse trisomies, for example, particularly those with death in early midgestation, death may be associated with very severe developmental and growth retardation. The genetic imbalance thus appears to have generally deleterious effects on basic cellular processes. But, in other trisomies that cause death only a few days later, development and growth appear to be much closer to normal, and the problem may lie with particular malformations or even with the placenta. Gropp (1981) has suggested that in some instances there may be relative placental insufficiency, with the aneuploid placenta being unable to serve the metabolic needs of the aneuploid fetus. Similar conclusions were reached by Boué *et al.* (1976) on the basis of their examination of human abortuses. Certain trisomies, particularly in groups A, B (some), C (some), and F, are lethal very early in embryogenesis, and they speculated that the early arrest is the result of the genetic imbalance. In other trisomies that are compatible with embryogenesis, there may also be growth or developmental retardation, although this is difficult to judge precisely because of the significant delay between the time of death in utero and expulsion of the fetus (estimated to be six weeks). However, a common feature of trisomic placentas was said to be hypoplasia, with few and frequently avascular villi and an underdeveloped trophoblast (particularly cytotrophoblast). This decrease in the number of cells in the placenta was thought to be capable of leading to a reduction of placental hormone secretion at a time when corpus luteum hormone production is not sufficient to maintain the pregnancy, thereby explaining the abortion of embryos with no or only minor malformations which would not themselves interfere with viability.

Polyploidy

Although not aneuploidy in the sense that this term is being used in this volume, polyploidy (the presence of one or more extra complete sets of the haploid chromosome complement) does raise issues that are relevant for consideration here. Despite the fact that genetic balance, at least in numerical terms, is apparently maintained, mammalian polyploids are inviable and, if they survive long enough in gestation, developmentally abnormal. This distinguishes them, for reasons which are not understood, from plants and amphibians in which functional and viable polyploids have long been recognized (Frankhauser, 1945; Lewis, 1980).

On theoretical grounds alone, polyploid cells or organisms, while possibly genetically balanced, are physiologically unbalanced. If we start with the assumption that gene dosage effects still exist in polyploid cells, then the volume of the cells will be directly proportional to the number of haploid chromosome sets present. Triploid (3n) cells will be 50% larger than the diploid (2n), and tetraploid (4n) cells will be twice as large. The validity of this relationship has been demonstrated for parenchymal cells in mouse liver, in which ploidies ranging from 2n to 32n normally coexist (Epstein, 1967; Epstein and Gatens, 1967). It has also been observed in tetraploid human fibroblasts (Chang *et al.*, 1983; Schmutz and Lin, 1983). However, while volume remains proportional to ploidy, surface area probably does not (Epstein, 1967; Snow, 1975). If, for the moment, we consider a cell to be a sphere, a doubling in volume, as would exist in going from diploidy to tetraploidy, would be accompanied by an increase in radius (r) of $2^{1/3}$ or 1.26 (since volume is equal to $4/3 \pi r^3$). Surface area, on the other hand, which is a function of r^2 (since $A = 4\pi r^2$), would increase by $(1.26)^2$ or 1.59. Therefore, the ratio between area and volume is reduced by a factor of 1.59/2.00, or 0.8. For a triploid cell, the surface area to volume ratio would be reduced by a factor of 0.87. Similar geometric considerations would apply to cells of any shape, since volume is ultimately a cubic function of the linear dimensions of the cell, while area is a square function.

Whether cell surfaces in polyploid cells actually follow these rules is unknown. It is conceivable that the cell could still make the normal total amount of surface membrane, especially since the synthesis of the protein constituents, which are themselves primary gene products, should continue at the normal (dosage-dependent) rate. But, if this were the case, then the cell would have to undergo a significant change in shape. For example, in the case of a spherical 2n cell, its 4n counterpart could maintain virtually the same surface to volume ratio if it became a cube (Fig. 9.3); a cell which started as a cube (when 2n) would have to elongate in some dimension in order to maintain the proper surface to volume ratio when 4n. For changes of 2n to 3n, the alternatives would, of course, not need to be as severe.

Thus, the options theoretically open to cells which become polyploid are to allow their surface to volume ratios to be reduced, to maintain the ratio by changing shape, or to effect a combination of the two. The first results in a physiological imbalance which could be deleterious, the second leads to problems with tissue architecture which certainly would be, and the third leads to both types of difficulties. Of course, the other way out of the problem would be for the cell to negate the gene dosage effects completely and reduce overall genetic activity by a half or a third (depending on whether it is 4n or 3n).

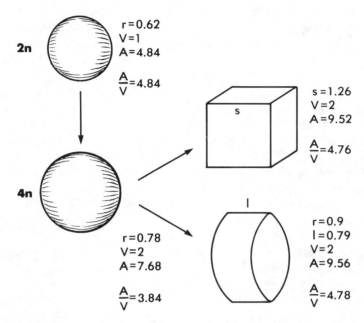

Fig. 9.3. Effect of changing ploidy of cell on surface to volume ratio. To restore its original (diploid) ratio, a spherical tetraploid cell would have to assume another shape, such as one closer to that of a cube or a cylinder.

While this type of mechanism may apply in some organisms, such as amphibians (Beçak and Pueyo, 1970; Audit *et al.*, 1976), there is no evidence that this can or does occur in mammals. However, there is some suggestion that individual gene products may depart from expectation (see below).

This brief analysis suggests that, if only geometric considerations applied, the consequences of triploidy should be significantly less severe than those of tetraploidy. As the material to be summarized below indicates, this does not seem to be the case, and other explanations for the severity of triploidy must be sought. These explanations may involve an imbalance between the autosomal and sex chromosomal complements, making triploidy, in addition to a problem of geometry, one of aneuploidy as well.

Among human spontaneous abortions, the frequency of tetraploidy is 6.3% and of triploidy 16.5% (Bond and Chandley, 1983). Among newborns, however, the cases of nonmosaic polyploidy are so few as to constitute material for individual case reports.

Tetraploidy

To date, only five cases of tetraploidy in liveborn human infants have been reported (Golbus *et al.*, 1976; Pitt *et al.*, 1981; Scarbrough *et al.*, 1984). Four of the five were males with a 92,XXYY chromosome complement. Most had prenatal growth retardation, and length of survival ranged from less than a day to a year. Common features in three or more affected infants included microcephaly, microphthalmia or anophthalmia with short palpebral fissures, beaked nose, apparently low-set ears with hypoplastic cartilage, prominent forehead with narrow biparietal diameter, high-arched or cleft palate, micrognathia, hypotonia, and microgyria (Scarbrough *et al.*, 1984). A few liveborns with mosaic tetraploidy (4n/2n) have also been reported, with the nervous system, as in full tetraploidy, being particularly vulnerable (Veenema, Tasseron, and Geraedts, 1982; Quiroz, Orozco, and Salamanca, 1985).

Considerably more has been learned about the effects of tetraploidy by studying animals, particularly mice, in which it can be produced experimentally. The principal techniques involve either suppression of cell division (cytokinesis) by cytochalasin B (Snow, 1973; Tarkowski, Witkowska, and Opas, 1977) or fusion of diploid cells with polyethylene glycol (Eglitis and Wiley, 1981). Snow's studies in random-bred Q-strain mice have been particularly interesting. When tetraploidy was induced at the two-cell stage with cytochalasin, the most immediate effect was a reduction of cell numbers in blastocysts from 54 to 20, and this difference persisted when the embryos were permitted to "outgrow" in culture (Snow, 1976). However, only about 60% of the 4n embryos produced outgrowths, and less than half of these contained inner mass cells capable of further embryonic differentiation. Similar results were obtained with 4n embryos allowed to implant in utero, with many of the 4n embryos forming trophoblastic vesicles incapable of yielding embryos. Snow (1976) postulated that the effect of reduction in cell number at the blastocyst stage was such as to make the number of interior cells marginal insofar as giving rise to a functional inner cell mass was concerned, so that the majority of 4n embryos were, for geometric reasons of another kind than those described earlier, unable to carry on embryogenesis.

When 4n embryos did develop in utero, a small proportion were capable of surviving to term and being liveborn (Snow, 1975). Those that did appeared externally normal but were rapidly eaten and could not be further examined. However, 4n fetuses examined at 14.5 to 16.5 days of gestation had numerous abnormalities, with the smallest fetuses showing the most severe developmental disturbances. Overall cell members were significantly decreased, and cell sizes were said to be increased (Snow, 1973). According to Snow (1975), the most common general defect was the presence of hemorrhages in a variety of tissues, especially the lungs, but also the brain, muscle, testis, and spinal cord. The hemorrhages were attributed to the mechanical stresses of circulating large (as much as four fold greater in volume than 2n cells), nucleated, yolk sac-derived blood cells through blood vessels no larger than those in 2n fetuses. Other severe developmental anomalies of the eyes and brain were also noted. To explain them, three possibilities could be considered: the hemorrhages themselves, reductions in cell numbers in tissue primordia, and metabolic disturbances in the tissues. Which, if any, was operative in any particular situation could not be determined.

The studies of Tarkowski, Witkowska, and Opas (1977), carried out in a different strain of mouse [(CBA × C57BL/10) F_1 × A] and with slightly different techniques, gave results which were somewhat at variance with those obtained by Snow (1973, 1975, 1976). The principal differences were the following. Survival during early postimplantation development was much greater, and the first abnormalities were not noted until the eighth day of pregnancy, when embryonic development begins to slow down. Formation of mesoderm was particularly affected, but development of the fetal membranes appeared to be spared. No embryos survived beyond the tenth day. Because 4n cells were virtually absent by days 10 and 13 from the embryonic portion of 4n/2n mosaics which they prepared, Tarkowski, Witkowska, and Opas (1977) speculated that 4n cells are intrinsically less viable and possibly have a significantly reduced rate of proliferation.

Although the mechanisms operative in producing the inviability and abnormalities in tetraploidy are unknown, the implications of the findings are clear. A change in the physiological balance of cells in embryos, brought about by altering cell size and the relationship between surface area and volume, can lead to a wide variety of developmental abnormalities. Since the original perturbation (the change in ploidy) is a specific one, it is not surprising that a reasonably consistent pattern of abnormalities should result. However, there is also a random element introduced by factors such as tissue hemorrhaging and the exact number of cells that end up in an organ primordium. Therefore, it is also not unexpected that there will also be a significant degree of variability so that, as Snow (1975) observed, only two tissues (blood and gonad) were always abnormal and the brain was abnormal most of the time.

Mention was made earlier of the possibility that gene dosage expectations might not always be met in tetraploid cells. Data on this point are mixed. On the one hand, Westerveld et al. (1972), in artificially created 4n cells, found that the activities per cell of both X-linked [glucose–6-phosphate dehydrogenase (G6PD) and phosphoglycerate kinase] and autosomal [6-phosphogluconate dehydrogenase (6PGD)] enzymes were increased 2 to 2.5 fold over 2n values. Similarly, Priest and Priest (1969) observed a 2-fold increase in collagen and total protein synthesis when cloned 4n cells of an established rat fibroblast-like cell line were compared with 2n cells from the same line. Chang et al. (1983) found proportional increases in volume, total protein, and the activities of G6PD and 6PGD in tetraploid human skin fibroblasts, but only a 1.4-fold increase in RNA per cell (which they attributed to earlier senescence of the 2n cell strain). By contrast, Schmutz and Lin (1983) found that the specific activity (in terms of protein) of peptidase 5 was reduced in half when a 4n fibroblast line derived from a spontaneously aborted fetus was compared with a 2n strain (source unspecified), and Eglitis and Wiley (1981) reported lower than expected increases (only 1.22- to 1.54-fold) in the total activity of malate dehydrogenase in 4n embryos containing the same number of cells as 2n control embryos. However, because of potential technical problems, most of these studies, particularly the latter ones, must be viewed with caution.

In plants, the expected dosage-dependent changes in protein content are generally found, although in some instances individual proteins were either higher or, less commonly, lower in amount than would have been expected (De Maggio and Lambrukos, 1974; Timko, Vasconcelos, and Fairbrothers, 1980; Birchler and Newton, 1981; Birchler, 1983*a*). Proportional changes in enzyme activity have also been found in yeast (Ciferri, Sora, and Tiboni, 1969).

Triploidy

Experience with human triploidy is considerably more extensive than with tetraploidy, particularly because a moderate number of late triploid abortuses, stillbirths, and liveborns, and of liveborn 3n/2n mosaics have been identified and reported (Niebuhr, 1974; Wertelecki, Graham, and Sergovich, 1976; Harris, Poland, and Dill, 1981; Blackburn *et al.*, 1982; Tharapel *et al.*, 1983). Complete triploidy is characterized by placental enlargement and hydatidiform degeneration, intrauterine growth retardation, and a variety of congenital malformations. A total of 210 external anomalies were found in 44 3n infants, and internal anomalies were found in all autopsied cases (Blackburn *et al.*, 1982). The most frequent anomalies are defects of the eye (colobomas, microphthalmia), micrognathia, facial clefting, low-set and malformed ears, syndactyly, genital anomalies (in males), abdominal wall defects, congenital heart disease, and abnormalities of cranial bone and brain formation. Gosden *et al.* (1976) felt that the anomalies associated with triploidy are less constant in appearance than are the anomalies in any of the common autosomal trisomies. Survival of liveborn triploids is no longer than a week (Niebuhr, 1974).

The organs of human triploids are generally reduced in size, with some (eyes, kidneys, adrenals, ovaries, pituitary, and lung) being disproportionately affected. [However, larger than normal eyes have been observed by Moen, Werner, and Bersu (1984)]. Reminiscent of the 4n mouse embryos, subectodermal hemorrhages were found more frequently in 3n abortuses than in chromosomally normal abortuses (Harris, Poland, and Dill, 1981). In 3n/2n mixoploid mosaicism, which is compatible with survival, the most common abnormalities involve the limbs (syndactyly) and eyes, and congenital assymetry and ambiguous genitalia (in males) are frequently noted (Blackburn *et al.*, 1982; Tharapel *et al.*, 1983). Mental retardation is noted as development proceeds (Niebuhr, 1974; Graham *et al.*, 1981; Blackburn *et al.*, 1982; Tharapel *et al.*, 1983).

Once again, experimental animals have been useful for the study of triploidy. Although spontaneous triploidy has been reported in a variety of animals, mostly fetal (Beatty, 1957; Bomsel-Helmreich, 1965), and even in a living chicken with an intersex appearance (Ohno *et al.*, 1963), most work has been carried out with rats and mice. Triploid rat embryos were produced either by inducing polyspermy (dispermy) or by treating eggs with a mitotic inhibitor at the time of in vitro fertilization to induce digyny (Piko and Bomsel-Helmreich, 1960). Although the 3n embryos appeared normal for the first ten days, development then slowed down, and no living 3n fetuses were found three days later. Similarly, in spontaneous triploidy in CBA mice, development appeared normal until the ninth day, at which time the embryonic part of the egg cylinder became disproportionately small in relationship to the extraembryonic part (Wróblewska, 1971). A comparable phenotype was found in triploidy (digynic) induced with cytochalasin in A strain mice, with development again being normal for the first 7 days (Niemierko, 1981). Other investigators also detected abnormalities as early as day 7 or 8 in superovulated mice of the same strain, with the principal defect again being underdevelopment of the embryonic portion and relative or complete sparing of the extraembryonic portion (Endo, Takagi, and Sasaki, 1982). Although some of the variation in results obtained by different investigators may be technical in origin, it is also likely that the genetic background of the strains being used does affect the development of the mouse triploids (Wróblewska, unpublished, quoted in Niemierko, 1981).

In their study of preimplantation 3n embryos, Takagi and Sasaki (1976) found that there was a delay in the rate of cleavage, with cell cycle time increased an estimated 10%. Other evidence for impaired cellular proliferation was also obtained by Kamoun *et al.* (1978) with 3n rabbit embryonic fibroblasts and by Boué *et al.* (1975, 1976) with cultured cells from human triploid abortuses, and Mittwoch (1977) has suggested that changes in mitotic cycle times may play a role in the abnormal development of triploid zygotes. However, as in the case of the trisomies, the results are controversial, and normal rates of cell proliferation were reported by Kuliev *et al.* (1975) and Hassold and Sandison (1983) in cells cultured from abortuses, and by Schmickel *et al.* (1971) and Graham *et al.* (1981) in cells from infants. Likewise, mitogen-induced proliferative responses, measured by ^{3}H-thymidine incorporation, were, if anything, greater (when corrected for DNA content) in lymphocytes from a premature triploid neonate (Pittard, Sorenson, and Stallard, 1983), although Fryns *et al.* (1980) speculated that 3n peripheral lymphocytes are selected against in 3n/2n mosaics.

Biochemical studies of triploid cells have been relatively limited. Schmickel *et al.* (1971) found 1.5- (as expected), 1.7-, and 2.1-fold increases, respectively, in the DNA, RNA, and protein contents of human new-born 3n skin fibroblasts. Using cells cultured from the products of early spontaneous abortions, Junien *et al.* (1976) found that 11 autosomal enzymes had normal specific activities when expressed on a per milligram protein basis, and Maraschio *et al.* (1984) also found five of nine enzymes to be normal. The overall mean for all nine enzymes in the latter experiments was 1.1, and the outlying results were attributed to individual culture variability. Although the protein content per cell was not measured, these results would be compatible with the expected gene dosage effect if it is assumed that protein content increases in proportion to ploidy. However, Kuliev *et al.* (1975), using a single 3n embryonic cell strain, found significant increases in three enzymes and a decrease in one [if it is assumed that the activity data are expressed on a per milligram protein basis, as in their earlier paper (Kuliev *et al.*, 1974)]. And, using 2n and 3n rabbit embryo fibroblasts, Kamoun *et al.* (1978) observed a 1.23- to 1.30-fold increase in cell diameter (which would correspond to about a 2-fold increase in volume), no significant change in cellular protein content, and no change or decreases (relative to protein content) in the activities of three autosomal enzymes. It is very difficult to know how to interpret any of these data, and more and better controlled studies are clearly required. In passing, it is worth noting that in triploid *Drosophila*, cell size is increased (surface area of cells with bristles increased 1.27 fold), and the activities of several enzymes, expressed relative to DNA content, are unchanged (Lucchesi and Rawls, 1973).

Although triploid and tetraploid individuals share geometric problems that differ in degree but not necessarily in kind, they clearly differ with respect to the state of their sex chromosomes. However we look at it, triploids have sex chromosome aneuploidy. This is the case, irrespective of whether one or two X chromosomes are active in the cell. Except for those X-chromosomal loci that are not inactivated (Shapiro and Mohandas, 1983), the normal ratio of autosomal loci to active X-chromosomal loci is 2:1. In the triploid situation, this ratio becomes either 3:1 (if only one X chromosome is active) or 3:2 = 1.5:1 (if two X chromosomes are active). Since neither is normal, it is difficult to know how to evaluate the suggestion of Weaver *et al.* (1975) that triploids should differ in the severity of their abnormalities, depending on whether they have one or two active X chromosomes. In the case of XYY triploids, this imbalance would occur even during the period prior to X inactivation, so that rather than the 2:1 (in males) and 1:1 (in females) ratios ordinarily observed (Epstein, 1983*a*) (see Chapter 13), the ratio would be 3:1. This may be an explanation for why XYY triploids are not detected in mice (Endo, Takagi, and Sasaki, 1982) and are so rare in humans (Beatty, 1978).

Considerable evidence has been accumulated to show that either one or two X chromosomes may be active in triploid cells. Using late replication of the X chromosome as a mark of inactivation, Jacobs *et al.* (1979) found that of

27 triploid abortuses that were 69,XXY, 14 had both X chromosomes active and 11 had a mixed cell population with either one or two active X chromosomes. Of 9 69,XXX triploids, 4 had two early replicating (active) X chromosomes, and 5 were mixed. In general, fetal tissues tended to have a greater proportion of cells with two active X chromosomes, while older specimens had fewer. In a study of a single 69,XXY infant, Yu, Chen, and Fowler (1983) found that all cells in a variety of tissues had two active X chromosomes, except for lymphocytes, of which 3% had only one. By contrast, Maraschio *et al.* (1984) found only 14% of cells (lymphocytes and fibroblasts) with two active X chromosomes, the remainder having only one.

The results in mouse triploids are somewhat at variance with those in humans. In general, two X chromosomes are inactivated in most of the 60,XXX embryonic cells, whereas only the paternally derived X chromosome is inactivated in the extraembryonic trophectoderm and primitive endoderm. The number of active X chromosomes in the latter thus depends on the sex of the triploid and whether it is digynous (two haploid sets from the egg) or diandrous (two haploid sets from sperm) (Endo, Takagi, and Sasaki, 1982). If it is assumed that a 3:2 autosome to X-chromosome ratio causes less severe imbalance than a 3:1 ratio – and there is presently no evidence to support or refute this – then the difference between human and mouse triploids (some of the former being capable of surviving to term) could lie in the finding that, in general, the human triploids tend to have two active X chromosomes while the mouse triploids have only one.

The cytological assessments of X-chromosome activity are confirmed by biochemical determinations of the activities of X-linked enzymes, an approach which appears to have validity for determining the functional state of X chromosomes (Epstein, 1983*b*). In human material, Weaver *et al.* (1975) found evidence for two active X chromosomes in cells from a 69,XXY triploid. Junien *et al.* (1976) showed 1.3- and 1.4-fold increases over diploid values in the specific activities of two X-linked enzymes in nine sets of embryonic fibroblasts, both 69,XXX and 69,XXY in constitution. These values are close to the 1.33-fold increase expected with two active X chromosomes, since, on a per cell basis, protein increases 1.5 fold and the X-linked enzymes increase 2 fold, giving a relative increase of $2/1.5 = 1.33$. However, Vogel *et al.* (1983), using a single cell assay for α-galactosidase, showed that cultured skin fibroblasts from a single 69,XXY triploid were in two classes, those with one and those with two active X chromosomes, with the latter increasing in frequency as the cells were carried in culture. In rabbit embryo fibroblasts, Kamoun *et al.* (1978) found 24% and 15% increases for two X-linked enzymes, but no change in a third. These findings are difficult to interpret.

Given the fact that triploid individuals do have sex chromosome aneuploidy, the real question is whether it plays any role in the pathogenesis of their problems. Opinions on this score are mixed. Some authors believe that it does not (Weaver *et al.*, 1975; Mittwoch, 1977), at least insofar as problems other than those concerned with sex differentiation are concerned (Edwards *et al.*, 1967). Others, however, believe that it may (Gosden *et al.*, 1976; Endo, Takagi, and Sasaki, 1982) or does (Jacobs *et al.*, 1979). In fact, Jacobs *et al.* (1979) have gone so far as to suggest that one active X chromosome in the human embryo may be incompatible with fetal development, as may two active X chromosomes in the placenta be incompatible with normal placental development. As has already been noted, this pattern is just the opposite of what is found in triploid mouse embryos (Endo, Takagi, and Sasaki, 1982). Jacobs *et al.* (1982) pointed out that the source of the extra chromosome complement also influenced the outcome, independently of the sex or number of active X chromosomes in the fetus. In all instances in which the extra complement was paternal in origin, the triploid embryos gave rise to a partial hydatidiform mole.

In view of the possibility discussed earlier that there may be a differential expression of autosomes which is dependent on parental origin (Cattanach and Kirk, 1985), another form of genetic imbalance could also be operative in triploid individuals. Irrespective of which parent the third chromosome complement was derived from, the quantitative relationships between the overall expression of such differentially expressed regions and of the other parts of the genome would be disturbed in a manner similar to that just described for the X chromosomes.

Clearly the available data do not permit an unequivocal answer regarding the basis of the difficulties in triploidy. However, a comparison of tetraploid and triploid mouse embryos does suggest that while the former have serious difficulties not attributable to genomic imbalance per se, the latter seem to do worse in that none survive until term. Therefore, in this species, it is not unreasonable to conclude that the genetic imbalance, superimposed on a less severe geometric imbalance, leads to as severe an outcome, if not a more severe one. Similarly, although human tetraploidy is less frequently recognized than is triploidy (probably because fewer tetraploids are conceived), the phenotypes are not so dissimilar, and complete tetraploidy is, curiously, more compatible with survival beyond term than is triploidy. Therefore, by the same reasoning, I infer from this that in man, as in the mouse, genetic imbalance in the triploids aggravates the problem of geometric imbalance.

Part IV

Experimental systems for the study of mammalian and
human aneuploidy

10

Generation and properties of mouse aneuploids

In the preceding part, several references were made to experimental systems for studying various aspects of the known or potential effects of aneuploidy and polyploidy. In addition to man himself, mention has been made of a number of organisms, ranging from bacteria, yeast, and plants to *Drosophila*, amphibians, and the mouse. Not already mentioned are also nematodes (Meneely and Wood, 1984), fish (Davisson, Wright, and Atherton, 1972), birds (Bloom, 1972; Blazak and Fechheimer, 1981; Macera and Bloom, 1981), and amphibians (with aneuploidy as opposed to polyploidy) (Guillemin, 1980*a,b*). However, in only three nonhuman organisms – plants, *Drosophila*, and the mouse – has the work been part of a systematic investigation of the effects of aneuploidy, and of these organisms only the mouse can be considered as being reasonably close to man genetically, developmentally, and physiologically. Therefore, while the other organisms, as well as others not mentioned or yet used for research on aneuploidy, will continue to be or may become quite useful for the investigation of various aspects of the aneuploidy problem, it is the mouse that promises to be the most useful in the short run. For this reason, separate consideration of the mouse as an experimental system for studying aneuploidy is warranted. Since this subject has been extensively reviewed quite recently (Epstein, 1981*a*, 1985*a,b*; Epstein *et al.*, 1984; Gropp *et al.*, 1983), the aim of this chapter will be to highlight and summarize the major findings.

Generation of autosomal aneuploids

Full trisomies and monosomies

The methods now used for the generation of full-chromosome aneuploidy in the mouse are based on the utilization of parental stocks heterozygous for two Robertsonian fusion (translocation, metacentric) chromosomes which share an arm in common (termed "monobrachial homology"). Earlier methods described by Baranov and Dyban (1972) and Gropp, Giers, and Kolbus (1974), which made use of males carrying a single translocation chromosome bred to karyotypically normal females, have for the most part been abandoned because of the low yields and lack of specificity of the aneuploid progeny. Both

207

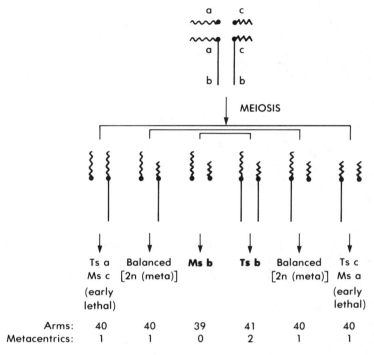

Fig. 10.1. Meiotic segregation of double Robertsonian fusion (metacentric) chromosome heterozygotes with monobrachial homology (for chromosome b). Of the six possible gametes produced, two result in Ts-Ms combinations that are early lethals; two result in balanced [2n(meta)] progeny that have 40 chromosome arms and one metacentric; one results in monosomy (Ms) b with 39 arms and no metacentrics, also an early lethal; and the last results in Ts b with 41 arms and two metacentrics. The Ts b and Ms b gametes are produced in equal numbers (Epstein and Travis, 1979). Reprinted by permission from Epstein (1985a).

of these problems are avoided by use of the breeding scheme, first described by White *et al.* (1974a) and Gropp, Kolbus, and Giers (1975), illustrated in Fig. 10.1. It has been used to produce all of the mouse trisomies and monosomies. Although a variety of segregation patterns are possible, trisomy and monosomy for the homologous chromosome (arm) are easily recognizable by the presence of two metacentrics and a total of 41 chromosome arms and of 39 chromosomes with no metacentrics, respectively. High frequencies of aneuploid progeny are generally obtained. The yields range from 14% to 22% each for both trisomics and monosomics when determined within two days after fertilization and from 15% to 42% for trisomics alone (the monosomics having died) in midgestation (Epstein and Travis, 1979; Epstein *et al.*, 1977) (see also Table 10.2). It would be expected from the breeding scheme (Fig. 10.1), that monosomic and trisomic embryos should be conceived in equal numbers. This has been directly shown to be the case for aneuploidy of several different chromosomes, indicating that nullisonic and disomic gametes are equally effective in fertilization (Epstein and Travis, 1979).

The laboratory mouse, *Mus musculus*, normally has 19 pairs of acrocentric autosomes and two acrocentric sex chromosomes, and only a few radiation-induced Robertsonian fusion metacentric chromosomes have been identified in laboratory animals. However, several groups of feral mice living in relatively circumscribed geographical areas have been found which carry large numbers of spontaneously occurring fusion chromosomes (Gropp and Winking, 1981). The first identified was *M. musculus poschiavinus*, which has seven pairs of fusion chromosomes. The existence of these metacentrics reduces the total chromosome number to 26, although the number of chromosome arms remains 40. Other feral mice have been found which have as many as nine pairs of metacentric chromosomes, with only the two chromosome 19s and the two sex chromosomes existing in the free state. Equally as remarkable as the fact that these fusion chromosomes should exist at all is the fact that different groups of mice have different combinations of chromosomes. Thus, about 60 different metacentric chromosome combinations are known at the present time, some found in more than one subspecies but others apparently being unique (Gropp and Winking, 1981). This is about a third of the total number of metacentrics involving dissimilar chromosomes that are theoretically possible. These metacentric chromosomes are now designated by symbols such as in the following example: *Rb(16.17)32Lub/Rb(11.16)2H* refers to an animal doubly heterozygous for two different Robertsonian fusion (*Rb*) chromosomes, one composed of chromosomes 16 and 17 and designated as number 32 in the Lübeck (*Lub*) series, and the other composed of chromosomes 11 and 16 and designated as number 2 in the Harwell (*H*) series. This combination, with monobrachial homology for chromosome 16, is one of the combinations in the male which, when mated to a normal female, produces trisomy and monosomy 16 conceptuses.

Fortunately, the feral mice carrying the translocation chromosomes are interfertile with laboratory mice, and it has therefore been possible to transfer these chromosomes into the laboratory strains. Individual stocks homozygous for different translocation chromosomes now exist, some on relatively highly inbred backgrounds. Lists of available breeding stocks are presented in Searle (1981), Gropp and Winking (1981), and Epstein (1985a), and are updated yearly in *Mouse News Letter*.

In using the double-heterozygote breeding scheme shown in Fig. 10.1, it is desirable, since the females must nearly always be sacrificed to obtain the aneuploid progeny, to have the males carry the translocation chromosome. However, this is not always possible, since such males may sometimes be limited in fertility or even completely sterile. This problem can usually, although not always, be overcome by choosing a different combination of metacentric chromosomes or even the same combination with one or both metacentrics being derived from different sources. When this does not work, doubly heterozygous females can be used, although this entails considerably more effort in breeding. It should be noted that tetrasomic embryos and fetuses can be generated when male and female double heterozygotes are bred with one another (Debrot and Epstein, 1985).

The importance of the development of the breeding scheme in Fig. 10.1 and of the isolation of the necessary translocation chromosomes and breeding of the stocks which carry them cannot be underestimated. These accomplishments have made it possible for the first time to use a mammal, the mouse, for systematic and controlled investigations of many aspects of aneuploidy.

Duplications and deletions (segmental aneuploidy)

While all of the full trisomies and monosomies can be generated by the method just described, the breeding of mice with duplications and deletions (partial trisomies and monosomies) is dependent upon the availability of appropriate sets of radiation-induced reciprocal translocation chromosomes. Unfortunately, only a small number of existing translocations (listed in Searle, 1981, and in *Mouse News Letter*) give rise to progeny which have only a single aneuploid segment. More usually, inviable embryos with combined duplications and deficiencies are the result. Thus, unlike the situation in *Drosophila*, in which segmental aneuploids covering virtually the entire genome can be prepared and examined (Lindsley, Sandler, *et al.*, 1972), there are still marked technical limitations in the generation of the mouse segmental aneuploids.

The two types of translocations that most commonly give rise to viable segmental aneuploids are illustrated in Fig. 10.2. And, once viable aneuploids (generally trisomics) have been produced, they can, if fertile, transmit the translocation chromosome in unbalanced form to their progeny with high frequency (Fig. 10.3). This, of course, greatly simplifies the production of segmental aneuploids and facilitates their use for experimental purposes.

Properties of autosomal aneuploids

All complete trisomies and monosomies of the mouse have been produced, as have a modest number of segmental aneuploids. Most attention has been directed to the trisomies and to the segmental aneuploids capable of surviving

Fig. 10.2. (a) Tertiary trisomy resulting from 3:1 meiotic segregation of a balanced reciprocal translocation with a small marker chromosome. Although labeled as Ts a2, it should be noted that a small region of b, b1, is also duplicated. (b) Partial trisomy (duplication) resulting from adjacent–2 segregation of a balanced reciprocal translocation in which virtually all of one chromosome (a) is translocated to the proximal region of the second (b). Reprinted by permission from Epstein (1985*a*).

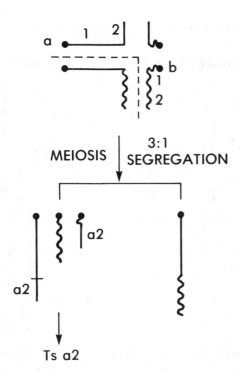

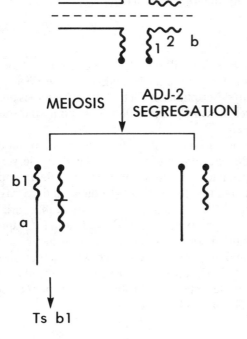

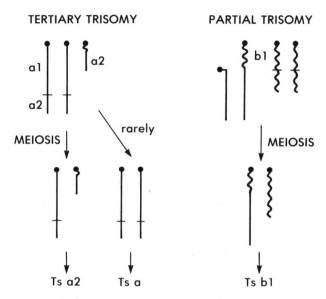

Fig. 10.3. Transmission of tertiary trisomy (left) and partial trisomy or duplication (right) at meiosis. Reprinted by permission from Epstein (1985*a*).

past birth. The monosomics, which die in the periimplantation period, have not yet received the attention they deserve.

Trisomies

As reference to Table 10.1 will reveal, the major share of the work in defining the phenotypes of the complete trisomies has been done by Gropp and his collaborators. Full descriptions of each of the phenotypes are contained in the original references and in Epstein (1985*a*), and only the salient features will be summarized here.

Although they do survive until at least midgestation, all of the full mouse trisomies, with the exceptions of trisomy 19 and to a lesser extent trisomies 16 and 18, die prior to parturition. The times of death appear to cluster in three periods: at about 9–11 days, 13–15 days, and 16 days to term (Table 10.1, see also Fig. 9.2), although there is considerable variability in the time of death of each (Gropp, 1981). As was noted earlier (Chapter 9), the precise cause of death of the trisomics is not known. In some instances, it appears to be related to extremely poor embryonic growth and development. In others, it has been suggested that it may be caused by relative placental insufficiency, with the aneuploid placenta being unable to serve the metabolic needs of the aneuploid fetus (Gropp, 1981).

Table 10.1. *Survival and phenotypes of complete mouse trisomies*

Trisomy[a]	Survival to (days of gestation)	Hypoplasia	Developmental retardation	Congenital malformations			Other abnormalities
				Heart[b]	Central nervous system	Other	
1	14–15	+ to +++	+ to +++		Brain maldevelopment: holoprosencephaly	Craniofacial dysplasia	Small placenta with hypoplasia of vascular labyrinth
2	10–13						
3	10–11	+++	+++				
4	9–12	++ to +++					
5	9–11						
6	12–15	+ to +++	+ to +++				
7	11–13	++	+++				
8	9–12	+++	+++				
9	10–11				Occasional exencephaly		
10	13–15	+ to ++	+	VSD in some			
11	9–12	+++	+++				
12	14–17 (NMRI background)[c]	++	0 to +	VSD (50%)	Exencephaly (100%); enophthalmia		Edema
13	13–birth	++	+	Pulmoni stenosis, VSD		Cleft Palate	Edema; hemopoietic stem cell defects; placental hypoplasia

Table 10.1. (cont.)

Trisomy[a]	Survival to (days of gestation)	Hypoplasia	Developmental retardation	Congenital malformations			Other abnormalities
				Heart[b]	Central nervous system	Other	
14	14–19	++	+	VSD, transposition of great vessels, DORV	Exencephaly (50%)		
15	9–11	+++	+++				
16	14–term	++	+	Great vessel anomalies; endocardial cushion defect	Brain hypoplasia; inner ear abnormalities	Open eyelids	Edema; hemopoietic stem cell defects; placental hypoplasia
17	10–12	+++	+++		Caudal hypoplasia; exencephaly in some		
18	to term[c]		+	VSD		Cleft palate	
19	1–4 weeks	++		VSD or DORV			Placental hypoplasia; small thymus

214

From Epstein (1985a).

[a]References: Trisomy 1 (Gropp, Giers, and Kolbus, 1974; Gropp, Kolbus, and Giers, 1975; Gropp, Putz, and Zimmerman, 1978; Gropp, 1978). Trisomy 2 (Gropp, 1982). Trisomy 3 (Gropp, Giers, and Kolbus, 1974). Trisomy 4 (Gropp, Giers, and Kolbus, 1974; Gropp, 1982. Trisomy 5 (Gropp, 1978; Gropp, Winking, and Gropp, 1981). Trisomy 6 (Gropp, Putz, and Zimmerman, 1978; Gropp, 1982). Trisomy 7 (Jüdes, Winking, and Gropp, 1980). Trisomy 8 (Baranov and Dyban, 1972; Gropp, Kolbus, and Giers, 1975). Trisomy 9 (Gropp, 1982).Trisomy 10 (Gropp, Giers, and Kolbus, 1974; Pexieder, Miyabara, and Gropp, 1981; Gropp, 1982). Trisomy 11 (Gropp, Giers, and Kolbus, 1974).Trisomy 12 (Gropp, Giers, and Kolbus, 1974; Gropp, Kolbus, and Giers, 1975; Putz et al., 1980, Pexieder, Miyabara, and Gropp, 1981; Herbst et al., 1981; Gropp and Grohé, 1981). Trisomy 13 (Gropp, Giers, and Kolbus, 1974; Hongell, Herbst, and Gropp, 1980, 1984; Gropp, 1982; Hongell and Gropp, 1982). Trisomy 14 (Gropp, 1978, 1982; Putz and Morriss-Kay, 1981; Pexieder, Miyabara, and Gropp, 1981). Trisomy 15 (Gropp, Putz, and Zimmerman, 1978). Trisomy 16 (Gropp, Giers, and Kolbus, 1974; Claussen, Gropp, and Herbst, 1981; Herbst et al., 1982a; Miyabara, Gropp, and Winking, 1982; Pexieder, Miyabara, and Gropp, 1981; Oster-Granite, Baker, and Ozand, 1983; Oster-Granite, et al., 1983; Ozand et al., 1984; Gearhart, Oster-Granite, and Hatzidimitriou, 1985; Oster-Granite and Hatzidimitriou, 1985; Msall, Oster-Granite, and Gearhart, 1985; Epstein, Cox, and Epstein, 1985; Epstein et al., 1985). Trisomy 17 (Baranov and Dyban, 1972; Gropp, Giers, and Kolbus, 1974; Gropp, Kolbus, and Giers, 1975; Winking and Gropp, 1978). Trisomy 18 (Evans, Brown, and Burtenshaw, 1980; Gropp, Gropp, and Winking, 1981). Trisomy 19 (White et al., 1972, 1974a,b; Gropp, 1978, 1981; Grohé and Gropp, 1980; Pexieder, Miyabara, and Gropp, 1981; Baranov, Vaisman, and Udalova, 1982; Bersu and Mieden, 1980; Bersu, Crandall, and White, 1982, 1983; Gropp et al., 1983; Bersu, 1984).

[b]VSD, ventricular septal defect; DORV, double-outlet right ventricle.

[c]See Chapter 9 for discussion of relationship of survival to genetic background.

With respect to their early times of death, the mouse trisomics are not significantly different from the human trisomics, which also nearly all die during gestation. Even in the case of trisomy 21, the most viable of the human trisomics, approximately 75% of the recognizable trisomic embryos and fetuses die during gestation (Boué *et al.*, 1981).

Growth retardation is almost invariably present, and congenital malformations are frequently detected. Among the interesting developmental lesions that have been recognized are holoprosencephaly (in trisomy 1), congenital heart disease (in trisomies 12, 13, 14, 16, and 18), cleft palate (in trisomies 13 and 18), and exencephaly (in trisomies 12 and 14). As is the case with the human trisomies, the phenotypes of the mouse trisomies, when development into the later part of gestation is possible, are often quite specific.

Because of its compatibility with survival beyond term, trisomy 19 is, of course, of particular interest. The longest survival observed thus far has been to day 27 after delivery. The viable animals lag considerably behind the normal in weight gain, especially after the ninth day, and apparently die of respiratory failure (G. Grohé and A. Gropp, personal communication). Unlike most of the other trisomies, trisomy 19 is not characterized by the presence of congenital malformations, and this may account to some extent for the ability of trisomy 19 mice to survive at all in the neonatal period. Even those mouse trisomies that might be metabolically capable of survival would not be able to live because of the impediments associated with their congenital defects (along with the maternal infanticide which occurs in such situations).

Studies of the anatomical aspects of the pathogenesis of the congenital defects found in many of the trisomies have been carried out. Of particular note in this regard is the work on congenital heart disease in several of the trisomies (Pexieder, Miyabara, and Gropp, 1981; Miyabara, Gropp, and Winking, 1982) and on exencephaly in trisomies 12 and 14 (Putz *et al.*, 1980; Putz and Morriss-Kay, 1981). Overall, the most intensive studies have been carried out with trisomy 16, because of its possible relevance to human trisomy 21. This work will be considered in detail in Chapter 12.

Monosomies

Unlike the mouse trisomics, which live until at least midway through gestation, monosomic embryos generally die in the perimplantation period. As was noted earlier, monosomic and trisomic embryos are conceived in equal numbers when the double-heterozygote breeding scheme is used, and equal frequencies have been directly demonstrated during the preimplantation period for aneuploidy of several chromosomes (Table 10.2) (Dyban and Baranov, 1978; Epstein and Travis, 1979; Baranov, Dyban, and Chebotar, 1980; Epstein, 1981*b*; Magnuson, Smith, and Epstein, 1982; Baranov, 1983*a*; Magnuson *et al.*, 1985). Therefore, by comparing the frequencies of monosomic and trisomic embryos early in gestation, it is possible to determine when

Table 10.2. *Frequency of aneuploidy in early embryonic progeny of doubly heterozygous males and normal females*

Chromosome	Cross[a]	Day of gestation[b]	Frequency of aneuploidy		References
			Monosomy	Trisomy	
1	ICR × Rb1Bnr/	3	16%	16%	Epstein and Travis
	Rb10Bnr	4	9	17	(1979)
	F₁ × Rb1Bnr/+	3	6.4	7.7	Baranov (1983a)
	ICR × Rb1Bnr/	2	14.9	20.9	Magnuson et al.
	Rb10Bnr	3	11.5	28.8	(1985)
		4	7.1	18.6	
		6	0	21.3	
2	F₁ × Rb4Iem/+	3	3.8	2.2	Baranov (1983a)
3	F₁ × Rb1Bnr/+	3	11.6	10.3	Baranov (1983a)
4	ICR × Rb2Bnr/	4	12.1	13.8	Magnuson et al.
	Rb4Rma	6	0	16.0	(1985)
5	Rb1Wh/Rb3Bnr	3	10.7	15.3	Baranov (1983a)
	× F₁				
10	ICR × Rb8Bnr/	3	15.0	15.0	Magnuson et al.
	Rb10Bnr	4	5.7	11.3	(1985)
		6	2.0	18.0	
11	ICR × Rb4Bnr/	3	13.2	26.5	Magnuson et al.
	Rb8Bnr	4	1.5	31.8	(1985)
12	ICR × Rb5Bnr/	4	13	13	Epstein and Travis
	Rb9Bnr				(1979)
	ICR × Rb5Bnr/	3	13.9	15.3	Magnuson et al.
	Rb9Bnr	4	15.4	11.5	(1985)
		6	0	32.7	
14	ICR × Rb6Bnr/	3	22.4	18.4	Magnuson et al.
	Rb16Rma	4	14.8	27.8	(1985)
		5	7.4	22.2	
		6	3.4	15.3	
15	Rb3Bnr/Rb1Ald	3	11.3	12.7	Dyban and Baranov
	× ?				(1978)
	ICR × Rb64Lub/	2	9.5	7.1	Magnuson et al.
	Rb4Rma	3	7.7	30.8	(1985)
		4	6.2	27.1	
		6	0	25.0	
	ICR × Rb3Bnr/	4	10.2	24.4	
	Rb4Rma				
16	F₁ × Rb7Bnr/+	2	6.6	4.4	Baranov (1983a)
	ICR × Rb32Lub/	2	17.0	17.0	Magnuson et al.
	Rb2H	3	6.2	12.5	(1985)
		4	8.0	18.0	
		6	0	20.0	
17	ICR ×	3	9	22	Epstein and Travis
	Rb11Rma/	4	8	13	(1979)
	Rb11em				

Table 10.2. *(cont.)*

Chromo-some	Cross[a]	Day of gestation[b]	Frequency of aneuploidy		References
			Monosomy	Trisomy	
	Rb7Bnr/Rb11em	2	11.5	16.0	Baranov (1983*a*)
	× F$_1$	3	0	22.8	
	ICR ×	2	7.6	17	Magnuson *et al.*
	Rb11Rma/	3	3.4	18.1	(1985)
	Rb11em	4	8.3	12.5	
	ICR ×	3	14.3	21.4	Magnuson *et al.*
	Rb11Rma/	4	20.8	20.8	(1985)
	Rb11em	6	0	30.0	
18	ICR × *Rb6Rma/*	3	15.8	19.7	Magnuson *et al.*
	Rb10Rma	4	13.5	15.4	(1985)
		6	0	20.8	
19	ICR × *Rb163H/*	3	18	20	Epstein and Travis
	Rb1Ct	4	2	24	(1979)
	C57BL/6 ×	3	16	20	
	Rb163H/	4	1	19	
	Rb1Ct				
	F$_1$ × *Rb163H/*	2	14.4	16.8	Baranov (1983*a*)
	Rb1Wh				
	ICR × *Rb163H/*	3	17	17	Magnuson, Smith,
	Rb1Wh	4	3	17	and Epstein
		9	0	21	(1982)
	ICR × *Rb163H/*	3	18.0	14.0	Magnuson *et al.*
	Rb1Wh	4	8.9	16.8	(1985)
		6	0	12.0	
	BALB/c ×	4	9.0	16.7	
	Rb163H/	6	0	22.6	
	Rb1Wh				
	DBA/2 ×	4	10.2	14.3	
	Rb163H/				
	Rb1Wh				
	C57BL/6 ×	4	17.3	19.6	
	Rb163H/				
	Rb1Wh				

[a]The genetic cross used in breeding the aneuploid progeny is represented as female strain × male strain. Although a doubly heterozygous parent was used in most instances, a singly heterozygous parent (*Rb/ +*) was occasionally used. F$_1$ refers to the F$_1$ generation of the cross CBA × C57BL/6. The full designations of the Robertsonian fusion chromosomes are as follows: *Rb(6.15)1A1d, Rb(1.3)1Bnr, Rb(4.6)2Bnr, Rb(5.15)3Bnr, Rb(11.13)4Bnr, Rb(8.12)5Bnr, Rb(9.14)6Bnr, Rb(16.17)7Bnr, Rb(10.11)8Bnr, Rb(4.12)9Bnr, Rb(1.10)10Bnr, Rb(8.19)1Ct, Rb(11.16)2H, Rb(9.19)163H, Rb(8.17)11em, Rb(2.6)4Iem, Rb(16.17)32Lub, Rb(15.17)64Lub, Rb(4.15)4Rma, Rb(2.18)6Rma, Rb(1.18)10Rma, Rb(2.17)11Rma, Rb(8.14)16Rma, Rb(5.19)1Wh.*

death of the former begins to occur. Studies of this type have been carried out in Leningrad and San Francisco, and the results are shown in Table 10.2. Two salient features emerge from this analysis: the different monosomies differ considerably in the time at which death occurs and, even for the same monosomy, the stage at which death occurs may differ between institutions or even, over time, at the same institution. The latter differences are presumably the result of genetic differences in the breeding stocks used. Even the same translocation-carrying stock will change over time as inbreeding is allowed to proceed. This is apparent in the results shown in Table 10.2 for monosomies 17 and 19.

Although death may not occur until several days after fertilization, it is sometimes possible to show significant effects of the monosomic state on cleavage and cell proliferation considerably earlier (Table 10.3). For some monosomies, such as 2 and 6, the number of cells in day 3 embryos may be between 30% and 50% of those present in normal control embryos or in trisomic embryos of the same gestational age. In others, there is no evidence for retardation of cleavage until at least day 4. Precisely what happens to monosomic embryos during implantation is less well understood. By day 6, nearly all examined monosomics have died, and viable monosomies are certainly no longer present by midgestation. Baranov (1983a) observed some viable cells in day 8 embryos with monosomy 16 (plus a segmental duplication of chromosome 17), although in most instances only trophoblastic vesicles without embryonic elements were observed. We observed a few marginally viable monosomy 14 embryos on day 6 (Magnuson *et al.*, 1985).

The cause(s) of death of the monosomic embryos are not known. By breeding the double-translocation heterozygous males to highly inbred females, it could be shown that the lethality of monosomy 19 is not the result of uncovering (hemizygosity for) recessive lethal genes (Epstein and Travis, 1979). In this situation, the only chromosome 19 present was the one derived from the inbred parent and was presumed to be devoid of such genes. Examination of two-dimensional polyacrylamide gel protein patterns did not reveal any marked differences between monosomy 19 and diploid control embryos, and no significant ultrastructural differences other than an increased number of cytoplasmic degradative bodies were found (Magnuson, Smith, and Epstein, 1982). Likewise, the only microscopic changes described by Baranov (1983a) were those directly attributable to cellular degeneration.

Notes to Table 10.2. *(cont.)*

*b*Gestational days are counted with the day following impregnation (which occurs during the night) as day 0.*

Table 10.3. *Cell numbers in aneuploid preimplantation embryos*

Chromo-some	Day of gestation[a]	Number of cells in			References
		Monosomic	Diploid	Trisomic	
1	3	41	38.5	39.6	Baranov (1983a)
2	3	14.5[b]	52.6		Baranov (1983a)
3	3	33.2	39.6	38.5	Baranov (1983a)
5	2	4.3	4.4	4.5	Baranov (1983a)
	3	14.1[b]	29.1	32.5	
6	3	27.4	52.3		Baranov (1983a)
14	3	30.9	42.9	39.5	Magnuson et al. (1985)
	4	68.9	95.6	80.3	
15	3	23.7	32		Dyban and Baranov (1978)
	4	57.3	110.2	70.2	Magnuson et al. (1985)
17	2	6.6[b]	8.3	8.2	Baranov (1983a)
19	2	13	12	13	Magnuson, Smith and
	3	46	56	60	Epstein (1982)
	3	35.6	34.0	29.8	Baranov (1983a)
	4	93.3[c]	151	112	Magnuson et al. (1985)
		85.5[d]	124.6	125.8	
		74.2[e]	82	78	

[a]Day 0 = day after impregnation.
[b]Death occurs during morula stage.
[c]BALB/c mother. See Table 10.2.
[d]DBA/2 mother. See Table 10.2.
[e]C57BL/6 mother. See Table 10.2.

Duplications and deletions (segmental aneuploidy)

Since the production of most segmental aneuploids depends upon specialized breeding schemes (Fig. 10.2) and is wholly dependent on the availability of appropriate translocations, the mouse duplications and deletion phenotypes have not, in general, been extensively studied. This is unfortunate since they deserve attention equal to if not greater than that given to the whole-chromosome aneuploids. Many of them are viable beyond the newborn period and have traits worthy of investigation – growth retardation, developmental abnormalities, and neurological impairment. While the males are frequently sterile or semisterile, the females are often fertile and may have aneuploid progeny (Fig. 10.3). In such instances, once the original aneuploids (usually duplications referred to as tertiary or partial trisomics) are generated, their perpetuation becomes a relatively easy matter. This occurs in *T(14;15)6Ca*, *T(5;12)31H*, *T(16;17)43H*, *T(1;13)70H*, *T(-;-)158H*; *T(-;-) 194H*, and *T(10;13)199H*.

The characteristics of the individual segmental aneuploidies compatible with postnatal viability are presented in Table 10.4. They are listed in numerical and alphabetical order of the designations of the reciprocal translocations used to generate them, and further detail may be obtained from the original references and from Searle (1981) and Epstein (1985*a*). The properties of some of the prenatally lethal segmental aneuploids have also been examined. Although many are midgestational lethals, some, such as deficiency of the middle segment ($\sim 10\%$) of chromosome 17 (Df17CD) or of the terminal segment of chromosome 7 (Df7F4), are lethal during the preimplantation period (Baranov, 1983*b*). Unfortunately, these results are difficult to interpret since other imbalances (duplications) were present simultaneously.

Chimeras and mosaics

In addition to the purely morphological studies described earlier in this chapter, the full trisomics and, to a lesser extent, monosomics have been used for a number of other studies. Some, on gene dosage effects and the cell cycle, have already been described in Chapters 4 and 9, respectively. Others, on trisomy 16, will be discussed in detail in Chapter 12.

The formation of chimeras or mosaics was originally undertaken as a means for rescuing aneuploid cells, since, as has already been demonstrated, none of the mouse monosomics and only one of the trisomics are compatible with postnatal survival. Having living juvenile and adult animals with aneuploid cells and tissues would make possible a variety of investigations, particularly functional in nature, which are not possible with fetal material. Two different types of chimeras or mosaics have been prepared. The first, radiation chimeras, have been made by implanting hematopoietic stem cells from fetal trisomic liver into irradiated hosts. This results in the repopulation of the hematopoietic (including lymphoid) system with trisomic cells, so that all blood and lymphoid elements are trisomic although the remainder of the animal is diploid. The second type of chimera (the term "mosaic" is sometimes used synonymously) has been prepared by aggregating aneuploid and diploid preimplantation embryos and then transferring these aggregates back into pseudopregnant females. This method can be used with both monosomic and trisomic embryos, and the resulting chimeras may have a mixture of aneuploid and diploid cells in any or all tissues. These chimeras are, therefore, formally equivalent to human aneuploid/diploid mosaics. For the mouse, the usual designations of radiation and aggregation chimeras are, respectively, aneuploid (Ts) $\rightarrow$ diploid(2n) and aneuploid (Ts or Ms) $\leftrightarrow$ diploid(2n).

Radiation chimeras

All of the work on radiation chimeras reported to date has been carried out by Gropp and his collaborators. Such chimeras have been prepared with he-

Table 10.4. Phenotypes of viable mouse segmental aneuploids

Translocation	Aneuploid segment	Hypoplasia	Fertility M	Fertility F	Other features	References
Duplications						
T(14;15)6Ca	15(cent→B3) + 14(telomere)	+	−	+	Nervous, shaky; reduced viability	Cattanach (1967); Eicher (1973)
T(1;17)190Ca	17(cent→B) + 1(telomere)	+	−			Lyon, Sayers, and Evans (1978)
T(7;X)1Ct	7(C→E3)		+	+	Males reduced in viability	Eicher (1967, 1970)
T(5;12)31H	15(cent→B) + 12(Fl→tel)	+	−	+		Beechey, Kirk, and Searle (1980)
T(X;1)38H	11(E1→tel) + X(cent→A2)		−	+	Abnormal gait; skeletal variability (exencephaly in utero)	Searle et al. (1983)
T(16;17)43H	17(cent→B) + 16(cent→A)	+	+	+		Forejt, Capková, and Gregorová (1980); Gregorová, Baranov, and Forejt (1981)
T(1;13)70H	1(cent→A4) + 13(D1→tel)	+	+	+	Malformations of skull (short snout, malocclusion); hypotonia	Boer (1973); Boer and Groen (1974); Boer, van der Hoeven, and Chardan (1976); Boer and Branje (1979); Buul and Boer (1982)
T(-;-)158H	?		−	→	Abnormal postural reflexes	Lyon and Meredith (1966)
T(-;-)194H	?		−	→		Lyon and Meredith (1966)

T(10;13)199H	10(cent→C1) – 13(cent→A1)	+	+		Lyon and Meredith (1966); Washburn and Eicher (1975)
T(2;4)1Sn	2(H3→H4)	+		Exencephaly	Washburn and Eicher (1975)
T(1;13)70H + *T(1;13)1Wa*	1(A4→C2 or C3)[a]			No phenotypic abnormalities (not karyotyped	Boer and Wauben-Penris (1984)
Deletions					
T(5;12)31H	15(cent→B) + 12(F1→tel)	+	–	+	Skeletal fusions	Beechey, Kirk, and Searle (1980)
T(-;-)194H	?	–	–	Broad head between eyes	Lyon and Meredith (1966)

From Epstein *et al.* (1984); Epstein (1985*a*).

[a]Males carrying two overlapping translocations involving the same chromosomes were used to generate progeny aneuploid for the region of overlap in chromosome 1.

matopoietic cells from the livers of trisomies 12, 13, 14, 16, 18, and 19 (Gropp *et al.*, 1983), and successful repopulation has been obtained with cells from trisomies 12, 14, 18, and 19 (Herbst *et al.*, 1981; Pluznik *et al.*, 1981; Dyban, 1982; Gropp *et al.*, 1983). However, successful repopulation does not mean that the trisomic hematopoietic systems are necessarily completely normal. Thus, engrafted trisomy 12 stem cells showed full histological maturation in erythroid, granulocytic, and lymphoid cell lines, and were even devoid of the very mild ultrastructural and functional abnormalities that characterized trisomy 12 stem cells in affected fetuses (Pluznik *et al.*, 1981; Gropp *et al.*, 1983). However, when trisomic stem cells from Ts12 → 2n radiation chimeras were secondarily transferred to new irradiated hosts along with normal diploid stem cells, a procedure which could be done twice serially, they appeared to be at a developmental or proliferative disadvantage (Hongell, Herbst, and Gropp, 1984). Furthermore, stimulation of the chimeric trisomy 12 lymphocytes in vitro with lipopolysaccharide or in vivo with sheep red blood cells resulted in lower than normal proliferative and antibody responses, respectively (Sellin, Schlizio, and Herbst, 1982). Although these lymphoid defects were not observed in trisomy 19 lymphoid cells from radiation chimeras (Pluznik *et al.*, 1981; Sellin, Schlizio, and Herbst, 1982), such chimeras took longer than normal to develop erythroleukemia after exposure to Rauscher leukemia virus (Herbst *et al.*, 1982*b*). Nevertheless, despite such relatively subtle functional defects, the Ts12 → 2n and Ts19 → 2n radiation chimeras were capable of surviving for longer than seven months after transplantation, as were also Ts1 → 2n and Ts18 → 2n chimeras (Gropp *et al.*, 1983).

In contrast to the effective repopulation of irradiated hosts with stem cells from the several trisomies just discussed, transfer of fetal trisomy 13 or 16 stem cells is much less successful. Fetal trisomy 13 liver cells are unable to restore hemopoiesis and lymphopoiesis normally and have a proliferative disadvantage when injected with normal fetal liver cells (Gropp, 1982; Herbst *et al.*, 1982*a*; Hongell, Herbst, and Gropp, 1984). Similar problems are also found with trisomy 16 stem cells, and these results will be discussed in detail in Chapter 12.

Aggregation chimeras

The second type of chimera that can be prepared with aneuploid cells has a chimeric or mosaic composition in most or all tissues. Such chimeras are generally prepared by aggregating preimplantation mouse embryos using the technique pioneered by Mintz (1971), and the chimeras to be discussed here were prepared in this manner. However, another method for preparing chimeras is based on the injection of early embryonic inner cell mass cells of one type, in this instance aneuploid, into the cavity (blastocele) of blastocysts of another type (diploid) and then transferring these injected blastocysts into

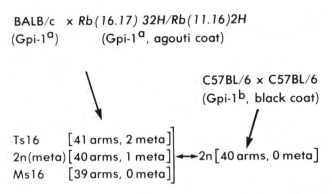

Fig. 10.4. The preparation of aneuploid ↔ diploid aggregation chimeras, illustrated here for aneuploids of chromosome 16. The figures in the brackets refer to the total number of chromosome arms (acrocentric chromosomes have one, and metacentric chromosomes two) and the number of metacentric chromosomes per cell.

pseudopregnant females (Gardner, 1968; Moustafa and Brinster, 1972). The only Ts ↔ 2n chimeras of this type were prepared with trisomy 12 cells (Ts12 ↔ 2n), but their properties have not been reported (A. Gropp, personal communication).

In the preparation of aneuploid ↔ 2n aggregation chimeras, embryos derived from the desired double-heterozygote mating (Fig. 10.1) are aggregated one to one and blindly, in so far as their chromosome constitution is concerned, with diploid embryos of an appropriate type. The resulting chimeras will, therefore, be of three types: Ts ↔ 2n, Ms ↔ 2n, and 2n(meta) ↔ 2n (Fig. 10.4). The last of these are chimeras in which both components are diploid, but the component from the double-heterozygote mating is recognizable by the presence of a single metacentric chromosome and a total of 40 chromosome arms. Both the chromosomally normal animal in the double-heterozygote mating and the known diploid component of the chimera are selected so that the two components of resulting chimeras will display genetic markers which make it possible to distinguish them from one another and therefore to quantitate their relative contributions. Such markers almost invariably include alleles for electrophoretic variants of an enzyme, such as glucose phosphate isomerase (GPI), which is present in all tissues and is readily quantifiable (Fig. 10.5). In some instances, coat color can also be used as a marker (Fig. 10.6), as can a variety of other enzymes, proteins, and antigens (such as the *H–2* products).

The preparation of aggregation chimeras is quite "labor intensive." Although the yields are reasonable, a much greater number of aggregates must be prepared than are finally recovered as chimeras late in gestation or after birth. An appreciation of the work involved may be gained by examining

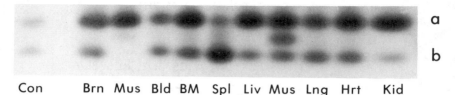

Con   Brn Mus Bld BM Spl Liv Mus Lng Hrt   Kid

Fig. 10.5 Glucose phosphate isomerase (GPI) patterns in tissues of a Ts17 ↔ 2n chimera. The trisomic component is represented by GPI–1B (designated b). The control column contains equal concentrations of GPI–1A and 1B. Note the hetero-dimeric GPI–1AB in the sample of muscle (Mus) on the right.

Fig. 10.6. Coat-color chimerism of Ts17 ↔ 2n chimeras. The dark pigmentation represents areas of trisomic pigment cells in the coat.

Table 10.5, in which data on the preparation of three types of Ts ↔ 2n chimeras are presented. It is clear from these data that the major difficulty is in obtaining a high rate of successful pregnancies after transfer of the aggregates. However, once this barrier is overcome, the frequency of chimerism is

Table 10.5. *Preparation of fetal Ts ↔ 2n aggregation chimeras*

	Ts15 ↔ 2n	Ts16 ↔ 2n	Ts17 ↔ 2n
Number of aggregates transferred	559	588	1538
Term or liveborn progeny (% of aggregates)	96(17%)	94(16%)	162(11%)
Chimeras (% of progeny)	40(42%)	53(56%)	56(35%)
Ts ↔ 2n chimeras (% of chimeras)	4(10%)	7(13%)	8(14%)

Data from Epstein *et al.* (1982*b*), Cox *et al.* (1984), and Epstein, Smith, and Cox (1984).

reasonably high (42% overall), but the number of Ts ↔ 2n chimeras, while not too far from expectation, may be somewhat low. The observed frequencies of the latter range from 10 to 14% (mean 13%) of all recognizable chimeras (Table 10.5), while the observed frequencies of embryos trisomic for the same chromosomes during preimplantation development ranged, in our own hands (Table 10.2), from 7 to 27% (mean 18.3%). Although the data in the two tables are not directly comparable, it does appear that fewer term and living Ts ↔ 2n chimeras are obtained, especially if it is recognized that all Ms ↔ 2n chimeras either die by this time or, more likely for many, lose their monosomic cell population.

Aggregation chimeras viable at term that incorporate cells trisomic for chromosomes 15, 16, 17, and 19 have been successfully prepared. The tissue compositions of the first three of these are summarized in Table 10.6. In all series, the overall proportions of trisomic cells in the Ts ↔ 2n chimeras are lower than the proportions of the genetically related (sib) 2n(meta) cells in the control 2n(meta) ↔ 2n chimeras. The difference is most striking for Ts17 ↔ 2n, in which the frequency of trisomy 17 cells is about half of that of the corresponding 2n(meta) cells. These figures are, of course, means, and thus hide the wide variations in tissue compositions among different chimeric individuals. An example of this variation is shown in Fig. 10.7 for Ts15 ↔ 2n chimeras.

These differences may come about in two ways. One is that there is a threshold above which the frequency of trisomic cells leads to developmental abnormality and embryonic or, more likely, fetal death. The existence of such a threshold has been recognized in aggregation chimeras prepared with embryonal carcinoma stem cells and diploid embryos (Fujii and Martin, 1983). However, the fact that recognizable phenotypic abnormality is not necessarily

Table 10.6. *Tissue compositions of aneuploid ↔ diploid chimeras*[a]

	Fetal chimeras						Adult chimeras			
Tissue	Ts15 (4)[b]	2n(meta) (3)	Ts16 (7)	2n(meta) (9)	Ts17 (2)	2n(meta) (5)	Ts16 (4)	2n(meta) (8)	Ts17 (4)	2n(meta) (5)
Kidney	40%	46%	60%	63%	30%	72%	40%	40%		
Heart	40	54	51	58			30	40		
Liver	38	58	53	62	30	68	40	50		
Brain	43	54	63	58	40	83	40	50		
Thymus	15[c]	60	36[c]	64			20[c]	60		
Spleen	38[c]	63	78	73			20[c]	70		
Blood	35	48					10[c]	70	31%[c]	74%
Limbs, tails, or fibroblasts	23	53	63	70						
Placenta	48	56								
Lung			53	61			30	50		
Coat							30[c]	60	50[c]	79
All tissues	36	55	57	64	33	74	29	54	41	77

Data from Epstein *et al.* (1982*b*); Cox *et al.* (1984); Epstein, Smith, and Cox (1984).

[a]Only tissues for which both Ts and 2n(meta) values were both available are included in this table. In each column, the proportion of the trisomic (Ts) or the matched 2n(meta) component, respectively, of sets of Ts ↔ 2n and 2n(meta) ↔ 2n chimeras is shown.

[b]Number of chimeras analyzed.

[c]Significant difference between trisomic and 2n(meta) proportions at $p \leq 0.05$ by Wilcoxon rank sum test.

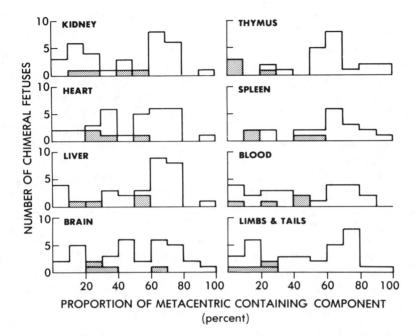

PROPORTION OF METACENTRIC CONTAINING COMPONENT
(percent)

Fig. 10.7. Left: Distribution of proportions of Ts 15 and 2n(meta) components in kidney, heart, liver and brain of Ts ↔ 2n and 2n(meta) ↔ 2n chimeras, respectively, as estimated by analysis of glucose phosphate isomerase compositions of tissue extracts. The Ts 15 and 2n(meta) distributions do not differ significantly. The Ts 15 values are represented by the crosshatched bars and the 2n(meta) values by the open bars. Right: Distribution of proportions of Ts 15 and 2n(meta) components in thymus, spleen, blood, and mesenchyme of Ts 15 ↔ 2n and 2n(meta) ↔ 2n chimeras. The Ts 15 proportions are significantly reduced in thymus, spleen, and mesenchyme. Ts 15, crosshatched bars; 2n(meta), open bars.

associated with a high frequency of aneuploid cells argues that mechanisms other than a general threshold effect of trisomic cells may be operative (Epstein, Smith, and Cox, 1984).

The other explanation for the low proportion of trisomic cells in the Ts ↔ 2n chimeras is that the trisomic cells in vivo are at a proliferative disadvantage relative to diploid cells. If this is the case, then the fact that the trisomic cells occur in the proportions as high as they do may be partially attributable to the fact that the genetic background of the mating which produces them is favorable. In all Ts ↔ 2n series in Table 10.6, the 2n(meta) proportions in the 2n(meta) ↔ 2n chimeras are higher than 50%. This is most striking in the trisomy 17 series in which the 2n(meta) proportion is 74–77%.

In addition to the fact that trisomic cells occur in lower than control frequencies is the finding that the decreases in proportion are not uniform for all tissues. As the data in Table 10.6 illustrate, certain trisomic cell populations may be particularly vulnerable. Thus, the trisomic contributions to lymphoid and hematopoietic tissues and coat in Ts16 ↔ 2n chimeras and the lymphoid tissues in Ts15 ↔ 2n chimeras are much lower than are their contributions to other tissues. This discrepancy becomes even more striking in chimeras alive postnatally, and it is sometimes impossible to detect the trisomic component in the red cells or lymphocytes of Ts15 ↔ 2n and Ts16 ↔ 2n chimeras. This issue will be discussed further in Chapter 12.

The compositions of the placentas in the Ts ↔ 2n chimeras are of interest. In the Ts15 ↔ 2n and control 2n(meta) ↔ 2n chimeras, the placentas tended to be composed almost entirely of only one cellular component. Two Ts15 ↔ 2n chimeras had placentas which were 90–100% trisomic in composition, and these placentas were significantly increased in weight [0.327 and 0.220 g in comparison to 0.124 ± 0.027 g (SD) for control 2n(meta) ↔ 2n placentas] (Epstein, Smith, and Cox, 1984). Since these chimeras were viable to term and phenotypically normal, it does not seem possible to ascribe the early demise of trisomy 15 fetuses (Table 10.1) primarily to placental insufficiency.

The general physical normality and freedom from gross neurological deficits of the Ts ↔ 2n chimeras is striking, since much of our experience with human mosaics is, of course, based on their ascertainment because of abnormalities. Of the Ts15 ↔ 2n chimeras examined at term, only one appeared abnormal and had a somewhat rounded appearance with a short snout and short neck (Epstein, Smith, and Cox, 1984). Similar characteristics of the face and head were also observed in two Ts15 ↔ 2n chimeras followed during postnatal life. Similarly, among the fetal Ts16 ↔ 2n chimeras, one was grossly edematous and had a phenotype indistinguishable from that of non-chimeric trisomy 16 fetuses (Cox *et al.*, 1984). Unexpectedly, the proportion of trisomy 16 cells in this fetus was the lowest of any of the fetal Ts16 ↔ 2n chimeras examined. The postnatal Ts16 ↔ 2n chimeras were normal in weight and, except for unilateral microphthalmia of uncertain significance in one, free of congenital abnormalities. None of the Ts17 ↔ 2n chimeras, fetal or liveborn, had any discernible abnormalities, except for a kink in the tail of one (Epstein *et al.*, 1982*b*). General growth in terms of weight was just in or below the lower range of their 2n(meta) ↔ 2n littermates (Epstein *et al.*, 1984).

The longevity of the Ts ↔ 2n chimeras has not been studied systematically, since most are killed at times convenient for analysis. However, some Ts16 ↔ 2n and Ts17 ↔ 2n animals have been followed for a year or more after delivery (Epstein *et al.*, 1982*b*; Cox *et al.*, 1984). Although some were able to breed successfully, none of our living Ts15 ↔ 2n, Ts16 ↔ 2n, or Ts17 ↔ 2n chimeras were able to contribute the genome from their trisomic cell population to progeny (Epstein *et al.*, 1982*b*; Cox *et al.* 1984).

Because of the way in which they were prepared (Fig. 10.4), several of the fetuses and mice in the three chimera series just described could have been monosomy ↔ diploid chimeras. However, of a total of 149 chimeras examined (Table 10.5), only one (0.7%) was a Ms ↔ 2n chimera, in contrast to 19 (12.8%) Ts ↔ 2n chimeras. This chimera, which was found in the Ts15 ↔ 2n series, contained a monosomy 3 component in addition to both 2n(meta) and 2n components (Epstein, Smith, and Cox, 1984). It should be noted that none of the metacentric chromosomes used contained chromosome 3 (Fig. 10.4), and the monosomic cell line is presumed to have derived from the 2n component of the chimera. In a culture of tail fibroblasts, 54% of the mitotic spreads were monosomic. Excepting this rare and still unexplained occurrence, Ms ↔ 2n chimerism has been demonstrated only during the first part of gestation and has been systematically investigated only for monosomy 19. Of 722 aggregates prepared, 12% implanted and grew and 82% were chimeric at day 9 (Magnuson, Smith, and Epstein, 1982). This frequency of chimerism is, of course, much greater than has been observed at term in any of the other series (Table 10.5). Of the karyotypable chimeras, 13% were Ts19 ↔ 2n and 7% Ms19 ↔ 2n. While the aneuploid component was usually >50% in the former, it was always less than 50% in the five Ms19 ↔ 2n chimeras analyzed, all of which were normal in appearance.

In addition to the several series of aneuploid ↔ diploid chimeras prepared in my laboratory, aneuploid ↔ diploid and polyploid ↔ diploid chimeras have also been prepared by other investigators. Lu and Markert (1980) described the formation of two viable putative tetraploid ↔ diploid chimeras. The presence of tetraploid cells was documented in one animal which was 50% chimeric in the coat and 3% in blood lymphocytes. This animal did not gain weight after weaning and was sacrificed at 27 days. The other possible 4n ↔ 2n chimera, which grew normally, was said to be 10% tetraploid in the coat. Several chimeras made by injecting teratocarcinoma cells into normal blastocysts have been shown to be aneuploid ↔ diploid in character. Most of the teratocarcinoma cells used in the preparation of such chimeras demonstrated both sex chromosome (XO) and autosomal aneuploidy, the latter including trisomies 6; 6, 8, and 10; 11 (with a deletion of 14); and 19 (also with a deletion of 14) (Iles and Evans, 1977; McBurney, 1976; Nicolas *et al.*, 1976; Dewey *et al.*, 1977; Cronmiller and Mintz, 1978; Papaioannou *et al.*, 1978). In most instances both the frequency and degree of chimerism were low (Papaioannou *et al.*, 1978).

These studies, particularly those involving trisomy 6,XO teratocarcinoma cells (Dewey *et al.*, 1977; Cronmiller and Mintz, 1978), are complementary to ours in demonstrating that a variety of trisomic cells can be "rescued" and become fully differentiated and functional. Therefore, even though some of the trisomies, especially 15 and 17, are associated with quite early lethality in utero, all of the results are compatible with the inference that even the most severe trisomies are not associated with generalized cell lethality. Such lethality is presumed to be the result of disruption of processes, such as intermediary metabolism and energy production, which are essential for individual cell survival (Magnuson *et al.*, 1982). If such defects do occur in trisomic cells, they are apparently capable of being corrected by normal diploid cells. Similarly, diploid cells are also capable of correcting the biochemical and physiological abnormalities which interfere with normal differentiation and morphogenesis in completely trisomic fetuses. Such correction is presumably effected by the passage of soluble substances from normal to trisomic cells, or perhaps by direct contact of these two classes of cells. However, as the work on trisomy 16 and on Ts16 ↔ 2n chimeras clearly shows (Chapter 12), such correction does not necessarily occur in every tissue even if contact is maintained throughout gestation.

The situation with monosomic cells is more problematic. Experience with monosomy 19 indicates that this condition, like the several trisomies studied, also does not cause generalized cell lethality, although it may seriously interfere with cell proliferation. However, in view of the very early time of death of all of the monosomies, it remains to be demonstrated whether this conclusion applies to all. In addition to the formation of other types of Ms ↔ 2n chimeras, another approach may be to prepare teratocarcinoma or embryonal carcinoma cell lines from monosomic preimplantation embryos. This approach was successfully used with embryos homozygous for the t^{w5} mutation which, like the monosomies, acts as an early embryonic lethal mutation (Magnuson *et al.*, 1982).

Teratocarcinoma cells

In the previous section, mention was made of the use of teratocarcinoma cells in the preparation of chimeras, either by aggregation or more commonly by injection. A considerable literature now exists on the potential usefulness of the mouse teratocarcinomas as systems for studying both embryogenesis and oncogenesis, and these concepts are reviewed by Mintz, Illmensee, and Gearhart (1975), Martin (1975, 1980), Strickland (1981), Mintz and Fleischman (1981), and Silver, Martin, and Strickland (1983). In addition to their ability to be successfully incorporated into chimeras, as described earlier, they can also be used to study differentiation in vitro. Under appropriate culture conditions, the undifferentiated teratocarcinoma cells, carried in the form of embryonal carcinoma stem cell lines, will give rise to a variety of differentiated cell types (Mintz, Illmensee, and Gearhart, 1975; Martin, 1980). However, tissue organization is far from normal, particularly as differentiation proceeds, and the particular pathway of differentiation is not predictable.

The latter problem has to some extent been overcome in two ways. The first is by the isolation of cell lines that differentiate in restricted paths. The second is by the use of specific agents, such as retinoic acid and cyclic AMP, which can direct differentiation along certain directions (Strickland, Smith, and Marotti, 1980; Adamson and Grover, 1983; M. Sherman *et al.*, 1983; McBurney *et al.*, 1983). However, given the many difficulties inherent in these types of studies, it does not appear that it will be fruitful, at least at present, to apply these approaches to a general study of the effects of aneuploidy.

How then might teratocarcinoma cells be useful in the investigation of mammalian aneuploidy? I can suggest two possibilities. The first is for the investigation of a variety of properties of aneuploid and normal control cells in permanent and presumably equivalent cell lines. A real difficulty in work with cells cultured from trisomic mouse embryos and fetuses is their lack of permanence, when compared with even human diploid skin fibroblasts, and their tendency, if they do survive, to transform rapidly and become heteroploid. Tetratocarcinoma cells, maintained in the form of embryonal carcinoma stem cells, could potentially overcome these difficulties, and the usefulness of such lines for studies on X-chromosome expression and inactivation has been well demonstrated (Martin *et al.*, 1978). Aneuploid and polyploid cells of all types could be analyzed by this approach, and a variety of issues relating to gene dosage effects and their effects on cellular metabolism, structure, and adhesion investigated. In the case of polyploid cells, questions relating to the relationships between the synthesis of gene products, cell volume, and cell surface area could be examined in detail, as could the process and effects of X-chromosome inactivation. In theory, it would be possible to prepare a series of teratocarcinoma cells covering the entire range of mouse autosomal trisomy so that imbalance of all parts of the genome could be examined. Coupled with knowledge of the genetic map, this would permit a systematic investigation of the effects of unbalancing any of the large and continuously increasing number of biochemical (structural gene) loci that have been mapped in the mouse.

The second possible area of usefulness of teratocarcinoma cells in the study of aneuploidy is for the investigation of the effects of autosomal monosomy. Although tissue differentiation and organogenesis are disorganized in teratocarcinomas growing in vivo or in vitro, the earliest stages of embryonic development are reasonably well duplicated in embryoid bodies formed by aggregated embryonal carcinoma stem cells (Martin, Wiley, and Damjanov, 1977; Martin, 1980). Since, as is discussed earlier in this chapter, this is the time in development at which most embryos with autosomal monosomies appear to die, monosomic teratocarcinoma cell lines could be quite useful for studying the nature of the deleterious effects of monosomy.

An approach of the type contemplated here has two principal requirements – that it be possible to prepare teratocarcinoma cell lines from aneuploid or polyploid preimplantation embryos and that these lines be chromosomally stable. Most of the teratocarcinoma lines that have been prepared and investigated to date have either appeared spontaneously or were induced by transfer of early embryos to ectopic sites. These lines are almost invariably aneuploid or heteroploid [see Appendix Table 1 in Silver, Martin, and Strickland (1983) for summary of karyotypes of teratocarcinoma cell lines], although one line (designated METT–1) considered to be euploid was prepared by Mintz and Cronmiller (1981). This line can be carried in the form of embryoid bodies and is capable of both somatic and germinal differentiation when injected into blastocysts to form chimeras.

Recently, a new approach to the preparation of teratocarcinoma cell lines was described, one in which preimplantation embryos are placed directly into culture (Martin, 1981; Evans and Kaufman, 1981). The cell lines that have been obtained are pluripotent and are generally (Evans and Kaufman, 1981) but not invariably (Martin, 1981; Magnuson *et al.*, 1982) euploid. However, it is likely that euploidy could be maintained by cytogenetically monitoring and cloning the cell line at appropriate times. While there should not be any difficulty beyond that associated with normal embryos in initiating such lines from trisomic or polyploid embryos, the situation with monosomic embryos is more in doubt. The new methods for establishing teratocarcinoma cells make use of late-stage blastocysts capable of attaching in vitro (Evans and Kaufman, 1981) or of inner cell masses derived from such blastocysts (Martin, 1981). Therefore, the question with regard to monosomic embryos is whether they would remain viable long enough to reach the stage in development necessary for teratocarcinoma stem cell imitation. The data presented in Table 10.2 suggest that at least some of the mouse monosomies might – particularly monosomies 12 and 18, monosomy 19 generated with the most recently used breeding stocks, and especially monosomy 14 for which some viable day 6 implants have been detected. Further work will be required to determine whether such stem cell lines can be made.

11

Models for human aneuploidy

In the previous chapter, I discussed mouse aneuploidy as a system in its own right, one capable of being used for a variety of studies of the effects of aneuploidy in mammals. At this point, it is worth considering explicitly the advantages of using such a system, many of which will have already become obvious. The major theoretical advantages to using animal models in general and the mouse in particular as experimental systems for studying the pathogenesis of aneuploid states have been discussed in earlier articles on the subject (Epstein, 1981a, 1985a; Epstein et al., 1984). Among the benefits expected to be derived from using animal models are the following:

1. Animal models facilitate the developmental analysis of the pathogenesis of abnormalities, particularly during crucial stages of organogenesis. Even if appropriate tissues were available in man, it would be impossible to carry out any investigations other than static biochemical or morphological examinations. In the mouse, studies of neurological and cardiac abnormalities have been of particular interest.
2. The in vitro and in vivo investigation of cells and tissues other than blood elements and fibroblasts becomes possible. This applies particularly to the central nervous system, the system most sensitive to the effects of aneuploidy in man but the most difficult to get at.
3. Aneuploid animals allow for the study of the preimplantation and early postimplantation stages of development which are particularly critical for an understanding of the effects of monosomy, as well as of trisomies and polyploidies which lead to early death in utero.
4. The effects of genetic and environmental factors on the phenotype of the aneuploid state can be assessed. A controlled analysis of these factors would not be possible in man.
5. By preparing and analyzing chimeras, it is possible to assess which abnormalities are cell autonomous and which are correctable by exposure to or contact with normal cells.
6. Chromosomally unbalanced animals facilitate the performance of gene dosage studies by permitting better control of experimental conditions, by making possible the use of genetic markers, and by permitting analysis of gene products that are restricted in cellular distribution. Such dosage stud-

ies are the first step in the investigation of the direct molecular consequences of aneuploidy.

7. It is possible to investigate aneuploidy of all parts of the genome in a
   systematic manner. In the mouse, this clearly applies to all whole-arm
   trisomies and, at present, to a limited number of duplications. This is in
   contrast to the situation in man, in which certain types of aneuploidy are
   rarely or never observed and, as a result, cannot be studied even in cultured cells except on an opportunistic basis.
8. Aneuploid animals may ultimately provide a test system for proposed
   therapeutic approaches. At present, no conceptual basis for therapy exists
   and empirical approaches have proved unrewarding (Share, 1976; Karp,
   1983; Smith *et al.*, 1984).

Information presented in the preceding chapters of this volume testify to
the fact that many of these theoretical expectations have already begun to be
realized. However, since two of the major reasons for studying aneuploidy
are to understand how it causes abnormalities in man and what, if anything,
can be done to prevent or ameliorate these untoward consequences, I shall
now turn to a consideration of the development of animal models for specific
human chromosomal disorders.

Comparative gene mapping

If a nonhuman system is to be used as a model for human aneuploidy, it is
necessary that there be homology between the unbalanced regions of the genome in the animal model and in the human disorder for which it is to be a
model. Demonstration of such homology is based on comparative gene mapping. In addition to the cytogenetic analysis and mapping of the human chromosomes, similar studies have been carried out in a variety of other mammals, most notably the primates, cattle, sheep, cat, dog, Chinese hamster,
rat, and, of course, the mouse (Minna, Lalley, and Francke, 1976; Lundin,
1979; Roberts and Ruddle, 1980; Dalton *et al.*, 1981; Stallings and Siciliano,
1981; Nash and O'Brien, 1982; O'Brien and Nash, 1982; D'Eustachio and
Ruddle, 1983). The most recent summary of the interspecific companions of
the gene maps of a large number of species is contained in Human Gene
Mapping 7 (Roderick *et al.*, 1984).

Nonhuman primates

Both evolutionary and cytogenetic considerations would suggest that nonhuman primates would be the closest species to man genetically, and the comparative mapping data are in agreement (Dutrillaux, 1979; Roderick *et al.*,
1984). Aneuploid primates have been observed, and, as might be expected,
their phenotypes were not dissimilar from those of humans with the homolo-

gous condition. Most notable was a chimpanzee named Jama, who was identified because of her low birth weight, slow rate of growth, and the presence of several congenital malformations and other physical abnormalities. These included bilateral, partial syndactyly of the toes, clinodactyly, prominent epicanthic folds, protruding tongue, a short neck with excess skin folds, hypotonia, hyperextensibility, a congenital heart defect (patent ductus arteriosus, atrial and ventricular septal defects), and developmental retardation (McClure *et al.*, 1969; McClure, 1972). As might be expected from the phenotype, this animal, who died at 17 months of age while being anesthetized, had the chimpanzee equivalent of Down syndrome. This was confirmed cytogenetically by the observation of 49 rather than 48 chromosomes (McClure *et al.*, 1969), and the extra chromosome was shown to be a small acrocentric chromosome homologous to the human chromosome 21 (Benirschke *et al.*, 1974).

A male gorilla with an extra chromosome 22 was reported by Turleau, Grouchy, and Klein (1972), and a female orangutan with an extra chromosome 22 by Andrle *et al.* (1979). The latter animal was retarded in growth and had a "more pronounced" upward slant to the lid axis, "imperfect" convolution of the ear, an open mouth, hypotonia, and reduced motor activity.

Other primates with chromosomal abnormalities have included a rhesus macaque with gonadal dysgenesis and a vertebral anomaly who had an XO karyotype (Weiss *et al.*, 1973) and three pigtailed macaques with either XXX or XX/XXX mosaicism (Ruppenthal *et al.*, 1983). These trisomy X animals had numerous developmental (skeletal), psychological (cognitive, social adaption), and neurological (motor, visual) abnormalities. They were considered to have the monkey equivalent of human mental retardation, which was apparently considerably more severe than is observed in 47,XXX human females (Robinson *et al.*, 1979; Stewart, Netley, and Park, 1982).

The similarities between the primates and humans with Down syndrome or its equivalent and with monosomy X (XO) indicate that, as expected, the genetic homologies are very great. They further speak to the issue of the specificity of the effects of at least autosomal aneuploidy, since the human phenotypes are remarkably preserved in the primates. Why the monkey version of XXX should be so much more severe than it is in humans is not obvious. However, despite their obvious genetic and phenotypic suitability as models for human aneuploidy, aneuploid primates are detected only rarely. Furthermore, primates are quite expensive to maintain and breed in captivity and are not always easy to handle. They do not, therefore, offer anything as a general model system for human aneuploidy.

Autosomal aneuploidy in the mouse

Of all of the mammalian species in which genetic mapping has been carried out, the mouse clearly offers the most to investigators interested in aneuploidy. It is low in cost and breeds rapidly. The genetic background and environment can be controlled, and appropriate genetic markers can be introduced as required. And, as has been discussed in the previous chapter, progeny with whole-chromosome aneuploidy can be produced in relatively high yields.

The ability to use the mouse for construction of specific human chromosome disorders is dependent upon the degree of homology between the human and mouse genomes. At first glance, there is little in the karyotypes to suggest that such homologies might exist (Fig. 11.1, see also Appendix). However, the comparative mapping of biochemical and other loci indicates that there are numerous sets of genes which are syntenic (present on the same chromosome) in both man and mouse. These syntenies, which are presumed to represent linked sets of genes (Nadeau and Taylor, 1984), contain between two and nine or more loci, and range in length from 1 to 24 centimorgans (cM). It has been suggested that as many as 60% of autosomal linkages may be conserved in mouse and man (Nadeau and Eicher, 1982). In a recent calculation based on data from 13 such syntenies, the average length of a conserved segment was estimated to be 8.1 ± 6.0(SD) cM (Nadeau and Taylor, 1984). These figures are likely to change as more syntenic groups are analyzed.

A list of the reported homologous syntenies in man and mouse is presented in Table 11.1. Examination of this table reveals, as might be expected, that with the possible exception of the X chromosome, there are no human and mouse chromosomes that are completely homologous with one another with regard to the genes that they carry. For example, genes from a single human chromosome arm, for example 1p, may be found on mouse chromosomes 3 and 4. Conversely, a single mouse chromosome may carry genes represented on two or more different human chromosomes – for example, mouse chromosome 7 and human chromosomes 11p, 15q, and 19. It is obvious, therefore, that there has been considerable intermixing of chromosome segments during evolution. In their calculations, Nadeau and Taylor (1984) estimated that 178 ± 39 exchanges distributed randomly through the autosomal part of the genome would explain the data available to them.

Consideration of the above facts make it clear that it will not be possible to duplicate in the mouse the genetic imbalances associated with specific human aneuploid states if we are restricted to working with whole-chromosome aneuploidy. While it may be possible to duplicate certain segmental aneuploidies reasonably well, the major limitation at present is in the availability of appropriate translocation stocks for generating the desired mouse segmental aneuploids. Nevertheless, even with the whole-chromosome aneuploidies – and in general this means only the trisomies – a model system would, under favorable circumstances, permit the generation of a mouse in which several

Fig. 11.1. Human (above) and mouse (below) karyotypes. Prepared by Drs. Cheng Zai-yu and Steven Schonberg.

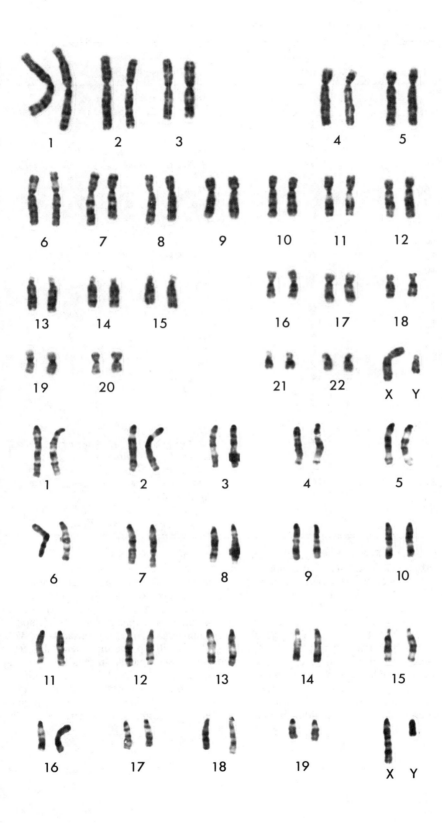

Table 11.1. *Homologous syntenies in man and mouse*

Human chromosome	Mouse chromosome	Loci (mouse symbols)[a]
1p	3	*Amy-1, Amy-2, Ngf*
1p	4	*Pgd, Gpd-1, Eno-1, Fuca, Ak-2, Pgm-2*
1q	1	*Pep-3, sph, Ren*
3p	9	*Acy, Bgl*
4	5	*Pgm-1, Pep-7, Alb-1, Afp*
6p	17	*H-2D, H-2I, C-4, C-2, H-2K, Glo-1, Bf, Sod-2*
6q	9	*Pgm-3, Mod-1*
7	5	*Gus, Mor-1, Asl, Psp*
7q	6	*Trp-1, Cpa*
9p	4	*Aco-1, Galt, Ifl*
9q	2	*Ak-1, Abl*
10	10	*Pyp, Hk-1,*
10	19	*Lipa, Got-1*
11	2	*Acp-2, Cs-1*
11p	7	*Ins-1, Hbb, Ldh-1, Hras-1*
11q	9	*Apl-1, Ups, Es-17*
12	15	*Ela-1, Int-1*
12p	6	*Gapd, Tpi-1, Kras-2, Ldh-2*
12q	10	*Cs, Pep-2, Ifg*
14q	12	*Igh, Pre-1*
14q	14	*Np-2, Tcra*
15q	2	*Sdh-1, B2m*
15q	7	*Idh-2, Fes*
15q	9	*Mpi, Pk-3*
16	8	*Got-2, Ctrb, Aprt*
17	11	*Myh* (3 genes), *Tk-1, Glk, Erba, Hox-2*
19	7	*Gpi-1, Pep-4, Lhb*
20	2	*Ada, Src, Itp*
21	16	*Sod-1, Ifrc, Prgs*
22q	15	*Arsa, Dia-1, Sis*
X	X	*G6pd, Hprt, Ags, Pgk-1, Tfm, Phk, Mdx, Sts, Spf, Hyp, Bpa, Mo*

Data from Roderick *et al.* (1984) and Buckle *et al.* (1984).
[a]Abbreviations: *Abl*, cellular homolog of Abelson murine leukemia virus oncogene; *Aco-1*, aconitase-1 (soluble); *Acp-2*, acid phosphatase-1 (kidney); *Acy*, aminoacylase-1; *Ada*, adenosine deaminase; *Afp*, α-fetoprotein; *Ags*, α-galactosidase; *Ak-1*, adenylate kinase-1 (soluble); *Ak-2*, adenylate kinase-2 (mitochondrial); *Alb-1*, serum albumin; *Amy-1*, amylase-1 (salivary); *Amy-2*, amylase-2 (pancreatic); *Apl-1*, apolipoprotein A-I; *Aprt*, adenine phosphoribosyltransferase; *Arsa*, arylsulfatase A; *Agl*, argininosuccinate lyase; *B2m*, β_2-microglobulin; *Bf*, properdin factor B; *Bgl*, β-galactosidase; *Bpa*, bare patches (chondrodysplasia punctata in man); *C-2*, complement factor 2; *C-4*, complement factor 4; *Cpa*, carboxypeptidase A; *Cs*, citrate synthase (mitochondrial); *Cs-1*, catalase-1; *Ctrb*, chymotrypsinogen B; *Dia-1*, NADH-diaphorase; *Ela-1*, elastase-1; *Eno-1*, enolase-1; *Erba*, esterase-17 (human esterase-A4); *Fes*, cellular homolog of feline sarcoma virus oncogene; *Fuca*,

genes present on a human chromosome of interest are similarly unbalanced in a single animal. At worst, it would permit several different animals to be generated, each of which reproduced the state of imbalance of at least one locus carried by an aneuploid human chromosome. Whereas the former would be preferable for studying the concomitant effects of unbalancing several genes at the same time, the latter would still allow the analysis of the effects of unbalancing individual genes of particular interest. In either situation, unless relatively short chromosome segments were involved (the size depending on the actual length of the conserved syntenies), there would of course also be imbalance of loci present on the mouse chromosome being manipulated that were not carried on the human chromosome for which the

Notes to Table 11.1. *(cont.)*

α-L-fucosidase-1; *G6pd*, glucose-6-phosphate dehydrogenase; *Galt*, galactose-1-phosphate uridylyltransferase; *Gapd*, glyceraldehyde-3-phosphate dehydrogenase; *Glk*, galactokinase; *Glo-1*, glyoxylase-1; *Got-1*, glutamate-oxaloacetate transaminase-1 (soluble); *Got-2*, glutamate-oxaloacetate transaminase-2 (mitochondrial), *Gpd-1*, glycerol-3-phosphate dehydrogenase; *Gpi-1*, glucose phosphate isomerase-1; *Gus*, β-glucuronidase; *H-2D*, histocompatibility locus region D (human *HLA-A*); *H-2I*, histocompatibility locus region I (human *HLA-I*); *H-2K*, histocompatibility locus region K (human *HLA-D*); *Hk-1*, hexokinase-1; *Hbb*, β-globin; *Hox-2*, homeo box sequence-2 (Rabin *et al.*, 1985; Joyner *et al.*, 1985); *Hprt*, hypoxanthine guanine phosphoribosyltransferase; *Hras-1*, cellular homolog of Harvey murine sarcoma virus oncogene; *Hyp*, hypophosphatemia; *Idh-2*, isocitrate dehydrogenase (mitochondrial); *Ifg*, interferon-γ (immune); *Ifl*, interferon-α (leukocyte); *Ifrc*, interferon receptor; *Ins-1*, insulin-1; *Int-1*, cellular homolog of mouse mammary tumor virus oncogene; *Itp*, inosine triphosphatase; *Kras-2*, cellular homolog of Kirsten murine sarcoma virus, *Ldh-1*, lactate dehydrogenase-1 (A); *Ldh-2*, lactate dehydrogenase-2 (B); *Lhb*, β-subunit of luteinizing hormone; *Lipa*, lysosomal acid phosphatase A; *Mdx*, muscular dystrophy; *Mo*, Menkes disease (mottled mutations in mouse); *Mod-1*, malic enzyme-1 (soluble); *Mor-1*, malate dehydrogenase (soluble, human *Mdh-2*); *Mpi*, mannose phosphate isomerase; *Myh*, myosin heavy chain; *Ngf*, nerve growth factor; *Np-2*, nucleoside phosphorylase-2; *Pep-2*, peptidase-2 (human peptidase B); *Pep-3*, peptidase-3 (human peptidase C); *Pep-4*, peptidase-4 (human peptidase D); *Pep-7*, peptidase-7 (human peptidase S); *Pgd*, 6-phosphogluconate dehydrogenase; *Pgk-1*, phosphoglycerate kinase-1; *Pgm-1*, phosphoglucomutase-1; *Pgm-2*, phosphoglucomutase-2; *Pgm-3*, phosphoglucomutase-3; *Phk*, phosphoglycerate kinase, liver; *Pk-3*, pyruvate kinase-3; *Pre-1*, α₁-antitrypsin; *Prgs*, phosphoribosylglycinamide synthetase; *Psp*, phosphoserine phosphatase; *Pyp*, inorganic pyrophosphatase; *Ren*, renin; *Sdh-1*, sorbitol dehydrogenase-1; *Sis*, cellular homolog of simian sarcoma virus oncogene; *Sod-1*, superoxide dismutase-1 (soluble), *Sod-2*, superoxide dismutase-2 (mitochondrial); *Spf*, ornithine carbamoyltransferase deficiency (sparse fur mutation in mouse); *sph*, spherocytosis (erythrocyte α-spectrin)(Huebner *et al.*, 1985); *Src*, cellular homolog of Rous sarcoma virus oncogene; *Sts*, steroid sulfatase; *Tcra*, T-cell receptor α-chain (Dembić *et al.*, 1985; Collins *et al.*, 1985); *Tfm*, testicular feminization; *Tk-1*, thymidine kinase-1; *Tpi-1*, triosephosphate isomerase-1; *Trp-1*, trypsin-1; *Ups*, uroporphyrinogen I synthase.

model is intended. This being the case, it is unlikely that the phenotype of an aneuploid mouse would be identical to that of the human aneuploid for which it is intended to be a model, especially if the morphological differences between the two species are also considered. However, this lack of phenotypic identity should not create a serious problem as long as the phenomena being studied in the model system are precisely identified and appropriately related to the human situation.

Three general approaches to using mouse models can be envisioned (Epstein *et al.*, 1984). The first is to investigate the metabolic, physiological, or developmental consequences of unbalancing specific loci. In the case of the full trisomies that are not liveborn, this can be carried out either in the trisomic fetuses, in trisomic hematopoietic and lymphoid cells in radiation chimeras, or in the trisomic cells present in liveborn aggregation chimeras. In the last instance it may actually be possible to isolate the aneuploid cells by building suitable genetic markers into the aneuploids (for example, surface antigens such as controlled by the *H–2* locus) and then separating the aneuploid from diploid cells with appropriate cell separation techniques.

The second use of the mouse models is to investigate phenomena that occur in both the human and the corresponding mouse aneuploids even though the specific locus or loci involved are unknown. In this approach, the assumption is made that if the phenomena are specific and reasonably similar in the human and the mouse – for example, a characteristic type of heart disease – then the underlying mechanisms are also likely to be the same or at least similar (Epstein, 1981*a*). This is, of course, a big assumption. It is based on the belief, the evidence for which has already been discussed, that the effects of aneuploidy are, in the main, specific consequences of the unbalancing of specific loci. Furthermore, even though a homologous synteny may at present be defined by as few as two loci, this does not mean that only two loci are unbalanced. If we take Nadeau and Taylor's (1984) figure of 8.1 cM as the mean length of homologous segment, and assume a total of 50,000 structural gene loci (McKusick and Ruddle, 1977) and a total map length of ~1600 cM or ~2700 cM, respectively, for mouse (Roderick and Davisson, 1981) and man (N. Morton *et al.*, 1982), the genetic content of the average syntenic segment would be of the order of 150–250 loci. For the longest segment, 24 cM, the number would be in the range of 450–750, and for the smallest, 1 cM, between 20 and 30. These estimates are, of course, quite crude, and may be serious underestimates. Chaudhuri and Hahn (1983) reported 150,000 different mRNA sequences in developing mouse brain. Nevertheless, the estimates do indicate that an aneuploid mouse model will reproduce the imbalance of substantial numbers of human loci. Therefore, if the same phenotypic effects are observed in the mouse as are found in the human, it seems reasonable as a working hypothesis to attribute them to the same mechanism. (It can of course be argued that the number of homologous loci will be only a small fraction of the number present on the unbalanced chromosome, thereby raising the possibility of spurious phenotypic associations rather than specific ones. I prefer the latter.)

Finally, mouse models are useful for the study of defects unique to themselves. In this instance they are no longer models for specific human disorders and would not illuminate the mechanisms operating in these conditions. However, their investigation would nevertheless be of great value for the general analysis of the genetic control of developmental processes.

For which human aneuploid conditions might the mouse trisomies provide models? In the following chapter, I shall discuss the development of mouse trisomy 16 as a model for human trisomy 21. This is at present the best example of the type of model I have been describing. At the phenotypic level, Gropp (1982) has suggested that mouse trisomy 1, with its holoprosencephaly and occasional cyclopia, may be a model for human trisomy 13 in which the same features may be observed. At present, there is nothing in the mouse maps (Table 11.1) to suggest that there is a biochemical basis for this pathological similarity (but nothing to deny it either). Aside from mouse chromosome 16, potential mouse homologs for the chromosomes involved in the other common full trisomies of man, chromosomes 13 and 18, have not been identified. Potential mouse homologs for human chromosomes 8 and 9 do exist, but trisomics for these chromosomes are unfortunately among the earliest to die in utero. Nevertheless, the presence of congenital heart disease in several of the mouse trisomies, as in many of the human aneuploidies, suggests that the mouse may be quite useful for studying the pathogenesis of this type of disease. As Gropp (1982) pointed out, although "no other 'homologous trisomies' in mouse and man are known at present . . . , the rapid progress of gene mapping may soon provide new hints. In the meantime it seems worthwhile to seek for insights into problems of human pathology, by the comparatively easy access to special organs in the murine trisomic systems."

Perhaps the most important use of the mouse models may be in the investigation of the development and function of the nervous system, the system that is ultimately of greatest concern in clinical practice. The question of whether all of the observed neurological and cognitive deficits in living aneuploid humans are anatomical in origin has never been satisfactorily answered, and may never be answered with human material alone. It is here that intensive study of the mouse aneuploids may be particularly valuable. These investigations should be directed not only at the neuropathology of the trisomies and viable segmental aneuploids, but also at the neurophysiology and neurochemistry of the living aneuploids (including trisomy 19) and of cells derived from both viable and inviable aneuploids. From the point of view of decreasing the burden of mental retardation, the identification of potentially correctable physiological (including metabolic and neurotransmitter) aberrations would be of paramount importance.

X-chromosomal aneuploidy in the mouse

The one exception to the high degree of genomic rearrangement in mammals is the X chromosome, which, although morphologically different in man and mouse, seems to have retained the same set of genes. This conservation of the X chromosome was originally stressed by Ohno (1967, 1979) and extends to both biochemical loci and loci related to specific disease states (Buckle *et al.*, 1984). One possible exception is the locus for steroid sulfatase (*Sts*) for which evidence both favoring (Gartler and Rivest, 1983) and opposing (Balazs *et al.*, 1982, no data presented) its location on the mouse X chromosome exists. The most recent suggestion is that there are functional *Sts* loci on both the X and Y chromosomes in the mouse (Keitges *et al.*, 1985). Further evidence in other species also supports [in wood lemmings, Ropers and Wiberg (1982)] or contradicts [in marsupials, Cooper *et al.* (1984)] its mapping to the X chromosome. Nevertheless, the high degree of homology between the human and mouse X chromosomes would suggest that useful mouse models of human sex-chromosome aneuploidy should exist.

The preparation and analysis of mouse 39,X (XO) females will be described in Chapter 13 and will not be further considered here. The remainder of this section will be concerned, therefore, with other sex chromosome anomalies in the mouse that could serve as models for human aneuploidy. Examples of X-chromosome aneuploidy in other mammalian species are discussed by Wurster-Hill, Benirschke, and Chapman (1983).

Unlike XO females, which are relatively common in mice (whether bred or occurring spontaneously), XXY males are distinctly uncommon. The first confirmed reports of such animals were by Cattanach (1961) and Russell and Chu (1961), who described, respectively, two and one sterile males which were capable of copulating with females. Their frequency in the progeny of normal matings was estimated as 0.02%, as opposed to 0.76% for XO (Russell and Chu, 1961; Cattanach, 1974; see also Russell, 1976). These animals are normal in size but have extremely small testes devoid of spermatogenesis (Russell and Chu, 1961). Interstitial cell hyperplasia develops during adult life, but the hyalinization and tubular atrophy typical of human XXY males does not occur in the mouse (Cattanach, 1974). Mouse XXY germ cells are apparently able to survive, as in man, for some time in the testes, but eventually do disappear (see Chapter 13).

A small number of normal-appearing XYY mice have also been described (Cattanach, 1974; Evans, Beechey, and Burtenshaw, 1978). Four were full XYY, two were XY/XXY mosaics, and the last was an XO/XYY mosaic with a XYY germ line (Evans, Ford, and Searle, 1969). Several of these animals were detected because of sterility, and the others were considered likely to be sterile (Cattanach, 1974). One, however, which was possibly an XY/XXY mosaic, may have sired two litters of embryos (Evans, Beechey, and Burtenshaw, 1978). In general, although spermatogonia were present, cell death occurred throughout spermatogenesis, so that few germinal elements survived meiosis. Three additional sterile XYY males were reported by Cacheiro and Generoso (1975). They were induced at a frequency of 0.5% by treatment of their fathers with 6-mercaptopurine.

The high rate of sterility in the XYY mice makes them analogous to humans with XYY, who were also found to have testicular abnormalities. In eight XYY males ascertained by institutional surveys (not because of infertility or problems with sexual development), various degrees of spermatogenic arrest were noted (Baghdassarian *et al.*, 1975); three had oligospermia. No reproducible hormonal abnormalities were found. The cause of the sterility in mice and the testicular abnormalities and presumed sterility in some cases of humans with XYY is unknown, but has been attributed by Burgoyne (1979) to spermatocyte death resulting from a disruption of sex-chromosome pairing.

Although spontaneously occurring XXY and XYY mice are, because of their very low frequencies, not very useful models for their human counterparts, there is another condition in the mouse which may provide a very useful model for these human conditions. The autosomal gene, sex reversal (*Sxr*), causes chromosomally XX (female) animals to develop as phenotypically normal but sterile males. Like the mouse XXY males, the sex-reversed females are capable of copulating normally, but the testes of older animals are generally devoid of germ cells and develop interstitial hyperplasia (Cattanach, 1974). Although primordial germ cells are present in fetal testes, many degenerate by the time of birth, and none is present by 10 days of age (Cattanach, Pollard, and Hawkes, 1971). XY males carrying *Sxr* were found to have smaller than normal testes and were occasionally sterile. XO males with *Sxr* are phenotypically normal males with reasonably normal spermatogenesis and spermiogenesis. However, they appear to be sterile and have grossly abnormal and immobile sperm (Cattanach, Pollard, Hawkes, 1971).

Although the possibility that *Sxr* involved the translocation of a portion of the Y chromosome to an X chromosome was considered at the time of its original description (Cattanach, Pollard, and Hawkes, 1971), cytological evidence for this could not be obtained for over ten years (Chandley and Fletcher, 1980). However, with the development of a recombinant DNA probe for a minor satellite DNA component (designated Bkm after the banded krait, an Indian snake, from which it was prepared) related to mouse Y chromosome sequences, it could be shown by in situ hybridization that one X chromosome of sex-reversed XX mice carries Bkm-reactive sequences at the distal end (Singh and Jones, 1982). In male (XY) carriers, however, the Y chromosome was shown to possess, in addition to a proximal Bkm-hybridizing region (the presumed normal site of the testis-determining region) (Singh and Jones, 1982) a second cytologically similar region at the distal end of the chromosome (Evans, Burtenshaw, and Cattanach, 1982). There-fore, at meiosis during spermatogenesis, the aberrant Bkm-reactive region on the distal part of the Y chromosome is translocated to the distal part of the X chromosome by crossing-over. In fact, Burgoyne (1982) postulated that such a crossover is obligatory even in normal male meiosis (see Fig. 13.3). The *Sxr* factor or region is thus identified with the Bkm-reactive and testis-determining region, Y(Td), of the Y chromosome.

On the basis of the genetic and cytologic findings, the simplest interpreta-tion is that XX (*Sxr*/ +), XY (*Sxr*) and XO (*Sxr*) males are equivalent, re-spectively, to XXY, XYY, and XY males insofar as the translocated segment of the Y chromosome is concerned. The only real difference is between XX (*Sxr*/ +) and XXY, in that the translocated segment of the Y present on the X chromosome is subject to X inactivation, at least some of the time. That this is the case was demonstrated when inactive *Sxr*-bearing X chromosomes were preferentially selected for, and the resulting XX (*Sxr*/ +) animals were found to be fertile females (Cattanach *et al.*, 1982; McLaren and Monk, 1982). Therefore, XX (*Sxr*/ +), while a reasonable and useful approximation of mouse XXY and secondarily of human XXY, is not to be considered as nec-essarily identical to either. The same considerations apply to the possible re-lationship between the fertile X(*Sxr*)Y(*Sxr*) male mice described by McLaren and Burgoyne (1984) and XYYY.

Taken together, the several X chromosome aneuploidies in the mouse would appear to represent useful although not exact models for sex aneu-ploidy in humans. The reason for the phenotypic differences, given Ohno's (1967) "law" regarding the conservation of the loci on the mammalian X chromosome, is not immediately obvious. It may, however, be related to dif-ferences between man and mouse in the nature of the noninactivated loci on the inactivated X chromosome. This issue will be considered in greater detail in Chapter 13.

Directed aneuploidy

In an article written nearly twenty years ago, Lederberg (1966) stated that

Human nuclei, or individual chromosomes and genes, will also be combined with those of other animal species; these experiments are now well under way in cell culture. Before long we are bound to hear of tests of the effect of dosage of the human twenty-first chromosome on the development of the brain of the mouse or the gorilla. . . . As bizarre as they seem, they are direct translations to man of classical work in the fruit-fly and in many plants. They need . . . just a small step in cell biology.

Clearly the goal of inserting a human chromosome 21 into the mouse to create a model for Down syndrome has not been attained, although it has to some extent been duplicated by the breeding of mice trisomic for the appropriate homologous chromosome (see Chapter 12). Nevertheless, recent developments in cellular and molecular biology suggest that it may in fact become possible to insert, if not a whole human chromosome into the mouse genome, at least one and likely a substantial number of loci from that chromosome. These developments can be considered in two categories: transgenic mice and chromosome-mediated gene transfer.

Transgenic mice

Transgenic mice are mice which carry genes transferred into them immediately following fertilization, while still at the pronuclear stage of development. Such transfer is accomplished by injection of the desired DNA sequences, usually incorporated into a plasmid, and is thus an extension of the technique of DNA-mediated gene transfer. The technical aspects of the production of transgenic mice were recently reviewed by Gordon (1983) and by Brinster *et al.* (1985).

Although a few adult mice carrying single copies of the injected genes in a nonintegrated form have been reported (Gordon, 1983), the injected DNA is integrated into the genome in the vast majority of transgenic mice. Such integrated sequences may be passed through the germ line, so that particular strains of transgenic mice can be perpetuated. Germ line transmission of integrated genes has been demonstrated for several genes: rabbit β-globin (Costantini and Lacy, 1981; Wagner *et al.*, 1981; Lacy *et al.*, 1983), human β-globin (Stewart, Wagner, and Mintz, 1982), leukocyte interferon (Gordon and Ruddle, 1981), herpes virus thymidine kinase (Palmiter, Chen, and Brinster, 1982), rat growth hormone (Palmiter *et al.*, 1982), chicken transferrin (McKnight *et al.*, 1983), a rearranged mouse immunoglobulin x-chain (Brinster *et al.*, 1983), rat myosin light chain 2 (Shani, 1985), porcine SLA histocompatibility antigen (Frels *et al.*, 1985), and human growth hormone-releasing factor (Hammer *et al.*, 1985). Integrated sequences may be present in anywhere from one to 150 copies per cell. When present in multiple copies, the donor DNA is nearly almost integrated into a single genomic site as head-to-tail concatamers (Gordon, 1983), although these tandem inserts are not obtained when the DNA is injected by iontophoresis (Lo, 1983). Although integration does not occur at homologous sites in the genome, the process does not appear to be a completely random one.

Despite the success in producing transgenic mice with foreign DNA integrated into their somatic and germ lines, reproducible expression of the integrated sequences has occurred much less frequently. The most consistently successful results have been obtained with the genes for either herpes thymidine kinase or rat growth hormone fused to the metallothionein–1 promoter (Palmiter, Chen, and Brinster, 1982; Palmiter *et al.*, 1982), with the mouse immunoglobin x-chain (Brinster *et al.*, 1983) and μ-chain (Grosschedl *et al.*,

1984) genes, and with the rat pancreatic elastase I gene (Swift *et al.*, 1984).

The methodology of transgenic mouse formation is clearly not yet at the point at which it can be used to produce mice carrying single extra copies of particular genes which are expressed in normal amounts in the appropriate tissues at the correct times. However, given the dramatic progress in this field, it is not inconceivable that such a goal might be attainable in time (Marx, 1983). If so, it would be possible to produce mice carrying three rather than two functional copies of genes of interest and to assess directly the effects of this imbalance on development and function. An analogous system using "adjustable" expression vectors has recently been described in bacteria (Walsh and Koshland, 1985).

Chromosome-mediated gene transfer

Another approach to the problem of creating animals aneuploid for specified segments of the genome would be to use chromosomes or centromere-containing chromosome segments rather than DNA as the vehicle for gene transfer. In this situation, true chromosomal aneuploids would be produced, and trisomy for regions considerably larger than one or just a few genes could be obtained. In theory, chromosomes could be injected into pronuclear eggs, in the same manner as the DNA sequences used to obtain transgenic mice. If such chromosomes could become appropriately incorporated into the mitotic apparatus and replicate normally, the problem of site of integration would be obviated and the loci on the inserted chromosome should be correctly regulated.

The only precedent for this approach comes from chromosome-mediated gene transfer into mammalian cells, recently reviewed by McBride and Peterson (1980), and it is unfortunately not a very hopeful one. In the early studies, the efficiency of chromosome uptake and of subsequent gene transfer was quite low. The efficiency of transformation could be improved by using dimethylsulfoxide in the treatment medium (Miller and Ruddle, 1978). However, neither this method nor the earlier ones result in the integration of intact chromosomes into the genome, although chromosome fragments are sometimes detected. McBride *et al.* (1982) found that while up to 15% of transcribed X-chromosome single-copy sequences could be detected in unstable transformants, the amount was below the level of detection in stable transformants.

If insertion and integration of chromosomes into mammalian embryos can be successfully effected, it may also become possible to create and insert artificial chromosomes carrying specific sets of loci. Such artificial chromosomes are being constructed in yeast and contain cloned genes, origins of replication (for DNA synthesis), centromeres, and telomeres (Murray and Szostak, 1983; Blackburn, 1985). If large enough (>20 kb), these artificial chromosomes are mitotically and meiotically stable, although considerably less so than natural chromosomes. However, this work is still in its early stages, and considerable improvements are to be expected. If these improvements are attained, it is reasonable to visualize that it might ultimately become possible, by using similar techniques, to apply the approach to the creation and investigation of aneuploidy in mammals.

Part V

Three major clinical problems of human aneuploidy

12

Trisomy 21 (Down syndrome)

Having presented a detailed review of both the theoretical and experimental aspects of our current understanding of aneuploidy, I now turn to a consideration of three important human clinical problems associated with aneuploidy. Two of these are specific disorders caused by chromosome imbalance: trisomy 21, with its phenotype of Down syndrome, and monosomy X (or XO), with its phenotype of Turner syndrome or gonadal dysgenesis. The third problem is not a single genetic disorder, but, rather, a whole group of acquired disorders in which aneuploidy appears to play a significant but poorly understood role – cancer. For the first two conditions, the intent, once again, is to focus on questions of mechanism and to relate what is known about the chromosomes involved to the phenotypes that are observed. If it is not possible to make such connections, the discussion will point up those issues which appear to offer the most with regard to arriving at mechanistic solutions.

The discussion of trisomy 21 in this chapter will begin with what is known about the genetic structure of human chromosome 21. It will proceed to consideration of first the implications of the imbalance of loci known to be on the chromosome and then of those aspects of Down syndrome the relation of which to imbalance of loci on chromosome 21 is not presently apparent. Finally, a newly developed animal model system for trisomy 21 will be discussed.

The structure of human chromosome 21

The genetic map

The most recent summary of the map of chromosome 21 is presented in the report of the committee on the genetic constitution of chromosomes 18, 19, 20, 21, and 22 (Westerveld and Naylor, 1984) in Human Gene Mapping 7 (1984), and the number of identified loci is very small indeed. The report lists one nucleolar organizing ribosomal RNA locus (*RNR*) (Henderson, Warburton, and Atwood, 1972) and three enzyme loci and one receptor locus that are confirmed as being present on chromosome 21: superoxide dismutase–1 (*SOD1*) (Tan, Tischfield, and Ruddle, 1973), the interferon-α/β receptor (*IFRC*) (Tan, Tischfield, and Ruddle, 1973), phosphoribosylglycinamide synthetase (*PRGS*) (Moore *et al.*, 1977; Patterson, Graw, and Jones, 1981; Cox *et al.*, 1983), and the liver isoenzyme of phosphofructokinase (*PFKL*) (Vora and Francke, 1981; Cox *et al.*, 1983). In addition, two enzymes are listed as provisionally mapped to chromosome 21: phosphoribosylaminoimidazole synthetase (*PAIS*) (Patterson, Graw, and Jones, 1981) and cystathionine-β-synthase (*CBS*) (Skovby, Krassikoff, and Francke, 1984; Chadefaux *et al.*, 1985). Also on the provisional list are genes for a renal β-amino acid transport system (*AABT*) (Connolly, Goodman, and Swanton, 1979) and for the control of 5-hydroxytryptamine (serotonin) concentration in the blood (5-hydroxytryptamine oxygenase regulator, *HTOR*) (Ternaux *et al.*, 1979). The latter situation will be discussed later in this chapter and does not warrant designation as a specific locus. The provisional list contains genes for a few surface antigens, *S14* (Chan, Ball, and Sergovich, 1979) and *MF13* and *MF14* (Schroder *et al.*, 1984), MF13 and MF14 being glycoproteins with molecular weights of 86 and 145 kilodaltons. The relationship of these antigens to one another or to others, such as that described by Miller *et al.* (quoted in Scoggin *et al.*, 1983), is unknown.

Attempts have been made to map unidentified polypeptides to human chromosomes by two-dimensional electrophoresis, either by using human × animal cell hybrids segregating human chromosomes or by looking at dosage effects in cells aneuploid for chromosome 21 (Van Keuren, Merril, and Goldman, 1983; Klose, Zeindl, and Sperling, 1982; Scoggin *et al.*, 1983). Although a very small number of such polypeptides have been tentatively assigned to chromosome 21, the results of these studies are still inconclusive.

In addition to the loci mentioned above, the report of Westerveld and Naylor (1984) also lists several unidentified DNA sequences, designated *D21S1, 6, 8,* and *15–20,* localized to chromosome 21 by Watkins *et al.* (1984) and Stewart *et al.* (1984). Other recombinant DNA probes for unknown chromosome 21 sequences have been described by Graham, Hall, and Cummings (1984) (repetitive sequences) and by Davies *et al.* (1984), Devine *et al.* (1984), Antonarakis *et al.* (1985), Brown *et al.* (1985), Gusella *et al.* (1985), Kittur *et al.* (1985), and Patterson *et al.* (1985b) (single-copy sequences). It is expected that the number of such probes will rapidly increase as more libraries of chromosome 21 DNA sequences are prepared and analyzed.

In addition to these anonymous sequences for which no specific functions are known, recombinant DNA probes for the superoxide dismutase–1 gene have been prepared and used by Lieman-Hurwitz *et al.* (1982) and Sherman *et al.* (1983) (see also Groner *et al.*, 1985). The existence of these *SOD1* probes will permit a detailed exploration of the region of the chromosome in the vicinity of the *SOD1* gene, a region thought to be associated with the Down syndrome phenotype.

Fine-structure mapping studies have been undertaken for all of the loci definitely mapped to chromosome 21, and the results of those studies are shown

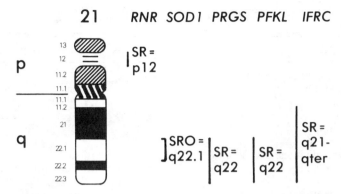

Fig. 12.1. Regional mapping of loci on chromosome 21. SR, smallest region; SRO, smallest region of overlap (of several partial duplications or deletions). See text for references. Redrawn with modifications from Westerveld and Naylor (1984). A tentative mapping of *PRGS* to 21q22.1 has recently been reported (Chadefaux *et al.*, 1984).

in Fig. 12.1 (see also Fig. 12.7). The three enzyme loci – *SOD1*, *PRGS*, and *PFKL* – have been placed in band 21q22, with *SOD1* and possibly *PRGS* further localized to band 21q22.1 (Sinet *et al.*, 1976; Poissonnier *et al.*, 1976; Tan, 1976; D.R. Cox *et al.*, 1983; Chadefaux *et al.*, 1984; Cox and Epstein, 1985). *IFRC* is present in the region 21q11→qter (Epstein and Epstein, 1976). While restricting the locations of these loci somewhat, the present mapping assignments still require considerably more refinement. Cystathionine-β-synthase has been assigned to the region between 21q21 and 21q22.1 (Chadefaux *et al.*, 1985).

The Down syndrome region

In addition to mental retardation, hypotonia, and, in about 40% of the cases, congenital heart disease, the physical phenotype of Down syndrome consists principally of features that are characterized as variants or minor malformations. A list of the salient physical features, with their frequencies, is presented in Table 12.1. Several characteristic dermatoglyphic features are also associated with Down syndrome. Although some of the features are capable of quantitative evaluation, others are more subjective. Nevertheless, the diagnosis of Down syndrome based on physical examination is generally quite accurate, and the accuracy can be improved, even without resorting to chromosome analysis, by using a diagnostic index which gives special weight to certain features with high discriminative power (Rex and Preus, 1982). However, it must be realized that the overall phenotype of Down syndrome consists of many features that are not included in Table 12.1. These features,

Table 12.1. *Physical characteristics of Down syndrome*

Feature	Frequency (%)[a]
Oblique (upslanting) palpebral fissures	82
Loose skin on nape of neck	81
Narrow palate	76
Brachycephaly	75
Hyperflexibility	73
Flat nasal bridge	68
Gap between first and second toes	68
Short, broad hands	64
Short neck	61
Abnormal teeth	61
Epicanthic folds	59
Short fifth finger	58
Open mouth	58
Incurved fifth finger	57
Brushfield spots	56
Furrowed tongue	55
Transverse palmar crease	53
Folded or dysplastic ear	50
Protuding tongue	47

[a]Means of frequencies in Table 5-1 of Pueschel (1982*c*).

including immune defects, susceptibility to leukemia, and the development of Alzheimer disease late in life, are discussed in later sections of this chapter.

If our intent is to understand the pathogenesis of Down syndrome, it is necessary to know which part or parts of chromosome 21 are responsible for the phenotype. Accordingly, cases in which only part of chromosome 21 was duplicated have been intensively studied, by the methods outlined in Chapter 5, to arrive at a phenotypic map of chromosome 21. It must be realized that in these studies the phenotype has been considered mainly on subjective grounds, and quantitative diagnostic approaches have not been used. The present consensus, summarized by Summitt (1981) and Rethoré (1981), is that the full Down syndrome phenotype appears when band 21q22 is duplicated, and of this, subbands 21q22.1 and probably 21q22.2 are required. As Fig. 12.1 indicates, this is the same general region that carries the genes for SOD–1, PRGS, and PFKL. Patterson *et al.* (1982) estimate that this region carries 50 to 100 genes.

At issue at the present time is the proximal margin of the zone, both in terms of its precise chromosomal location and its relationship to *SOD1*. Although all classical cases of Down syndrome were originally thought to have an elevated SOD–1 activity, recent reports (summarized in Fig. 12.2) have

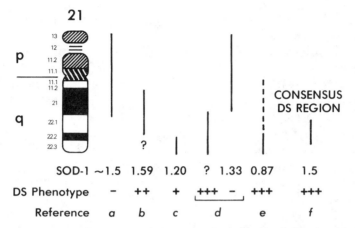

Fig. 12.2. Exceptional cases of partial trisomy 21 which are cited as calling into question the commonly accepted relationship between the region responsible for the Down syndrome phenotype and the location of *SOD1*. The consensus location for the Down syndrome phenotype (21q22.1→22.2) is shown at the left. The dashed bar indicates the region claimed to be inactivated by spreading of X inactivation from the X chromosome of an X;21 translocation. See text for full details. SOD–1, red cell SOD–1 activity. DS phenotype: -, absent; +, some features present; + +, many features present but some typical features absent; + + +, typical phenotype. References: *a* (Cottoni-Durand, 1979); *b* (Jenkins *et al.*, 1983); *c* (Habedank and Rodewald, 1982); *d* (Leschot *et al.*, 1981); *e* (Taysi *et al.*, 1982); *f* (Summitt, 1981; Rethoré, 1981).

raised questions about this association. Thus, Cottoni-Durand (1979) [quoted in Rethoré (1981)] observed a mentally retarded but otherwise phenotypically non-Down syndrome boy with dup(21pter→q21) and an elevated SOD–1 activity. Similarly, Jenkins *et al.* (1983) reported a girl with elevated SOD–1 activity and dup(21q21→q22.?2) who had some but not all of the features of Down syndrome. She had mild mental retardation, brachycephaly, low posterior hairline, Brushfield spots, small square ears, brachydactyly, and widened interdigital spaces between the first and second toes, but lacked slanting palpebral fissures, epicanthic folds, simian creases, and typical dermatoglyphics. From these reports, I would infer that triplication of the *SOD1* locus and/or of the immediate region surrounding it is not sufficient in itself to produce the full phenotype of Down syndrome if enough of band q22 is not duplicated. In a reciprocal but still compatible manner, Habedank and Rodewald (1982) reported three moderately retarded adult siblings with dup (21q22.2→qter) and some phenotypic features of Down syndrome: upward-slanting palpebral fissures (not impressive in the published photograph) and scrotal (furrowed) tongue. However, the ears were described as "jug handle," the nose as "relatively differentiated," and the dermatoglyphic indices as normal or in the overlap region between control and Down syndrome distributions. SOD–1 activity was considered as normal, although by my calculation it was somewhat elevated [mean 1.20, just above the control mean + 2(SD)]. It may be noted in passing that, by "subtraction," these cases and those of Jenkins *et al.* (1983) and Poissonier *et al.* (1976) would place the region for the typical dermatoglyphic changes of Down syndrome in band 21q22.1 or the margin between 21q22.1 and 21q22.2.

Taken at face value, and assuming that SOD–1 activity is really not increased, the cases of Habedank and Rodewald (1982) would suggest that duplication of the region of 21q22 that carries *SOD1* is required for the full phenotype of Down syndrome. However, an apparently contradictory case has been reported by Taysi *et al.* (1982). They observed a child with clinically typical Down syndrome resulting from an unbalanced X;21 translocation involving 21q11→qter. Late replication studies of (presumably) the lymphocytes indicated that the t(X;21) chromosome was inactivated, and measurements of red cell SOD–1 activity gave normal values which were attributed to spreading of inactivation into the proximal part of the autosomal segment of the translocation chromosome. Taysi *et al.* (1982) interpreted these results to mean that the region of 21q22.1 that contains *SOD1* is not involved in the Down syndrome phenotype, or that the "clinical effect or damage" may have occurred prior to the time of X inactivation. I very much doubt the latter, and am somewhat reluctant to accept the former, principally because it seems risky to infer the state of activity of the translocated *SOD1* locus and surrounding region of chromosome 21 in a variety of tissues during and subsequent to gestation only from measurements of red cell SOD–1 activity. It would have been preferable to have measured SOD–1 activity in several organs and tissues.

Another report which came to a similar conclusion is that of Leschot *et al.* (1981). They described two sibs, one with dup(21q21.4→qter) and del (9pter→p24), who had several typical features of the Down syndrome phenotype (with some extra anomalies as well), including an endocardial cushion defect; and the other with dup(21pter·q21.4) and dup(9pter·p24), who may have been retarded but was otherwise phenotypically normal. The latter child had a red cell SOD–1 activity 1.33 times the normal range, and on the basis of this it was suggested that *SOD1* is in or proximal to band 21q21.4. In the absence of SOD–1 studies in the first child and considering the intermediate level of SOD–1 activity in the second, it is difficult to know how to evaluate these findings. In fact, at this point in time, it would be better to determine the number of *SOD1* loci present by molecular hybridization, using an *SOD1* probe, rather than by enzyme assay, which, in the present case, gives somewhat equivocal results.

Rethoré (1981) also argues that the chromosomal region containing *SOD1* is not causally related to the Down syndrome phenotype because "the diminution of its activity (by deletion) does not suffice to induce the malformation syndrome in the contre-type to that of trisomy 21." This reasoning seems tenuous at best (see Chapter 8).

The reason for much of the interest in SOD–1 and its relationship to the phenotype of Down syndrome is, as will become apparent in the next section, that much has been written about the possible pathogenetic significance of the increase in SOD–1 activity. In addition, *SOD1* is being used as one of the loci to define a chromosome segment in the mouse which is homologous to human chromosome segment 21q (see below). Nevertheless, however the matter is eventually resolved (and at present I consider it unresolved), it should be kept in mind that the *SOD1* locus is triplicated in virtually all cases of trisomy 21 and that mental retardation still occurs if the region carrying this gene is present in an extra dose, irrespective of whether the full Down syndrome phenotype is present. Therefore, imbalance of the chromosomal region around *SOD1*, and perhaps of *SOD1* itself, does have phenotypic effects worthy of consideration. Furthermore, duplication of the region 21q11→q22 also results in mental retardation, even in the absence of other gross phenotypic abnormalities, indicating that it should also be considered when the pathogenesis of the mental retardation of trisomy 21 is being studied.

The effects of unbalancing loci on chromosome 21

Earlier in this volume, the gene dosage effects associated with aneuploidy for chromosome 21 were summarized (Table 4.1), and these included changes in the expression of *IFRC* (interferon-α receptor), *PFKL* (phosphofructokinase, liver), *PRGS* (phosphoribosylglycinamide synthetase), *SOD1* (superoxide dismutase–1), and *CBS* (cystathionine-β-synthase). In each case, the mean activities or concentrations of the gene products were close to 1.5 times the diploid value in trisomic cells and 0.5 times the diploid in monosomic cells. Gene dosage effect data are not yet available for the one other locus provisionally mapped to chromosome 21 (*PAIS*) or for chromosome 21-determined surface antigens.

As has already been discussed in great detail in Part III, gene dosage effects by themselves are significant in the causation of the abnormalities associated with aneuploidy only insofar as they lead to secondary effects which interfere with developmental processes and metabolic functions. In this section I shall discuss what is known or has been speculated about the consequences of unbalancing the few loci known to be on human chromosome 21. (Cystathionine-β-synthase will not be discussed since no data are as yet available). Given the much greater number of loci likely to be on even that part of the chromosome involved in the pathogenesis of Down syndrome and the basic assumption that many loci are likely to be involved in the generation of the overall phenotype, I am not suggesting that the loci discussed here have major roles in the generation of the Down syndrome phenotype (or that they necessarily have any role at all). Rather, the intent in discussing them is to point out what effects their imbalance does or might have and how, in the light of the earlier theoretical discussion, these effects could contribute to developmental and functional abnormality.

Superoxide dismutase–1

"There is a bizarre enzymatic activity universally present in respiring cells. The substrate is an unstable free radical that can be present only in miniscule amounts at any instant, and the reaction catalyzed proceeds at a rapid rate even in the absence of the enzyme. Yet the enzyme is essential for the survival of aerobic cells." So wrote Fridovich (1975) in introducing a review of the biochemistry of the superoxide dismutases, enzymes which catalyze the dismutation of superoxide radicals, O_2^-, to H_2O_2 by the reaction

$$O_2^- + O_2^- + 2H^+ \rightarrow H_2O_2 + O_2$$

These enzymes occur in mammals in two forms, one a soluble (cytoplasmic) enzyme, SOD–1, containing copper and zinc, and the other a mitochondrial manganese-containing enzyme, SOD–2. It is the former that is determined by human chromosome 21. As suggested by Fridovich (1975), SOD–1 is generally regarded as a protective enzyme, and numerous examples exist of the beneficial effects of exogenously added enzyme. These include, for example, prevention of oxygen damage to cultured endothelial cells (Freeman, Young, and Crapo, 1983; Freeman *et al.*, 1985); reduction of radiation-induced chromosome breaks in lymphocytes (Nordenson, Beckman, and Beckman, 1976), of radiation-induced damage to erythrocytes (Petkau *et al.*, 1976) and bone marrow stem cells (Petkau *et al.*, 1975), and of spontaneous and mitomycin C-induced chromosome breaks in normal and Fanconi anemia fibroblasts (Sudharsan Raj and Heddle, 1980); inhibition of paraquat-stimulated lipid peroxidation in mouse lung microsomes (Bus, Aust, and Gibson, 1976) and of radiation-induced peroxidation of phospholipid membranes (Petkau and Chelack, 1976); impairment of the ability of phorbol myristate acetate-stimulated monocytes to lyse erythrocytes (Weiss, LoBuglio, and Kessler, 1980); and protection against alloxan- and streptozotocin-induced breaks in pancreatic islet cell DNA (Uchigata *et al.*, 1982). If SOD–1 is indeed protective, it would be expected that an increase in its concentration, if it had any effect at all, would be beneficial to the individual, but just the opposite has been suggested (Sinet, 1982; Björkstén, Marklund, and Hägglöf, 1984). Why?

The most direct mechanism for an adverse effect was postulated by Björkstén, Marklund, and Hägglöf (1984), who drew a comparison between the increased SOD–1 in polymorphonuclear leukocytes (Table 4.1) and reported decreases in the bactericidal capacity of trisomy 21 leukocytes (Gregory, Williams, and Thompson, 1972; Kretschmer *et al.*, 1974; Costello and Webber, 1976; Barkin *et al.*, 1980). Since superoxide is believed to be involved in the bactericidal process, possibly as a precursor to the highly reactive hydroxyl radical (OH·) (Babior, 1978), they reasoned that an increase in SOD–1 might reduce the amount of available O_2^- and therefore interfere with the process. However, there is a problem with this argument. The enzyme system which generates O_2^- is located in the outer membrane of polymorphonuclear leukocytes (Green, Schaefer, and Makler, 1980) and releases the O_2^- to the outside of the cell (Salin and McCord, 1975) or into the bacteria-containing phagocytic vacuoles which, although within the cell, are separated from the cytosol by the plasma membrane. In both instances, the superoxide generated would not be accessible to the cytoplasmic SOD–1 and should not, therefore, be affected by a change in its activity. Nevertheless, Annerén and Björkstén (1984) have reported that the mean superoxide concentration in phorbol myristate acetate-stimulated polymorphonuclear leukocytes is reduced by 21% in trisomy 21. Whether this reduction results from increased degradation or decreased synthesis or from some other more general alteration in cell physiology remains to be determined.

A more general set of reasons for suggesting an adverse effect from increased SOD–1 activity was put forward by Sinet, Lejeune, and Jerome (1979) and Sinet (1982), and the logic seems to go as follows. As a result of increased SOD–1 activity, the concentration of O_2^- is reduced in trisomic cells while the concentration of H_2O_2 is increased. The latter, by a series of reactions, some also involving O_2^-, serves as a precursor of OH·, the hydroxyl radical. This radical is an extremely powerful oxidant and initiates injurious peroxidation reactions leading to the formation of organic hydroperoxides (lipid peroxides). In support of this proposal, it has been claimed that trisomic fibroblasts are more sensitive to the toxic effects of high oxygen tensions (95% oxygen, 5% carbon dioxide) than are diploid cells and produce higher quantities of lipid peroxides (Mayes, Muneer, and Sifers, 1984). It is also possible that SOD–1 itself can use H_2O_2 to cause peroxidation reactions and might be inactivated by O_2^- and H_2O_2. The last of these effects would, according to Sinet (1982), result in a paradoxical reduction of SOD–1 activity in trisomy 21, but there is no evidence to support this possibility, and all available in vivo data (Table 4.1) are contrary to this expectation.

Except for the fibroblast data just cited, the evidence in support of an increased level of tissue peroxidation is, in the main, indirect. Several authors have reported increased activities of glutathione peroxidase, an enzyme which catalyzes the reduction of H_2O_2 and of organic hydroperoxides in trisomy 21 erythrocytes (Frischer *et al.*, 1981; Sinet, 1982; Nève *et al.*, 1983;

Annerén *et al.*, 1984; Björkstén, Marklund, and Hägglöf, 1984) and fibroblasts (Sinet, Lejeune, and Jerome, 1979), with the peroxidase activities in the trisomic cells averaging, respectively, 1.42 and 1.91 times the control values. My own laboratory (Feaster, Kwok, and Epstein, 1977) did not find an increased activity in fibroblasts.

The glutathione peroxidase results have been considered to be significant because this enzyme is believed to be induced when cells are exposed to oxidative stress (Sinet, 1982), the inference being that trisomic cells are being subjected to such stress as a result of the increased SOD–1 activity. In fact, based on the finding of a positive correlation between *red cell* glutathione peroxidase activity and measured IQ (r = 0.58), Sinet, Lejeune, and Jerome (1979) appear to be suggesting that the ability of trisomic individuals to respond to the putative oxidative stress caused by the increased SOD–1 activity is a determinative factor in their ultimate intellectual development. The ultimate level of glutathione peroxidase is thus a function of the individual rather than a measure of the degree of oxidative stress and associated tissue damage. That other genetic (or environmental) factors might be involved in determining the level of red cell glutathione peroxidase activity is suggested by the finding of Annerén *et al.* (1984) that there is a positive correlation (r = 0.42) between the activities found in children with Down syndrome and in their normal siblings.

The significance of the suggested relationship between glutathione peroxidase activity and IQ is brought into question by the one piece of direct evidence presently available. Brooksbank and Balazs (1984) directly measured SOD–1 activity, in vitro lipid peroxidation, and glutathione peroxidase activity in cerebral cortex homogenates from control and Down syndrome fetuses. As expected, SOD–1 activity was increased. However, although in vitro lipid peroxidation was increased by 53% in the Down syndrome brain, glutathione peroxidase activity was not. Nevertheless, these results are consistent with the conjectured increase in lipid peroxidation in trisomy 21 brain. More data bearing on this point with both human and mouse trisomy 16 (see below) material are clearly required.

One additional point concerning SOD in trisomy 21 deserves mention. In contrast to the increased activity of SOD–1, the activity of SOD–2, the mitochondrial manganese-containing enzyme, is decreased in platelets (Sinet *et al.*, 1975; Baret *et al.*, 1981) fibroblasts (Crosti *et al.*, 1983), and possibly in lymphocytes [Björkstén, Marklund, and Hägglöf (1984) found a decrease; Baret *et al.* (1981) did not], but not in granulocytes or serum (Baret *et al.*, 1981). The average reported decreases for the first three cell types were, respectively, to 59%, 64%, and 67% of control values. However, in terms of total activities, the mitochondrial SOD–2 activities were only 1% (lymphocytes) to 16% (platelets) of the SOD–1 activities. The cause and functional significance of reduced SOD–2 activity is unclear, although Sinet *et al.*

(1975) have suggested that it might contribute to mitochondrial damage and impaired cell function in trisomy 21. The generality of the finding for other tissues requires further exploration.

At the present time, a pathogenetic relationship between increased SOD–1 activity and either impaired intelligence or the development of Alzheimer disease (see below) has not been established. I personally find the suggested relationships extremely tenuous and, in the absence of directly relevant data, am reluctant to implicate the 50% increase in SOD–1 activity in the etiology of the abnormalities associated with Down syndrome. Despite the suggested mechanisms, it is difficult for me to visualize how this amount of increase in a supposedly protective enzyme converts it into an injurious one. Nevertheless, while the finding of Brooksbank and Balazs (1984) of increased in vitro lipid peroxidation in trisomic fetal brain could have explanations unrelated to SOD–1, they are certainly provocative and suggest that further investigation is warranted. Insofar as the role of SOD–1 is concerned, the critical questions are whether, in brain and other tissues, O_2^- concentrations are decreased, H_2O_2 concentrations are increased, and SOD–1 concentration is rate-limiting for the dismutation of O_2^-. Only if the latter were shown to be the case would it even be possible to visualize such a role for SOD–1. However, irrespective of whether SOD–1 itself is ultimately implicated, the numerous reports of elevated glutathione peroxidase activity in trisomic cells suggest that further efforts to understand its basis are warranted.

The interferon-α receptor

The locus for the interferon-α (IFN-α) receptor, now designated *IFRC*, was originally mapped to human chromosome 21 as the gene (called *AVP*) that determines the species-specific response of hybrid cells to human IFN-α (Tan, Tischfield, and Ruddle, 1973) and IFN-β (Slate *et al.*, 1978). Although several investigators suggested that the locus does govern the synthesis of a cell surface receptor for interferon (for summary, see Epstein and Epstein, 1983), direct proof did not exist until the demonstration that the number of specific cell surface receptors on human fibroblasts binding ^{125}I-labeled human IFN-α is directly proportional to the number of chromosomes 21 present in the cells – a primary gene dosage effect (Epstein *et al.*, 1982a) (Table 4.1). Further evidence is provided by the results of Shulman *et al.* (1984), who found that antibodies to human chromosome 21-coded cell surface components block the binding of IFN-α to human cells.

Because of the numerous biological effects reported for interferon, my wife, Professor Lois B. Epstein, I, and our collaborators, especially Dr. Jon Weil, have been particularly interested in examining the consequences of the increased concentration of IFN-α receptors on trisomic cells – the secondary effects of imbalance of this locus (see Fig. 5.1). These studies may be divided into two major categories: the effect of trisomy 21 on the synthesis of inter-feron-induced proteins and its effect on the various biological responses brought about by interferon (for a complete review, see Epstein and Epstein, 1983). Although the latter were actually the first studied, the most direct re-lationship between changes in receptor number and the action of interferon is observed by examining the synthesis of IFN-induced intracellular polypep-tides and enzymes. Induction of these proteins is presumably intermediate in the pathway between binding of interferon to its receptors and the develop-ment of the several biological responses.

The synthesis of eight different IFN-induced intracellular polypeptides has been examined in matched sets of trisomy 21 and diploid fibroblasts, some over a wide range of interferon concentrations, and the results have been quite consistent. In the early studies using natural IFN-α, the mean induction of three of these peptides was 1.67 times greater in trisomy 21 fibroblasts than in diploid fibroblasts (Weil, Epstein, and Epstein, 1980; L. B. Epstein *et al.*, 1981). In the most recent studies, involving all eight polypeptides and using both natural and recombinant IFN-a, the relative polypeptide inductions in the trisomic cells were 1.4 $\pm$ 0.3(SD) and 1.8 $\pm$ 0.5 times greater, respec-tively, than in the control diploid cells (Weil *et al.*, 1983*a*). Given the exper-imental scatter of the data, all of these values are compatible with the ratio of 1.5, expected on the basis of gene dosage. It seems quite clear, therefore, that the synthesis of the intracellular polypeptides we have examined follows the dosage relationship that would be predicted from the trisomy 21-determined increase in the number of *IFRC*-coded receptors. Results compatible with this conclusion were also obtained with cells monosomic for chromosome 21 (Weil, Epstein, and Epstein, 1980).

The identities and functions of the several IFN-α-induced polypeptides are not known. However, similar studies have also been carried out on the induc-tion of the enzyme, (2'-5')oligoisoadenylate synthetase (2–5A synthetase). This enzyme, by forming (2'-5')oligoisoadenylate chains, initiates a se-quence of events which, through activation of endonuclease E, result in in-hibition of protein synthesis in vivo and in vitro and are thought to inhibit viral replication and release (Baglioni, 1979). Again, when matched trisomy 21 and diploid cells were compared, the mean induction of 2–5A synthetase was 1.70 fold greater in the trisomic cells, and the response was proportional to the log of the IFN-α concentration, with no saturation detectable at the highest concentration tested, 2000 IU/ml (Fig. 12.3) (Weil *et al.*, 1983*c*).

The theoretical implications of changes in receptor number have been dis-cussed in detail in Chapter 5, and the applications of these considerations to the interferon response system are worthy of note. Since, as has already been discussed, trisomic cells have an increased (150%) number of receptors per cell, they are capable of having more occupied receptors than are normal cells. Thus, with IFN concentrations that result in a saturation of two-thirds

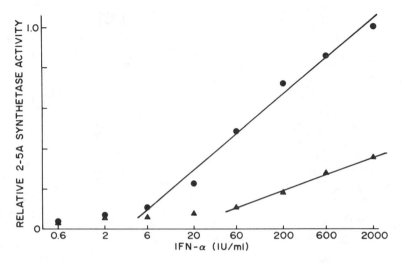

Fig. 12.3 Induction of 2–5A synthetase by IFN-α in matched diploid (●) and trisomy 21 (▲) fibroblasts. The values depicted are relative to a value of 1.0 for trisomic cells treated with 2000 IU/ml IFN-α. Redrawn from Weil *et al.* (1983c).

of the receptors on diploid cells, the trisomic cells will have as many occupied receptors as a completely saturated normal cell (Fig. 5.7). Higher IFN concentrations will therefore result in the attainment of receptor occupancies by trisomic cells that would be completely unattainable by normal cells. Since, for both the several polypeptides and 2–5A synthetase, the degree of induction appears to be closely related to the number of receptors occupied, trisomic cells should, at high IFN concentrations, be capable of levels of biological responses to the interferon of which normal cells are wholly incapable. Conversely, the trisomic cells should be capable of reaching functionally relevant threshold levels of polypeptides and 2–5A synthetase at much lower IFN concentrations than will diploid cells. Both points should be apparent from the data depicted in Fig. 12.3.

Starting shortly after the original mapping of *IFRC* to chromosome 21, a large number of studies have been carried out on the sensitivity of trisomy 21 cells to the antiviral and other biological effects of IFN-α. The results of these studies are summarized in Table 12.2. It is quite apparent that trisomic cells are highly sensitive to interferon in the development of several of these effects, requiring from 3 to 15 times less IFN than equivalent diploid cells. Thus, whereas the protein induction responses, which are presumed to be closely related to the binding of interferon to its receptors, show gene dosage effects, the more complex biological actions of interferon do not. How can these enhanced responses, graphically illustrated in Fig. 12.4, be explained?

Table 12.2. *Responsiveness of trisomy 21 cells to interferon-α[a]*

Effect	Responsiveness	References
Polypeptide induction[b]		
Three polypeptides	1.67	Weil, Epstein, and Epstein (1980); L. B. Epstein *et al.* (1981)
Eight polypeptides	1.4 1.8* }	Weil *et al.* (1983a)
(2'-5') oligoisoadenylate induction[b]	1.7	Weil *et al.* (1983c)
Protection of cells against viral challenge[c]	5.0	Tan *et al.* (1974)
	3.7	Tan, Chou, and Lundh (1975)
	10.	DeClercq, Edy, and Cassiman (1975)
	3.1	Epstein and Epstein (1976)
	3.9	Tan (1976)
	6.7	DeClercq, Edy, and Cassiman (1976)
	2.1	Lubiniecki *et al.* (1978)
	8.0 11.3 }	Weil, Epstein, and Epstein (1980)
	6.6 14.7 }	L. B. Epstein *et al.* (1981)
	3.1	Weil *et al.* (1983c)
	6.5 3.0* }	Weil *et al.* (1983a)
Inhibition of proliferation of fibroblasts	6.7	Tan (1976)
Inhibition of maturation of monocytes to macrophages	3.7	Epstein, Lee, and Epstein (1980)
Inhibition of lectin-induced lymphocyte mitogenesis	2.9 5.0[d]	Cupples and Tan (1977)

[a]Natural IFN-α was used in all experiments except those denoted by an asterisk in which recombinant DNA-produced IFN-α was used.
[b]Results are expressed in terms of quantities of proteins synthesized or enzymes induced in response to the same concentration of IFN-α in trisomic and diploid cells.
[c]Result are expressed in terms of the ratio of IFN-α concentrations (diploid/trisomic) required to produce the same biological effect (50% inhibition) in diploid and trisomic cells. (See Fig. 12.4.)
[d]IFN-β.

If the antiviral and other biological effects of interferon being assayed require high degrees of receptor occupancy to achieve the 50% end point used in these assays, then it might be possible to explain some or even all of the enhanced sensitivity of the trisomic cells by the properties of IFN-receptor

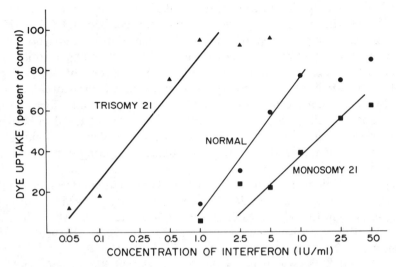

Fig. 12.4 Relative sensitivity of trisomic, diploid, and monosomic fibroblasts in an antiviral assay. The degree of response to interferon is proportional to the dye uptake, and 100% represents complete protection. Redrawn with corrections from Weil, Epstein, and Epstein (1980).

interactions. The relevant calculations are summarized in Fig. 5.10 and indicate that it would be necessary to postulate receptor occupancies of 80% or greater for diploid cells to explain the observed threefold and greater increases in sensitivity demonstrated by trisomic cells. However, our own data (Weil, Epstein, and Epstein, 1980) and those of Aguet and Blanchard (1981) indicate that the 50% inhibition of viral cytopathogenicity or synthesis (which is the end point of the assay) occurs at interferon concentrations considerably lower than are required to occupy half of the diploid number of receptors. Enhancement at the level of IFN-receptor interaction therefore seems unlikely. Furthermore, the observation that nonamplified gene dosage effects are preserved for polypeptide and 2–5A synthetase induction, which also depends on binding of IFN to its receptors, also argues against such an explanation.

Unfortunately, our lack of understanding of the details of the interferon response system makes it impossible to provide a satisfactory answer to the question of the mechanism of the amplified sensitivity of trisomic cells. However, a variety of conjectures are possible. One is that there is nonlinearity in the activation of endonuclease E by the oligomers of 2–5A synthesized by the IFN-induced 2–5A synthetase (Weil *et al.*, 1983c). This could result from the fact that only trimers and larger oligomers are capable of activating the endonuclease. Another possibility is that there is a multiplicative interaction of several different induced proteins, each induced to a 1.5-fold greater extent in trisomic cells. Were this the case, a threefold enhanced sensitivity would require the participation of at least three different factors, assuming each to contribute to the reaction proportionally to its concentration [$(1.5)^3 = 3.375$], or of one factor acting proportionally to the cube of its concentration, or of some combination of the two types of effects. Yet another explanation might be that a 50% increase in a process acting at an early stage of, for example, viral replication might significantly enhance the effectiveness of a later process by decreasing the number of viral target structures on which the late process acts (Weil *et al.*, 1983*c*). Further knowledge of the details of the mechanisms of interferon action will be required to test these possibilities and to arrive at an understanding of what is actually transpiring in trisomic cells exposed to interferon.

The discussion of the response of trisomy 21 cells to interferon has to this point focused on IFN-α, since *IFRC*, the only formally mapped IFN-receptor gene, is thought to code for the IFN-α or IFN-α/β receptor, which does not recognize IFN-γ. However, trisomy 21 cells also show an enhanced responsiveness to IFN-γ in terms of polypeptide induction [1.9 fold (L. B. Epstein *et al.*, 1981)], 2–5A synthetase induction [1.33 fold (Weil *et al.*, 1983*c*)], and antiviral effect [3.1 fold (Epstein and Epstein, 1976), 28.6 fold (L. B. Epstein *et al.*, 1981), and 17.5 fold (Weil *et al.*, 1983*c*)]. Once again the difference between the first two (gene-dosage related) and the third (amplified) responses are noted. These results indicate that the action of IFN-γ, which is mediated by a different receptor and has a variety of different peptide-inducing (Weil *et al.*, 1983*b*) and biological (for summary, see Epstein and Epstein, 1983) effects is also mediated in some way by chromosome 21. The simplest explanation would be that the IFN-γ receptor is also coded for by this chromosome, but it does not seem to be.

Does the increased sensitivity to interferon in trisomy 21 contribute to the development of the phenotype of Down syndrome? The straightforward answer to this question is that we do not know. We had previously suggested (Epstein and Epstein, 1980; C. J. Epstein *et al.*, 1981) that this increased sensitivity might somehow interfere with the development of an immune response, because of the inhibitory effects of IFN-α on lymphocyte mitogenesis and other aspects of the immune response (L. B. Epstein, 1977; Epstein, 1979). However, given the large number of immune response-enhancing effects recently described for IFN-γ (Leibson *et al.*, 1984; L. B. Epstein, 1984; Sidman *et al.*, 1984), which is itself a product of stimulated lymphocytes, and the enhanced sensitivity of trisomic cells to IFN-γ as well as IFN-α and IFN-β, it is not clear that the overall effect of the increased number of interferon receptors in trisomy 21 on the development or maintenance of immune responses would be a negative one. Therefore, the relationship, if any, of *IFRC* to the immune defects in Down syndrome (see below) remains to be elucidated. However, it does seem somewhat paradoxical that given their enhanced sensitivity to interferon, individuals with Down syndrome are, as will be discussed shortly, still at higher risk of developing viral infections and leukemia.

Although the interferon response system has not as yet been definitively implicated in the pathogenesis of Down syndrome, the observations discussed earlier have considerable theoretical importance. First, they demonstrate how primary gene dosage effects, in this case involving the concentration of IFN-α receptors, are translated into secondary dosage effects involving several elements of the molecular response to the binding of interferon. Second, of even greater interest, they show that a dosage-dependent process can, in appropriate circumstances, become greatly amplified so that the ultimate biological differences between aneuploid and diploid cells or organisms are much greater than would be predicted from simple gene dosage considerations. If this occurs with the interferon response system, could it not also occur for other metabolic systems as well?

Phosphoribosylglycinamide synthetase and phosphoribosylaminoimidazole synthetase

Using somatic cell hybrids between Chinese hamster ovary cells with mutations making them dependent on exogenous purines and human cells, Moore *et al.* (1977) and Patterson, Graw, and Jones (1981) mapped two of the enzymes of the purine biosynthetic pathway to human chromosome 21. These enzymes, phosphoribosylglycinamide synthetase (PRGS) and phosphoribosyaminoimidazole synthetase (PAIS or AIRS), catalyze, respectively, the third and sixth steps of the purine biosynthetic pathway:

$$\text{PRGS}$$
$$\text{Phosphoribosylamine} + \text{glycine} + \text{ATP} \rightarrow$$
$$\text{phosphoribosylglycinamide} + \text{ADP} + \text{Pi}$$
$$\text{PAIS}$$
$$\text{Phosphoribosylformylglycinamidine} + \text{ATP} \rightarrow$$
$$\text{phosphoribosylaminoimidazole} + \text{ADP} + \text{Pi}$$

Patterson, Graw, and Jones (1981) have shown the two enzymes to be coordinately regulated. These enzymes, as well as GAR formyltransferase (the fourth enzyme in the pathway), may also be physically associated or perhaps even present together on a multifunctional polypeptide, as is the case in yeast and *Drosophila* (Patterson *et al.*, 1985*a*).

Gene dosage effects have been demonstrated for PRGS in trisomy 21 cells (Table 4.1), and Patterson *et al.* (1982) have speculated that the increased activity of PRGS may result in increased de novo purine biosynthesis in trisomic individuals. In support of this notion, they point to decreased purine biosynthesis in a Chinese hamster ovary cell mutant with decreased PRGS activity and to the elevation of serum purine levels in Down syndrome. Increases in serum uric acid levels have been reported by several investigators (for summary, see Pueschel, 1982*a*), with the increases ranging from 16% to 44%. A particularly thorough study was carried out by Pant, Moser, and Krane (1968), who found mean increases of 43% and 44% in trisomic males and females, respectively, with significant differences being found in all age groups. Since urinary uric acid excretion was, if anything, increased, they inferred that the hyperuricemia was the result of purine overproduction. Other suggestions regarding the etiology of the hyperuricemia have been made, including increased degradation of purines resulting from enhanced leukocyte turnover (Mellman *et al.*, 1968) or associated with an increased adenosine deaminase activity (Puukka *et al.*, 1982). The increased enzyme activity could also be an effect of increased synthesis rather than a cause of increased degradation. Furthermore, the issue of overproduction (or increased purine degradation) versus diminished excretion has still not been fully resolved. Nevertheless, I would agree with Patterson, Graw, and Jones (1981) that "the basis for elevated serum purines in these patients should be reinvestigated." At the same time, it is also necessary to determine whether trisomy 21 cells do actually synthesize increased amounts of purines. Once again, as with SOD–1, and as is discussed in detail in Chapter 5, the issue is whether the cellular concentrations of PRGS and/or PAIS do control the flux of metabolites through the purine biosynthetic pathway.

Phosphofructokinase, liver

Phosphofructokinase (PFK), a rate-limiting enzyme in glycolysis, catalyzes the reaction

$$\text{PFK}$$
$$\text{Fructose–6-phosphate} + \text{ATP} \rightarrow \text{Fructose–1,6-diphosphate} + \text{ADP}$$

A tetrameric enzyme, PFK is composed of three types of subunits: M (muscle), L (liver), and P (platelet). Of these, it is the L type of subunit that is coded for by human chromosome 21 (Vora and Francke, 1981). The subunit makes up virtually all liver PFK, as well as substantial amounts of the PFK of red cells, lymphocytes, polymorphonuclear leukocytes, and fibroblasts (Vora, 1982). As is summarized in Table 4.1, a dosage effect for red cell PFK has been demonstrated in trisomy 21. However, it is not clear why the increase in PFK activity should reach ~50%, as it does in so many reports, since PFKL comprises only about half of red cell PFK subunits (Vora, 1982). Conversely, it is not clear why fibroblasts and leukocytes, which are predominantly PFKL [50–70% (Vora and Francke, 1981)], do not show any dosage effect (Conway and Layzer, 1970; Layzer and Epstein, 1972). In fact,

Sorbi and Blass (1983) reported that mean trisomy 21 fibroblast PFK activity was inexplicably reduced to 74% of control values. Given the greatly expanded state of knowledge concerning the genetic control of PFK isoenzymes since the original dosage experiments were performed, it would be of interest to see these experiments repeated with better control of experimental conditions. To date, no data are available on PFK activity in trisomic liver or brain.

In view of the supposed centrality of PFK in metabolic regulation, it is of considerable interest to know whether changes in PFK activity in trisomy 21 have any metabolic effect in vivo. Unfortunately, this does not appear to have been directly studied. Analysis of glycolytic intermediates in erythrocytes showed a variety of changes, including decreases in ATP and 2,3-diphosphoglyceric acid (Kedziora *et al.*, 1972). Strangely, these results are not unlike those found in red cells of individuals with decreased red cell PFK activities (Vora, 1982). However, considering the numerous non-dosage-related changes in enzyme activities in trisomic red cells (Hsia *et al.*, 1971), it is difficult to know how to judge changes in red cell metabolism in trisomy 21. A recent study, which may be relevant, demonstrated that young adult (19 to 27 years of age) subjects with Down syndrome have increased (up to 40%) rates of brain glucose metabolism as demonstrated by positron emission tomography (Schwartz *et al.*, 1983). While there are numerous potential explanations for this finding, it is of interest that Lowry and Passonneau (1964) have suggested that PFK is the rate-limiting enzyme of glycolysis in the brain. Unfortunately, the fact that the PFKP subunit predominates in adult brain (Vora, 1982) would argue against a role for increased PFK activity in explaining the increased cerebral glucose metabolism in trisomy 21.

It would seem that if a trisomy 21-related increase in PFK activity is to have any chance of having an effect, it would be in the liver that such an effect should be detectable. Only direct study of glycolysis in this tissue will reveal whether this is the case.

The cellular phenotype of trisomy 21

Trisomy 21 and its clinical manifestation, Down syndrome, have been the subjects of intensive investigation since the original description by Down in 1866. The accumulated literature is extremely voluminous and is too extensive to review in detail here. Recent summaries will be found in Smith and Berg (1976), Burgio *et al.* (1981), Tolksdorf and Wiedemann (1981), and Pueschel and Rynders (1982), and a comprehensive bibliography has been published by Pueschel and Steinberg (1980). Given the intent of this volume to consider issues of mechanism, only certain aspects of the Down syndrome phenotype will be discussed (the physical phenotype has already been listed in Table 12.1). Emphasis will be placed on those phenotypic abnormalities which appear to offer some promise of providing information about the mechanisms underlying their development. For the most part, these abnormalities involve cellular and physiological processes, some of which are presumably continually being influenced by the genetic imbalance associated with the trisomic state. The first four to be discussed are concerned with the properties of cultured trisomic cells.

Cell proliferation

In 1970, Kaback and Bernstein (1970), following up on an observation by Mittwoch (1967), published a provocative report of a decreased rate of DNA synthesis in trisomy 21 fibroblasts when compared with diploid controls. The $t_{1/2}$ for the decline in specific activity of labeled cellular DNA, a measure of the cell doubling time, was 2.51 days in trisomic cells and 1.65 days in the controls. An increased cellular concentration of RNA was also found in growing trisomic cells. Unfortunately, the cells being compared were not well matched, with both skin biopsy and foreskin-derived fibroblasts being used. The same problem attached to the work of Segal and McCoy (1973), who reported mean population doubling times of 1.48 and 0.97 days for trisomic and diploid fibroblasts, respectively. However, using only skin biopsy-derived fibroblasts, Paton, Silver, and Allison (1974) found that the G_2 phase of the cell cycle was prolonged three hours in trisomic cells. Porter and Paul (1973), using cells of undescribed origin, found an increased length of G_1, as did Cummings, All, and Baro (1981) who also found an elevated concentration of cellular RNA, probably ribosomal RNA, resulting from a decreased rate of turnover.

Because of the difficulties in interpreting data obtained with unmatched cells, we (Schneider and Epstein, 1972) repeated the experiments of Kaback and Bernstein (1970) using four pairs of cell lines matched for tissue site of origin, donor age, and handling in vitro. In continuous cell culture, the rate of cell population doubling decreased linearly with time, with the trisomy 21 cells always having a slightly lower rate (about 10–15%). Furthermore, the cumulative number of population cell doublings was reduced by 20% (from a mean of 51.5 to 40) in the cultured trisomy 21 cells.

In contrast to these results, all suggesting some degree of impairment of fibroblast proliferation in vitro, Ruble et al., (1978) and Hoehn et al. (1980), also using fibroblasts matched for site of origin and for culture and storage conditions, found that the mean number of cumulative population doublings was not reduced (trisomic, 49.3; diploid, 48.6) and that doubling times were not significantly increased (trisomic, 1.31 days; diploid, 1.27 days). Each of these means had a substantial standard deviation, and it was concluded that "small cohorts of fibroblast strains can be quite variable, but without apparent systematic correlation to genome size. . . . Fibroblast cultures from patients with constitutional aneuploidy display wide ranges of longevities and a pattern of cell cycle kinetics similar to those observed among euploid cultures."

Where does the truth lie? Research on cell proliferation is clearly beset by many technical problems, and, at the present time, the data favoring or denying a difference in proliferative rates and cumulative population doublings seem to be about equally persuasive. From our own studies (Schneider and Epstein, 1972), it would appear that a difference in proliferative rate, if present, is likely to be much smaller in magnitude than claimed by others. However, it would still be of interest to know whether trisomic and diploid cells are truly different. While, as has already been discussed in detail in Chapter 9, I do not believe that a decreased rate of cellular proliferation, especially if small, can provide a general explanation for the development of the aneuploid phenotype, it is nevertheless possible that an altered rate of cellular proliferation can affect the development of specific organs or tissues. Unfortunately, virtually all of our experience is with fibroblasts or lymphocytes (whose behavior may be even more difficult to interpret – see below), and the properties of these cells may not actually be representative of the properties of other types of cells that may have important roles in differentiation and morphogenesis. This investigational problem has not as yet been surmounted.

Sensitivity of cells to radiation and to mutagenic and carcinogenic chemicals

Because of the increased susceptibility of children with trisomy 21 to the development of leukemia, considerable interest has been focused on the sensitivity of trisomic cells to mutagenic and carcinogenic agents, including radiation, chemicals, and viruses. The object of these studies has been to determine whether there is an intrinsic defect in trisomic cells which might make them more susceptible to oncogenic transformation. Following the initial reports of Dekaban, Thron, and Steusing (1969), Chudina, Malyutina, and Pogosyane (1966), and Kučerová (1967), Sasaki and Tonomura (1969) carried out a detailed study of the sensitivity of cultured trisomic lymphocytes to γ-irradiation in the unstimulated (G_0) state. With 160 rads of radiation, the trisomic cells had approximately two times as many dicentric and ring chromosomes (after 50 hr of culture) as did age-matched control cells. This finding was independent of the age of the donor, and the same relative sensitivities were also found in trisomic and diploid cells in the blood of a patient with mosaic trisomy 21. Sasaki and Tonomura (1969) therefore concluded that cells with trisomy 21 show an "intrinsically heightened chromosomal radiosensitivity."

Since the time of these initial reports, there have been numerous other reports also describing an increased radiosensitivity of trisomic lymphocytes as measured by a variety of techniques and at different stages of the cell cycle. Data from several of these reports are summarized in Table 12.3. What is

Table 12.3. *Sensitivity of trisomy 21 cells to radiation*

Treatment	Stage in mitotic cycle	Aberrations scored[a]	Sensi- tivity[b]	Reference
Lymphocytes				
γ-rays (300 rad)	S	All	1.5	Kučerová (1967)
γ-rays (160 rad)	G_0/G_1	D + R	1.7–2.1	Sasaki and Tonomura (1969)
X-rays (150 rad)	G_0/G_1	D	1.4	Lambert *et al.* (1976)
X-rays (400 rad)	G_0/G_1	Micronuclei	1.8	Countryman, Heddle, and Crawford (1977)
γ-rays (200 rad)	G_0/G_1	SCE	1.3	Biederman and Bowen (1977)
X-rays (400 rad)	G_0/G_1	All	1.2	Kučerová and Polikova (1978)
X-rays (50 rad)	S	SCE	1.3	Crossen and Morgan
(100 rad)			1.0	(1980)
X-rays (200 rad)	G_0/G_1	D	1.1	Preston (1981)
X-rays (200 rad)	G_0/G_1	D + R	1.2	Morten, Harnden, and
(400 rad)			1.5	Taylor (1981)
X-rays (150 rad)	S	E	1.3	Athanasiou and Bartsocas (1982)
γ-rays (200 rad)	G_0/G_1	Del + E	~2	Leonard and Merz (1983)
(100 rad)	G_2		1.0	
γ-rays (2 Gy)	G_0/G_1	D + R	1.6	Morimoto *et al.* (1984)
Lymphoblastoid cell lines				
γ-rays (100–500 rad)	–	B	NS[c]	Huang *et al.* (1977)
Fibroblasts				
X-rays (100 rad)	–	D + R	2.8[d]	Higurashi and Conen (1972)
X-rays (150–300 rad)	–	Mut	NS	Yotti *et al.* (1980)
			NS	Leonard and Merz (1983)

[a]Abbreviations: B, breaks; D, dicentric chromosomes; Del, deletions; E, chromosomal exchanges; Mut, mutations; R, ring chromosomes; SCE, sister chromatid exchanges.
[b]Aberrations in trisomy 21 cells/aberrations in diploid cells.
[c]No significant difference.
[d]Single cell-strain used, unmatched for site of origin. Similar increased sensitivity also found in trisomy 13, del(4p), del(5p), r(B), and reciprocal translocation cell strains.

perhaps most striking about the pooled data is that the effect, when present, is relatively small – the increased production of aberrations by trisomic cells being of the order of 1.1 to 2.1 fold. In some experiments, when the dose of irradiation or stage of the cell cycle was changed, the effect was no longer detectable.

What significance can be ascribed to these observations? Sasaki and Tonomura (1969) speculated that "the abnormally elevated chromosomal radiosensitivity in Down's syndrome might be brought about by an altered enzymatic condition which occurred, possibly in reflection of a derangement of genetic homeostasis, in the cells having chromosomes in excess." In other words, the trisomic state was viewed as causing an intrinsic increase in cellular radiosensitivity. However, recent opinion has turned against this interpretation, and it now appears that the increased sensitivity may be a product of alterations in certain aspects of the cell kinetics of trisomic lymphocytes. Thus, Morimoto *et al.* (1984) now state that the "hypersensitivity might be explained by a reduced average life span of circulating blood lymphocytes in trisomy 21 patients," with unstimulated G_0 trisomy 21 lymphocytes being in a state similar to that of phytohemagglutinin (PHA) -stimulated normal lymphocytes. Similarly, Leonard and Merz (1983) assert that the "increased chromosomal radiosensitivity previously reported for the trisomy 21 lymphocyte is related to an apparent rapid response to stimulation by PHA." Therefore, "the metabolic imbalances produced by trisomy 21 do not directly reduce the ability of a trisomic cell to sustain and to repair damage which would otherwise result in a chromosome aberration." These differences between trisomic and diploid individuals in the state of their unstimulated (G_0/G_1) lymphocytes may account for the more rapid repair of chromosome aberrations which Countryman, Heddle, and Crawford (1977), Preston (1981). and Athanasiou and Bartsocas (1982) suggest is the cause of the higher aberration frequencies in trisomic cells.

In support of the conclusion that there is no intrinsic defect in the repair of chromosome aberrations in trisomy 21, Leonard and Merz (1983) point to their results with fully stimulated trisomic lymphocytes (in the G_2 stage of the cell cycle) which no longer show a difference from normal (Table 12.3) and to the results obtained by themselves and others with cultured fibroblasts and lymphoblastoid cells (Table 12.3). On the basis of these results, it seems reasonable to conclude that trisomic lymphocytes may, under certain experimental conditions, show a modestly increased production of ionizing radiation-induced chromosomal aberrations. However, this increase seems to be the result of physiological changes in the trisomic cells and not to a specific alteration in the cellular mechanisms involved in the repair of radiation damage. It is also of interest to note, apropos the earlier discussion of SOD–1, that the finding of, if anything, an enhanced sensitivity to radiation may be construed as evidence against the elevation in SOD–1 having a functional effect in vivo in trisomy 21.

Table 12.4. *Sensitivity of trisomy 21 cells to mutagenic and carcinogenic chemicals*

Treatment	Stage in mitotic cycle	Aberrations scored[a]	Sensitivity[b]	References
Lymphocytes				
Dimethylbenz(α)-anthracene (10^{-4} M)	S	All	3.8	O'Brien *et al.* (1971)
Mitomycin C (10 ng/ml)	S	SCE, all	NS	Kučerová and Polikova (1978)
Trenimone (10^{-8} M)	S	SCE	1.6	Aldenhoff, Wegner,
	S	B	2.8	and Sperling (1980)
Mitomycin C (10 ng/ml)	S	SCE	1.1 (NS)	Crossen and Morgan (1980)
Busulfan (0.5–2 μg/ml)		SCE	?NS	Serra, Bova, and Neri (1981)
Bleomycin (10^{-8} M)	G_0	D + F	1.4	Vijayalaxmi and Evans (1982)
Bleomycin (100 μg/ml)	G_0	D + R	2.2	Morimoto *et al.* (1984)
Bleomycin (20–100 μg/ml)	G_0	SCE	NS	Iijima *et al.* (1984)
Lymphoblastoid cell lines				
Mitomycin C (1–5 μg/ml)		B	NS	Banerjee, Jung, and
Caffeine (0.25–1 mg/ml)		B	NS	Huang (1977)
Fibroblasts				
N-Methyl-N-nitrosourea (1mM)	G_1 + early S	Trans	6.5[c]	Kaina *et al.* (1977)
Trenimone (10^{-9} M)	–	B	3.9–4.6[d]	Aldenhoff, Wegner, and Sperling (1980)

[a]Abbreviations as in Table 12.2. Trans, translocations; F, fragments.
[b]Aberrations in trisomy 21 cells/aberrations in diploid cells.
[c]Cells not matched for site of origin.
[d]Single pair of cells from trisomy 21/2n mosaic.

Data on the effects of mutagenic and carcinogenic chemicals, generally alkylating agents, on the induction of chromosome aberrations in trisomic cells are more limited than for radiation (Table 12.4). As they now stand, little can be inferred from them. In the best analyzed work on lymphocytes (Morimoto *et al.*, 1984), the considerations that apply to radiation seem to hold for the chemical agent, bleomycin, as well. The available fibroblast results are flawed by the inadequate matching of cells or the limited number of cell strains studied, although the finding of a difference in response to the alkylating agent, trenimone, between trisomic and diploid cells in the same sample from a trisomy 21/2n mosaic are of interest (Aldenhoff, Wegner, and Sperling, 1980). For the present, however, the published data are insufficient to support a conclusion that trisomic cells are inherently more sensitive to mutagenic or carcinogenic chemicals.

Virus-induced chromosome aberrations and cell transformation

Another type of procedure used to probe the sensitivity of trisomy 21 lymphocytes to chromosomal damage has been viral infection, either spontaneous or induced by vaccination. Measles infection has been reported to increase the number of chromosome breaks per cell 4.8 fold in trisomic individuals, while the increase in normals was only 1.2 fold (Higurashi, Tamura, and Nakatake, 1973). By contrast, measles vaccination decreased the frequency of sister chromatid exchanges in cultured trisomic cells by 13% from the level in untreated cells (Knuutila *et al.*, 1979). Unfortunately, no values for control cells were given, and it is impossible to evaluate this result. As with measles infection, the number of chromosome breaks per cell immediately after chicken pox infection increased 2.8 fold in trisomy 21 but only 1.8 fold in normal individuals (Higurashi *et al.*, 1976). These data are too few to know what significance to attach to them.

Of greater interest than studies of virus-induced chromosome aberrations have been the investigations of cell transformation induced by transforming viruses. Simian virus 40 (SV40) has been the principal virus studied, and tranformation and infection have been assessed in two ways – the establishment of transformed cell foci in a cloning assay and the induction of the SV40 T-antigen in individual acutely infected cells. The latter is considered to be a measure of likelihood of transformation (Aaronson, 1970). The results of these assays are summarized in Table 12.5. By both assays, trisomy 21 cells appear to be more sensitive to transformation than do diploid controls, the increase in sensitivity being of the order of two to three fold. However, the significance of these observations is uncertain. First, as with many similar types of studies, the selection and matching of cells is frequently open to question. When the matching was good, as, for example, when trisomic and diploid cells were cloned from trisomy 21/2n mosaic cell populations, the trisomy 21-diploid difference did not necessarily hold up. Lubiniecki *et al.* (1979) studied three such pairs of cells and found that the *diploid* cells were significantly more sensitive than were the trisomic cells in two of the three cases. Second, the results were sometimes quite dependent on the choice of assay conditions. Thus, when Potter, Potter, and Oxford (1970) looked at the induction of SV40 T-antigen in nondividing cells, as opposed to the dividing

Table 12.5. *Sensitivity of trisomy 21 fibroblasts to SV40 transformation and infection*

Trisomy 21	Diploid	Ratio[a]	References
Transformants (colonies per 10⁴ cells)			
7.8(5)[b]	3.5(4)	2.2	Todaro and Martin (1967)
7.0(4)	2.2(3)	3.2	Potter, Potter, and Oxford (1970)
2.1(4)	0.78(3)	2.7	Young (1971)
Induction of SV40 T-antigen (percent of cells)			
2.34(3)	0.97(2)	2.4	Potter, Potter, and Oxford (1970)
11.2(1)	1.8(5)	6.2	Aaronson (1970)
18.1(2)	8.8(3)	2.1	Payne and Schmickel (1971)
16.0(17)	11.5(76)	1.4	Lubiniecki et al. (1979)

[a]Trisomy 21/diploid.
[b]Number of different cell strains used in parentheses.

cells used to derive the values in Table 12.5, the diploid cells were more, rather than less, sensitive than the trisomic cells (by a factor as great as eight fold).

Finally, the observations on SV40 transformation, whatever they do mean, do not appear to be unique to trisomy 21. In their initial report, Todaro and Martin (1967) obtained the highest level of transformation, 14.8 colonies/10⁴ cells, in a single trisomy 18 cell strain. In much more extensive studies, Lubiniecki, Blattner, and Fraumeni (1977) found elevated frequencies of T-antigen expression in cells from individuals with a variety of different aneuploid states, including 45,X, 47,XXY, trisomy 13, trisomy 18, del(4p), del(18q), and monosomy 21. These results suggested to them that aneuploidy per se, rather than trisomy 21, is responsible for the increased transformation by SV40 virus or expression of SV40 T-antigen. If so, this would appear to constitute a real example of a nonspecific effect of aneuploidy. However, before coming to such a conclusion, I think that it will be necessary to confirm these observations in appropriately controlled experiments and then, if such verification is obtained, to determine the mechanism or mechanisms responsible for the increased sensitivity. The limited data on this last point indicate that the difference between trisomy 21 (one cell strain only) and controls disappears when SV40 DNA rather than whole SV40 virus is used (Aaronson, 1970) and that aneuploid cells do not absorb virus more efficiently than normal cells (Lubiniecki, Blattner, and Fraumeni, 1977). It must also be remembered that all of the foregoing discussion is based on work with only one virus.

The situation with trisomy 21 and SV40 transformation, as it presently stands, is not dissimilar from that described earlier for cell proliferation and sensitivity to radiation. There seems to be something different between trisomic and diploid cells, generally small in magnitude, but the results are not always consistent, their interpretation is unclear, and their significance is uncertain. Despite the great amount of effort devoted to them, the outcome of these investigations must be considered as disappointing.

The β-adrenergic response system

A most intriguing report of an enhanced sensitivity of trisomy 21 fibroblasts to the induction of cyclic AMP was published in 1981 by McSwigan *et al.* (1981). When skin fibroblasts obtained from a variety of sources were treated with 1 μM isoproterenol, the cAMP content of trisomy 21 cells increased to 29 times the initial level in 10 minutes, in comparison to the 2.5- to 3.2-fold increase observed in diploid cells and in cells trisomic for chromosomes other than 21, a differential of 9 fold. Monosomy 21 cells were less responsive than diploid cells.

The response of trisomic cells was, as expected, exaggerated by addition of an inhibitor of phosphodiesterase, but the trisomy 21/nontrisomy 21 differential still persisted (4.5 to 6.1 fold). Epinephrine had a similar although less pronounced effect, and epinephrine-induced platelet aggregation (supposedly an α-adrenergic function in platelets) was produced at a considerably lower concentration with trisomy 21 than with control platelets (Sheppard *et al.* 1982). Similar results for fibroblast cAMP induction, but with somewhat smaller differentials, were also obtained by Reed *et al.* (1983) for both isoproterenol and epinephrine.

Since no differential was observed when prostaglandin E_1 or cholera toxin was used as an inducer, it was concluded that the enhanced sensitivity could not be attributed to increased adenylate cyclase or decreased phosphodiesterase activity (McSwigan *et al.*, 1981). Furthermore, the provisional mapping of the β-adrenergic receptor to chromosome 5 (Sheppard *et al.*, 1982) argues against a gene dosage effect at the receptor level such as that found for the interferon-α receptor.

In contrast to the very short-term induction of cAMP by isoproterenol, Hösli and Vogt (1979) examined the induction of alkaline phosphatase activity in fibroblasts treated for several days with isoproterenol, theophylline (a phosphodiesterase inhibitor), and ascorbic acid. Although the mean noninduced activity was only 1.8 times higher in trisomy 21 than in diploid cells, it was 6.9 times greater after induction. However, these increases in both noninduced and induced levels were also found for cells from individuals with other forms of both autosomal and sex chromosomal aneuploidy, casting doubt on the connection between these observations and the work on short-term cAMP induction. Nevertheless, it must be noted that once again an apparently nonspecific biochemical effect of aneuploidy is being described, one which Hösli and Vogt (1979) have speculated "could contribute to developmental breakdown of chromosomally unbalanced individuals."

Returning to the observations of McSwigan *et al.* (1981) and Sheppard *et al.* (1982), it is clear that much remains to be done to confirm them and to understand their basis. They bespeak some alteration in the function of a membrane receptor-controlled system which could have important implications for a variety of physiological and possible developmental processes in which catecholamines and cAMP are involved. Unfortunately, related work has not yet been carried out in cells other than fibroblasts, so it is difficult to know how generalizable the results are. Although Tam and Walford (1980) found that cAMP concentrations were lower and cGMP concentrations higher than normal in unstimulated trisomy 21 T lymphocytes, and that adenylate cyclase and guanylate cyclase were, by contrast, higher and lower, respectively, these results do not bear on the question of responsiveness to adrenergic agonists. Moreover, they are more likely to be a reflection of differences in the overall state of the lymphocytes themselves rather than of a specific differential responsiveness to adrenergic stimulation.

Platelet serotonin

A quite considerable literature has accumulated on abnormalities of tryptophan metabolism and platelet serotonin (5-hydroxytryptamine) in trisomy 21, with the latter being of particular interest because of its possible relevance to understanding central nervous system neurotransmitter function. The earlier work in these areas is summarized in Pueschel (1982a). The principal observation has been that the concentration of serotonin is decreased in whole blood to about 65% of normal (Tu and Zellweger, 1965; Berman, Hultén, and Lindsten, 1967), a decrease attributable to a reduction in the level of platelet serotonin (Jerome, 1968). In platelets, levels as low as 35–40% of normal were observed (Lott, Chase, and Murphy, 1972; Bayer and McCoy, 1974). Studies of serotonin uptake by platelets demonstrated a reduced rate of influx, with a maximal uptake velocity, V_{max}, about 42% of control (Bayer and McCoy, 1974). Following the observation of an increased sodium content and decreased ATPase activity in trisomic platelets, McCoy *et al.* (1974) suggested that, since serotonin is actively cotransported with sodium (sodium out and serotonin in), these factors might be responsible for the reduced rate of serotonin uptake. When measured on a per platelet basis, (Na^+/K^+)-ATPase activity in trisomic platelet membranes was reduced by 30% and ouabain-binding (to the ATPase) by 20% (McCoy and Enns, 1978). The ouabain-sensitive (ATPase mediated) efflux of sodium was reduced by 50%. It has also been suggested that abnormalities in potassium transport might also be involved (McCoy and Enns, 1980; Enns, McCoy, and Sneddon, 1983), and low concentrations and uptakes of platelet calcium have also been described (More *et al.*, 1982; McCoy and Sneddon, 1984). None of these proposals has as yet been unequivocally proven, and no specific mechanisms have been suggested to explain how the trisomic state affects platelet function. The possibility that many of the effects are secondary to an overall change in platelet physiology is raised by the findings that trisomic platelets are reduced in volume (More *et al.*, 1982) and that serotonin content may be correlated with volume (Shuttleworth and O'Brien, 1981). A reduction in platelet volume could also explain the ~35% decrease in the activity of monoamine oxidase, again expressed on a per platelet basis (Fowler *et al.*, 1981), as well as the decreased ATPase activity. However, such an explanation would not apply to the reported 60–70% decreases in erythrocyte membrane Na^+/K^+- and Mg^{++}-ATPase activities (Xue, Shen, and Dong, 1984), nor to the decreased binding of imipramine to trisomic platelets (Tang *et al.*, 1985).

It has been suggested that the platelet, with its uptake and storage of amines, can serve as a model for synaptosomes in the central nervous system (Paasonen, 1968; Abrams and Solomon, 1969; Boullin, 1975). However, although direct studies of serotonin concentration and uptake in central nervous system neurons have not been conducted, Lott, Murphy, and Chase (1972) have demonstrated that the cerebrospinal fluid concentration of 5-hydroxy-indoleacetic acid, the principal catabolite of serotonin, is not significantly reduced in trisomy 21. Therefore, in view of both this finding and the questions raised by the possible reduction in platelet volume, it remains to be shown whether the platelet abnormalities have any relevance for central nervous system function.

Cellular aggregation

As an approach to developing a model for explaining the pathogenesis of the characteristic cardiac malformations of trisomy 21 (see below), Wright *et al.* (1984) measured the rates of divalent cation-independent aggregation of fibroblasts derived from the lungs and endocardial cushion-determined structures of trisomic and normal midtrimester fetuses. The rate of aggregation was considered to be a measure of cellular adhesiveness. With the fetal lung fibroblasts, the rate of aggregation was clearly greater for all (five) of the trisomic cell strains assayed than for the eight control and one trisomy 13 strains (Fig. 12.5). With cardiac fibrobasts, which were more difficult to obtain, two out of three trisomic strains had increased rates of aggregation (Fig. 12.5). No major differences were seen between the small numbers of trisomic and control fetal skin fibroblasts studied.

The differences in the rates of aggregation do not seen to be attributable to differences in hyaluronidate concentration or metabolism or in gross membrane lipid composition. To explain the differences between normal and

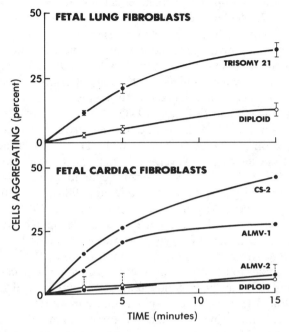

Fig. 12.5. Divalent cation-independent aggregation of trisomy 21 and diploid fetal lung and cardiac fibroblasts. Aggregation is expressed as percent of cells in aggregates of two or more cells. In the upper panel, the mean ± SEM amounts of aggregation for five trisomic and eight diploid cell strains are shown. In the lower panel, the mean + 2(SD) for five diploid cell strains from two donors is shown, along with three individual trisomic strains from two donors. ALMV, anterior leaflet of mitral valve; CS, conal septum; 1 and 2, trisomic donors. Plotted from the data in Tables 1 and 2 of Wright *et al.* (1984).

trisomic fibroblasts, Wright *et al.* (1984) have favored a chromosome 21-determined gene dosage effect for the concentration of a cell surface macromolecule or macromolecules involved in divalent cation-independent cellular aggregation. The fact that differences are not observed with all cell types is considered to reflect differences among the embryonic tissues from which the cells are derived or differences in the developmental status of these organs.

Whether the differences, if they are confirmed, in rates of aggregation and, inferentially, adhesiveness, between trisomy 21 and diploid lung and cardiac fibroblasts do explain the genesis of cardiac malformations (see below) and pulmonary hypoplasia (Chi and Krovetz, 1975; Cooney and Thurlbeck, 1982) in trisomy 21 remains to be determined. Nevertheless, the observations are certainly of interest in themselves since they suggest that an effect of aneuploidy on cellular interactions, such as has been proposed in Chapter 7, has actually been observed. Further work in this area is clearly essential.

Hematological, immunological, and cardiac defects

In this section, I shall discuss three pathological processes which involve the development, control, and/or function of tissues or organs. The consequences of these abnormal processes include leukemia, immune defects, and congenital cardiac malformations. Abnormalities of the central nervous system will be considered in the following section.

Leukemia

In 1970, Miller surveyed the incidence of leukemia in individuals with Down syndrome. Based on his own data and those of several other investigators [for a review of the early literature, see also Sassaman (1982)], he concluded that the relative risk of leukemia in Down syndrome is at least 18 times that of normal children under 10 years of age and is considerably higher in trisomic children under five years of age. Evans and Steward (1972) arrived at a similar estimate – a 10- to 18-fold increase over the risk to the normal childhood population under 15 years of age. Based on mortality figures, Scholl, Stein, and Hansen (1982) concluded that Down syndrome adults between 20 and 34 years of age, and probably older as well, also have an excessive rate of leukemia mortality. Other malignancies did not appear to have contributed to excessive rates of mortality.

The types of leukemia associated with Down syndrome were extensively reviewed by Rosner and Lee (1972), using both their own data and data obtained from a survey of the literature. In both instances they found that congenital and newborn cases were more frequently acute nonlymphoblastic leukemia (ANLL), whereas cases in older infants and children were predominantly acute lymphoblastic leukemia (ALL) (Table 12.6). Similar results were obtained by Stiller and Kinnier Wilson (1981), who observed an increase in the frequency of ALL and a decrease in the frequency of ANLL during the first two years of life (Table 12.6). From three years of age on, the distribution of types was the same for both Down syndrome and non-Down syndrome cases. Robison *et al.* (1984) found that the age of onset of ANLL was younger in trisomic than in diploid patients, but the ages were the same for ALL. The length of survival of Down syndrome individuals after diagnosis of ALL was less than that of diploid individuals. Although major emphasis has been placed on ALL and ANLL, it should be noted that the latter may include other forms of leukemia, such as acute megakaryoblastic leukemia, in addition to acute myelocytic leukemia (Lewis *et al.*, 1983).

In their review, Rosner and Lee (1972) also pointed to the high frequency of "transient" acute leukemia or leukemoid reactions in newborns with Down syndrome. In these cases, there was apparently a complete remission, and it is now believed that the affected individuals never had true leukemia at all

Table 12.6. *Types of leukemia in Down syndrome*

Source of cases	Age	Total cases	Proportion of types (%)[a]			Reference
			ALL	ANLL	AUL	
Acute Leukemia Group B	<1 mo	5	20.0	80.0		Rosner and Lee (1972)
	1 mo–19 yr	41	69.8	30.2		
Literature	<1 mo	47	42.1	57.9		
	1 mo–19 yr	229	69.1	30.9		
Oxford Study of Childhood Cancer	<1 yr	10	10.0	80.0	10.0	Stiller and Kinnier Wilson (1981)
	1–2 yr	25	44.0(82.2)[b]	52.0(13.7)	4.0(14.1)	
	3–14 yr	35	80.0(79.2)	17.1(18.2)	2.9(2.6)	

[a] Abbreviations: ALL, acute lymphoblastic leukemia; ANLL, acute nonlymphoblastic leukemia; AUL, acute undifferentiated or unspecified leukemia.
[b] Proportions in non-Down syndrome population in parentheses.

[although Rogers *et al.* (1978) reported a case which they believe did]. Rather, they appear to have had what Ross, Moloney, and Desforges (1963) termed "ineffective regulation of granulopoiesis masquerading as congenital leukemia." Unlike true acute leukemia, there are normal numbers of granulocytes and macrophage stem cells (CFU-GM) in the bone marrow (Heaton *et al.*, 1981; Denegri *et al.*, 1981; Barak *et al.*, 1982; Rogers *et al.*, 1983). However, that the presence of an extra chromosome 21 is of importance in this condition is borne out by the fact that, in several cases of such leukemoid reactions recently described in phenotypically normal trisomy 21/2n mosaics, the abnormally proliferating cell population was always trisomic for chromosome 21 (Brodeur *et al.*, 1980; Heaton *et al.*, 1981; Seibel, Sommer, and Miser, 1984).

The relationship between leukemoid reactions and true leukemia still remains to be defined. While it is, of course, possible that they are unrelated, it is tempting to try to visualize some significant relationship between two such infrequent aberrations affecting leukocyte proliferation. However, while a relationship between leukemoid reactions and ANLL can be visualized since they affect the same cell lineage and since occasional cases of the former proceeding to the latter have been described (Lin *et al.*, 1980), it is more difficult to understand a relationship with ALL. However, this difficulty might be resolved if the ontological relationship between early lymphoid and granulocytic stem cells is recalled (Quesenberry and Levitt, 1979; Cline and Golde, 1979). Of interest in this regard is the report by Muchi *et al.* (1980) describing a case of Down syndrome with a type of leukemia in which both myeloid and lymphoid cell surface markers were present. Furthermore, the finding of other hematological abnormalities in trisomic newborns, especially an increased hematocrit, but also including either thrombocytosis or thrombocytopenia (Miller, Sherrill, and Hathaway, 1967; Miller and Cosgriff, 1983), is compatible with the notion of a generalized abnormality in stem cell regulation. In fact, Weinberger and Oleinick (1970) suggested that there is an extensive congenital defect of bone marrow function in Down syndrome newborns, possibly due to the presence of a marrow-stimulating humoral substance. I am not aware that this suggestion has been followed up, but it is an interesting one. Even more intriguing would be an increased sensitivity of trisomic stem cells to the stimulating effects of normal concentrations of trophic (growth) factors, analogous to the increased sensitivity of trisomic cells to interferon.

The only direct study of stem cells in hematologically normal Down syndrome adults was carried out by Standen, Philip, and Fletcher (1979). Contrary to what the foregoing discussion would lead us to expect, they found significantly reduced (by 73%) numbers of granulocytic stem cells (CFU-C) in the peripheral blood. They suggested that there may be a corresponding diminution of marrow stem cells or an impaired release of stem cells from the marrow to the circulating blood. In either event, they proposed that the stem cell population may be "defective" in trisomy 21 and this may be one factor "which encourages the development of a leukemic clone." Based on a poor response to bone marrow transplanted from a Down syndrome donor, Storb *et al.* (1980) also suggested that trisomic stem cells are defective. However, Denegri *et al.* (1981), in their study of trisomic infants with a leukemoid reaction, found normal granulocytic colony formation from peripheral blood (although quantitative control data were not given).

What then is the etiologic relationship between trisomy 21 and leukemia? That trisomy 21 should itself be regarded as the predisposing factor would appear to be true, but Rowley (1981) nevertheless felt compelled to develop this argument at great length. The evidence which she amassed includes the data cited above on the incidence of leukemia and on the relationship of leukemoid reactions to the presence of trisomic cells in trisomy 21/2n mosaics. In addition, she pointed to the fact that trisomy 21, alone or with other types of chromosome imbalance, is the most common acquired change in the leukemic cells in ALL (31% in children, 7% in adults) and one of the most common in ANLL (7% in children, 5% in adults) occurring in constitutionally diploid individuals. She thus etiologically equated the leukemia with leukemoid reactions and constitutional with acquired chromosome imbalance. Hecht (1982) objected quite strongly to the latter and, in so doing, totally rejected the idea that constitutional trisomy 21 is directly responsible for acute leukemia in Down syndrome. I would agree with him that the etiologic equivalence of constitutional and acquired chromosome abnormalities is open to debate and shall take up this issue again in Chapter 14. However, I do not see how the relationship between the existence of the trisomy and the increased susceptibility to leukemia can be denied. The relationship clearly exists.

The crux of the matter, of course, is the mechanistic nature of the relationship. In her discussion, Rowley (1981) compares the situation in trisomy 21 to that which exists with the two chromosomal deletions, of 11p and 13q, which predispose individuals to the development, respectively, of Wilms tumor and retinoblastoma (Figs. 3.3 and 3.4; see Chapter 14). The present thinking, based on the ideas of Knudson and on more recent experimental evidence (see Chapter 14 for discussion), is that the addition of a second deletion or similar mutation affecting the same locus or loci to the deletion already present constitutionally is involved in the genesis of these two tumors, and probably others as well. Whether similar "two-hit" considerations also apply in the same sense to leukemia in Down syndrome, with the trisomy 21 being considered as the first "hit," is certainly open to question. I personally do not think that there is presently any real evidence to support such an argument. Furthermore, unlike the malignancies associated with deletions, with frequencies of one-third or higher in chromosomally abnormal individuals, the frequency of leukemia in Down syndrome, however high the relative risk, is still one or two orders of magnitude lower, suggesting that another type of mechanism might be operative.

The possibility that mechanisms such as increased susceptibility to radiation, chemical, and/or virus-induced chromosomal/genetic damage and transformation are involved in the susceptibility to leukemogenesis provided the rationale for the studies in these areas described earlier in this chapter. As has already been noted, the evidence concerning an increased sensitivity of trisomic cells is still far from persuasive. Even if accepted at face value, the relative increase in chromosomal damage to trisomic cells is only 1.5 or 2 fold, and very high doses of radiation or mutagens are often required to demonstrate an effect. This sensitivity, if it occurs at all, seems greatest in resting mature lymphocytes, and the applicability of any of these considerations to stem cell populations has not been tested. Therefore, claims for a causal relationship between increased susceptibility to chromosomal damage and increased susceptibility to leukemia (Todaro and Martin, 1967; Countryman, Heddle, and Crawford, 1977; Athanasiou and Bartsocas, 1982) must be viewed with caution. Furthermore, as pointed out by Lubiniecki *et al.* (1979) in their discussion of SV40-induced cellular transformation, "not all patients with Down syndrome had elevated T-antigen expression, and a patient . . . who developed leukemia had a normal level of T-antigen. . . . From the available data, it is difficult to conclude that elevated values of T-antigen expression in patients with Down syndrome . . . are related to cancer risk or particular cytogenetic defects." Based on the finding that all of 11 trisomic individuals with ANLL had additional karyotypic abnormalities, as opposed to only 41% of 64 constitutionally normal patients with ANLL, Alimena *et al.* (1985) have suggested that these findings are "another reflexion of increased susceptibility and/or decreased resistance of such cells to chromosome damaging effects of environmental agents."

Another possibility that might be considered is an altered host defense or tumor surveillance system related to abnormalities of lymphocyte function (see below) or of the response to interferon. Insofar as the latter is concerned, the result would be a paradoxical one, since it might be expected (although there is no real evidence pro or con) that enhanced sensitivity to interferon should be protective against malignancy rather than increasing susceptibility. With regard to the immune system itself, Ugazio (1981) has drawn an analogy between the situation in Down syndrome and the association of leukemia with a variety of primary immunodeficiency disorders and has speculated that the leukemoid reactions might be the result of "lack of homeostatic control within the lymphoid system" which results, in turn, from a deficiency of suppressor T lymphocytes. Finally, it has even been suggested that the increase in SOD–1 activity might, by mechanisms described earlier in this chapter, also be involved (Teyssier *et al.*, 1984).

We are thus left with many possibilities, none proven. In the end, the explanation might derive from some still undefined factor or from a combination of factors, both of the type considered here and yet to be discovered. This factor or factors might operate alone or be superimposed on the potentially enhanced proliferative capacity of trisomic hemopoietic stem cells. However the increased susceptibility to leukemia is eventually explained, I expect that it is more likely to be related to a trisomy 21-produced change in some aspect(s) of cellular physiology rather than to the presence of a specific leukemia susceptibility locus on chromosome 21 analogous to the loci involved in Wilms tumor and retinoblastoma. I suspect that equating an extra chromosome 21 ($+21$) with a constitutional mutation (Rowley, 1981) or deletion, as is present in constitutional forms of the latter conditions, will not prove to be fruitful. Nevertheless, I do agree with Rowley (1981) that a specific gene or genes on chromosome 21 will ultimately be shown to be involved in the process of leukemogenesis, however indirect the connection may ultimately turn out to be.

Immunological defects

The immunological status of individuals with trisomy 21 has been the subject of intensive investigation for many years, primarily because of clinical observations suggesting that they are more susceptible to a variety of infectious diseases (Siegel, 1948a) and, as has just been discussed, to the development of leukemia. Infection still constitutes the leading cause of death of trisomic individuals (Øster, Mikkelsen, and Nielsen, 1975; Scholl, Stein, and Hansen, 1982). While the early studies [summarized in Rigas, Elsasser, and Hecht (1970); Smith and Berg (1976); and Nurmi (1982)] were mainly concerned with immunoglobulin levels and specific antibody responses, later investigations have focused on various aspects of lymphocyte function. In general, the literature has been characterized by considerable disagreement and contradiction, stemming, in large part, from differences in subject selection (age, institutionalized versus noninstitutionalized), choice of control subjects, and the methodologies employed. Nevertheless, a picture of immunological impairment associated with trisomy 21 still emerges.

Numerous reports on serum immunoglobulin concentrations have appeared during the past decade and a half and are summarized in Table 12.7. Although these reports are difficult to compare with one another because of numerous differences in the populations studied, the general consensus is that serum IgG is elevated, particularly in older subjects. However, there are reports of normal IgG levels and, in newborns, of decreased concentrations (Miller *et al.*, 1967). The consensus with regard to IgM and IgA is less clear. Serum concentrations of IgM have been variously reported as decreased, normal, and even increased, and the differences do not appear to be age related. Similarly, IgA has been reported to be decreased, normal, and increased. A transition from normal levels in children to elevated levels in adults has been

Table 12.7. *Serum immunoglobulin concentrations in Down syndrome*

| Immuno- | Concentration | | |
globulin	Increased	Normal	Decreased
IgG	Adinolfi, Gardner, and Martin (1967); Bernard et al. (1976); Björkstén et al. (1980); Burgio et al. (1975); Dyggve and Clausen (1970); Gershwin et al. (1977); Greene, Shenker, and Karelitz (1968); Jacobs et al. (1978); Rosner, Kozinn, and Jervis (1973); Seger, Buchinger, and Ströder (1977); Stiehm and Fudenberg (1966); Watts et al. (1979); Whittingham et al. (1977); Yokoyama et al. (1967).	Agarwal et al. (1970); Chen, Walz, and Carroll (1977); Griffiths, Sylvester, and Baylis (1969); Gulliya and Dowben (1983); Kitani, Abo, and Kokubun (1977); Lopez et al. (1975); Reiser et al. (1976); Sutnick, London, and Blumberg (1969).	Miller et al. (1967).
IgM	Adinolfi, Gardner, and Martin (1967).	Björkstén et al. (1980); Dyggve and Clausen (1970); Gershwin et al. (1977); Greene, Shenker, and Karelitz (1968); Griffiths, Sylvester, and Baylis (1969); Kitani, Abo, and Kokubun (1977); Miller et al. (1967); Reiser et al. (1976); Rosner, Kozinn, and Jervis (1973); Seger, Buchinger, and Ströder (1977); Watts et al. (1979); Whittingham et al. (1977); Yokoyama et al. (1967).	Agarwal et al. (1970); Bernard et al. (1976); Burgio et al. (1975); Chen, Walz, and Carroll (1977); Gulliya and Dowben (1983); Lopez et al. (1975); Sutnick, London and Blumberg (1969); Stiehm and Fudenberg (1966).
IgA	Björkstén et al. (1980); Greene, Shenker, and Karelitz (1968); Griffiths, Sylvester, and Baylis (1979); Gulliya and Dowben (1983); Stiehm and Fundenberg (1966); Watts et al. (1979).	Adinolfi, Gardner, and Martin (1967); Agarwal et al. (1970); Bernard et al. (1976); Jacobs et al. (1978); Kitani, Abo, and Kokubun (1977); Lopez et al. (1975); Reiser et al. (1976); Sutnick, London, and Blumberg (1969).	Gershwin et al. (1977).

claimed (Dyggve and Clausen, 1970; Rosner, Kozinn, and Jervis, 1973) and may explain some of the discrepancy.

Several immunological studies on Down syndrome have focused on the numbers and types of lymphocytes, and the results have again been quite contradictory. Quantitative studies of T lymphocytes in institutionalized subjects with Down syndrome have demonstrated a reduction, often quite small, in the proportion or absolute number of T lymphocytes (Burgio *et al.*, 1975; Levin, Nir, and Mogilner, 1975; Franceschi *et al.*, 1978; Levin *et al.*, 1979; Handzel *et al.*, 1979; Duse *et al.*, 1980; Park, 1981; Gulliya and Dowben, 1983; Church *et al.*, 1983; Yachie *et al.*, 1984), although we (Epstein and Epstein, 1980) and others (Reiser *et al.*, 1976; Whittingham *et al.*, 1977; Walford *et al.*, 1982) found normal proportions or numbers of T and B lymphocytes in trisomic children (see also the recent studies listed in Table 12.8 in which normal proportions of OKT3[+] or 9.6[+] T cells were found).

With the development of monoclonal antibodies directed against T-lymphocyte subsets, the T cells of Down syndrome subjects have been subjected to further analysis. The results of several recent studies are summarized in Table 12.8. With the exception of Gupta *et al.* (1983), the consensus appears to be that the proportion of helper cells (OKT4[+], Leu–3a[+]) is decreased, resulting in a decreased, perhaps even reversed (<1.0) ratio of helper to suppressor (OKT8[+], Leu–2a[+]) cells. That this is not a uniform finding is shown by the work of Karttunen *et al.* (1984), who found that the half of their subjects who had normal proliferative responses to phytohemagglutinin (PHA) and were, in general, younger in age, also had OKT4[+]/OKT8[+] ratios closer to normal (Table 12.8). (Unfortunately, no normal values were actually given in the paper.)

Despite the abnormalities in PHA-stimulated proliferation and OKT4[+]/OKT8[+] ratios observed by the last group of investigators, they were unable to document any defect in the production of interleukin–2 (IL–2) by PHA-stimulated lymphocytes. In this study, IL–2 was being used as a measure of lymphocyte function. Another measure of lymphocyte function is the production of interferon by mitogen-stimulated cells. In our own studies (Epstein and Epstein, 1980), PHA-stimulated interferon production was normal, but Funa *et al.* (1984) reported increased production of interferon, presumably IFN-γ, by leukocytes after stimulation with concanavalin A (Con A) and *Lens culinaris* lectin. Conversely, Nair and Schwartz (1984) found a ~50% decrease in the production of interferon by trisomic lymphocytes stimulated with K562 cells in vitro.

The most commonly used approach to the study of lymphocyte function in trisomy 21 has been the assessment of PHA-stimulated lymphoycyte transformation or proliferation in vitro. As with other immunological investigations, quite discrepant results have been obtained. Using blast transformation or the fraction of cells incorporating ³H-thymidine as a criterion, it has been

Table 12.8. *T-lymphocyte subsets in Down syndrome*

Population studied	Proportion of T cells (%)				Reference
	OKT4+ OKT3+)	OKT8+ (or Leu-3a+)	OKT4+ (or Leu-2a+)	OKT4+ OKT8+	
3-54 yr				<1.24 in 34%	Park and Averian (1982)
4 mo–12 yr		42(50)[a,b]	27(28)	1.9(2.0)	Church *et al.* (1983)
Children	71(74)	38(43)	44(23)	0.9(1.9)	Burgio *et al.* (1983)
2–12 yr	73(72)[c]	55(58)	20(31)	2.8(1.9)	Gupta *et al.* (1983)
Low responders[d] (14–37yr)	77	29	44	0.7⎤	Karttunen *et al.* (1984)
Normal responders[d] (5–20 yr)	65	33	26	1.3⎦	
7–16 yr	52(67)	23(42)	21(23)	1.1(1.9)	Yachie *et al.* (1984)
2–42 yr	72(74)	43(43)	45(25)	1.0(1.7)	Maccario *et al.* (1984)
6–37 yr		28(33)[e]	15(13)[f]	1.9(2.5)	Philip *et al.* (1985a)

[a]Control values in parentheses
[b]OKT4+ significantly reduced
[c]9.6+ cells
[d]Low and normal proliferative responses to PHA stimulation
[e]Leu-3a+3b+
[f]Leu-2a+15+

claimed that trisomic cells have a normal (Hayakawa *et al.*, 1968; Matte, Sasaki, and Obara, 1969; Agarwal *et al.*, 1970) or decreased (Smith *et al.*, 1976; Serra, Arpaia, and Bova, 1978) response to PHA at optimal doses and an enhanced response at very low concentrations (Hayakawa *et al.*, 1968). With ³H-thymidine incorporation, generally after 72 hours in culture, as a measure, it has been reported that, at optimal PHA concentrations, the response of trisomic cells is either normal (Nadler, Monteleone, and Hsia, 1967; Fowler and Hollingsworth, 1973; Cupples and Tan, 1977; Epstein and Epstein, 1980; Church *et al.*, 1983) or, in most reports, decreased (Mellman, Younkin, and Baker, 1970; Rigas, Elsasser, and Hecht, 1970; Burgio *et al.*, 1975; Bernard *et al.*, 1976; Whittingham *et al.*, 1977; Seger, Buchinger, and Ströder, 1977; Gershwin *et al.*, 1977; Serra, Arpaia, and Bova, 1978; Franceschi *et al.*, 1978; Björkstén *et al.*, 1980; Park, 1981; Walford *et al.*, 1982; Yachie *et al.*, 1984). Although some authors have described an enhanced response by trisomic cells at low PHA concentrations (Rigas, Elsasser, and

Hecht, 1970; Bernard *et al.*, 1976), this was not confirmed by other investigators (Serra, Arpaia, and Bova, 1978). Of note, however, are the reports demonstrating an age-dependence of the proliferative response to PHA in trisomic lymphocytes but not in normal lymphocytes. Thus, both Burgio *et al.* (1975) and Seger, Buchinger, and Ströder (1977) found normal responses to PHA in cells from trisomic subjects up to 10 years of age and decreased responses in subjects over 10 years, and Karttunen *et al.* (1984) found a strong negative correlation (r = −0.69) between proliferative response and age.

With regard to other mitogenic agents, Gershwin *et al.* (1977) and Franceschi *et al.* (1978) described a decreased ³H-thymidine incorporation response to Con A, a finding parallel to what they observed with PHA. McCoy *et al.* (1982) also examined Con A-stimulated lymphocytes and, using polyamine content as a measure of cell growth, found a diminished response in trisomic cells. However, our own studies (Epstein and Epstein, 1980) and those of Funa *et al.* (1984) with Con A did not reveal a significant difference in ³H-thymidine incorporation between trisomic and normal cells. Gershwin *et al.* (1977) reported decreased response to pokeweed mitogen, a B-cell mitogen, as did Smith *et al.* (1976), Franceschi *et al.* (1978), Church *et al.* (1983), and Yachie *et al.* (1984). Funa *et al.* (1984) observed a decreased response to *Lens culinaris* lectin.

Of considerably greater interest than the response to nonspecific mitogens has been the response of trisomic T lymphocytes to specific antigens. The earliest studies were carried out by Fowler and Hollingsworth (1973), who demonstrated a normal in vitro proliferative response to PPD in cultures of trisomic lymphocytes obtained from donors having positive PPD skin tests. Funa *et al.* (1984) also found normal responses to staphylococcal, streptococcal, and Sendai virus antigens. By contrast, our work with tetanus toxoid-stimulated T lymphocytes revealed a significantly reduced proliferative response by trisomic cells (Fig. 12.6), suggesting that there is a defect in immunological memory, and these results have recently been confirmed and extended (Philip *et al.*, 1985a). Again, the proliferative response of trisomic lymphocytes to tetanus toxoid, as well as to influenza virus antigens, was significantly reduced, as was antigen-induced interleukin–2 production. Furthermore, although circulating levels of antibodies against tetanus toxoid and the viral antigens were normal, the in vitro secondary antibody responses to these antigens were markedly reduced.

The results with tetanus toxoid are not incompatible with some of the earlier reports on the in vivo antibody responses to specific antigens. Siegel (1948b) studied antibody response to tetanus toxoid and typhoid vaccine in institutionalized subjects with Down syndrome and other forms of retardation. The trisomic subjects had significantly reduced antibody titers against the two antigens, both following primary inoculations and after booster injections. Furthermore, a greater proportion of trisomic than normal subjects

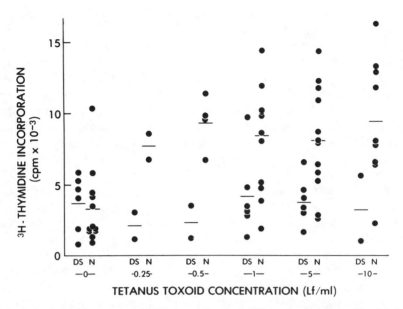

Fig. 12.6. The incorporation of ³H-thymidine into DNA of tetanus toxoid-stimulated lymphocytes from trisomic (DS) and normal (N) subjects. The response of trisomic lymphocytes was lower at all concentrations of tetanus toxoid tested. Redrawn with modifications from Epstein and Epstein (1980).

were characterized as poor reactors. Using influenza A/PR8 vaccine to which the subjects had not been previously exposed, Gordon, Sinha, and Carlson (1971) found a poorer response, in terms both of the mean titers and of the proportion of subjects developing a rise in titer greater than four fold, in institutionalized trisomic subjects over 14 years of age. A similar decrease in the mean titers developed by trisomic subjects also occurred after vaccination with influenza A_2/Japan. Lopez *et al.* (1975) studied the response of institutionalized 12- to 18-year-old subjects to bacteriophage ϕX174 and found diminished primary and secondary responses in trisomic subjects. However, half of the trisomic subjects were able to develop normal tertiary antibody responses, although the other half remained poor responders. In a different type of study, Hollinger *et al.* (1972) investigated the development of antibodies to Australia antigen by institutionalized subjects exposed to hepatitis. While 45% of control subjects developed antibodies, only 18% of exposed Down syndrome subjects did so.

Recently, Nurmi *et al.* (1982*b*) studied the antibody response of adults with trisomy 21 to pneumococcal vaccine. Before vaccination, serum antibodies to all six pneumococcal polysaccharides were lower than normal in the trisomic subjects, the difference being most striking in males. After vaccination, the trisomic subjects had lower antibody levels than their institutionalized controls, with males again having lower titers than females. However, once again, as with so many of the other studies, contradictory results have been published. Leibovitz and Yannet (1941) reported normal antibody responses to immunization of institutionalized subjects with typhoid vaccine, although examination of their table suggests that the response of the trisomic subjects may actually have been less than normal. And, in a carefully controlled study of institutionalized subjects, there was no statistically significant difference between the antibody titers to tetanus toxoid developed by trisomic and control subjects after a primary immunization and two booster injections (Griffiths and Sylvester, 1967).

Ugazio *et al.* (1978) reported decreased titers of "natural" antibodies to *E. coli* O antigens and to rabbit erythrocytes. Similarly, Fekete *et al.* (1982) reported that although a higher proportion of Down syndrome children (mainly 6 to 10 years of age) exhibited cellular proliferative responses to herpes simplex antigen, suggesting that they had previously been infected, a significantly lower proportion (about one-third) had circulating antibodies against herpes simplex virus, and in them the titers were lower than in controls. Reduced proportions of antibody-positive subjects, again with low titers, were also found for adenovirus and influenza virus antigens.

An interesting connection between the immune system and the interferon response system discussed earlier is provided by the natural killer (NK) cell, but once again the data are contradictory. Natural killer cells have cytotoxicity, which is independent of antibody and complement, toward a variety of tumor target cells, and this cytotoxicity is enhanced by exogenous interferon. Nurmi *et al.* (1982*a*) reported that trisomic lymphocytes (from institutionalized patients of unknown age) have a higher spontaneous NK-cell activity than do cells from normal individuals, but respond less well to interferon with regard to the proportional increase in NK activity. Nevertheless, both trisomy 21 and control cells reach the same activity after interferon stimulation. Nurmi *et al.* (1982*a*) speculated that these observations could be reconciled by assuming that trisomic cells, being inherently more sensitive to interferon, are stimulated to a greater degree by circulating interferon in vivo and therefore attain a higher basal level. Addition of exogenous interferon is then able to increase NK activity only to the same maximum in both trisomic and control cultures. Similar results were obtained by Philip *et al.* (1985*b*), who also showed that the trisomic cells are more sensitive to both interferon-α and interferon-γ. Interleukin–2 had no effect on the trisomic NK activity, presumably because the activity was already at a high level. In contrast to these results, Funa *et al.* (1984) found that samples from trisomic children had normal unstimulated NK-cell activities, but, as would be expected from the *IFRC* dosage, were more responsive to the NK-activity-stimulating effect of interferon in vitro. In fact, the ratio between the trisomic and control cells of the mean slopes of the dose-response curves of NK activity versus IFN concentration was about 1.5. The basis of the discrepancy between the two sets of results is not obvious, but each in its own way is compatible with an enhanced responsiveness of NK cells to interferon. Spina *et al.* (1981) and Nair and Schwartz (1984) also found normal levels of unstimulated NK-cell activity in trisomic subjects, although the latter found in addition a diminished response of trisomic lymphocytes to the NK-enhancing activity of interleukin–2.

Despite all of the contradictions, a picture of immunodeficiency still emerges from the numerous data obtained on Down syndrome. The problems seem to increase with age and are not present in all subjects and to the same degree in the individuals who have them. Abnormalities of T lymphocytes appear to play a major role, not only in T-cell functions themselves but also in the regulation of humoral immunity (antibody responses) as well. How can all of this be put together into a coherent picture, and what is the basic or principal effect of trisomy 21 on the immune system?

In their analysis of the problem, Burgio *et al.* (1983) pointed to a central role for the thymus in the pathogenesis of the immunodeficiency of Down syndrome. To support this argument, they cited both anatomical and functional abnormalities of the thymus. Levin *et al.* (1979) described thymic abnormalities in a series of 12 infants ranging in age from 1 day to 15 months. In comparison to age-matched controls, the trisomic thymuses were characterized by marked lymphoid depletion, resulting in a thin cortex and poor corticomedullary demarcation. The Hassall's corpuscles were *increased* in size (by a factor of seven fold in area) and were cystic in two-thirds of the cases, including newborn infants. The same abnormalities of Hassall's corpuscles had also been described earlier by Müller-Hermelink (1977) and by Benda and Strassman (1965), again in Down syndrome infants as young as a few months of age. The latter authors regarded the changes in the thymus as being so characteristic as to be virtually diagnostic of trisomy 21. Although the true significance of these abnormalities of Hassall's corpuscles could not be established, it was suggested by Burgio *et al.* (1983) that they represented degenerative changes in the thymic epithelium. If this were the case, the putative hormonal products of the epithelium would be expected to be reduced, and claims to this effect were made by Duse *et al.* (1980), Franceschi *et al.* (1981), and Fabris *et al.* (1984). Duse *et al.* (1980) found a significantly lowered (mean = 39% of normal) activity of "thymus-dependent human serum factor" in 18 trisomic patients between 8 months and 4 years of age. Franceschi *et al.* (1981) likewise found a marked dimunition in the activity of "facteur thymique serique" (FTS), especially in the younger age groups when the control activities were much higher than in older individuals. Later, this group (Fabris *et al.*, 1984) reported mean activities of FTS about eight fold lower than normal in trisomic individuals between 0 and 20 years of age.

However, to complicate matters, they also reported that addition of zinc sulfate to the plasma samples abolished an endogenous FTS-inhibitory activity and restored the measured FTS activity to normal. They concluded, therefore, that the decreased FTS activity was the result of plasma FTS-inhibitory activities and not of a primary failure of the thymus. And, to explain the higher concentration of the inhibitor, which appeared to be inactivated by binding zinc, they postulated that an observed reduction in mean plasma zinc concentrations (86.3 μg/dl in trisomy 21 versus 105 μg/dl in young controls) was responsible for the increased inhibitory activity in trisomic plasma.

The story with zinc is not an entirely new one. Björkstén *et al.* (1980), following up an earlier report by Milunsky, Hackley, and Halsted (1970), also found a decreased serum zinc in Down syndrome, the reduction averaging about 24%, along with several immunological defects. They therefore administered zinc therapeutically to their Down syndrome subjects and observed improved delayed hypersensitivity responses in the skin and a somewhat improved lymphocyte proliferative response to a high dose of PHA. Yachie *et al.* (1984) also examined zinc levels in trisomic individuals and found a 27% reduction, along with reduced numbers of OKT4[+] T cells and impaired mitogen-stimulated lymphocyte proliferation. A few control subjects with low plasma zinc concentrations also had depressed OKT4[+] numbers, but did not have impaired proliferative responses. Dietary administration of zinc (along with several other trace metals) resulted, in a period of four to six weeks, in significant increases in the proportions of OKT3[+] and OKT4[+] cells in both trisomic and zinc-deficient control subjects, and enhanced proliferative responses in trisomic lymphocytes. The latter increases were reversible when zinc administration was discontinued, but the changes in OKT-reactive cells persisted for several months.

Despite these results, it does not seem likely that all of the immunological difficulties in trisomy 21 can be attributed to a deficiency of zinc. The findings on the neonatal thymus, in particular, argue against such an interpretation. Returning then to Burgio *et al.* (1983), it is their concept that, as a consequence of a primary defect in the thymus, the ultimate maturation of thymocytes into fully competent T lymphocytes does not take place. This abnormality of T-cell differentiation is viewed, in turn, as setting up a vicious cycle of recurrent infection, further stress on the immune system, and progressive "exhaustion" or "stress deficiency" of T- and B-cell functions. The appearance of autoimmune phenomena and malignancy is viewed as part of this process, and institutionalization, with greater chances for infection, is regarded as an accelerating factor. A similar proposal has also been made by Whittingham *et al.* (1977). The recurrent infections presumably lead to the elevated levels of IgG in adult life, and the decreased ratio of helper to suppressor T lymphocytes leads to impaired antibody responses and to decreased lymphocyte proliferative responses. Although Nair and Schwartz (1984) reject the concept of immunoregulatory dysfunction in favor of intrinsic defects in effector cells, they do agree that abnormalities of T lymphocytes are central to the immunological problems in Down syndrome.

Although I am not comfortable with the notion of immune system exhaustion, the concept of an evolution of abnormalities in immune function is an attractive one and can probably help explain many of the disparate immunological observations listed earlier. The idea that environmental factors – in this case, infection – affect the phenotypic expression of whatever defect(s) are the direct result of the trisomic state is compatible with the concepts discussed earlier in this volume, and it is certainly not necessary to invoke notions such as "exhaustion" to visualize how control of immune responses might become altered. However, the relationship with malignancy suggested by Burgio *et al.* (1983) must be considered as highly speculative, particularly since the leukemia occurs quite early and there is no evidence for an increased frequency of other malignancies later in life.

At the present time, it appears to me that the most promising direction for further research on the problem of the immune system in Down syndrome will be along the lines of investigation of thymic epithelium and lymphocyte development and function in trisomic newborns and, if possible, trisomic fetuses. By reducing the influence of exogenous environmental factors, it should be possible to define more precisely those developmental and functional abnormalities that result directly from trisomy 21 and thereby to come closer to establishing the true nature of the relationship between trisomy 21 and the immunological abnormalities which ultimately ensue. The work on the immune defects in mouse trisomy 16 (see below) may also have considerable relevance.

Congenital cardiac malformations

In two reviews, Rehder (1981) and Pueschel (1982*b*) summarized the literature on the incidence and types of congenital heart disease found in individuals with Down syndrome. Depending on the methods of ascertainment and diagnosis, a wide range of figures were cited. The most unbiased postnatal clinical or autopsy figures gave a frequency of congenital heart disease of 30–40%. However, Rehder (1981) reported an incidence of 45%, not including lesions that spontaneously correct themselves after birth (such as patent ductus arteriosus and some atrial septal defects), among unselected midtrimester trisomy 21 fetuses aborted after prenatal diagnosis. The characteristic cardiac lesion in trisomy 21 has long been recognized as the persistent atrioventricular (AV) canal (also known as an endocardial cushion defect), but again the estimates of the proportion of all cardiac anomalies this represents have varied, from as low as 11% to as high, in Rehder's (1981) series, as 73%. Why such a range?

Rehder (1981) points out that failure of normal fusion of the endocardial cushions results in a spectrum of lesions, ranging from atrial septal (ostium primum) and membranous ventricular septal defects to the complete atrioventricular canal. When interpreted on this basis, the great majority of cardiac defects in trisomy 21 can be regarded as endocardial cushion defects. Furthermore, many cases without gross abnormalities have minor cardiac developmental anomalies which occur at the same location, the venous inflow tract of the heart, and involve the endocardial cushions of the atrioventricular region and the atrial septa, the same tissues involved in the gross malformations. Rehder (1981) therefore believes that trisomy 21 affects the process of endocardial cushion formation as a whole, with the degree of abnormality that results being determined either by "different penetrance," developmental compensation [reminiscent of B. Shapiro's (1983a) proposed developmental homeostasis], and/or endogenous or exogenous environmental factors. The important inference, therefore, is that the effect of trisomy 21 on cardiac morphogenesis is highly specific, even though the exact pathology may differ to some extent among individuals. In light of the discussion on mouse trisomy 16 which is to come, it is also worth mentioning that another class of lesions, anomalies of the aortic arch and vessels (not including patent ductus arteriosus), are found in 15% of fetuses with trisomy 21 (Rehder, 1981).

What is the relationship of trisomy 21 to abnormalities of endocardial cushion formation? Earlier in this chapter, I described the studies of Wright *et al.* (1984) on the aggregation of lung fibroblasts and of cardiac fibroblasts derived from the endocardial cushion region of aborted human fetuses. It will be recalled (see also Fig. 12.5) that they found higher rates of aggregation in trisomy 21 cells and inferred from this that the cells are more adhesive. To illustrate how an increase in adhesiveness could affect cardiac morphogenesis, Kurnit *et al.* (1985) have developed a computer-based model to simulate the influence of several aspects of cell behavior, including adhesiveness, on the formation of the endocardial cushions. Adhesiveness, as defined in their work, is the probability that two cells lying adjacent to one another "recognize" each other and adhere so that each no longer migrates nor divides. The principal outcome of their calculations is that, by use of a threshold model, a moderate increase in adhesiveness can completely prevent fusion of the endocardial cushions, whereas smaller increases can significantly reduce the likelihood of complete fusion. Thus, an increase in the value of adhesion, A, from 0.15 to 0.20 or 0.25 (in arbitrary units) results in successful fusion of the endocardial cushions in 100%, 92%, and 52% of instances, respectively. Therefore, as Kurnit *et al.* (1985) point out, a change in the concentration of a single factor, such as a cell surface molecule involved in cellular adhesion, can result in a situation in which abnormal development occurs some but not necessarily all of the time. This comes about, in their model, because of certain assumptions about the stochastic nature of cellular migration and proliferation. The theoretical significance of this conclusion is that it is not necessary to invoke such concepts as decreased developmental buffering or multifactorial causation to explain the occurrence, in only a proportion of susceptible individuals (in this case those with trisomy 21), of a specific group of developmental abnormalities. A reasonable explanation based on considerations of gene dosage involving a single developmentally important factor can readily be developed. Kurnit *et al.* (1985) state their conclusion as follows: "In our model, stochastic effects modify penetrance, resulting in an apparently non-Mendelian inheritance pattern for congenital heart defects following Mendelian inheritance of an allele that predisposes to the heart defect. Thus, individuals with the same genotype and fetal cardiac environment could have different clinical results solely on the basis of chance." One need only to substitute the term "gene dosage dependent" for "Mendelian inheritance" to make this statement directly applicable to trisomy 21. This approach is clearly consistent with the general philosophy set forth in this volume and the mechanistic considerations regarding cell surface functions set forth in Chapter 7. More work to establish its validity is certainly warranted.

The nervous system

While the several pathological conditions just discussed are of clinical significance, the one condition of overriding clinical importance in Down syndrome is, of course, mental retardation. Clearly, the brain, insofar as it is involved in cognition and other highly integrated mental functions, seems to be the organ most vulnerable to the deleterious effects of autosomal aneuploidy, whichever chromosome is involved. Having said this, it must immediately be pointed out that we do not know how these deleterious effects arise. While gross abnormalities often do occur in certain aneuploid states, they do not in themselves provide a complete explanation for the very severe functional abnormalities which are almost invariably present.

In considering the possible effects of aneuploidy on the nervous system, two categories of abnormalities need to be considered: anatomical and functional. The former includes, in turn, abnormalities of neuronal organization and of neuronal structure. Together these constitute what might be considered as the "wiring" of the brain, and their abnormality is the direct *developmental* consequence of the aneuploid state. However, it is also possible to visualize that while the development of the brain may be normal, its function may not be, because of a combination of major and/or minor derangements of metabolism, membrane properties, neurotransmitter synthesis and release, and electrical properties which are not associated with visible defects. The many precedents from the inborn errors of metabolism testify to the severe effects of metabolic derangement on intellectual development and function. Were the major defects to turn out to be functional rather than anatomical, there might be greater hope for the development of therapeutic strategies to improve central nervous system function.

In addition to its effects on intelligence and control of muscle tone early and throughout life, trisomy 21 is also associated with a process of neuronal degeneration during the adult years. This process, which is analogous or identical to Alzheimer disease (presenile/senile dementia), results in significant pathological changes in the brain and may possibly further compromise the already impaired mental functioning. In this section, I shall review what seems to be known about the anatomical and functional consequences of trisomy 21 in the nervous system and will examine the relationship between the trisomic state and the development of the pathology of Alzheimer disease.

Neuropathology

Like so many aspects of Down syndrome, there has been considerable disagreement about the frequency and nature of anatomical abnormalities of the brain. However, the best summary seems to be that of Rehder (1981) and of Gullotta, Rehder, and Gropp (1981):

> It is, indeed, tempting to suspect morphological alterations to be responsible for the clinical picture of severe mental retardation. However, no specific neuropathologic pattern has, so far, emerged from intensive anatomic and histologic studies, and in the majority of cases with Down's syndrome the results of neuropathologic study are normal. . . . It thus seems reasonable to assume that structural alterations of the central nervous system are neither constant nor obligatorily accompany a specific chromosomal syndrome. However, their appearance follows patterns of semi-specificity. . . . Mental retardation as a frequent symptom of most chromosomal disorders is very rarely associated with structural anomalies of the CNS which could explain this symptom per se. Therefore, the clinician who expects from the neuropathologist an interpretation of the pathogenesis of mental retardation will be disappointed.

A similar sentiment is expressed by Warkany (1971): "One can conclude that the abnormalities of the central nervous system of patients with Down syndrome are varied and inconstant . . . . It is futile to look for a description of the 'mongoloid brain.'"

Scott, Becker, and Petit (1983) agree that the histological findings have been inconsistent, but they do accept the findings [attributed to Crome, Cowie, and Slater (1966)] that brain weight is low normal and the size of the cerebellum and brainstem may be reduced to an even greater extent. They also describe a "characteristic" narrowness of the superior temporal gyrus. Among the semispecific findings in Down syndrome are nerve cell heterotopias in the white layers of the cerebellum and vermis which were found in 16% of infantile and fetal Down syndrome brains (Gullotta, Rehder, and Gropp, 1981). These heterotopias are attributed to disturbance or retardation of embryonal cell migration, perhaps as a consequence of a "general developmental retardation."

Detailed neuronal architecture, studied with Golgi-type preparations, has been analyzed in a handful of cases. Because of the limitations in numbers, all reports must be viewed with caution. Suetsugu and Mehraein (1980) studied the brains of seven individuals with Down syndrome who died between the ages of 3 and 23 years and compared them with brains of nonretarded persons of the same age. They observed a significant diminution, of the order of 25–30%, of the number of spines in the middle and distal segments of the apical dendrites of the pyramidal neurons. They believe this finding to be rather specific for Down syndrome and have postulated that it might represent an early change in the process of dendritic degeneration. Similarly, Marin-Padilla (1976) conducted a detailed analysis of the brain of a 19-month-old infant with Down syndrome and reported a variety of abnormalities of the dendritic spines, altered axospinous synapses, and a peculiar form of intrinsic vacuolar change in the dendrites with scattered neuronal fragmentation and necrosis. The specificity of the last of these abnormalities was considered uncertain because of the child's clinical history. Nevertheless, when Takashima *et al.* (1981) analyzed the numbers of visual cortical neurons, dendritic arborization, and numbers of dendritic spines in 14 fetuses (14–40 weeks) and infants (to 12 months) with Down syndrome, they found changes consistent with those just described. While the fetuses had no abnormalities, the newborns had short, thin spines and older infants had decreased number of spines, thin spines of unusual length, and short basilar dendrites. These results were interpreted as suggesting that in Down syndrome, unlike the situation in normal infants, the growth of dendrites and of dendritic spines (in numbers and length) plateaus after birth. The abnormalities of spine morphology and number are considered to be nonspecific, and it is not clear whether they are a cause or an effect of, or are unrelated to, the mental retardation in Down syndrome (Takashima *et al.*, 1981). Nevertheless, they are reported as being more constant in Down syndrome than in other conditions leading to mental retardation (Suetsugu and Mehraein, 1980).

Other evidence for defects in brain histogenesis have also been reported. Thus, Sylvester (1983) described delayed development of the dentate gyrus and hippocampus in the brain of a single trisomy 21 fetus, but felt that this is related to the smaller sizes of these structures in adult brains. Colon (1972) reported a decreased number of occipital cortical neurons per unit volume in Down syndrome and concluded that while the overall architecture of the central cortex is normal, neuronal differentiation is not. Ross, Galaburda, and Kemper (1984) also described a poverty of granular cells throughout the cortex and suggested that the affected cell is the aspinous stellate cell, a type of GABAergic neuron. Wisniewski, Laure-Kamionowska, and Wisniewski (1984, 1985) have reported decreased neuronal densities in layers II and IV of the occipital cortex (area 17) of trisomic individuals ranging in age from newborn to 14 years. These decreases averaged 30% during the first year of life and 20–25% thereafter. However, in the frontal cortex (area 10) and the parahippocampal gyrus (area 28), the reductions were only 10–20% during the first year of life. In a single case of an infant with Down syndrome studied at 5.5 months, cholinergic cells were decreased ~50% in the substantia innominata and locus coeruleus (McGeer *et al.*, 1985).

Studies of synapse formation in Down syndrome fetal brains have not revealed striking differences (Scott, Becker, and Petit, 1983). Synaptic density in the neocortex was normal, although the frequency of primitive synaptic contacts was possibly increased and pre- and postsynaptic width and length were possibly reduced. Scott, Becker, and Petit (1983) believe that these changes could reduce the frequency of synaptic transmission and thereby contribute to mental retardation. Wisniewski *et al.* (1985) also found that although synaptic density in the visual cortex of trisomic brains was decreased only slightly, the surface area of synaptic contact was 20–30% lower.

It is obvious that a clear picture of the state of the "wiring" does not emerge. Although there is accumulating evidence for abnormalities of neuronal differentiation and migration in fetal and infant brains, it is not certain, if the developmental plasticity of the young brain is taken into account, how many of the changes are permanent or functionally significant. Considering when they appear (after birth) and their nonspecificity, the abnormalities of dendrites and of dendritic spine formation can be interpreted as representing secondary changes in neurons which are metabolically or otherwise affected by the trisomic state. While possibly contributing to the neurological difficulties, they do not appear to be the cause.

Since one of the principal neurological manifestations of trisomy 21 is hypotonia, a brief departure from consideration of the brain is in order to mention the findings of Landing and Shankle (1982) on the structure of skeletal muscle in Down syndrome. These investigators reported that muscle fibers are increased in diameter (1.26 fold) and the number of nuclei per unit length is decreased (to 0.85 of normal), resulting in increases in fiber area/nucleus and fiber volume/nucleus, of 1.60 and 1.96 fold, respectively. They suggested that a "simple explanation for the facts . . . is that a biochemical 'signal' emanating from, or controlled by genetic information on, chromosome 21 determines the number of muscle fiber nuclei by controlling the rate of increase of these nuclei. Since $\frac{2}{3} \times \frac{3}{2} = 1$, the same level of 'signal' could be maintained by two-thirds the number of nuclei with trisomy 21 as by a (50% greater) number of nuclei with a normal chromosomal complement." In this calculation they made area/nucleus the critical factor, since this is what is increased by a factor close to $\frac{3}{2}$. The numerology, however interesting it might appear from the point of view of gene dosage, is far from convincing, and it is not clear why surface area/nucleus should be the regulated developmental parameter. The idea of looking at muscle fibers from individuals with monosomy 21 to ascertain whether they have opposite findings (Landing, Dixon, and Wells, 1974) may be a good way to test the hypothesis.

In addition to their proposals regarding fiber development, Landing and Shankle (1982) also suggested that the hypotonia, without weakness, typical of Down syndrome may reflect muscle fiber abnormality rather than a cerebral dysfunction. This is an interesting notion which requires further investigation, since the issue of a neurological versus a muscular basis of the hypotonia is still unresolved (Smith and Berg, 1976; Scola, 1982). The reported beneficial effect of 5-hydroxytryptophan, a precursor of serotonin, on muscle tone (Bazelon *et al.*, 1967) does not settle the issue, since both peripheral and central mechanisms can be postulated and the drug had no other significant effects on the central nervous system or the development of intelligence (Bazelon *et al.*, 1968; Weise *et al.*, 1974).

Neurobiology

If the mental retardation and possibly hypotonia of Down syndrome cannot be attributed to gross anatomical abnormalities of the brain, and probably not to primary microscopic abnormalities either, what is the source of the difficulty? Lejeune (1977, 1979) has suggested that the problem is essentially a metabolic one. Arguing both by analogy with various inborn and acquired errors of metabolism, each with either similar or opposing ("countertype") abnormalities, and from observations on various aspects of intermediary metabolism, oxygen metabolism (see discussion of superoxide dismutase, above), and the properties of the cholinergic system, he has proposed that there is a series of interrelated abnormalities principally affecting one-carbon and folate metabolism. These proposals extend also to morphological abnormalities: "It seems difficult to consider as purely fortuitous the fact that three mental deficiencies [hypothyroidism, dup(12p), and trisomy 21], correlated with short nose, short stature and adiposity, and clinically sometimes very similar, are related to enzymatic changes so close to one another" (Lejeune, 1977). To the extent that these proposals are heuristic and "open the way to further investigations," as Lejeune (1977, 1979) states, they are worthy of consideration. However, to this point they do not appear to have added significantly to our understanding of the pathogenesis of either the mental or physical abnormalities of Down syndrome. The attempt to reduce most or all forms of mental retardation to a common schema which ultimately becomes the whole of intermediary metabolism does not really serve to clarify the pathogenetic mechanisms operative in and unique to each. It also obscures the differences of mechanism likely to apply to conditions with modestly altered concentrations of several gene products (as in trisomy 21) as compared with disorders associated with a complete lack of a single enzyme activity (as in the inborn errors of metabolism).

While rejecting the concept of a unitary overall metabolic explanation for all of the abnormalities in Down syndrome, I certainly do not reject the possibility that specific metabolic aberrations do have a significant role in the pathogenesis of the condition and, in particular, on the cerebral dysfunction. However, at the present time there is no convincing evidence for any, and the myriads of studies carried out on various aspects of metabolism peripherally do not shed a light on what may be happening in the brain.

Considerable interest has focused on peripheral neurotransmitter function in Down syndrome, and mention has already been made of the studies on plasma and platelet serotonin. In his discussions, Lejeune (1977, 1979) pointed specifically to the cholinergic system. Citing evidence of hypersensitivity to the peripheral effects of atropine, he postulated a constitutional deficiency of cholinergic activity because of less rapid synthesis or release of acetylcholine. However, it must be noted that while hypersensitivity to the mydriatic effect of atropine has been repeatedly confirmed (Berg, Brandon, and Kirman, 1959; O'Brien, Haake, and Braid, 1960; Priest, 1960; Mir and Cumming, 1971; Lejeune, Bourdais, and Prieur, 1976), there is still debate about its cardioacceleratory effects. While Harris and Goodman (1968), in an apparently well-controlled study, observed an increased sensitivity of trisomics to the cardiac effect of atropine, Mir and Cumming (1971) and Wark, Overton, and Marian (1983) did not (although the latter commented that the matching of their Down syndrome and control group subjects was not optimal). Moreover, Harris and Goodman (1968) suggested that the increased cardioacceleratory response to atropine is mediated, at least in part, through β-adrenergic receptors. This idea is, of course, of considerable interest in view of the greatly increased response of fibroblasts to isoproterenol discussed earlier in this chapter.

The adrenergic system has also received other attention. The activity of the enzyme, dopamine-β-hydroxylase, which converts dopamine to norepinephrine, is significantly decreased in the plasma of individuals with trisomy 21 (Wetterberg *et al.*, 1972). Nevertheless, although urinary epinephrine excretion was decreased, urinary norepinephrine excretion was not, and plasma concentrations of both epinephrine and norepinephrine were normal (Keele *et al.*, 1969).

While somewhat confusing, the results of studies on peripheral neurotransmitters do suggest that abnormalities are present in Down syndrome. The basis of these abnormalities has yet to be defined, but is certainly worthy of investigation. Perhaps of even greater importance is to ascertain whether neurotransmitter abnormalities are present in the central nervous system. Although abnormalities of the cholinergic system are present in adult life (see discussion of Alzheimer disease in next section), the real question is whether these or any other types of neurotransmitter abnormalities are present in infants and children. If such abnormalities are present and if they contribute to the development of mental retardation, they could provide a rational point for pharmacological intervention. While the trial with 5-hydroxytryptamine had no beneficial effect on the development of intelligence (Weise *et al.*, 1974), this certainly does not rule out the possibility that other agents may ultimately be found that will.

Another biochemical approach to investigating the effects of trisomy 21 on the central nervous system has been to look at the structure of myelin. This subject has been reviewed by Shah (1979), who concluded that in Down syndrome all of the phospholipids of myelin contain reduced amounts of mono-unsaturated fatty acids but do not have unequivocally abnormal amounts of polyunsaturated fatty acids. However, the opposite seems to be true of synaptosomal phospholipids, with reduced proportions of polyunsaturated fatty acids and normal amounts of monounsaturated fatty acids being found. Monounsaturated fatty acids are reduced in sphingomyelin. The basis for these abnormalities, which are also found in phenylketonuria, is unknown. Shah (1979) speculated that a putative immaturity of the Down syndrome brain results in a deficiency of desaturase activity. He further speculated that the changes in fatty acid composition may affect membrane fluidity and transport processes at the nerve endings and, in turn, "may at least in part be responsible for functional abnormalities" present in this condition. However, Banik *et al.* (1975), who found decreased amounts of myelin in brains from trisomic individuals between 18 and 50 years of age, concluded just the opposite: "It seems unlikely that deficiency in myelin explains the mental retardation found in Down's syndrome but rather that amyelination reflects a more general structural change affecting neuronal growth and synaptogenesis."

One reason for the differences in interpretation is, of course, that biochemical analyses, while providing valuable information about effects of trisomy 21 on the chemical structure of elements of the nervous system, do not directly speak to the question of alterations in function. Unfortunately, there are few data that really do bear on this question, and they principally involve electrophysiological studies of one type or another. Electroencephalographic (EEG) studies have not been very revealing. Ellingson, Eisen, and Ottersberg (1975) observed a significantly, but not strikingly increased overall rate of EEG abnormality, especially in the age groups between 1 month and 10 years. There were no specific EEG patterns associated with Down syndrome, and the abnormalities were not well correlated with specific behavioral or neurological signs and symptoms. However, Elul, Hanley, and Simmons (1975), using a statistical analysis of a small number of EEG patterns, suggested that synaptic contacts on cortical neurons are "impoverished" in Down syndrome, resulting from an incomplete postnatal development of neuronal interconnections. These conclusions were considered to be consistent with the independently described histological observations of dendritic abnormalities mentioned earlier. Based on an excess of theta wave-activity, Tangye (1979) also referred to an immaturity of cerebral development.

Somewhat closer to the issue of neuronal function in Down syndrome are the studies of visual and auditory evoked potentials. Significant differences in patterns were found between Down syndrome and other retarded and nonretarded subjects (Gliddon, Busk, and Galbraith, 1975; Squires *et al.*, 1980; Folsom, Widen, and Wilson, 1983; St. Clair and Blackwood, 1985), the meaning of which still remains to be established. As expressed by Folsom, Widen, and Wilson (1983), "clearly our understanding of neural development in Down's syndrome and/or the underlying cochlear and neural mechanisms that generate the auditory brain-stem response in Down's syndrome is incomplete."

Perhaps the most intriguing functional studies of Down syndrome neurons to date are those of Scott *et al.* (1982; also Scott, Becker, and Petit, 1983) on fetal and infant dorsal root ganglion neurons cultured in vitro. Virtually all calculated electrical parameters of Down syndrome neurons were found to be altered, often by as much as 20–40%. Although changes were also found in cells from a single culture of trisomy 18 neurons, the pattern was not the same as in trisomy 21, implying some degree of specificity. To explain the "bewildering array" of electrical alterations, Scott, Becker, and Petit (1983) postulated a combination of alterations in potassium permeability, both in resting (unexcited) and excited neurons, and in membrane lipid constitution. Although there was no evidence to prove it, they also postulated that similar abnormalities may also occur in central nervous system neurons, and calculated a 7% increase in action potential conduction velocity in Down syndrome nerve fibers and a 15.5% decrease in the rate of conduction along Down syndrome dendrites. The latter, they suggested, "might critically slow down the integration of stimulatory and inhibitory postsynaptic potentials so as to profoundly disturb CNS function and therefore may be the neurological basis of the mental retardation in Down syndrome."

All of these inferences are, of course, highly speculative. However, what must be kept in mind is that the changes which have been described were found in midtrimester fetal neurons (16–23 weeks of gestation), when axonal and dendritic development in Down syndrome is indistinguishable from normal. While confirmation and extension of these results to central nervous system neurons are certainly required, this type of in vitro cellular approach appears to be an important one which may ultimately be able to provide considerable insight into the effects of trisomy 21 on neuronal function and, if consistent abnormalities are found, into their biochemical and ultrastructural basis.

Alzheimer disease

While the first two parts of this section have been concerned with neuroanatomical, neurochemical, and neurophysiological changes potentially related to the mental retardation and hypotonia of Down syndrome, this part is concerned with what appears to be quite a different effect of the trisomic state on the central nervous system – presenile/senile dementia. Recent reviews of this area were published by Sinex and Merril (1982), Katzman (1983), and Epstein (1983c).

The possibility of a relationship between Down syndrome and dementia has been recognized for over 100 years (Fraser and Mitchell, 1876) and between Down syndrome and Alzheimer disease (as defined pathologically) for over 50 years (Struwe, 1929). Numerous authors have described the pathological hallmarks of Alzheimer disease in the brains of adults with Down syndrome [for summary, see Whalley and Buckton (1979)]. While the initial observations were based on light microscopy, a few recent reports have been based on ultrastructural analysis of a total of eight Down syndrome brains (Ohara, 1972; Schochet, Lampert, & McCormick, 1973; Ellis, McCulloch, & Corley, 1974). The consensus of these reports is that the brains of adults with Down syndrome possess granulovacuolar cytoplasmic changes, senile plaques, and neurofibrillary tangles structurally indistinguishable from those found in the brains of individuals with Alzheimer disease. In addition to these qualitative similarities, it has also been reported that the number of neurofibrillary tangles and the loss of hippocampal pyramidal neurons in Down syndrome brains are quantitatively similar to those observed in the brains from patients with Alzheimer disease (Ball and Nuttall, 1980) and that the same regions of brain are preferentially affected (Ball and Nuttall, 1981). The latter assertion has, however, been questioned (Ropper and Williams, 1980).

Chemical similarities between the Alzheimer disease and Down syndrome brains have also been observed. These include decreases of choline acetyltransferase (ChAT) and acetylcholinesterase (AChE) in various parts of the brain (Yates *et al.*, 1980), of norepinephrine in the hypothalamus (Yates *et al.*, 1981), and of dopamine and 5-hydroxytryptamine in several parts of the brain (Nyberg, Carlsson, and Winblad, 1981). Glenner and Wong (1984) have shown that the partial amino acid sequences of β_2 proteins from cerebrovascular amyloid fibrills obtained from the brains of individuals with Alzheimer disease and from elderly subjects with Down syndrome are virtually identical.

The time of appearance and the frequency of the lesions of Alzheimer disease on Down syndrome brains have been matters of particular concern, since the interest in the relationship between the two conditions has been as much a function of the early and generalized appearance of the lesions as of the nature of the lesions themselves. In this extensive autopsy series, which included 347 cases of Down syndrome, Malamud (1972) found that 5 out of 312 (1.6%) brains from individuals dying below the age of 40 (20–38 years) and 35/35 (100%) from individuals over the age of 40 (42–69 years) had the gross pathological changes of Alzheimer disease. Of the latter, 60% were described as severe and comparable to the most advanced cases of Alzheimer disease, and 40% were mild to moderate. From their review of the literature, Whalley and Buckton (1979) concluded that the neuropathological changes are almost universal over the age of 35 years, but they did point out that exceptions have occasionally been noted. In a study of 100 brains of institutionalized individuals with Down syndrome, Wisniewski, Wisniewski, and Wen (1985) found senile plaques and tangles in all cases over 30 years of age, and brain weights were generally below the mean -2 SD from the second decade on.

Of principal interest in the context of the present discussion is the pathogenetic relationship between Down syndrome and the development of the pathological changes of Alzheimer disease. Several possibilities have or may be considered. The first is that common or related genetic factors lead to both conditions. The facts that Down syndrome is the result of a genetic disorder and that Alzheimer disease can in some instances be familial and apparently hereditary (Heston and Mastri, 1977; Heston *et al.*, 1981) have led to speculation that the two conditions may share certain genetic factors. This speculation has been further stimulated by the observation that there may be an excess of individuals with Down syndrome (and with myeloproliferative and autoimmune disorders) among the relatives of patients with familial Alzheimer disease (Heston *et al.*, 1981), and by reports that individuals with familial Alzheimer disease demonstrate an increased frequency of aneuploid cells or of acentric fragments in culture of mitogen-stimulated peripheral blood lymphocytes (Ward *et al.*, 1979; Nordenson *et al.*, 1980). The former was not found in the extensive family studied by Nee *et al.* (1983), and the latter was also not confirmed in several studies, some double-blind (White *et al.*, 1981; Buckton *et al.*, 1983; Moorhead and Heyman, 1983; Smith, Broe, and Williamson, 1983, 1984).

To explain the associations they observed, Heston and Mastri (1979) considered a variety of possible genetic explanations centered on an abnormality of microtubular organization and function and involving the existence of an abnormal gene or genes. However, despite the originally intriguing nature of the reported familial associations between familial Alzheimer disease and Down syndrome, pursuit of these associations has not, in my view, been fruitful. It has led to an attempt to consolidate the causes of the nondisjunction which leads to trisomy 21, the mechanisms of how trisomy 21 leads to the abnormalities associated with Down syndrome, and the etiology of the morphological abnormalities of Alzheimer disease into a single, unifying entity which is presently unsupportable. Therefore, I think that it is more appropriate to look specifically at how trisomy 21 might lead to the development of Alzheimer disease.

Earlier in this chapter, the known and possible effects of imbalance of the loci known to be on chromosome 21 were discussed, particularly with regard to their potential role in the pathogenesis of the mental retardation of Down syndrome. Similar considerations, especially involving SOD-1, have also been applied to Alzheimer disease, and Sinet, Lejeune, and Jerome (1979) and Delahunty, Kling, & McCormack (1982) have postulated that elevated SOD-1 and glutathione peroxidase activities, by generating toxic intermediates, produce neuronal damage. As before, while I do not personally find this hypothesis attractive, partly because of lack of sufficient evidence for it and partly because of its attempt to provide a unitary explanation for the major neurological defects in Down syndrome, it nevertheless has the virtue of being a testable one (see below) and may therefore be worth pursuing.

It is possible to speculate how the imbalance of the other known chromosome 21 loci might be implicated – for example, how the enhanced sensitivity of trisomic cells to interferon might paradoxically affect immune functions in an adverse manner (Epstein and Epstein, 1980; C. J. Epstein *et al.*, 1981) and thereby increase susceptibility to extrinsic or intrinsic infectious agents. But, in view of our ignorance about the role of infectious agents in the pathogenesis of Alzheimer disease, such speculations would, for the moment, be premature. At present, therefore, there is no evidence to support the notion that the specific genetic abnormalities in Down syndrome – the increase in the synthesis of one or an unknown number of chromosome 21 gene products – directly cause the Alzheimer disease changes characteristic of this condition. On the other hand, the near universality of these findings still makes a direct causal relationship a real possibility, and Schweber (1985) has argued that Alzheimer disease results from a minute duplication of chromosome 21.

As was also discussed earlier, individuals with Down syndrome are highly susceptible to infection, both bacterial and viral, presumably as a result of deranged immunological function. However, as was just mentioned, the significance of these immunological aberrations to an understanding of the pathogenesis of the Alzheimer disease changes depends, of course, on whether the possibility that Alzheimer disease is the result of an infectious process is seriously entertained. This question has been the subject of numerous reviews [for example, Roth (1982)], but the answer is still unknown.

Another highly speculative explanation suggested for the development of Alzheimer disease in Down syndrome is that of Aziz *et al.* (1982), who observed a modest increase (overall mean, 116% of normal) in concentration of serum dihydrobiopterin in Down syndrome subjects. They speculated that this might be the result of increased oxidation of cell tetrahydrobiopterin (because of elevated SOD–1, etc.), which would in turn lead to reduced tetrahydrobiopterin levels, consequent reduced neurotransmitter formation, and thereby ultimately to mental retardation early in life and Alzheimer disease later.

In the literature on Alzheimer disease in Down syndrome, it is frequently asserted that the pathological changes are a consequence of the premature aging which characterize the syndrome. The implication here is that Alzheimer disease is itself a disease of aging and that trisomy 21, by causing early aging, thereby leads to the development of the neurological changes. At present, none of the assumptions underlying this reasoning can be considered as proven. The arguments concerning the relationship of Alzheimer disease to normal aging of the brain, recently summarized by Roth (1982), are outside the purview of this discussion, but it is clear that a substantial case can be made for the conclusion that Alzheimer disease is not an inevitable consequence of aging and does indeed represent disease. Similarly, it is not at all clear that Down syndrome does cause premature aging. The definition of premature aging is extremely difficult, since there is still little agreement on what constitutes aging per se as opposed to age-related disease. Several years ago, in a discussion of another condition thought to represent premature aging (Epstein *et al.*, 1966), I quoted the idealized criteria of Casarett (1964) for premature aging, and they are worth briefly considering here. Premature aging leads to (1) an earlier increase in mortality, without alteration of the shape of the mortality curve; (2) proportional advancement in time of all diseases or causes of death, without alterations of degree, sequence, or absolute incidence, and without induction of disease; and (3) proportional advancement in time of all morphological and physiological manifestations of the aging process.

It is quite clear that Down syndrome is associated with an earlier increase in mortality, but the age-related mortality figures have been changing quite rapidly during the past several decades, and the true shape of the mortality curve is uncertain (Øster, Mikkelsen, and Nielsen, 1975; Masaki *et al.*, 1981). The changes in survival are principally the result of improved methods for treating infection, and the reported causes of death, especially in individuals dying at a young age, are heavily biased in this direction. Therefore, data are not presently available to determine whether Down syndrome meets the first two criteria for premature aging. With regard to the third criterion – advancement in time of the morphological and physiological manifestations of aging – the data are mixed. Older individuals with Down syndrome look prematurely aged, with graying of the hair and dryness and fine wrinkling of the skin. Changes in behavior and personality are common (Wisniewski *et al.*, 1978). However, when attempts were made to use objective criteria of aging, the results were surprising. Cardiac valve fenestrations, supposedly an age-related change, occur earlier and with greater frequency in Down syndrome subjects (Sylvester, 1974). By contrast, the mechanical behavior of the skin of individuals with Down syndrome, as measured by a quantity defined as "limit strain," is within usual limits (Murdoch and Evans, 1978) although it can be interpreted as consistent with 10 years' advanced aging (Edwards, 1978). Their blood pressure, both systolic and diastolic, is lower than normal (Murdoch *et al.*, 1977; Richards and Enver, 1979). And, based on findings in a limited number of cases, it is claimed that trisomic individuals are remarkably free of atheroma (should this be interpreted as aging or disease?)

(Murdoch *et al.*, 1977), although preatherosclerotic lesions are seen in younger individuals (Moss and Austin, 1980). Therefore, despite repeated assertions that Down syndrome is associated with premature aging, it must be concluded that it is not at all clear that this is the case. In fact, careful reading reveals that most such assertions are based on equating the Alzheimer disease changes with premature aging – a process of circular reasoning to say the least. As an explanation for the Alzheimer disease in Down syndrome, the concept of premature aging is seriously wanting at the present time.

There is one other possibility that should also be considered, which is that the neurons of individuals with Down syndrome are intrinsically defective from the beginning. The neuropathological evidence on this point has already been summarized earlier in this section. If, as was suggested earlier, the dendritic and related changes are interpreted as being secondary to metabolic imbalance or structural abnormality, the question again becomes one of the relationship between trisomy for chromosome 21 and the nature of these abnormalities. Moreover, if these early changes are considered as being related to the later pathology of Alzheimer disease, it is necessary to explain the lag time of some three decades, without progressive neurological impairment, before the latter becomes fully manifest. However, if the problem is one of an age-related diminution of neurons which are already quantitatively deficient at birth, thereby permitting neuronal densities to fall below some critical threshold, a relationship between Alzheimer disease and early developmental abnormalities might exist.

Clearly, we are far from understanding the nature of the relationship between Down syndrome and Alzheimer disease. Therefore, while it has been suggested that "Down syndrome, with its partially characterized genotypic and phenotypic abnormalities, is an appropriate model for the study of the pathogenesis of the lesions (of Alzheimer disease)" (Burger and Vogel, 1973) and that "these cytoarchitectonic predilections reinforce the search for the pathogenesis of Alzheimer disease through further examination of the 'Alzheimer-type' lesions found in adult cases of Down syndrome" (Ball and Nuttall, 1981), the necessary insights have yet to be developed. Nevertheless, the constancy of the relationship between the two conditions gives us reason to expect that a better understanding of the effects of trisomy 21 could play an important role in solving the puzzle of the etiology of Alzheimer disease.

An animal model of trisomy 21 (Down syndrome)

The foregoing discussion in this chapter points up the sparseness of our knowledge about the pathogenesis of the diverse abnormalities which characterize Down syndrome, and it is clear that new approaches to the problem are required. One such approach in which my collaborators and I have been particularly interested is in the development of an animal model of trisomy

21 in the mouse. The rationale for such a model and the technical aspects of generating aneuploid mice have already been extensively discussed in Chapters 10 and 11.

Comparative mapping

To create a mouse model for human trisomy 21, it is necessary to map the genes known to be present on chromosome 21 onto the mouse genome and to determine whether there is a chromosome segment or segments in the mouse homologous to regions of chromosome 21. As was discussed earlier, that this could be the case was suggested by the demonstration of syntenic groups of genes which, despite the obvious differences in chromosome structure, are present in both man and mouse (see Table 11.1). Therefore, three of the genes known to be on human chromosome 21 – *SOD1*, *IFRC*, and *PRGS* – were mapped by somatic cell genetic techniques, and all were found to be on mouse chromosome 16 (Francke and Taggart, 1979; Cox, Epstein, and Epstein, 1980; Epstein, Cox, and Epstein, 1980; Cox, Goldblatt, and Epstein, 1981). Further, the two genes, *SOD1* and *PRGS*, which in humans are known to be on band 21q22 (Sinet *et al.*, 1976; Cox *et al.*, 1983), the region which in the trisomic state is associated with the Down syndrome phenotype (Fig. 12.2), have been mapped to the distal one-fifth of mouse chromosome 16 (Cox and Epstein, 1985). The present status of the comparative maps of human chromosome 21 and mouse chromosome 16 is shown in Fig. 12.7.

The existence of the conserved synteny of *SOD1*, *IFRC*, and *PRGS* provides a basis for hoping that all three genes, as well as many others still to be defined, are in the distal segment of mouse chromosome 16 and that mouse chromosome 16 does carry a chromosome segment wholly or to a great extent homologous to the "Down syndrome region" of human chromosome 21. However, that mouse chromosome 16 is not completely homologous to human chromosome 21 – which would not in any case have been expected – is demonstrated by the fact that the gene (*IGLC* in man, *Igl* in mouse) for the immunoglobulin λ light chains has been mapped to mouse chromosome 16 and to human chromosome 22 (Francke *et al.*, 1982), and the gene for somatostatin (*SST*) to mouse chromosome 16 and human chromosome 3q (Roderick *et al.*, 1984). The size of mouse chromosome 16 is about 35% larger than that of human chromosome 21 in terms of total genetic length in centimorgans [assuming total lengths of 1600 cM and 2700 cM for mouse and man, respectively (Roderick and Davisson, 1981; N. Morton *et al.*, 1982) and relative haploid lengths of 3.86% and 1.7%]. If chromosome 21 has a lower than average density of euchromatin (Hoehn, 1975; Korenberg, Therman, and Denniston, 1978), the disproportion will be even greater.

Therefore, trisomy for all of mouse 16 will result in a degree of genetic imbalance more extensive than in human trisomy 21, and a more appropriate

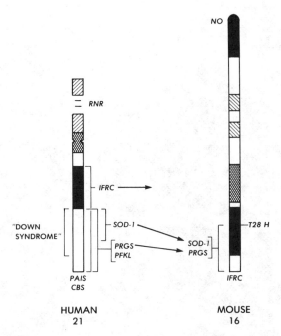

Fig. 12.7. Comparative maps of human chromosome 21 and mouse chromosome 16 drawn approximately to scale and showing G-band pattern. A tentative mapping of *PRGS* to 21q22.1 has recently been reported (Chadefaux *et al.*, 1984).

model would therefore be an animal with a duplication of just the distal part of chromosome 16. Such animals are just now in the process of being made and examined, and the present discussion will be confined to the properties of mice trisomic for all of chromosome 16.

The phenotype of mouse trisomy 16

Some aspects of the phenotype of fetuses with mouse trisomy 16 have already been listed in Table 10.1, and the most extensive studies are those of Gropp, Giers, and Kolbus (1974) and of Miyabara, Gropp, and Winking (1982). Although death begins to occur after day 14, trisomy 16 fetuses can survive to or near term. Cattanach's (1964) claim of viability to adulthood has not been confirmed. The trisomic fetuses are moderately growth retarded, with weights about 10–25% less than those of normal littermates. There is also a modest diminution in placental weight. Between days 14 and 17, a very striking generalized edema is present (Fig. 12.8), which nearly resolves by day 19. However, late-gestation trisomy 16 fetuses have short, wide necks which may represent the residual of the edema (Fig. 12.9). In addition, they appear

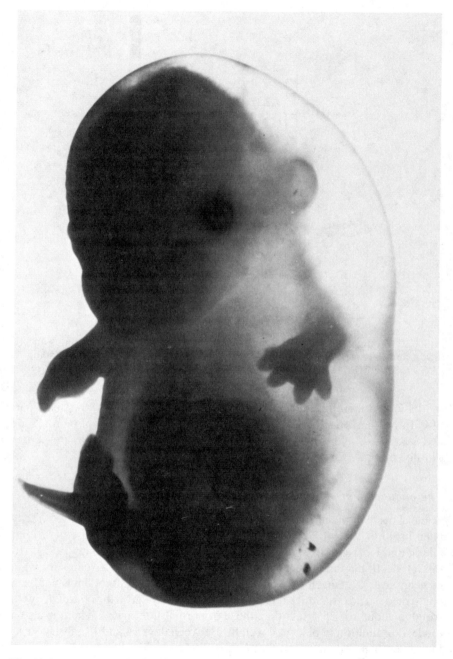

Fig.12.8. Trisomy 16 fetus at day 14 of gestation. The massive edema is visible. Reprinted by permission from C. J. Epstein (1984).

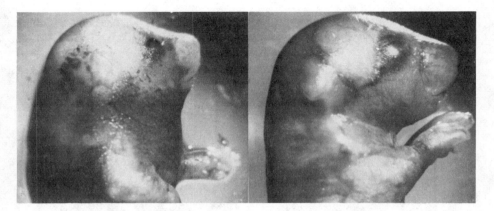

Fig. 12.9. Facial appearance of trisomy 16 (left) and diploid (right) sib fetuses at day 18 of gestation. Note the short thick neck, small ear, shortened snout, and open eyelid of the trisomy 16 fetus. Reprinted by permission from Epstein (1985a).

somewhat brachycephalic and have short snouts. Unlike normal fetuses, which have closed eyelids late in gestation, the trisomic fetuses have open eyelids, frequently accompanied by lenticular dysplasia or degeneration (Cox *et al.*, unpublished; Miyabara, Gropp, and Winking, 1982; Gropp, personal communication) and defects of the retina (Oster-Granite, Baker, and Ozand, 1983). The fetal vasculature of the placenta is hypoplastic (Gearhart, Oster-Granite, and Hatzidimitriou, 1985).

Congenital heart disease is a nearly invariant (96%) feature of trisomy 16 (Pexieder, Miyabara, and Gropp, 1981; Miyabara, Gropp, and Winking, 1982; Msall, Oster-Granite, and Gearhart, 1985). Anomalies of the great vessels, including overriding aorta, double-outlet right ventricle, complete transposition, and persistent truncus arteriosus, represent 85% of the cardiovascular anomalies present. However, more than half of the trisomy 16 fetuses with cardiac defects also have an endocardial cushion defect. Other lesions, usually found in combination with those listed above, include right-sided aortic arch, interrupted aortic arch (Fig. 12.10), coarctation of the aorta, aberrant subclavian artery, and tricuspid valve stenosis (Miyabara, Gropp, and Winking, 1982). Renal anomalies, with a reduction in the number of glomeruli and tubules and with hydronephrosis and hydroureter by day 18, have also been observed (Miyabara, Gropp, and Winking, 1982).

The total weight of the brain of trisomy 16 fetuses is significantly reduced (to as low as 32–38% of controls), as are the weights of the telencephalic and diencephalic/brainstem regions (Oster-Granite *et al.*, 1983; Ozand *et al.*, 1984; Singer *et al.*, 1984). Overall, there appears to be a decrease in cell proliferation in the ventricular zone, resulting in impaired radial and tangential growth of the pallium and decreased numbers of cortical neurons (Blue *et*

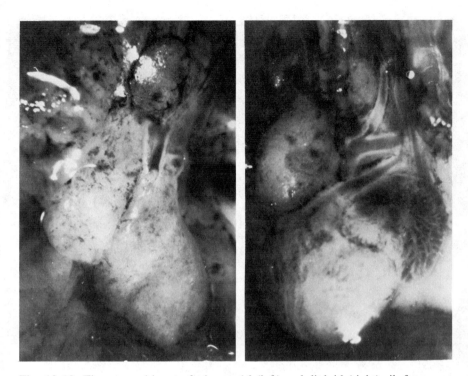

Fig. 12.10. Thymus and heart of trisomy 16 (left) and diploid (right) sib fetuses at day 18 of gestation. The left lobe of the thymus of each animal has been removed. Note the small thymus (upper-right corner) high in the neck of the trisomy 16 fetus, as well as the thick ventricular wall and absence of the aortic arch. Reprinted by permission from Epstein (1985*a*).

al., 1984). However, the incremental rates of brain growth, in terms of both weight and protein, are normal (Ozand *et al.*, 1984). There is a delay in cerebellar foliation and in development of the hippocampal formation, along with structural alterations of the cochlear and vestibular portions of the inner ear (Oster-Granite, Baker, and Ozand, 1983; Oster-Granite and Hatzidimitriou, 1985). Development of the basicranium and craniofacial apparatus is also retarded, and there is hypocellularity of the facial mesenchyme throughout the fetal period (Oster-Granite, 1983).

Detailed neurochemical analyses of the trisomy 16 fetal brain have been carried out by Gearhart *et al.* (1983), Singer *et al.* (1984), and Ozand *et al.* (1984). Their results are summarized in Table 12.9. Ozand *et al.* (1984) interpret their results as indicating "an arrested development of the cholinergic neurons and altered expression of catecholaminergic functions . . . reminiscent of those [changes] described in the nervous system of Down syndrome

Table 12.9. *Neurochemical studies of fetal trisomy 16 mouse brain*

Enzyme or metabolite	Specific activity or concentration relative to controls[a]		
	Whole brain[b]	Hemispheres[c]	Diencephalon/ brainstem[c]
Cholinergic system			
Choline acetyltransferase	↓	nl	↓
Acetylcholinesterase	↓		
Muscarinic receptors	↓	nl	nl
Catecholaminergic system			
Tyrosine hydroxylase	↓	↓ (NS)	nl
Dopamine-β-hydrozylase	↓ [d]		
Dopamine	↓	↓	↓ (NS)
Norepinephrine	↓	↓	↓
DOPA decarboxylase	↑		
Catechol-O-methyltransferase	↑		
Monoamine oxidase	nl		
Serotonergic system			
Serotonin (5-HT)		↓ (NS)	↓ [e]
5-Hydroxyindoleacetic acid		↓ (NS)	↓
GABAergic system			
Glutamate decarboxylase		nl	nl

[a]All differences are statistically significant ($p \leq 0.05$) unless marked NS (not significant).
[b]Ozand *et al.* (1984).
[c]Singer *et al.* (1984).
[d]Significantly decreased in 3 of 5 experiments.
[e]Day 17 of gestation.

patients obtained at autopsy." Similarly, Singer *et al.* (1984) suggest that their studies point to a delay in the neurogenesis and growth of the brain, but not to aberrant neuronal development, with certain neuronal systems being more severely affected than others. Particularly affected, both pre- and postsynaptically, is the cholinergic system, but the catecholaminergic and serotonergic systems are also impaired. The findings with regard to the cholinergic system are considered to be of particular interest because a specific deficit in this system has been found to characterize Alzheimer disease, both by itself and in association with Down syndrome (see above).

It is, of course, intriguing that an endocardial cushion defect is present in about half of the mouse trisomy 16 fetuses and in about 32% of Down syndrome fetuses and that aortic arch defects, which are so characteristic of mouse trisomy 16, are also found in human trisomy 21 (see above). Aside from these quite striking defects, and perhaps the retarded development of the brain (cholinergic deficits have not been shown to be characteristic of Down syndrome in early life), the transient edema, and the shortened neck and snout, the gross phenotype of complete mouse trisomy 16 does not, as expected, precisely mimic that of human trisomy 21 (Table 12.10). However, even without straining credulity, those similarities which *are* present are not trivial. Considering the genetic differences, the prenatal and perinatal lethality of mouse trisomy 16, possible differences in the formation of the limbs and face, and the absence of very many striking gross phenotypic features in Down syndrome, it is not obvious in what ways the human and mouse phenotypes could be more similar morphologically. However, there may be significant functional similarities in the immunological and hematological systems, and these will be discussed in the following subsection.

Immunological and hematological development of mouse trisomy 16

When attempts were made to use trisomy 16 fetal liver cells for the repopulation of lethally irradiated hosts, major difficulties were observed. Although radiation chimeras could be successfully prepared with cells from several types of trisomics (see Chapter 10), survival of the trisomy 16 radiation chimeras (Ts16 → 2n) was quite limited, with only 10% surviving beyond two months (compared to 50% survival in the controls) (Herbst *et al.*, 1982a). Although circulating granulocytes reached normal levels, red cell counts were moderately (~20%) and lymphocytes were severely (≥80%) reduced. Ultrastructural examination revealed disturbances in erythropoiesis and hemoglobin biosynthesis (Claussen, Gropp, and Herbst, 1981).

The abnormalities displayed by the transplanted trisomy 16 hematopoietic cells in the Ts16 → 2n chimeras are also reflected by the behavior of trisomy 16 cells in Ts16 ↔ 2n aggregation chimeras. The relevant data have already been presented in Table 10.6. Unlike most trisomy 16 cell types, which comprise 50–60% in chimeras examined prior to term and 30–40% in juvenile and adult chimeras, there are marked deficiencies of trisomy 16 cells in the blood, spleen, thymus, and bone marrow, particularly in the adult chimeras (Cox *et al.*, 1984). In this regard, it is of interest that Richards (1969) and Ford (1981) have summarized data that show that in human trisomy 21/2n mosaics, which are formally homologous to mouse Ts16 ↔ 2n chimeras, the human trisomy 21 cells are nearly always more prevalent proportionally in fibroblast cultures than in lymphocyte cultures from the same individuals.

Taken together, these observations point to abnormalities of trisomy 16 hemopoietic precursors and have stimulated further investigations of the trisomic stem cell populations. Herbst *et al.* (1982a) demonstrated reductions in two types of stem cells: the multipotential myeloid stem cell (designated

Table 12.10. *Comparison of phenotype of mouse trisomy 16 with that of human trisomy 21 at birth*

	Mouse trisomy 16[a]	Human trisomy 21[b]
Survival beyond term	None	~30% (or less)[c]
Growth in utero (weight)	Decreased 10–25%	Birthweight reduced ~10%[d]
Edema in utero	Massive but transient	Transient edema of neck,[e] rarely generalized[f]
Facies	Flat snout, short neck, open eyelids	Flat face, short neck, epicanthic folds
Brain development	Retarded, with reduced weight, enzyme activities, and metabolite concentrations	? Retarded maturation Head circumference decreased ~2% at birth[g]
	Structural alterations of cochlear and vestibular portions of inner ear	Anomalies of inner ear[h]
Congenital heart disease	Present in 96% with aortic arch anomalies in >80%, endocardial cushion defect in ~50%	Present in ~45% with endocardial cushion defect in ~32% (of all) and aortic lesions in ~15%
Immunological and hematological	Severe thymic hypoplasia and reduction in many stem cell populations	Thymic hypoplasia Leukemoid reactions and infantile leukemia
	Poor lymphoid and erythroid cell survival in radiation and aggregation chimeras	Reduced proportion of trisomic lymphocytes in blood of trisomy 21/2n mosaics

[a]See text for references.
[b]References in text except as noted.
[c]Hook (1982); Boué et al. (1981).
[d]Matsunaga and Tonomura (1972); Pueschel, Rothman, and Ogilby (1976).
[e]Personal observation of trisomy 21 abortuses (Dr. Bryan Hall, personal communication).
[f]Fujimoto et al. (1983).
[g]Hall (1964).
[h]Igarashi et al. (1977).

CFU-S, colony forming unit – spleen) and the granulocyte-macrophage stem cell (designated CFU-C, colony forming unit – culture). On a relative basis (in terms of the number of fetal liver cells used in the assay), the former was found to be reduced by about 50% and the latter by ≥75%. The relative reduction in trisomy 16 CFU-C was most pronounced between 14 and 16 days of gestation when it was >90%, but decreased to only 75% on day 18 as the number of CFU-C in the control fetal livers began to drop.

My collaborators and I (Epstein *et al.*, 1983, 1985) have confirmed these results and have also examined several other aspects of lymphopoiesis and hemopoiesis. The results are summarized in Table 12.11 and are expressed on an absolute basis (reduction per fetus). Since the number of cells in the trisomy 16 fetal liver is about half that in control fetal livers, the relative numbers of the various stem cell populations assayed are about twice those that would be inferred from the data in the table. In brief, there is a profound hypoplasia of the thymus (Fig. 12.10) and spleen, profound reduction in CFU-C, in the primitive erythroid precursors (BFU-E) and in the ability of fetal liver cells to be transformed by the Abelson virus. Marked reductions are also noted in CFU-S, in the later erythroid stem cell (CFU-E), in hematocrit, and in pre-B and B lymphocytes. Despite the great deficiency of thymocytes, in vitro cultures of thymuses from day 14 fetuses reveal that the proliferation and maturation of these cells occur at normal rates. This suggests that the thymic hypoplasia may result from a deficiency of prothymocytes in the fetal liver or a defect in the migration of these cells to the primitive thymus. A similar precursor cell defect may also explain the lack of transformation of trisomy 16 liver cells by Abelson virus. This experiment was originally done to determine, because of the increased frequency of leukemia in human trisomy 21 (see above), whether the trisomic cells are more susceptible to malignant transformation. The finding of vastly decreased susceptibility was, therefore, a great surprise.

Currently under investigation is the question of whether trisomic T lymphocytes, in addition to being numerically deficient, are also functionally abnormal. Work with the fetal thymus culture system suggests that trisomic thymic lymphocytes mature more slowly than normal fetal cells insofar as their ability to respond to the proliferative stimulus of Con A is concerned, but they do eventually attain a degree of responsiveness comparable to normal (Epstein *et al.*, 1985).

These immunological findings, particularly the ones concerning the thymus and T lymphocytes, are of course of great interest because of the thymic and T-lymphocyte abnormalities in Down syndrome which were described in detail earlier in this chapter. Furthermore, there is evidence that the thymus plays a role in the normal development of erythroid and granulocytic cells as well as lymphoid cells (Wiktor-Jedrzejezak *et al.*, 1977), but it is not known

Table 12.11. *Immunological and hematological defects in mouse fetuses with trisomy 16*[a]

	Time of gestation (day)	Change relative to littermate controls[b]
Lymphoid system		
T lymphocyte ontogeny		
MICG-bearing cells in liver (?pre-T)	14	− 20%
Thymocytes	14	− 80%
T lymphocytes (in thymus)	18	− >80%
B lymphocyte ontogeny in liver		
Pre-B lymphocytes (by staining)	17–18	− ~60%
B lymphocytes (by staining)	18	− ~40%
B lymphocytes (as CFU-B)	17	− ~60%
Transformation with Abelson virus	17–18	− >95%
Myeloid system		
Multipotential stem cells (CFU-S) in liver	16	− 70%
Granulocyte-macrophage stem cells		
(CFU-C) in liver	16	− 70–95%
Spleen cells	18	− 80%
Erythroid progeny		
BFU-E (liver)	14	− ~90%
CFU-E (liver)	14	− 50%
Hematocrit	18	− 50%
Liver cells	14–18	− 50–60%

Taken from Epstein *et al.* (1985).
[a]Abbreviations: MICG, macromolecular insoluble cold globulin; CFU-S, colony forming unit – spleen; CFU-C, colony forming unit – culture; CFU-B, colony forming unit – B lymphocyte; CFU-E, colony forming unit – erythroid; BFU-E, burst forming unit – erythroid.
[b]Reductions are expressed on an absolute basis. For those assays carried out with fetal liver, an absolute reduction of 50% in cell number would correspond to a normal relative number of cells since the liver itself is reduced 50–60% in cell number.

if the abnormalities of erythroid and granulocytic proliferation often observed in children with trisomy 21 are secondary to abnormal thymic development or if they represent an additional primary developmental defect. In either event, the evidence for thymic deficiency in Down syndrome, coupled with the finding of dramatic thymic hypoplasia in the trisomy 16 mouse fetus, suggests that similar mechanisms may result in abnormal lymphoid and hematopoietic development in both mice with trisomy 16 and humans with trisomy 21.

Uses of the model

Throughout this chapter, I have alluded to questions about the effects of trisomy 21 that might be suitable for investigation in the mouse model system just described, either in its fetal whole-trisomy form or in the form of Ts16 ↔ 2n mosaics. For those loci known to be on both human chromosome 21 and mouse chromosome 16, the animal model will permit a direct examination of the possible effects of the trisomic state. In particular, it will allow for a detailed examination of whether an elevated concentration of SOD–1, which is known to be increased in trisomy 16 fetuses (Cox, Tucker, and Epstein, 1980), results in any or all of the abnormalities of glutathione peroxidase and oxygen metabolism that have been postulated to be the basis of both the mental retardation and Alzheimer disease changes of Down syndrome. Given the fact that endocardial cushion defects are found both in the human and mouse trisomies, the latter will also make possible a detailed study of the pathogenesis of this condition, including a test of Kurnit *et al.*'s (1985) hypothesis regarding the role of increased adhesiveness of the cells which give rise to the endocardial cushions.

As has just been described, studies of the effect of trisomy on the immune system and other stem cell populations are well under way, as are investigations of neurochemical and neuroanatomical changes. Electrophysiological studies on cultured neurons, of the type described by Scott *et al.* (1982), will certainly be facilitated by the ready availability of fetal neuronal tissue, and it will be most interesting to see if abnormalities similar to those described in trisomy 21 are found. It will also be of great interest to extend both types of investigations to Ts16 ↔ 2n chimeras so that the functional consequences of the aneuploidy in living animals, rather than just in fetuses, can be analyzed. Of special interest will be changes in learning and behavior that might be correlated with structural or, more likely, chemical aberrations. If such are found, it will then be possible to test, in a systematic way, the effects of agents which alter brain chemistry and neurotransmitter activity.

The Ts16 ↔ 2n chimeras, coupled with the experimental scrapie system of Dickinson, Fraser, and Bruce (1979), could provide an approach to the study of the pathogenesis of the Alzheimer disease changes. While not an exact model of Alzheimer disease, the development of senile plaques in the brains of scrapie-infected mice (Bruce and Fraser, 1975; Wisniewski *et al.*, 1978) provides a specific marker and permits the following question to be asked: Will Ts16 ↔ 2n chimeras exposed to the scrapie agent develop such plaques earlier or more extensively than control animals? If so, it might be reasoned that there may be an intrinsic defect in the trisomic neurons which makes them more likely to develop plaques in response to an exogenous agent. If not, investigation of other agents known to be capable of inducing Alzheimer disease-type changes would then be warranted, including aluminum and extracts of Alzheimer disease-affected brain (Crapper-McLachlan and De Boni, 1982).

It is possible, once the actual degree of homology between human chromosome 21 and mouse chromosome 16 is established by conventional and molecular genetic techniques, that animals trisomic for only the truly homologous region can be constructed. This would serve to exclude the effects of loci external to the region and to better reproduce the state of genetic imbalance present in humans with trisomy 21. While it would not necessarily guarantee that the human phenotype would be more faithfully reproduced in the mouse, this is a possibility that could be hoped for. As a step in this directon we (D. R. Cox and C. J. Epstein) have been attempting to prepare fetuses trisomic for just part of mouse chromosome 16, the region distal to the T28H breakpoint (see Fig. 12.7). Work carried out by Kirk (1984) has suggested that fetuses with a duplication of this region (along with a deletion of the distal 5% or less of chromosome 2) have certain defects characteristic of trisomy 16 – in particular, generalized edema and congenital heart disease with left ventricular hypertrophy and atresia of the great vessels. Other abnormalities not characteristic of trisomy 16 and possibly related to the terminal deletion of chromosome 2 were very severe developmental retardation, kyphoscoliosis, anencephaly, and exencephaly. Once fetuses with only the duplication of distal chromosome 16 are prepared, it will be possible to determine which defects are related to imbalance of this region, and whether the other abnormalities, such as the immunological, hematopoietic, and neurological ones already described for trisomy 16, are present.

13

Monosomy X (Turner syndrome, gonadal dysgenesis)

Of all of the literature relating to a specific clinical area in human cytogenetics, the most voluminous is probably that which deals with abnormalities of the X chromosome, particularly monosomy X (45,X or 45,XO) and X-chromosomal deletions. In line with the general thrust of the discussion of aneuploidy in this volume, this chapter will be principally concerned with one aspect of the problem: the influence which the regulation of X-chromosome expression has on the phenotype of X-chromosomal aneuploidy. By virtue of the existence of mechanisms for X-inactivation in human females, interpretation of the effects of X-chromosomal aneuploidy involves considerations which go beyond those applicable to autosomal aneuploidy. These considerations and their application to an understanding of the phenotypes associated with monosomy X and del(X) will be discussed here. Except for the infertility associated with 47,XXY, the other common sex chromosome abnormalities (47,XXY, 47,XXX, and 47,XYY) will not be considered since no specific information beyond that available from analysis of 45,X is available. For a comprehensive review of the cytogenetics of the X chromosome, the reader is referred to the two volumes edited by Sandberg (1983*a,b*).

Properties of the human X chromosome

The life cycle of the X chromosome

Virtually all work relevant to our understanding of the cycle of X-chromosome inactivation and reactivation during ontogeny has been carried out in the mouse, and most of the data to be presented here are based on experimental work with that species. However, as will be shown, the available human data are consistent with the overall picture derived from work with the mouse.

The details of the cycle of inactivation and reactivation of the second X chromosome in female cells have been extensively discussed in two recent reviews (Epstein, 1981*c*, 1983*a*), and only the salient features will be presented here. A variety of methods have been used to assess the status of X-chromosome activity during development. These include cytological approaches based on the observation of a visual difference in the behavior of one of the X chromosomes (such as greater chromatin condensation and allocyclic replication – replication at a time different from that of the other chromosomes) and biochemical determinations of X-linked gene products. The latter approaches may be of two kinds: quantitative, which make use of gene dosage-effect relationships between the number of active X chromosomes and the activities of X-chromosomal gene products in the cell or embryo; and qualitative, in which each X chromosome is marked by a specific variant allele (Epstein, 1981c, 1983b). The relevant data derived from the use of these approaches in the mouse are summarized in condensed form in Table 13.1 and Fig. 13.1.

By cytological criteria, there is no evidence for an allocyclic, heteropyknotic, or late-replicating X chromosome prior to the blastocyst stage, 3.5 to 4 days after coitus. Conversely, by gene dosage-effect determinations, in which a 2:1 female/male ratio is considered evidence that both X chromosomes are actually functional, both X chromosomes do appear to be expressing in the entire embryo at the late morula-early blastocyst stage (3.5 days after conception), in the inner cell mass a day later, and probably in at least some of the cells of the epiblast (the embryo proper) 6 days after conception.

Inactivation is first observable in the trophectodermal cells of the mouse embryo at the full blastocyst stage (40–50 cells), four days after coitus. At this time, allocyclic replication and heteropyknosis of one of the two X chromosomes are detectable. One or two days later, in the 5.5- to 6.5-day embryo, an allocyclic and late-labeling X chromosome is detectable in embryonic cells. Biochemical gene dosage determinations give parallel results, and it appears that inactivation occurs first in the trophectodermal cells at the time of their differentiation from the inner cell mass, then in the primary endoderm after differentiation from the embryonic ectoderm, and finally in the epiblast (the embryo proper). It should be noted that it is the paternal X chromosome that is preferentially inactivated in all of the extraembryonic tissues. In view of the timing of the cellular and X-chromosomal changes, Monk and Harper (1979) have suggested that each inactivation event is coupled to an event in cellular differentiation, perhaps by necessity.

Once inactivation has occurred, it appears to be irreversible under normal conditions in all somatic and extraembryonic cells. Furthermore, both cytological and biochemical evidence are consistent with one of the X chromosomes being inactive in primordial and premeiotic germ cells. However, with the onset of meiosis, a condensed or heteropyknotic X chromosome is no longer observed, the XX/XY or XX/XO gene dosage-effect ratio progressively increases from 1 to 2, and both X-chromosome marker alleles become expressed. Thereafter, all stages of mouse oocyte growth following entry into meiosis and continuing even into adult animals show an X-chromosome gene dosage-effect when XX and XO oocytes (oocytes from XX and XO females)

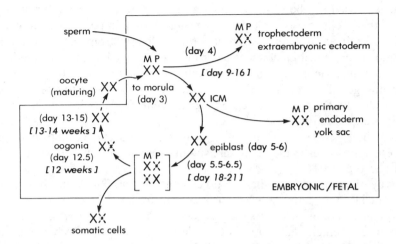

Fig. 13.1 The cycle of X-chromosome activity in the female mouse and human. The times in parentheses and brackets indicate the durations of, respectively, mouse and human embryonic or fetal development. All stages within the box are embryonic or fetal; those outside the box are present in juvenile and adult animals. M and P indicate maternally and paternally derived X chromosomes, respectively. Active X chromosomes are indicated by the solid *X*'s, inactive X chromosomes by the broken *X*'s. ICM, inner cell mass. Figure redrawn with modifications from Epstein (1981*c*).

are compared. The observed ratios between the activities of X-chromosome-determined products in oocytes from XX and XO mothers are, as might be expected from what has been found with autosomal dosage effects (Chapter 4, Table 4.1), very close to 2.0, ranging from 1.89 to 2.24 (mean 2.04) for three different enzymes (Epstein, 1969, 1972*a*; Kozak and Quinn, 1975; Mangia, Abbo-Halbasch, and Epstein, 1975; Monk and Harper, 1978). Monk (1981) has suggested that, as with X inactivation, X-chromosome reactivation and cellular dedifferentiation (return to totipotency of the germ line) are also linked events.

In general, the human data are consistent with the mouse data, both with regard to inactivation occurring first in the trophoblast (by 13–14 days) and later in the embryo itself (by 18 days) and to reactivation of the second X chromosome in germ cells at about 13 weeks of gestation (Epstein, 1983*a*) (Fig. 13.1 and Table 13.2). However, the relationship of reactivation to entry into meiosis remains uncertain.

Noninactivation of X-chromosome loci

The discussion of the life cycle of the X chromosome has been presented in terms of inactivation and reactivation of the entire second X chromosome in

Table 13.1 *X-chromosome expression during female embryogenesis and oogenesis in the mouse*

Time after coitus (days)	Stage/tissue	Enzyme dosage (XX/XY or XX/XO)[a]	Cytologic and other criteria	References[b]
Both X chromosomes active				
2–3	8-cell to early blastocyst	~2	No condensed, allocyclic or late replicating X	Epstein *et al.* (1978); Kratzer and Gartler (1978)
X-inactivation in trophpectoderm but not in inner cell mass				
4	Blastocyst	~1	Allocyclic, late-replicating X	Epstein *et al.* (1978); Monk and Kathuria (1977);
	Inner cell mass	~2	Transfer of cells with 2 active X chromosomes to chimeras	Gardner and Lyon (1971)
X-inactivation in epiblast				
5.5–6.5	Epiblast	~1	Allocyclic, late-replicating X	Monk and Harper (1979); Rastan (1982); Takagi, Sugawara, and Sasaki (1982)
Only one X chromosome active before meiosis				
11–13	Primordial germ cells	~1	Heterochromatic X Single allele expressed	Gartler, Rivest, and Cole (1980); Monk and McLaren (1981)
Both X chromosomes active throughout oogenesis				
(1–2 days postnatal to adult)	Oocytes	~2	No condensed X (Two alleles expressed in human oocytes)	Ohno (1963); Epstein (1969; 1972a); Mangia, Abbo-Halbasch, and Epstein (1975); Kozak and Quinn (1975); Monk and Harper (1978)

[a]Ratios of enzyme activities in XX and XY embryos or fetuses and in occytes from XX and XO females.
[b]Selected references are presented here. Full references are given in Epstein (1983a).

Table 13.2. *X-chromosome expression during female embryogenesis and oogenesis in the human*

Period of gestation	Stage/tissues	Cytologic and other criteria	References
Both X chromosomes active			
≤9 days	Embryo trophoblast	No sex chromatin	Park (1957)
X-inactivation in trophectoderm but not in inner cell mass			
?9–16 days	Trophoblast, yolk sac	Sex chromatin	Glenister (1956); Park (1957)
X-inactivation in embryo			
18–~21 days	Embryo	Sex chromatin	Glenister (1956); Park (1957)
Only one X chromosome active before meiosis			
12 weeks	Germ cells	No G6PD AB heterodimer	Gartler, Andina, and Gant (1975)
Both X chromosomes active throughout oogenesis			
13–14 weeks gestation to adult	Germ cells, oocytes	G6PD AB heterodimer	Gartler et al. (1972); Gartler, Andina, and Gant (1975); Migeon and Jelalian (1977)

somatic cells of human and mouse females. While this was originally believed to be the case, evidence now exists which indicates that all X-chromosome loci are not necessarily subject to inactivation. The possibility that the *Xg* locus is not subject to inactivation was first suggested by Fialkow *et al.* (1970) and Lawler and Sanger (1970). Shapiro and Mohandas (1983), whose work in this area has been of great importance, have summarized the data which indicate that at least two loci in the region Xp22.3→pter escape inactivation: *Xg* (the Xg^a blood group) and *STS* (steroid sulfatase). Qualitative data indicative of noninactivation are available for both loci, and quantitative data are available for steroid sulfatase. The latter, while supporting the concept of noninactivation, suggest that the process is not fully understood (see below). A putative H-Y antigen suppressor locus has also been mapped to Xp22.3 (Wolf *et al.*, 1980).

On the basis of the gene dosage considerations discussed throughout this book, it would be expected that a locus on the X chromosome, if not inactivated, should show dosage effects of the type displayed by autosomal loci and found in oocytes from XX and XO females. This has not, however, proven to be the case. While the activity of steroid sulfatase (STS) is greater in female cells than in male cells, the observed ratio (XX/XY) has not been 2.0. Rather, mean ratios of 1.4 and 1.6 have been found in placental tissue (Bedin *et al.*, 1981; Shapiro, Mohandas, and Rotter, 1982), 1.6 in hair follicles (Chance and Gartler, 1983), and from 1.4 to 1.7 in fibroblasts (Müller *et al.*, 1980; Ropers *et al.*, 1981; Bedin *et al.*, 1981; Lykkesfeldt, Lykkesfeldt, and Skakkebaek, 1984). A similar ratio was found when XX and XO fibroblasts were compared (Lykkesfeldt, Lykkesfeldt, and Skakkebaek, 1984), although Chance and Gartler (1983) found an XX/XO ratio of 2.7 (2.0 when standardized for lactate dehydrogenase activity). However, the latter investigators found that XXX and XXXX cells did not have activities higher than did XX cells, and Vogel *et al.* (1984) showed that the ratio of activities in 47,XXX and 45,X fibroblast clones derived from the same culture was only 1.8. Clearly, the activity of steroid sulfatase is not linearly related to X-chromosomal gene dosage.

Another observation relevant to a consideration of expression of the *STS* locus is that of Migeon *et al.* (1982). Using cloned cells from females heterozygous for steroid sulfatase deficiency and for two G6PD electrophoretic alleles, they were able to show that the activity of the *STS* locus on the active X chromosome was from 1.75 to 4.6 fold greater than that of the locus on the otherwise inactive X chromosome. Based on these results and on consideration of X inactivation in other species, such as *Drosophila*, they suggest that "it is possible that mechanisms have evolved to compensate for the single copy of X genes [in male cells] by enhancing the product output of the active X chromosome . . . through enhanced transcription or processing." As an alternative explanation, I would raise the possibility that the *STS* locus, rather than operating at reduced activity on all "inactive" X chromosomes, actually operates at full activity on some (the *STS* locus fully escaping inactivation) and is totally inactivated on others. The net effect would be the same. However, neither explanation satisfactorily explains why the XXX and higher X states do not have STS activities higher than does 46,XX.

Despite these still poorly understood aspects of noninactivation, it does seem clear that 45,X individuals do differ from 46,XX females with regard to the expression of the *Xg* and *STS* loci and presumably of other loci in the same region. However, it must also be kept in mind that although they are functionally monosomic in comparison to other females, they do not differ from normal males (Lykkesfeldt, Lykkesfeldt, and Skakkebaek, 1984).

In addition to the noninactivated region defined by *Xg* and *STS* (and also *MIC2X*, see below), it has been suggested that a second region of the short arm of the X chromosome also escapes inactivation (Therman *et al.*, 1976, 1980). This region, designated "*b*," is in the proximal third of Xp (the Q-band dark region, Xp11). It has been characterized by cytological criteria, including the time of replication relative to the rest of the chromosome and its degree of condensation in Barr (sex chromatin) bodies formed by abnormal X chromosomes. Therman *et al.* (1980) have also postulated that its degree of activity is inversely related to the number of X chromosomes in the cell. However, to date, no specific loci appropriate for gene dosage effect studies of the type carried out with steroid sulfatase have been unequivocally localized to this region (Miller, Drayna, and Goodfellow, 1984), and the assignment of noninactivation status to the *b* region remains hypothetical.

Mechanisms leading to phenotypic abnormalities

The phenotype of monosomy X, originally defined cytogenetically by Ford *et al.* (1959*b*), can be divided into three major components: gonadal dysgenesis, with failure of development of secondary sexual characteristics; shortness of stature; and congenital malformations (the Turner syndrome stigmata) which most frequently include a shield chest, short webbed neck, low posterior hairline, congenital lymphedema, short fourth metacarpals, pigmented nevi, cubitus valgus, hyperconvex nails, congenital heart disease (especially coarctation of the aorta), and renal anomalies (Haddad and Wilkins, 1958; Engel and Forbes, 1965; Ferguson-Smith, 1965; Therman, 1983). As with the autosomal aneuploidies, the several congenital malformations listed are not invariably present in every affected individual.

An understanding of the mechanisms leading to the phenotypic abnormalities of monosomy X must be based on two types of considerations. The first type includes those considerations relevant to chromosome imbalance in general, and the principles set forth in the earlier chapters on autosomal aneuploidy are equally applicable here. As with the latter conditions, no specific connection has as yet been established between imbalance of specific genetic loci and particular abnormalities. However, what distinguishes monosomy X from all of the autosomal aneuploid states is the role played by the X-inactivation process in determining where and when one or both X chromosomes are active. Only by coupling this information with the considerations applicable to aneuploidy in general will it be possible to understand the mechanistic relationship between monosomy X and the phenotype which it produces.

Gonadal dysgenesis

The mere existence of abnormalities associated with X-chromosomal aneuploidy indicates that the X-inactivation process is more complicated than was originally thought, and the previous section has described some of these complexities. However, it is probably not necessary to invoke any such complexities to explain the gonadal dysgenesis which is characteristic of monosomy X. As has been described earlier (Fig. 13.1; Tables 13.1 and 13.2), there is a significant period of time in oogenesis during which X inactivation is not operative – in the germ cell and oocyte from about the initiation of meiosis at midgestation (in the mouse) to the time of ovulation in adulthood. Since this is the normal state of affairs, it must be assumed that loss of an X chromosome during female germ cell development and oogenesis constitutes a situation of true genetic imbalance with regard to the functions determined by the X chromosome. In the oocytes, X-chromosome monosomy is thus functionally analogous to autosomal monosomy, a condition which, as has been discussed in Chapter 10, is highly lethal to early embryos, with death usually occurring about the time of implantation. While autosomal monosomy may not necessarily create a state of cell lethality, the evidence on monosomy ↔ diploid chimeras suggests that the survival of monosomic somatic cells, even in an otherwise normal environment, is compromised.

On the basis of such considerations, it has been suggested that the monosomic state present in the germ cells and oocytes of human female embryos lacking an X chromosome may, during the course of gestation, lead to oocyte degeneration and secondarily to ovarian dysgenesis (Ferguson-Smith, 1965; Lyon, 1970; Epstein, 1972*b*; Kennedy, Freeman, and Benirschke, 1977; Burgoyne, 1978; Therman *et al.*, 1980; Epstein, 1981*c*). Although liveborn humans lacking an X chromosome do have gonadal dysgenesis, with "streak" ovaries devoid of eggs, these individuals actually do form eggs in embryonic life which degenerate prior to birth (Singh and Carr, 1966; Carr, Haggar, and Hart, 1968; Weiss, 1971). In addition, female mice lacking an X chromosome, although initially fertile, have a significantly shortened period of reproductive capability (Lyon and Hawker, 1973; see below). If the duration of the gestation of a human female lacking an X chromosome (9 months) and that of the gestation and period of fertility of a mouse without an X chromosome (about 12 months) are compared, it is noted that these times are not very different on an absolute scale. It has also been recognized that defects in oocytes or their failure to migrate to the ovary may lead to maldevelopment of the ovary (Witschi, 1951), as occurs in mice homozygous for the W^v mutation (Mintz, 1959). In addition, despite their fertility, 39,X (XO) female mice have small litter sizes, and even chromosomally normal preimplantation embryos of 39,X mothers do very poorly in vitro (see below). Taken together, these observations suggest, as has already been stated, that the genetic imbalance present in the germ cells and oocytes of 45,X human female embryos may lead to oocyte degeneration and secondarily to ovarian dysgenesis, with its concomitant effects on sexual development. It is of interest to note that in deletions of the X chromosome, a period of fertility may precede a premature menopause, and Fitzgerald, Donald, and McCormick (1984) have speculated that the greater the number of relevant genes involved, the more rapid the rate of germ cell attrition.

As an aside, it is worth noting that it may also be the case that the absence of viable germ cells in the human Klinefelter syndrome (47,XXY) may represent just the reverse situation of that obtaining in monosomy X. Thus, it may be hypothesized that reactivation of the second X chromosome also takes place at the time of male meiosis. Therefore, the presence of two X chromosomes in male germ cells which ordinarily have just one is functionally analogous, in terms of dosage relationships, to tetrasomy for one of the autosomes, a condition likely to be highly deleterious to cells with this form of genetic imbalance. [A relevant model for this may be the case of a dysmorphic female with severe growth retardation who was found to have an active duplicated segment of the distal part of the long arm of the X chromosome (Magenis *et al.*, 1982).] This imbalance may then lead to progressive death of the spermatogonia and, in a manner similar to that postulated for gonadal dysgenesis, to the secondary changes in testicular structure. The presence of significant but reduced numbers of normal-appearing spermatogonia in the testes of infant boys in the first year of life (Mikamo *et al.*, 1968) is consistent with this hypothesis. Citoler and Aechter (1979) also described the presence of germ cells in the testes of fetuses therapeutically aborted in midgestation and found a higher than normal frequency of heterotopic germ cells in the 47,XXY fetal testis.

Short stature and congenital malformations

Although the nature of the cycle of X inactivation and reactivation in female germ cells provides a reasonable explanation, at least in part if not in whole, of the gonadal dysgenesis component of monosomy X, it does not explain the growth retardation and the diversity of specific somatic and functional abnormalities which constitute the Turner syndrome phenotype. If vague concepts such as altered cell cycle and a nonspecific heterochromatic effect are disregarded (see Chapter 9), two principal explanations for these manifestations can be considered [see also discussion of Therman *et al.* (1980) and Therman (1983)]. The first is related to the normally occurring period of genetic imbalance between males and females which lasts until the egg cylinder stage (Fig. 13.1). If this period represents a time of necessary genetic difference between males and females, rather than merely the time required for the inactivation process to go into operation, the lack of the second X chromosome might be expected to have adverse effects. However, for such effects to occur during and even beyond the time of somatic morphogenesis, it would have to be postulated that either there is a long period of carry-over of the X-chromosome gene products (or at least their effects) beyond the time of inactivation, or, alternatively, inactivation does not actually occur in some tissues until much later than is presently believed. While it is conceivable that such mechanisms could be operative, and the second of these was considered very early by Lyon (1962) and later by Steele (1976), there is little evidence to support them. Furthermore, the time scale of human development would make such delayed effects quite unlikely. The somatic anomalies associated with the Turner syndrome appear to have their genesis six or more weeks after conception [see Gray and Skandalakis (1972) for tables], at least 20 days after the time of observable sex chromatin in human embryos (Table 13.2).

The other and more likely possibility is that the problem lies with those regions of the X chromosome that do not undergo inactivation. If the Y chromosome in the 46,XY genome of the male is ignored, the somatic manifestations of monosomy X could be attributed to the deletion of that region or regions that normally escape inactivation. However, for this to be the case it would have to be postulated once again, since males have only one X chromosome, that there is a necessary difference between male and female somatic development in the role of X-chromosomal loci, both before and after birth. At present, this notion is difficult to accept. The alternative approach is to postulate that at least some functionally significant loci in the noninactivated region of the X chromosome are also represented by functional homologous loci on the Y chromosomes (Epstein, 1981c; Burgoyne, 1981). If this were the case, sex chromosome aneuploidy, whether for the X or the Y chromosome, would result in a state of imbalance which would potentially affect development and function.

Two possible models for how imbalance of such homologous or analogous sites on the X and Y chromosomes might be implicated in the genesis of abnormalities are shown in Fig. 13.2 (Epstein, 1981c). The first is a simple dosage mechanism, such that absence of a second sex chromosome, whether X or Y, produces a state of partial monosomy. The second is a regulatory model which could explain how loss of a single sex chromosome, or even deletion of part of one, could lead to total failure of some essential function or functions.

The suggestion of homologous regions on the X and Y chromosomes is not a new one and was raised by Lyon (1962, 1971) herself shortly after advancing the original X-inactivation hypothesis. Although this notion has been rejected by others (Ohno, 1967, 1979), it still must be considered an open question for several reasons. First, Y-chromosome aneuploidy is not without somatic manifestations; for example, both 47,XXY and 47,XYY males are taller on the average than chromosomally normal males (Price *et al.*, 1976). In view of the shortness of stature associated with monosomy X, this could be indicative that genes on both the X and Y chromosomes are involved in controlling longitudinal somatic growth. Moreover, human X;Y translocations have been described in which nullisomy for the distal part of Xp would have to be present unless functional homologous loci were also present on the Y chromosome (Pfeiffer, 1980b). Unless this region of the chromosome is relatively empty, it is difficult to visualize how such nullisomy could be tolerated at all.

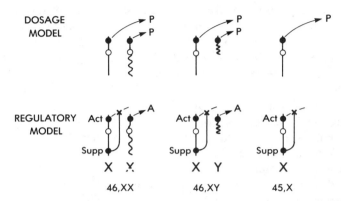

Fig. 13.2 Potential role of homologous sites on X and Y chromosomes in the pathogenesis of the 45,X phenotype. Dosage model (top) predicts a decrease by half in the synthesis of the homologous X and Y gene products, P. Regulatory model (bottom) invokes an interaction between "activator" loci (Act) on the homologous regions of the X and Y chromosomes which produce an active gene product (A) and a "suppressor" locus (Supp) which is present only on the X chromosome. The product of Supp blocks synthesis of A when the number of Act loci is equal to or less than the number of Supp loci. This model indicates how a dosage phenomenon can be translocated into an all-or-nothing effect. The wavy line and broken X indicate the inactivated X chromosome. Redrawn with modifications from Epstein (1981c).

Homologous pairing of the X and Y chromosomes at meiosis has been observed in a large number of mammalian species, with an X-Y synaptonemal complex involving the short arms of both chromosomes being described in human spermatogenesis (Solari and Tres, 1970; Moses, Counce, and Paulson, 1975; Solari, 1980). The existence of such pairing implies, although it certainly does not prove, that homologous (though not necessarily functional) coding sequences are present on Xp and Yp (Fig. 13.3). In fact, Polani (1982) has calculated that the X-Y pairing segment encompasses ≥27% of Xpter and ≥95% of Yp, and both he and Burgoyne (1982) regard the existence of such a pairing segment as obligatory for normal meiosis. Since crossing-over may (Polani, 1982) or must (Burgoyne, 1982) occur in the pairing region, loci distal to the crossover would behave as though they were autosomal. Burgoyne (1982) refers to them as pseudoautosomal. It should be noted, however, that Ashley (1984) has argued against the existence of homology between the X and Y chromosomes and believes that synapsis between the X and Y is nonhomologous.

Structural evidence is now beginning to accumulate that there are indeed homologous segments on the short arms of the human X and Y chromosomes (Miller, Drayna, and Goodfellow, 1984). Goodfellow *et al.* (1983) have identified a gene, *MIC2X*, in the region Xp22.3→pter, which controls the cell-

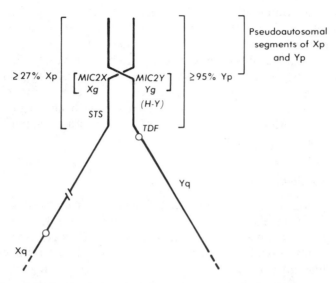

Fig. 13.3 Model of X-Y pairing in human meiosis according to Burgoyne (1982) and Polani (1982). According to Burgoyne (1982), the inactivated X and Y pairing segments are split into two regions by a single obligatory crossover, with genes distal to this point behaving as though autosomally inherited (pseudoautosomal). *TDF*, testis determining factor, is placed in the nonhomologous segment of the Y chromosome by Burgoyne (1982), but Polani (1982) places *H-Y*, which may be the same, in the homologous region. The lengths of the segments are as given by Polani (1982). The model shown here does not take into account the possible existence of a second noninactivated region, as proposed by Therman *et al.* (1976, 1980).

surface expression of an antigen defined by the monoclonal antibody, 12E7. A structurally homologous locus, *MIC2Y*, has been localized to the short arm of the Y chromosome in the region Ypter→q11. The X-chromosomal homolog, *MIC2X*, appears to escape inactivation (Goodfellow *et al.*, 1984), and it is postulated that this gene may be related to or even identical with *Xg* (Goodfellow *et al.*, 1983). If so, *MIC2Y* would then be related to or identical to the postulated Y-chromosome locus, *Yg*, homologous to *Xg*, for which evidence of functional expression has been obtained (Goodfellow and Tippett, 1981; Tippett *et al.*, 1984). A 25,000- to 30,000-dalton cell-surface glycoprotein defined by the monoclonal antibody AbO13 has been mapped to the X and Y chromosomes, with the X chromosome locus not being subject to X inactivation (Dracopoli *et al.*, 1985). The relationship between this antigen and those coded by *MIC2X* and *MIC2Y* has not yet been established, although it is quite possible that they are related or even identical.

Several other homologous loci or sequences present on both the X and Y chromosomes have been identified with recombinant DNA probes (Page *et al.*, 1982; Camerino *et al.*, 1984; Weissenbach *et al.*, 1984; Miller, Drayna, and Goodfellow, 1984; Wolfe *et al.*, 1984). Virtually all of these map to the long arm of the X chromosome and do not, therefore, support the notion of homologous X and Y short-arm segments (Erickson and Goodfellow, 1984). However, the existence of these homologous X and Y segments on Xq does not negate the possible existence or significance of active loci on the Y chromosome which are homologous to those in the noninactivated region(s) of the X chromosome. More data should be forthcoming as the mapping of the X and Y chromosomes proceeds.

Phenotypic mapping of the X chromosome

One test of the possible role of homologous noninactivated X- and Y-chromosome segments on the production of the monosomy X phenotype is provided by an examination of the clinical consequences of deletions of different segments of the X chromosome. Numerous authors have attempted to correlate different features of the phenotype with specific regions of the X chromosome, using the principles of phenotypic mapping outlined in Chapter 3. The results have been, to say the least, confusing and often contradictory, and several authors suggest that it is not possible to make firm phenotype-genotype correlations (Maraschio and Fraccaro, 1983; Fryns, Kleczkowska, and van den Berghe, 1983). These difficulties may stem, in part, from technical problems, including difficulties in assigning breakpoints and distinguishing interstitial from terminal deletions and inclusion of mosaic [del(X)/XO] with nonmosaic cases. Nevertheless, Therman (1983) has divided the phenotype of X-chromosome deletions into three main groups, Xp-, i(Xq), and Xq-, and the last of these has been further subdivided, as shown in Table 13.3. Perhaps the most important single conclusion is that, as stated by Teyssier, Bajolle, and Caron (1981), "gonadal dysgenesis, short stature, and physical signs of Turner syndrome are dissociate entities." They also suggested "that breakpoint localisation might not be the major factor influencing the deleterious effect of a deleted X chromosome," a conclusion that I would be hesitant to accept at this time. However, since our interest here is in the question of overall mechanisms, not fine-structure phenotypic mapping, the issue will not be further pursued at this point.

From the point of view of assigning possible consequences to imbalance of the noninactivated region at the distal end of Xp (Xp22→pter), it would appear from Table 13.3 that short stature is the only feature of monosomy X that can be directly attributed to deletion of this region, and many other investigators have arrived at the same conclusion. However, Therman (1983) has further suggested that other quantitative traits, such as total finger ridge count and palatal index (height/width), might also be influenced by imbalance of this region, since these traits appear to be linearly dependent on the total number

Table 13.3. *Phenotypic features of X-chromosome deletions*

Deleted region	Stature	Gonads	Somatic stigmata
Short arm			
Xpter→~p21	Short	Normal	None
All of Xp [as in i(Xq)]	Short	Dysgenesis	Most or all
Long arm			
Xqter→~p22	Normal	Dysgenesis	None to few
Xqter→~q13	Normal[a]	Dysgenesis	None to some
All of Xq [as in i(Xp)][b]	Normal[a]	Dysgenesis	Several

Note: This is a very rough breakdown of the features of supposed nonmosaic X-chromosome deletions, based on the outline of Therman (1983) and on the data of Grumbach and Conte (1981); Fryns, Petit, and van den Berghe (1981); Wyss *et al.* (1982); Goldman *et al.* (1982); Larizza *et al.* (1982); L. Shapiro (1983); Kaffe, Hsu, and Hirschhorn (1983); Maraschio and Fraccaro (1983); Schinzel (1983); Kaiser *et al.* (1984); and Skibsted, Westh, and Niebuhr (1984), not all of whom agree with one another.
[a]Some cases with reduction in stature have been reported.
[b]The existence of i(Xp) has been questioned by Therman and Sarto (1983) and Schinzel (1983).

of X chromosomes and, to a lesser extent, on the number of Y chromosomes (Barlow, 1973). A possible mechanism for such dependence, particularly for the total finger ridge count, has already been discussed in Chapter 8.

While it thus seems possible to attribute much or all of the growth problem ⌐to imbalance of the noninactivated distal segment of Xp [this was suggested over twenty years ago by Ferguson-Smith (1965)], and the gonadal dysgenesis to the effects on oocyte integrity of deletions of the proximal part of Xp and distal part of Xq (Table 13.3), the etiology of the somatic stigmata of Turner syndrome is still unexplained. If Therman *et al.* (1980) are correct in their identification of a second noninactivated segment, the *b* region, in the proximal part of Xp, it would be possible to attribute the congenital malformations to imbalance of this region if it were indeed a functional one. However, as has already been noted, this would require males and females to be fundamentally different in terms of the normal expression of loci in this segment, since no corresponding region is present on the Y chromosome. Moreover, this explanation, even if true, would still not explain why deletions of Xq should result in somatic abnormalities, if in fact they really do [see Teyssier, Bajolle, and Caron (1981); Skibsted, Westh, and Niebuhr (1984)], unless yet another noninactivated region is present on the long arm. Clearly, despite all we know about the X chromosome, we are still discouragingly far from understanding the origin of these abnormalities.

The lethality of monosomy X

A striking feature of monosomy X is its extreme prenatal lethality, with >95-99% of 45,X conceptuses dying during gestation (Simpson, 1976; Hook and Warburton, 1983). The recognizable and liveborn cases of monosomy X, with which the preceding discussion has been concerned, thus represent only a small and probably biased sample of the effects of this chromosome abnormality. The early mortality is much less in 46,Xi(Xq) cases, suggesting that imbalance of loci on the long arm of the chromosome may be particularly involved in the production of the lethality (Hook and Warburton, 1983). While many types of autosomal aneuploidy may also be associated with high rates of prenatal mortality, the difference from monosomy X is that, in the former, the individuals that do survive are generally quite severely affected, while in monosomy X the defects are functionally much less severe. Perhaps the closest analog is trisomy 21, with its moderately high rate (~75%) of prenatal loss (Boué et al., 1981) and its only mild or moderately severe congenital malformations (Chapter 12).

The time of death of 45,X embryos and fetuses appears to extend throughout the period of gestation, with the vast majority of 45,X conceptuses, perhaps 95%, dying during the first trimester and another 3–4% (~70–75% of those identifiable at the time of a midtrimester amniocentesis) dying during the second half of pregnancy (Hook and Warburton, 1983). The mean developmental age of the 45,X fetuses recognized in the first trimester abortions is about 6 weeks (Boué et al., 1976), and it is this high rate of early lethality that is of interest here.

Two general types of problems leading to early fetal demise can be visualized. The first is that the 45,X embryo or fetus is so defective as to be unable to survive even the early period of gestation. While pathological descriptions of 45,X abortuses do not indicate any unusual degree of developmental abnormality (although congenital malformations including cystic hygromas and horseshoe kidneys are certainly present in recognizable abortions) (Singh and Carr, 1966; Singh, 1970; Boué et al., 1976), it is still conceivable that those embryos eliminated earlier are more defective. One suggestion for a possible mechanism is that of Gartler and Sparkes (1963), who postulated that random inactivation applied to the single X chromosome of 45,X embryos so that, on the average, half of the cells of a 45,X embryo had no functional X chromosome. At present, there are no data bearing directly on this issue, although Santesson, Böök, and Kjessler (1973) did examine the status of the X chromosome in cultured skin cells from a 15-week 45,X abortus. None of the cells had sex chromatin, and no late-labeling X chromosome was found on DNA replication studies. If taken as representative of the situation in 45,X abortuses in general, it would seem that those embryos that do reach about 6 weeks of gestation do not have cells without a functional X chromosome. It is still conceivable, although – I would agree with Therman *et al.* (1980) – unlikely, that such would occur and lead to the death of 45,X conceptuses at an even earlier stage. Another possibility is that death occurs prior to the time of embryonic X-chromosome inactivation (Table 13.2) when the 45,X embryo is truly functionally monosomic for the X chromosome. This possiblity has already been considered earlier with regard to abnormalities of morphogenesis. While it is of course possible to make the analogy with autosomal monosomies, which die very early in gestation (see Chapter 10; Hassold *et al.*, 1978; Epstein and Travis, 1979), the situation is actually quite different for the X chromosomes. Male embryonic cells at the same stage of development have only one X chromosome and seem to tolerate this situation well without any known dosage-compensation mechanism being operative, at least in the mouse, to adjust X-chromosome gene-product levels to the female range (Table 13.1).

The other general explanation for the early death of 45,X embryos and fetuses is that the difficulty is not with *them* but with the placenta. The aneuploid state is visualized as interfering with placental growth and function so that the placenta is unable to sustain an otherwise reasonably normally functioning fetus. A similar suggestion has been made to explain the early death of both human and mouse trisomies and human triploids (Boué *et al.*, 1976; Ornoy *et al.*, 1978; Gropp, 1981) (see Chapters 9 and 10). Burgoyne, Tam, and Evans (1983) call attention to what they term "different strategies for the maintenance of pregnancy" in man and mouse. In the mouse, ovarian steroids appear to be involved throughout pregnancy; in man, steroids produced by the placenta become essential near the end of the first trimester. Therefore, any abnormality or immaturity of 45,X placental development producing a placental hormone deficiency could result in an inability to maintain an otherwise viable embryo and consequent spontaneous abortion (see also Boué, Boué, and Spira, 1974; Boué *et al.*, 1976). Burgoyne, Tam, and Evans (1983) further suggest that placental insufficiency not so severe as to lead to embryonic death could, either by so severely interfering with early embryonic growth as to set up a situation of maldevelopment during a later period of catch-up growth (Snow, Tam, and McLaren, 1981) or by a direct teratogenic action, lead to the production of congenital malformations. While conceivable, I do not consider these possibilities highly likely, mainly on the grounds of specificity of the pattern of somatic abnormalities associated with monosomy X.

If a placental mechanism is involved in the early lethality of monosomy X, why then might the placenta be abnormal in the first place? Two possibilities may be considered. The first is that, as in autosomal aneuploidy, the same or similar mechanisms, whatever they might be, that produce the somatic growth impairment and other abnormalities also affect placental growth and/ or function. The other possibility is that, as the result of inactivation of the only X, there could be a population of placental cells in which no X chromosome is active. In the mouse, there is preferential inactivation of the paternally derived X chromosome in extraembryonic tissues (Fig. 13.1), including the extraembryonic ectoderm, primary endoderm, and yolk sac, chorion, and allantois (Takagi and Sasaki, 1975; West *et al.*, 1977; Papaioannou and West, 1981; Harper, Fosten, and Monk, 1982; Takagi, 1983). The same may also be true, but has not been unequivocally proven, in man (Ropers, Wolff, and Hitzeroth, 1978; Migeon and Do, 1978), although contrary evidence (including evidence that X inactivation may not occur in all placental tissues) has been reported (Migeon *et al.*, 1985).

Assuming for the moment that preferential paternal X-chromosome inactivation does occur in man, what might be expected in the placentas of 45,X embryos? Data on the origin of the X chromosome on liveborns with 45,X are not helpful. Although they indicate that only 23% of cases have a paternally derived X chromosome (Sanger *et al.*, 1977), this is not necessarily representative of the status of 45,X conceptuses, especially if embryos with such paternally derived X chromosomes were more likely to die early in gestation (Warburton, 1983). However, a more serious objection to the notion of a role for inactivation of the single X chromosome in cells of the placentas of monosomy X embryos comes from work with XO and parthenogenetically derived female mice. In the latter, random X inactivation takes place in the extraembryonic tissues, even though no paternal X chromosome is present (Kaufman, Guc-Cubrilo, and Lyon, 1978). Conversely, in XO embryos in which the only X chromosome is paternally derived, inactivation of this chromosome does not occur (Frels and Chapman, 1979; Papaioannou and West, 1981). Therefore, if the mouse data are applicable, inactivation of the single X chromosome in 45,X human placentas is unlikely to occur, whatever the source of the X chromosome. As a result, if placental abnormality is involved in early lethality, it probably arises from genetic imbalance of the same type associated with other manifestations of the disorder rather than from some vagary of the X-inactivation process.

The 39,X mouse

At several points in this chapter, I have made reference to what is known from studies of the 39,X or XO mouse, and these animals are of course of interest as potential models for human monosomy X. Female mice with just one X chromosome were first described by Welshons and Russell (1959). Since these animals are fertile females, it was possible to establish a breeding stock for generating large numbers of XO progeny. In this stock, the X-linked gene, tabby (*Ta*), is used as a marker for the genetic state of the progeny (Cattanach, 1962). "Tabby" is a coat mutation recognizable in both hemizygous or heterozygous forms. Thus when, XO females are mated with *Ta*/Y males, the offspring are either full tabby females (and therefore *Ta*/O = XO), heterozygous tabby females (*Ta*/X = XX), or nontabby males. Schematically,

XO × *Ta*/Y → *Ta*/O + *Ta*/X + XY + YO
         (female,   (female,   (lethal)
          full   heterozygous
          tabby)   tabby)

In the next generation, the *Ta*/O females are mated with nontabby males:

Ta/O × XY → XO + *Ta*/X + *Ta*/Y + YO
          (female,   (female,   (male,   (lethal)
           nontabby) heterozygous full
                      tabby)   tabby)

The nontabby females in this cross are the desired XO animals, and they can in turn be mated to *Ta*/Y males to continue the alternating breeding scheme. Despite the fact that equal numbers of XO and XX females and of XY males are predicted by this breeding scheme, there is in fact a deficiency of the XO class, with ratios of 0.33 to 0.54:1:1 being observed at 3 weeks after birth among all progeny of XO (of either type) × XY matings (Cattanach, 1962; Brook, 1983).

Another method for producing XO females is by mating normal males with female carriers of a large radiation-induced X chromosome inversion, *In(X)1H* (Evans and Phillips, 1975). Because of the inversion, there is a high rate of X-chromosome loss, with the production of about 25% XO progeny (Evans and Phillips, 1975). That the process is not one of simple nondisjunction is demonstrated by the fact that XXY and XXX progeny are never observed (Phillips, Hawker, and Mosely, 1973). Unlike the first XO-breeding scheme described, the mothers of the XO daughters obtained by this method are XX [actually *X In(X)*] rather than XO in sex chromosome constitution. The importance of this for studies of early embryonic development was pointed out by Burgoyne and Biggers (1976). They showed that preimplantation embryos of all genotypes from XO mothers developed poorly in vitro, presumably as a result of the deficiency of X-chromosomal gene products in the egg which results from the gene dosage effects summarized in Table 13.1.

The phenotype of the XO mouse bears both similarities to and differences from that of 45,X human female. Clearly, the fact that it is fertile and free of congenital malformations (Welshons and Russell, 1959; Russell, Russell, and Grower, 1959) distinguishes the XO mouse from the XO human female with her characteristic pattern of congenital anomalies and gonadal dysgenesis. Nevertheless, XO mice do have impairments in fertility, with a significantly shorter reproductive life-span and litter sizes about half of normal (Lyon and

Hawker, 1973). The ovaries of the older (12 months) XO mice are devoid of small follicles and have a large amount of hypertrophied interstitial tissue. Even earlier in life, from 12 to 200 days after birth, the population of oocytes in XO ovaries is reduced by about 50% (Burgoyne and Baker, 1981a,b). This deficit is not present in fetal ovaries at the time of meiosis, although a developmental retardation in oogenesis was found. Because of the existence of both the prenatal and postnatal abnormalities in the ovaries of XO mice, it has been suggested, as has already been noted above, that the difference between XO humans and XO mice in this regard is more a matter of timing than a real qualitative difference (Burgoyne and Biggers, 1976). Since the ovaries of XO human fetuses and infants do have oocytes which gradually disappear (Carr, Haggar, and Hart, 1968), it may be that the length of gestation and the postnatal life-span of XO mice are just too short for the same process to occur in them.

Early in gestation (7.25 days), XO embryos are considerably smaller than their XX sibs and remain behind both developmentally and in weight until midgestation (Burgoyne, Tam, and Evans, 1983). The XO placentas are normal or slightly larger than normal in size. At birth, the XO neonates are slightly smaller than normal and have a decreased rate of growth until weaning, at which time the rate of weight gain normalizes (Burgoyne, Evans, and Holland, 1983). Since the preweaning deficit in weight is not corrected, the mature XO females are 1.5 to 2.0 g lighter than their XX sisters. This diminution in size, of the order of about 6–7%, is far from the degree of growth retardation seen in human XO females. The only exception to this are the occasional mouse XO "runts" which are extremely retarded during fetal development and have a severe germ cell deficiency, making them the closest mouse analog to both the somatic and gonadal state of the XO human (Burgoyne and Baker, 1981a; Burgoyne, Tam, and Evans, 1983).

As has already been discussed, perhaps the most striking feature of the human 45,X genotype is its very high degree of prenatal lethality. That there is a deficit in the mouse in the number of XO offspring born to XO mothers has also been noted, but opinion has been divided on whether this is the result of abnormal meiotic segregation, with preferential segregation of the X chromosome to the egg rather than the polar body, prenatal loss, or both. Evidence for and against the former possibility is summarized by Brook (1983), whose own investigation showed a random segregation of the X chromosome at meiosis. The loss of XO progeny during gestation was calculated to be 36.2%. This can be considered a maximal estimate, since any preferential segregation of the X chromosome to the oocyte would reduce the amount of prenatal loss necessary to account for the observed frequencies at birth. However, Burgoyne and Evans (unpublished, quoted in Burgoyne, Tam, and Evans, 1983), used the *In(X)1H* technique for generating XO progeny and observed only a negligible loss during pregnancy.

The XO mouse and 45,X human

It is clear from the previous discussion that human and mouse XOs differ in many respects. Unlike the human with monosomy X, the XO mouse is fertile, does not display a high degree of prenatal death, does not have congenital malformations, and, after weaning, is not significantly growth retarded. A possible explanation for the difference in fertility, relating to the length of gestation and of the reproductive period, has already been discussed earlier in this chapter. A suggested explanation for the difference in prenatal death based on placental function has also been considered, but other possibilities for this and the other differences between man and mouse are also conceivable. The most likely is that, despite the many genetic similarities between the human and mouse X chromosomes, there are also fundamental differences between them. Although several mapped genes appear in the same linear order on the two X chromosomes, an order not affected by the central location of the centromere in the human X chromosome, at least one is out of order (Buckle *et al.*, 1984). *Hprt*, which on the human X chromosome is in the distal part of Xq (Xq26–27) near *G6PD* and *GLA* (α-galactosidase), is close to the centromere of the mouse chromosome, near *mdx* (X-linked muscular dystrophy) and *Tfm* (testicular feminization). In addition, the putative meiotic pairing segment, which in humans is at the end of Xp (Fig. 13.3), is visualized as being at the distal end of the mouse X chromosome (Burgoyne, 1982; Evans, Burtenshaw, and Cattanach, 1982).

However, perhaps more to the point is the fact that the existence of a noninactivated segment on the mouse X chromosome has not been demonstrated. In fact, Lyon (1966) used the fact that the XO mouse is "so phenotypically normal" as "positive evidence" that the whole of the second X chromosome in the XX female is inactive and concluded that the evidence then available suggested species differences in the completeness of X inactivation. Unfortunately, this reasoning becomes circular in the present context. Nevertheless, no other evidence has been brought forward in support of the existence of a noninactivated segment of the mouse X chromosome. If its nonexistence could be unequivocally demonstrated, Lyon's (1966) argument would be strengthened, as would the general inference that the explanation for the somatic abnormalities and growth retardation in human monosomy X does lie with the noninactivated region or regions of the human X chromosome.

14

Cancer

Virtually every review of the cytogenetics of cancer begins with a reference to Boveri (1914), and this is certainly not inappropriate. He anticipated a relationship between aneuploidy and malignancy several decades before a relation between aneuploidy and any human disease was first established. His principal ideas, as quoted by Hauschka (1961), were the following:

The essence of my theory is . . . in general, a definite abnormal chromosome complex . . . asymmetrical mitosis could be chiefly considered for the origin of tumors. . . . My theory is able to explain above all the defective histological form and altered biochemical behavior of tumor cells. Are there any means of reaching a trustworthy decision as to the correctness of the views presented here? The most obvious would be to devote renewed attention to the counting of chromosomes, if possible with better techniques.

The required improvements in techniques did not occur for over forty years, but when they did, the validity of Boveri's insight was finally established. As viewed by Sager (1983), "Boveri's contribution to clear thinking about cancer ranks nearly with Mendel's contribution to clear thinking about genes. Written close to 70 years ago, Boveri's view of the origin of cancer anticipates much of what we know today." What still remains to be accomplished, as will be discussed in this chapter, is to discover the meaning of the relationship between aneuploidy and malignancy.

A comprehensive review of the cytogenetics of malignancy would be out of place in this volume, and the reader interested in pursuing the cytogenetic issues in further detail is referred to the books or articles by Sandberg (1980), Mitelman and Levan (1981), Rowley and Testa (1982), German (1983), Mitelman (1983), Rowley and Ultmann (1983), Yunis (1983), and de la Chapelle and Berger (1984), and to the reports of the First, Second, Third and Fourth International Workshops on Chromosomes in Leukemia (1978, 1980, 1981, 1984). The earlier literature is reviewed by Hsu (1961). In this chapter, I shall briefly summarize the information presently available on the association of aneuploidy with cancer. I shall then consider the mechanisms which have been proposed to explain the observed relationships, both in terms of the causation of malignancy and of the potential effects of the aberrations on cellular regulation. Of particular interest will be whether mechanisms based on gene dosage effects, such as have been discussed for the other developmental and functional consequences of aneuploidy, are applicable to malignancy.

Nonrandom associations of aneuploidy and cancer

When first studied cytogenetically, most malignant cells were found to have a bewildering array of cytogenetic aberrations which did not appear to have any consistency to them. This situation changed with the discovery (in chronic myelogenous leukemia) of the Philadelphia chromosome (Ph[1]), [originally thought to be a deletion of chromosome 22q (Nowell and Hungerford, 1960) and later shown to be a balanced 9;22 translocation (Rowley, 1973)], and of the nonrandom association of trisomy 8 and, to a lesser extent, trisomy 9 and del(7q) or monosomy 7 with several hematological malignancies (Rowley, 1975). Since then, a large number of more or less specific, or at least nonrandom, associations between chromosome aberrations and certain neoplasms have been described. As Rowley (1983*a*) and Yunis (1983) have pointed out, the possibility of discovering these associations depended on the development of the techniques of chromosome banding.

The chromosome aberrations associated with malignancy can be divided into two main groups: the apparently balanced translocations, as in the t(9;22) of the Philadelphia chromosome, and the aneuploid states (trisomies, monosomies, duplications, and deletions). In a large number of instances, more than one chromosome abnormality is present in a single malignant tumor or cell line. However, analysis of large numbers of tumors or cell lines from many individuals has made it possible to infer which cytogenetic aberrations are most consistently associated with certain tumors, and these associations are listed in Tables 14.1, 14.2, and 14.3.

The human data have been divided into two sections. Those relatively few instances in which the chromosome abnormality is constitutional and precedes the development of the tumor are listed in Table 14.1. However, the larger number of chromosome aberrations identified in man are the acquired ones (Table 14.2). Although some unbalanced aberrations are associated with just one type of malignancy, others, such as dup(1q23→qter), monosomy 5 or del(5q12→q32), monosomy 7 or del(7q22→qter), trisomy 8, and i(17q), appear in many diverse types of tumors. Greater specificity seems to be attached to the apparently balanced translocations listed in Table 14.2. The animal data (Table 14.3), often obtained from induced tumors as opposed to the generally spontaneous origin of the human tumors, are generally consistent with the human results, although a higher frequency of whole-chromosome trisomy and monosomy seems to be present.

As was noted earlier, a given tumor may display more than one cytogenetic aberration at a time, and the question has been raised as to whether any specific regions are of "decisive importance" in tumor development (Mitelman,

Table 14.1. *Constitutional chromosome aberrations associated with human malignancy*[a]

Chromosome aberration[b]	Malignancy
Aneuploidy	
del(11p13)	Wilms tumor
*del(11q14→q21)	Lymphoma[c]
del(13q14)	Retinoblastoma
del(20p12)	Multiple endocrine neoplasia, type 2[d]
trisomy 21	Acute leukemia[e]
Translocation	
t(3;8)(p1;q24)	
t(3;11)(p13 or 14; p15)]	Renal carcinoma

[a]For references see de la Chapelle and Berger (1984) and Yunis (1983).
[b]Aberrations marked with asterisk are among those identified by Mitelman (1984) as being regularly involved in human neoplasia (see text).
[c]Single case (Rudd and Teshima, 1983).
[d]Conflicting results reported, but original findings held up in double-blind study (Babu, Van Dyke, and Jackson, 1984).
[e]See discussion in Chapter 12.

1984). To answer this, Mitelman (1984) has surveyed a large number of reported aberrations in tumors of all types (Mitelman, 1983) and has identified 54 which are present by themselves in at least two independent cases. When the breakpoints of the aberrations so identified were compared with those in the entire group of 3844 cases surveyed, 96.1% of the tumors studied displayed breakpoints involving the chromosomal segments under consideration. From these results, Mitelman (1984) concluded that "only a limited number of breakpoints are involved regularly in chromosomal aberrations of human neoplasia." Many of the abnormalities listed in Tables 14.1 and 14.2 fall into the regions identified in this analysis and are marked with an asterisk. The other aneuploid states not included in the tables are i(1q), dup (3p25→pter), dup(3q25→qter), del(9q13→qter), del(11q14→qter), del(11q23→ qter), del(13q13→qter), and del(13q21→qter). The possible significance of these regions will be discussed in a later section.

Constitutional aneuploidy

Although they represent just a small proportion of all cases of malignancy, those associated with constitutional aneuploidy have been the subject of intense interest because of the opportunity they provide for relating oncogenes to a specific genetic defect (Table 14.1). The best studied of these conditions are two associated with deletions – retinoblastoma and Wilms tumor. The mapping of the relevant regions, 13q14 and 11p13, respectively, has already been discussed in Chapter 3 and is illustrated in Figs. 3.3 and 3.4. Work on these conditions, both in their chromosomally abnormal forms and in their hereditary, nonaneuploid forms, has provided the basis for the "two-hit" theory of carcinogenesis proposed by Knudson and discussed below.

Retinoblastoma and the "two-hit" theory of carcinogenesis

In an effort to explain the characteristics of the hereditary and spontaneous forms of retinoblastoma, Knudson and his collaborators (Knudson, 1971; Knudson, Hethcote, and Brown, 1975; Hethcote and Knudson, 1978) developed a model which has since been extended to other tumors, predominantly those of childhood, including Wilms tumor (Knudson and Strong, 1972a), neuroblastoma, and pheochromocytoma (Knudson and Strong, 1972b), but also including later-onset diseases such as medullary thyroid cancer (Jackson et al., 1979), carcinoma of the colon, neurofibrosarcoma, carcinoma of the breast, and malignant melanoma (Moolgavkar and Knudson, 1981). In this model, cancer is regarded as "the end result of two discrete, specific, and heritable events at the cellular level" which are regarded as being, "for all practical purposes, irreversible" (Moolgavkar and Knudson, 1981). While the first event is believed to involve a genetic alteration, the nature of the second is not specified. Although generally regarded as also being a genetic abnormality, Matsunaga (1979, 1980) has suggested that it could be an error in the process of differentiation, rather than a mutation.

In applying this theory to spontaneous tumors, the two events ("hits") are regarded as random occurrences which both happen to affect the same cell. For the dominantly inherited conditions, such as retinoblastoma (Schappert-Kimmijser, Hemmes, and Nijland, 1966; Carlson et al., 1979) and Wilms tumor (Matsunaga, 1981), the first event is presumed to be an inherited mutation. Therefore, when the association of retinoblastoma with a 13q14 deletion was first established, it required just a small extension of the model to propose that the deletion represents the first event and that the mutant locus (now designated $rb-$) characteristic of the inherited nonaneuploid form of retinoblastoma is in the 13q14 region (Knudson et al., 1976). Further evidence for this was obtained from detailed linkage and mapping studies carried out by Sparkes et al. (1980b, 1983). Although it was suggested that the mutation present in individuals with dominantly inherited retinoblastoma might increase the sensitivity of the cells to radiation and thereby increase the predisposition to the development of malignancy (Weichselbaum, Nove, and Little, 1980), it appears that the observed difference between normal and mutant somatic cells is too small to have any major effect (Morten, Harnden, and Taylor, 1981).

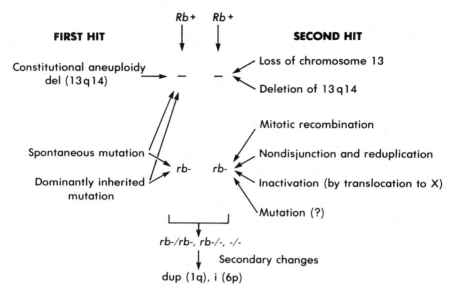

Fig. 14.1. Genetic basis of retinoblastoma by two-hit model, with secondary cytogenetic changes. Adapted from Murphree and Benedict (1984).

Recent work, based on both conventional and recombinant DNA types of genetic analysis, has cast considerable light on the nature of both the first and second hits in retinoblastoma, each of which appears to be mutational in nature. The general scheme has been summarized by Murphree and Benedict (1984). As before, the first event is a mutation or deletion of the *Rb +* locus, making it either *rb −* or *−*, respectively. The second event, involving the normal homolog of *Rb +* on the other chromosome 13, can be either a mutation (*rb −*), inactivation by translocation to an inactive X chromosome (functionally rb −), or deletion of either region 13q14 or of the entire chromosome 13 (monosomy 13). Additionally, and perhaps least expected, it can be the loss of the normal chromosome 13 and duplication of the mutant (*rb −*) one, or a mitotic recombination event bringing *rb −* onto the otherwise normal chromosome 13 (Fig. 14.1). Evidence for all but the first (*Rb + → rb −* by mutation) of these mechanisms has been obtained by Mohandas, Sparkes, and Shapiro (1982), Balaban *et al.* (1982), Benedict *et al.* (1983*b*), Cavenee *et al.* (1983), Godbout *et al.* (1983), Murphree and Benedict (1984), and Dryja *et al.* (1984).

The general interpretation of the results that have been obtained is that the retinoblastoma gene acts as a recessive character, with the cellular genotype of the tumor being either *rb − /rb −*, *rb − / −*, or *− / −* (Fig. 14.1). How then does this cause the tumor? The leading suggestion, based on the hypothesis

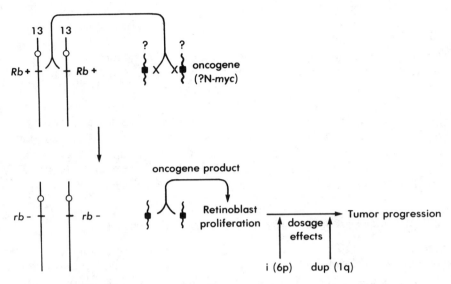

Fig. 14.2. Model for control of oncogene expression by *Rb* + locus on chromosome 13. Based on proposals of Comings (1973) and analysis of Murphree and Benedict (1984) and Lee, Murphree, and Benedict (1984). See text for details.

of Comings (1973), which was greatly ahead of its time, is that the *Rb* + locus acts as the suppressor of a cellular oncogene (proto-oncogene) which, by removal of the suppressive influence, becomes activated or overexpressive (Benedict *et al.*, 1983*b*; Godbout *et al.*, 1983; Murphree and Benedict, 1984) (Fig. 14.2). Presumably, the activation process resulting from loss of *Rb* + is specific for some type of primitive retinal cell, such as that proposed by Kyritsis *et al.* (1984). One candidate for the relevant oncogene is N-*myc* which has been found to be amplified (increased in gene number) 10 to 200 fold in two primary tumors and in a retinoblastoma cell line and, more importantly, to be overexpressed with or without amplification in ten tumors and a cell line (Lee, Murphree, and Benedict, 1984). In neuroblastomas, the amplification of the N-*myc* oncogene, as opposed to its overexpression, may be a late-stage event in tumor development (Brodeur *et al.*, 1984). The chromosomal location of N-*myc* has recently been discovered to be 2p23–24 (Schwab *et al.*, 1984).

   In addition to the chromosome 13 changes, retinoblastomas have also been found to have nonrandom increases in the frequency of dup(1q) and i(6p) (Benedict *et al.*, 1983*a*; Wang *et al.*, 1983; Murphree and Benedict, 1984; Chaum *et al.*, 1984). Murphree and Benedict (1984) have suggested that i(6p), which creates a state of tetrasomy for 6p, could result in the dosage-dependent overexpression of an "expressor" gene which enhances tumor formation (Fig. 14.2). In fact, they suggest that this might conceivably occur after only one chromosome 13 (*Rb+* locus) hit, but there is no evidence to support this. Dup(1q), which is found in many types of malignancy, is thought to regulate malignant progression rather than tumor production per se. If the suggestions with regard to the chromosome 1 and 6 changes are correct, and if these chromosomes are unaltered (see discussion of trisomy 15 and mouse leukemia below), it would mean that dosage-dependent mechanisms could be involved in the progression of retinoblastoma. However, it must be emphasized that the mechanisms proposed for the role of chromosome 13 itself in the pathogenesis of retinoblastoma do not depend on a dosage-dependent mechanism, and as such are presumed to be qualitatively different from what has been discussed in earlier chapters.

Wilms tumor and other forms of constitutional aneuploidy

Recent work on the genetic basis of Wilms tumor has run closely in parallel with that on retinoblastoma, and evidence has been obtained for homozygosity of chromosome 11 in 11 of 21 tumors, presumably in the region carrying the inactive or deleted *wt* allele (Koufos *et al.*, 1984; Orkin, Goldman, and Sallan, 1984; Reeve *et al.*, 1984; Fearon, Vogelstein, and Feinberg, 1984). The mechanisms to account for Wilms tumor formation, therefore, are the same as have been suggested for retinoblastoma. However, the constitutional 11p13 deletion associated with Wilms tumor is also mani fested by aniridia, genital abnormalities, and mental retardation (Riccardi *et al.*, 1980; Turleau *et al.*, 1984). It is not clear whether these anomalies are just the result of the deletion of loci close to the Wilms tumor locus, by the mechanisms discussed earlier in the book, or the result of some process analogous to that involved in tumor formation, which requires inactivation or loss of loci on both chromosomes 11. I would still prefer to think that the former is the case.

The other forms of tumors associated with constitutional aneuploidy – in particular, multiple endocrine neoplasia, type 2 with del(20p12) and familial renal carcinoma with t(3;8) and t(3;11) (Table 14.1) – may also arise by mechanisms similar to those proposed for retinoblastoma and Wilms tumor (Murphree and Benedict, 1984). In the case of renal carcinoma, the translocation, even though it appears balanced, is presumed to cause a deletion or at least inactivation of the relevant locus. Another interpretation will be discussed below. Reasoning somewhat in reverse, it has also been suggested (Murphree and Benedict, 1984) that the locus involved in neuroblastoma is on chromosome 1, since del(1p32→pter) is a nonrandomly associated, acquired chromosome defect in this tumor (Brodeur *et al.*, 1981). Hulten (1984) has made a similar suggestion with regard to carcinoma of the breast.

The situation with trisomy 21 and acute leukemia has already been discussed in Chapter 12. The frequency of leukemia in trisomy 21 is quite different from the frequencies of retinoblastoma and Wilms tumor in individuals with chromosomal deletions. Furthermore, the pathogenesis of the leukemia is probably mechanistically different from that of the other conditions as well. The arguments for and against consideration of leukemia in trisomy 21 as a malignancy associated with constitutional aneuploidy have already been considered.

At this point it is also worth noting that although other chromosome abnormalities are frequently associated with malignancy (Table 14.2), their prior (constitutional) presence in the genome does not itself necessarily lead to the development of malignancy. A case in point is trisomy 8, which is found as a secondary change in a large number and variety of malignancies. However, among 74 individuals with congenital mosaic trisomy 8, only 3 (4%) had malignancy, one each with acute nonlymphoblastic leukemia, Wilms tumor, and rhabdomyosarcoma (Riccardi and Forgason, 1979). A single case of trisomy 8 mosaism, neurofibromatosis, and chronic granulocytic leukemia has also been reported (Palmer *et al.*, 1983). While these findings could be interpreted as indicating that trisomy 8 does not by itself predispose to the development of malignancy, Riccardi and Forgason (1979) have taken the opposite tack and have suggested that there is a "definite, though nonspecific correlation between congenital . . . chromosome 8 abnormalities and the development of certain types of neoplastic growth." Although the correlation is certainly not a strong one, it still may be the case that given the appropriate initiating factor, the constitutional presence of trisomy 8 may serve to enhance or accelerate the progression of the tumor to a fully malignant lesion.

Acquired aneuploidy

Having presented the encouraging progress of research on malignancy associated with constitutional aneuploidy, we now come to a consideration of the much more common cytogenetic problems associated with malignancy – acquired aneuploidy. However, before dealing with aneuploidy per se, a brief review of the present thinking with regard to apparently balanced translocations in malignancy is in order, since much of current work is influenced by the findings in this area.

Translocations

As Tables 14.2 and 14.3 indicate, a large number of human malignancies and some animal neoplasms are characterized by the presence of apparently balanced translocations rather than frank aneuploidy, and some commentators have taken the position that such translocations or transpositions and related rearrangements represent the principal (Cairns, 1981; Radman, Jeggo, and Wagner, 1982; Sager, 1983; Sandberg, 1983c) or at least a common (Chaganti, 1983; Yunis, 1983; Neri, 1984) cause of cancer. Although early proposals (Comings, 1973) regarded these translocations as acting through a

Table 14.2. *Acquired chromosome aberrations associated with human malignancy*[a]

Chromosome aberration[b]	Malignancy
Aneuploidy	
del(1p32→pter)[c]	Neuroblastoma
del(1p21→pter)	Breast carcinoma[d]
del(1p11→p22)[c]	Malignant melanoma
*i(1q)	Endometrial carcinoma
*dup(1q23→qter)	Several malignancies
del(3p14→p23)	Small cell carcinoma of the lung
del(3p21→pter)	Renal cell carcinoma[d]
*del(5q12→~q32) or monosomy 5	Several malignancies
i(6p)	Retinoblastoma[e]
del(6p21→pter)[c]	Acute nonlymphoblastic leukemia
del(6q11→q31)[c]	Malignant melanoma
del(6q11–21→qter)	Renal cell carcinoma[f]
del(6q15→q21)[c]	Ovarian carcinoma
*del(6q21→qter)	Lymphoma, acute lymphoblastic leukemia
trisomy 7	Adenocarcinoma of the large bowel
*del(7q22→qter) or monosomy 7	Several malignancies
trisomy 8	Several malignancies
*del(11q13→q25)[c]	Acute nonlymphocytic leukemia
trisomy 12	Lymphoma, chronic lymphocytic leukemia
i(12p)	Seminoma
del(12q13→q15)[c]	Mixed salivary gland tumor
i(16q)	Bladder carcinoma
del(16)(p21→pter) or monosomy 16	Breast carcinoma[d]
del(16q22)	Acute nonlymphoblastic leukemia
*i(17q)	Several malignancies
*del(17q12→q21)	Acute nonlymphoblastic leukemia
*del(20q11→qter)[c]	Polycythemia vera and other hematological malignancies
trisomy 21	Acute lymphoblastic leukemia
*del(22q11→qter) or monosomy 22[g]	Meningioma
Translocations	
t(1;14)(q23-qter;q32)	Lymphoma
*t(2;8)(p11-p13;q24)	Burkitt lymphoma or acute lymphoblastic leukemia
*t(3;8)(p25;q21)	Parotid gland tumor (benign)
*t(4;11)(q21;q23)	Acute ?lymphocytic leukemia
t(6;12)(q15;p13)	Prolymphocytic leukemia
t(6;14)(q21;q24)	Ovarian carcinoma
*t(8;21)(q22;q22)	Acute nonlymphoblastic leukemia
*t(8;14)(q24;q32)	Burkitt lymphoma or acute lymphoblastic leukemia
*t(8;22)(q24;q11)	
*t(9;11)(p21;q23)	Acute nonlymphoblastic leukemia

Table 14.2. *(cont.)*

Chromosome aberration[b]	Malignancy
*t(9;22)(q34;q11)	Several malignancies
t(10;14)(q24;q32)	Lymphoma
t(11;14)(q13-q23;q32)	Lymphoma, acute lymphoblastic leukemia
t(11;19)(q23;p12 or q12)	Acute nonlymphoblastic leukemia
*t(11;22)(q24;q12)	Ewing sarcoma
t(14;14)(q24;q32)	Lymphoma
*t(14;18)(q32;q21)	Lymphoma
inv(14)(q11;q32)	Chronic lymphocytic leukemia (T-cell)[h]
*t(15;17)(q22;q12-q21)	Acute nonlymphoblastic leukemia
*inv(16)(p13;q22)	Acute nonlymphoblastic leukemia
t(4;14)(q12;q24?)	Lymphoma

[a]For references see de la Chapelle and Berger (1984), Sandberg (1983c), Yunis (1983), and Rowley (1984).
[b]Aberrations marked with asterisk are among those identified by Mitelman (1984) as being regularly involved in human neoplasia (see text).
[c]Translocations involving the same region also reported.
[d]Both del(1) and del(16) in some cases (Hulten, 1984).
[e]Squire *et al.* (1984).
[f]Wang *et al.* (1983); Benedict *et al.* (1983a).
[g]Frequent (8–10%) loss of a sex chromosome in addition to monosomy 22 (Zang, 1982).
[h]Zech *et al.* (1984).

deletion mechanism, similar to that postulated for retinoblastoma and Wilms tumor and also suggested for the 3;8 and 3;11 translocations associated with renal cell carcinoma, more recent thinking has taken quite a different direction. Thus, it is now thought by many that the leading effect of translocations is to cause the activation of loci which were previously unexpressed or were expressed at low levels, the chief candidates for such loci being the oncogenes (Klein, 1981; Radman, Jeggo, and Wagner, 1982; Gilbert, 1983). [For a less positive view, see Duesberg (1983).] Other mechanisms of oncogene activation also exist, but will not be considered here (see Willecke and Schäfer, 1984).

The cellular oncogenes are presumed to be intimately involved in the control of cellular proliferation, and some have recently been shown to code for proteins directly involved or related to those involved in growth regulation: v-*erb*-B for the epidermal growth factor receptor (Downward *et al.*, 1984), v-*sis* for platelet-derived growth factor (Robbins *et al.*, 1983; Deuel *et al.*, 1983), and possibly *Blym–1* for transferrin (Diamond, Devine, and Cooper, 1984) [for a general review of the products of oncogenes, see Hunter (1984)].

Table 14.3. *Acquired chromosome aberrations associated with malignancy in experimental animals*

Species	Chromosome aberration	Malignancy[a]	References
Mouse	trisomy 6	Bladder epithelial cell lines	Cowell (1980)
	trisomy 13	Mammary tumors	Dofuku, Utakoji, and Matsuzawa (1979)[b]
	trisomy 15	T-cell lymphomas and leukemias	Dofuku *et al.* (1975)
	del(15D2→D3)	Plasmocytoma	Wiener *et al.* (1984b)
	T(6;15)(C2;D2/3) *T(12;15)(F2;D2/3)*	Plasmocytoma	Wiener *et al.* (1984a)
	T(6;10)	Plasmocytoma	Perlmutter *et al.* (1984)
Rat	trisomy 2	DMBA-induced sarcomas and carcinomas	Ahlström (1974); Levan, Ahlström, and Mitelman (1974)
	del(2q24→q26)	MSV-transformed glial cells	Kano-Tanaka and Tanaka (1982)
	trisomy 4	ENU-induced neurogenic tumors	Haag and Soukup (1984)
	dup(4q11→q22)	MSV-transformed glial cells	Kano-Tanaka and Tanaka (1982)
	trisomy 7	RSV-induced sarcomas	Levan and Mitelman (1970)
	trisomy 12	MSV-transformed glial cells	Kano-Tanaka and Tanaka (1982)
	monosomy 15	Iodine deficiency-tumors induced thyroid	Al-Saadi and Beierwaltes (1967)
	T(6;7)(q32;q33)	Plasmocytoma	Wiener *et al.* (1982a)
Chinese hamster	dup3q	AzaC-transformed and plasmid-transfected CHEF/18 cells	Harrison *et al.* (1983); Lau *et al.* (1985)
	trisomy 11	DMBA-induced sarcoma	Mitelman, Mark, and Levan (1972)
Syrian hamster	monosomy 15	Virus-induced tumors and transformed cell lines	Pathak *et al.* (1981)

[a]Abbreviations: azaC, 5-azacytidine; DMBA, 7,12-dimethylbenz(α)anthracene; ENU, ethylnitrosourea; MSV, murine sarcoma virus; RSV, Rous sarcoma virus.
[b]Quesitoned by Spira and reaffirmed by Dofuku (Spira and Dofuku, 1980).

The reason for considering the oncogenes is that many translocations have been shown to occur in the region of such oncogenes, with a common finding being that a chromosomal segment carrying an oncogene is brought into juxtaposition with a region coding for an immunoglobulin sequence (Heisterkamp *et al.*, 1983; Klein, 1983; Leder *et al.*, 1983).

Several mechanisms for oncogene activation after translocation have been proposed. These include movement of the gene away from an adjacent cis-acting suppressor of transcription or next to an enhancer or active promoter of transcription (Gilbert, 1983; Klein, 1983; Sandberg, 1983c; Hayday *et al.*, 1984). Another possibility is the deletion or alteration of a region of the oncogene which confers sensitivity to a trans-acting suppressor (Leder *et al.*, 1983), which could be the oncogene product itself (Rabbits *et al.*, 1984). While the details of these mechanisms are not of concern to us here, the postulated outcome – enhanced oncogene expression – is. The evidence that, in the absence of gene amplification, such enhanced expression actually occurs is still quite limited, but reports to this effect have appeared for c-*myc* expression in non-Hodgkins and Burkitt lymphomas with t(8;14) (Nishikura *et al.*, 1983; Croce *et al.*, 1983; Hayday *et al.*, 1984) and in mouse plasmocytomas with *T(12;15)* (Marcu *et al.*, 1983), and for c-*abl* in chronic myelogenous leukemia cells with t(9;22) (Collins *et al.*, 1984). In the last case, the increase in oncogene expression was of the order of four to eight fold, although it occurred only after acute blast transformation and not in chronic myelogenous leukemia cells. In fact, several laboratories have failed to show an elevation of c-*abl* expression in chronic myelogenous leukemia (Yunis, 1983). For the mouse plasmacytomas, the increase was 10 to 20 fold. The same appears to be true for the expression of oncogenes in tumors in which translocations are probably not involved (Slamm *et al.*, 1984). Therefore, the increases in oncogene expression, when they are detected, are considerably greater than would be likely to result from a simple triplication of an oncogene locus or from trisomy of a chromosome carrying such a locus. The relevance of this will become apparent in the next section, in which a specific example of acquired aneuploidy with trisomy is considered.

Before closing this section, it is worth noting that Klein (1983) has suggested the t(3;8) associated with familial renal carcinoma does not act by a deletion-two hit mechanism such as was suggested earlier, but by activation of the oncogene c-*myc* located on chromosome 8 near the q24 breakpoint of the translocation. While certainly a plausible explanation, there is presently no evidence to support it.

Trisomy 15 in mouse lymphoma and leukemia – a prototype

A relationship between mouse T-cell lymphomas and leukemias and trisomy 15 was first documented by Dofuku *et al.* (1975), and numerous articles confirming and extending their observations have appeared in the succeeding decade. This relationship has probably been more intensively studied than any other between a specific acquired aneuploid state and spontaneous or induced malignancy, and as such can serve as a prototype for exploring the nature and meaning of the relationship. T-cell malignancies arise spontaneously in AKR strain mice and can be induced with a variety of agents (including viruses, carcinogens, and radiation) in this and other mouse strains. The striking finding is that, however they arise, the resulting tumors are quite consistently trisomic for chromosome 15 (Chan, Ball, and Sergovich, 1979; Spira *et al.*, 1979; Chan, Ens, and Frei, 1981). A similar although less uniform finding, has also been made for murine B-cell lymphomas (Fialkow, Reddy, and Bryant, 1980; Wiener *et al.*, 1981) and for Rauscher-virus-induced erythroleukemia and myeloid leukemia (Hagemeijer *et al.*, 1982). The notion that the trisomy for chromosome 15 has a critical role in the development of T-cell malignancy was supported by the finding that when chromosome 15 was coupled to another chromosome, as in a homozygous Robertsonian translocation, the malignant cells were trisomic for both chromosome 15 and the other coupled chromosome (Spira *et al.*, 1979; Herbst, Gropp, and Tietgen, 1981; Herbst and Gropp, 1982). This was interpreted as arguing against the possibility that trisomy 15 "is due merely to secondary selection from amongst a variety of potential trisomies" and for the conclusion that "duplication of some critical 15-associated element is causally involved in thymic leukemogenesis. . . . Apparently, the increased gene dosage of the critical segment is so important for the development of thymic leukemia that duplication of attached genetic material will have to be 'tolerated,' even if it involves the largest chromosome" (Spira *et al.*, 1979). On the basis of results obtained with DMBA-induced lymphomas [DMBA = 7,12-dimethylbenz(α)anthracene] in SJL strain mice, in which only the segment 15(D3→ter) is present in a triple dose as the result of an X;15 translocation, it has been inferred that the critical region is in this distal segment of chromosome 15 (Spira *et al.*, 1980).

Subsequent work has indicated that the situation is more complicated. First, it was found that when malignancy occurs in an animal carrying chromosomes 15 of different origins (as in F_1 hybrids between two strains), there is a hierarchy of likelihood that a specific chromosome 15 will be duplicated in the formation of the trisomic state. From a series of such experiments, the following order of susceptibility (highest to lowest) can be deduced: AKR > *Rb(4.15)* > CBA = C3H > C57BL > *T(9;15)9H* (Wiener *et al.*, 1979, 1982*b*). Therefore, not all chromosomes 15 are the same with regard to their participation in the oncogenic process, possibly because of different susceptibilities to mutation or proviral insertion (Spira *et al.*, 1981). Second, when somatic cell hybrids were prepared from AKR T-cell leukemia cells and diploid CBA cells marked with the *T6* [*T(14;15)6Ca*] marker, the highly tumorigenic segregants had an increased number (5.5 rather than 3) of AKR chromosomes 15 and decreased number (0.9 rather than 2) of CBA *T6* chromosomes 15 (Spira *et al.*, 1981). Although a difference between AKR and CBA chromosomes 15 had been shown in the susceptibility experiments just mentioned, these results were interpreted to indicate that one or more chromosomes 15 had been "changed by mutation or by insertion of proviral DNA" and that "the homologous non-duplicated chromosome 15 carries the normal counterpart of the changed gene that counteracts tumorigenic behavior." Further evidence for a qualitative change in the chromosome 15 was adduced from the observation of an altered temporal replication pattern in chromosome segment 15E (Somssich *et al.*, 1982), but the significance of this is unclear since all three chromosomes 15 were altered. Therefore, rather than merely acting by a change in gene dosage, as had originally been suggested, the trisomy was thought to act by duplication of a qualitatively changed gene in the distal part of chromosome 15. Klein (1981) summarized his thoughts as follows:

For tumour-associated non-random chromosome changes such as chromosome 15 trisomy it is simplest to invoke gene dose effects. It may be objected that a change in gene dosage of 50% is a relatively mild change. On the other hand, the relationship between the normal cell and its many growth controlling signals must be extremely finely poised, so that small changes may tip the balance and favour unlimited proliferations. However, the chromosomal analysis of high and low tumorigenicity segregants of somatic cell hybrids indicates that there also exists a qualitative difference between the relevant gene(s) carried by the tumour- and the normal cell-derived chromosome 15. This is probably due to a mutation or some other change, like proviral DNA insertion. Duplication of the altered chromosome has probably overcome *trans*-acting regulation by the normal homologue, resulting in the leukemic phenotype.

By analogy with the 12;15 and 6;15 translocations characteristic of the murine plasmocytomas (Table 14.3, see preceding section), in which the chromosome 15 breakpoint is in the region of the c-*myc* oncogene [which is translocated to the other chromosome (Crews *et al.*, 1982)], it was further speculated that the relevant qualitatively altered locus on the duplicated chromosome 15 in the trisomy 15 lymphomas/leukemias might also be c-*myc* (Klein, 1983; Wiener *et al.*, 1984a). On the basis of this, it was also theorized that mice with the *T(9;15)9H* translocation are resistant to lymphoma induction (see above) because the translocation has inactivated the oncogene (Wiener *et al.*, 1982b).

Recent evidence supports the suggestion that c-*myc* is indeed involved in the trisomy 15 malignancies. Wirschubsky *et al.* (1984) have shown that an EcoRI restriction fragment carrying c-*myc* is altered in size in the spontaneous trisomy 15 lymphoma designated TKA, and that all three copies of the region were altered in the subline, TIKAUT. Furthermore, retroviral sequences were found to be inserted immediately 5' to c-*myc* in the latter, as

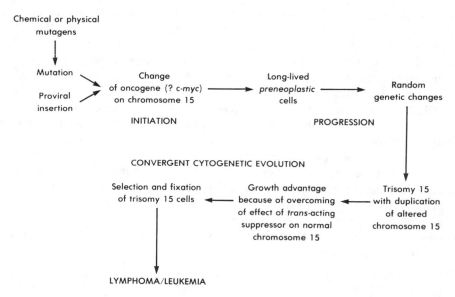

Fig. 14.3. Scheme of development of trisomy 15 lymphomas and leukemias in mice as envisioned by Klein (1979, 1981, 1983). See also Haas *et al.* (1984).

well as two other lymphomas, and c-*myc* was found to be highly transcribed in TIKAUT (Cory *et al.*, 1983). However, 18 other virus-induced lymphomas, both diploid and trisomic for chromosome 15, did not have a demonstrable change in the EcoRI c-*myc*-containing restriction fragment (Wirschubsky *et al.*, 1984), so the generality of the finding remains to be determined. Furthermore, another oncogene, c-*sis*, has also been mapped to mouse chromosome 15 (Kozak, Sears, and Hoggan, 1983), and its possible involvement in trisomy 15 lymphomas needs also to be considered.

The overall process of T-cell oncogenesis, as envisioned by Klein and his collaborators, is summarized in Fig. 14.3. However, this proposal, as ingenious as it is, is not without difficulties. In addition to its admittedly speculative and still unproven assumptions regarding the preneoplastic changes affecting chromosome 15 and its oncogenes, the true significance of the association of the malignant state with trisomy 15 has been brought into question by Boggs *et al.* (1983). They reasoned that if trisomy 15 is necessary for either tumor initiation or, as Klein (1979, 1981) suggests, tumor progression, it should be present in all cells of the tumor. However, both their own work and that of others indicate that the latter is not always the case. Thus, Herbst and Gropp (1982) found that about half of AKR mice with spontaneous leukemia did not have trisomy 15 tumor cells. Similarly, Chan, Ens, and Frei (1981) showed that diploid cells predominated in carcinogen-induced tumors in adult CFW/D mice, with the reverse being true in neonatal mice, and suggested that "the emphasis on the importance of trisomy 15 as a primary chromosome abnormality in the progression of murine lymphomas may be eclipsed by the finding that some lymphomas are characterized by normal diploid karyotypes." In addition to finding also that about half of spontaneous lymphomas in AKR mice did not have trisomy 15 cells, Boggs *et al.* (1983) showed that transplanted trisomy 15-negative tumors, while sometimes developing trisomy 15 cells, did not invariably do so and that serially transplanted lymphomas that were originally trisomic for chromosome 15 sometimes lost the extra chromosome 15. On the basis of these observations they concluded that, "despite the impellingly high frequency and nonrandom occurrence of trisomy, it may be secondary to the development of lymphoma."

It is possible to reconcile the observation of disappearance of trisomy 15 in the serially transplanted tumors with the Klein (1979, 1981) (Fig. 14.3) hypothesis by assuming that the chromosome 15 that is lost is the normal homolog, which, by its loss, further enhances the malignancy of the tumor by the removal of the putative trans-acting suppressor. A similar explanation might also be applicable to the finding of a diploid karyotype in primary tumors, the assumption being that loss of the normal chromosome 15 occurred concurrently with duplication of the activated one. Alternatively, it could be postulated that the initial oncogenic event(s) produced a sufficient degree of oncogene activation so that selection of a derivative trisomy cell line was not required for tumor progression. This would be analogous to what Klein *et al.* (1980) proposed in order to explain why mouse leukemias induced with the Abelson murine leukemia virus are and remain diploid. All of these suggestions are, of course, sheer speculation, which awaits further evidence regarding the structure and state of the chromosome 15 in a large number of the murine lymphomas and leukemias.

In an earlier chapter (Chapter 10), I made reference to our work on Ts15 ↔ 2n chimeras (Epstein, Smith, and Cox, 1984). The principal reason for preparing these animals was to see whether they could be used to determine if the prior existence of trisomy 15 would accelerate the rate of tumor appearance. If Klein's (1981) hypothesis were correct, this would not be expected, since it is the duplication of an altered chromosome 15 – not just any chromosome 15 – which is considered to be essential. Unfortunately, we could not obtain a satisfactory answer since, as is shown in Table 10.6 and Fig. 10.7, the trisomy 15 thymic lymphocyte population is disproportionately low in fetal Ts15↔2n chimeras, and is virtually or completely undetectable in adult Ts15↔2n chimeras (unpublished data). These findings, to the extent that they indicate that trisomy 15 itself does not confer a growth advantage to T lymphocytes, can be considered as at least being compatible with the Klein (1981) hypothesis.

Duplications and trisomies

Is the situation of trisomy 15 in mouse malignancies truly representative of the role of an increase in chromosome number in oncogenesis, or is it merely a special case? At present, it is impossible to give an unequivocal answer to the question. A brief glance at Table 14.2 reveals that most trisomies and duplications are nonspecific in their associations and are found in a wide variety of malignancies. However, Yunis (1983) points out that there are two types of disorders in which a nonrandom trisomy can be found as the only abnormality: trisomy 12 in about one-quarter to one-third of cases with chronic lymphocytic leukemia [see also Han *et al.* (1984)] and trisomy 8 in 5–10% of cases with acute nonlymphoblastic leukemia. He suggests that neoplasms such as these might arise through a mechanism of gene dosage, but again invokes the Klein (1981, 1983) hypothesis of duplication of a chromosome carrying an already activated oncogene. A similar proposal has been made to explain dup(3q) in Chinese hamster cells treated with 5-azacytidine (Harrison *et al.*, 1983), and a possible example of this mechanism has been found in a colon carcinoma cell line (Alitalo *et al.*, 1984). These cells often carried three copies of a marker chromosome on which the c-*myb* oncogene was amplified and overexpressed.

On the other hand, Gilbert (1983) suggests that duplications and trisomies, while not likely to be directly responsible for carcinogenesis, alter the division of both normal and malignant cells through gene dosage effects that are not cell type specific. These chromosomal changes presumably arise at random but are then selected for and tend to accumulate as the tumor spreads and becomes more refractory to therapy. The ways in which an extra dose of a specific locus or set of loci could enhance growth are not known, but a few possible mechanisms have been offered. As Chaganti (1983), following Klein (1981), points out, "the simplest hypothesis . . . would be of gene dosage, with even the slightest elevation of the oncogene product due to trisomy possibly being sufficient to cause an imbalance of some regulatory function related to the development or proliferation of target cells." At present there is no evidence to support this notion, and what evidence there is is against it. For example, induction of overexpression of the cellular oncogene, c-*src*, did not result in malignant transformation of chicken fibroblasts even though expression of the structurally somewhat different viral oncogene, v-*src*, did (Iba *et al.*, 1984). Similarly, while injection of a mutant Ha-*ras* oncogene product, a p21 protein, could efficiently transform NIH 3T3 cells, much larger amounts (at least 5 times as much) of the normal cellular Ha-*ras* product were required to produce the same effect (Stacey and Kung, 1984). In a related hypothesis, Murphree and Benedict (1984) advanced the concept of a balance between suppressor and expressor loci, and the application of this notion to dup(1q) and i(6p) in retinoblastoma has already been discussed. Such a mechanism would require a quite tightly balanced control system.

It is, of course, quite possible and probably even likely that duplications and trisomies do not always exert their effects on tumors through mechanisms affecting the expression of oncogenes or similar regulatory systems specifically concerned with growth. Also to be considered are mechanisms by which dosage effects alter other aspects of cellular metabolism or surface structure and, without affecting the regulation of growth per se, confer some selective advantage on cells which are so altered. In an attempt to define what such dosage effects might be, Rowley (1977) examined the relationship between the location of genes affecting carbohydrate and nucleic acid metabolism and certain regions commonly present in three [as in dup(1q)] or even four doses [i(17q)] in malignant cells. Unfortunately, the data at that time and even now (Human Gene Mapping 7, 1984) are still too few to permit this type of analysis to be carried out. Furthermore, virtually no data are available on whether the expected gene dosage effects are actually observed in aneuploid tumor cells. [The only report I could find is that of de la Chapelle, Vuopio, and Icén (1976) concerning glutathione reductase in trisomy 8 red cells (Table 4.1).] In fact, the metabolic changes that are observed in malignant cells seem to be quite unrelated to known loci on the specific chromosomes that are unbalanced, and the patterns often seem to more closely resemble the state of corresponding fetal tissues than of the original adult tissues. As such, they may merely be a reflection of the altered growth pattern of the tumor cell (Honey and Shows, 1983).

Despite the failure so far to make specific associations between loci on the unbalanced chromosomes and defined changes in some aspect of cell metabolism or structure, the premise that the affected chromosome is maintained in the unbalanced state, presumably by selection, because of the effects of the genes it carries seems to me to be a highly reasonable one (Mitelman and Levan, 1981; Rowley and Testa, 1982; Nowell, 1982; Wolman, 1984). So does the idea that certain changes, such as dup(1q) and trisomy 8 are quite general in their effects insofar as different types of tumors are concerned, while others are more restricted (Nowell, 1982; Gilbert, 1983). The challenge, then, is to identify the specific loci that are involved, and efforts in this direction will be aided by fine-structure gene mapping to localize the relevant genes (Rowley, 1977; Mitelman and Levan, 1981). While some of the unbalanced chromosomes may indeed carry qualitatively altered loci, as proposed by Klein (1981, 1983), it is likely, as I have already suggested, that this will not always be the case and that simple gene dosage considerations will be applicable.

Deletions and monosomies

Except for deletions and monosomies of human chromosomes 5 and 7, other acquired chromosomal deletions seem to have a higher degree of tumor specificity than do many of the duplications just discussed (Table 14.2) (Gilbert,

1983). One hypothesis to explain their role in oncogenesis follows the reasoning already outlined for the constitutional deletion-associated tumors. Therefore, as an example, while a single visible deletion may by itself be insufficient to produce a malignant tumor (Gilbert, 1983), it may, when coupled with a qualitative although invisible lesion of the homologous chromosome, result in oncogene activation because of the removal of a trans-acting suppressor. Another variant of this mechanism would be that a deletion permits recessive mutations, possibly those involving oncogene suppressors, to be expressed. This type of proposal was made by Ohno (1971) to explain the role of the Ph¹ chromosome at the time when it was still thought to be a deleted chromosome 22. However, it is not really known whether a second lesion on the homologous chromosome is actually required to advance the process of malignancy. For example, in preleukemia associated with del(5q) – the "5q– syndrome," there is a gradual expansion of the preleukemic clone to the level of clinically recognizable leukemia (Nowell, 1982). Furthermore, it could be visualized that, in some delicately balanced suppressor-expressor system, loss of even one suppressor locus could be sufficient to allow to oncogene activation and expression (Evans, 1983).

In the case of interstitial deletions, a special mechanism for activating an oncogene has been described. Wiener *et al.* (1984*b*) observed three murine plasmocytomas in which part of the D band of chromosome 15 was deleted, resulting, it is suggested, in the joining of the 5' end of the c-*myc* oncogene to an actively transcribed chromatin region in 15D1 and the consequent enhanced expression of c-*myc*. While certainly not a general mechanism, a process such as this could be operative for any of a number of reported interstitial deletions for which corresponding terminal deletions with the same proximal breakpoints are not also found (Table 14.2).

Sandberg (1983*c*) has made the opposite proposal, that the critical initiating factor is removal by deletion (or translocation) of a cis-acting suppressor. By this hypothesis, both interstitial and terminal deletions would have the same effect, that of permitting the expression of a previously suppressed locus. This locus could be a proto-oncogene which is normally active during embryogenesis, but is then turned off.

Once again, as with duplications and trisomies, it may not always be necessary to involve increased oncogene expression to explain the role of deletions and monosomies, and more straightforward gene dosage effect considerations might also be applicable. Unfortunately, this possibility is still hypothetical, and good examples do not exist. However, two instances of what appear to be relatively specific consequences of chromosomal deletion or monosomy are worthy of mention. One is the reported loss of the ability of herpes simplex antigen-stimulated leukocytes from most patients with preleukemia or acute lymphoblastic leukemia and del(5q) or monosomy 5 to produce interferon (type unspecified) (Pedersen-Bjergaard *et al.*, 1980). Al-

though an attempt was made to relate this finding to the presence of an interferon production gene on chromosome 5, this mapping is questionable. The
interferon-α, -β, and -γ structural genes are on chromosomes 9, 9, and 12,
respectively (Human Gene Mapping 7, 1984), and the observed result would
not be compatible with the expected gene dosage effect for deletion of a structural locus. The other example involves defective chemotaxis of granulocytes
from several patients with hematological malignancy or premalignancy associated with monosomy 7, and a demonstrated reduction of the surface glycoprotein, GP130, in three (Ruutu *et al.*, 1977; Gahmberg *et al.*, 1979). The
relationship between the chemotactic lesion and the surface protein changes
is unclear, as are their connections with monosomy 7, and the presented dosage effect data do not justify the conclusion (Gahmberg *et al.*, 1979) that the
gene(s) for these two functions can be mapped to chromosome 7.

A particularly striking example of a deletion or monosomy associated with
a specific tumor is that of monosomy 22 and meningioma. In a series of 140
meningiomas, a total of 103 (74%) were monosomic for chromosome 22,
sometimes in mosaic form and sometimes in association with aneuploidy of
other chromosomes (Zang, 1982). Translocations involving chromosome 22
have not been recognized. As described by Zang (1982), "The chromosome
aberrations in meningiomas show a uniformity observed in only very few
human tumors. The initial loss of one small acrocentric chromosome, occurring in over 95% of meningiomas showing chromosomal aberrations in all
cells or in mosaic form with chromosomally normal cells, suggests a uniform
pathogenetic mechanism, perhaps even a uniform etiology. However, it does
not prove both [? either]." Zang (1982) also pointed out that in many meningiomas, only a small minority of the cells display monosomy 22.

To explain these observations, as well as the findings that simian virus 40
(SV40) antigens were often detectable in meningiomas and that all patients
with meningiomas had antibodies against SV40 antigens, Zang (1982) suggested that SV40 either induces loss of chromosome 22 after infection or
more readily infects and transforms cells from which chromosome 22 has
already been lost. In either event, virus transformation induces proliferation
of the monosomy 22 cells and, by a humoral effect (release of a growth factor), of adjacent diploid cells as well. These proposals are, of course, purely
hypothetical, but the notion of chromosomally normal cells being stimulated
to divide by the abnormal cells is an intriguing one and potentially applicable
to other mosaic situations as well.

Primary versus secondary chromosomal changes

Throughout this chapter I have referred to acquired aneuploidy as representing a secondary change or changes following some other initiating event
which begins the transition of a normal diploid cell to a malignant one. This

concept, which follows the clonal evolution hypothesis of Nowell (1976), and the convergent cytogenetic hypothesis of Klein (1979), is expressed by numerous workers in the field of cancer research. The essence of this approach has already been discussed with regard to trisomy 15 and the mouse lymphoid malignancies, and has been illustrated for this specific case in Fig. 14.3. Central to the concept is the notion that the premalignant cell, once the initiating event has occurred, is capable of sustained growth and becomes genetically unstable and that this genetic instability eventually leads to oncogene activation or some other event which confers a selective advantage on the now malignant cell.

It is as yet unclear at what point the relatively specific translocations occur. By the formulation just presented, they probably occur quite early in the oncogenic process, but whether they actually represent the initiating event is unclear. The only commentators who appear to have suggested that this class of abnormalities, as well as the other cytogenetic abnormalities that are highly nonrandom in occurrence, may be primary initiating events are Mitelman and Levan (1981). On the basis of experimental work in animals (Mitelman, 1981), they suggest that certain aberrations may represent the result of a direct interaction between the inducing agent and the genetic material of the target cell. The secondary changes which then occur by chance disturbance of mitosis are thought to amplify the effect of the primary changes. On the basis of extensive analysis of the tumor cytogenetics literature, Mitelman (1984) believes, as was mentioned earlier in this chapter, that only a limited number of chromosome changes (mainly translocations but also deletions) are of "critical importance" and are sufficient in themselves to produce the malignant state. One observation consistent with this notion is the finding that deletions or monosomies of chromosomes 5 and 7 are quite characteristic of individuals with acute nonlymphoblastic leukemia who are occupationally exposed to mutagens, and are detected in almost 90% of aneuploid patients in whom this form of leukemia occurs secondary to cytotoxic therapy (Rowley, 1983b). However, a mechanism to explain how this could occur has not been proposed.

For the moment, then, it seems reasonable to conclude that certain cytogenetic abnormalities, particularly translocations, may occur quite early in the formation of a malignant cell, while others, mainly involving aneuploidy are later events involved in tumor progression. In each instance, the abnormalities that are observed are the ones that become fixed because they confer a selective (growth) advantage on the tumor cells. While many of the early abnormalities may be involved either in the activation of the expression of oncogenes or of other loci with a significant effect on cellular growth, or in an escape from normal control mechanisms, the later abnormalities presumably serve to further enhance the selective advantage of the tumor cells. While these latter events could also involve the activation of oncogenes and related events, perhaps of oncogenes different from those involved early in the process [see Land, Parada, and Weinberg (1983) for a discussion of the two-oncogene hypothesis], they could also involve other types of mechanisms related to simple gene-dosage effects involving cellular processes affected by genes on the affected chromosomes. At present there is no evidence to support the idea that simple dosage effects are involved in the early pathogenesis of malignancies. However, there is also no evidence that proves that, in some special instances, they may not be.

Part VI

Conclusion

15

Principles, Mechanisms, and Models

It gradually dawned on us that what we eventually came to identify as a "differentiated" cell of a given type, shape, structure, colour, and behavior, was really just the last scene of a long play of interactions, the earlier stages of which had been hidden from us. We came to realize that, in dynamic terms, the so-called "stages" of the development of a cell, an embryo, or a disease, represented simply cross-sections through a continuous stream of processes, recorded by whatever methods of assay and portrayal – whether in pictures or in graphs – we had at hand. The picture of an overt "character" thus emerged as merely the residual index of prior covert *dynamics*; "form," as the registered outcome of formative processes. (Weiss, 1973)

So also with aneuploidy and its effects. What we observe as the phenotypes associated with unbalanced chromosome abnormalities represent but the end results of the lifelong impacts of specific types of genetic imbalance on the normal processes of development and function. Therefore, complete understanding of how aneuploidy produces its effects will ultimately require a full knowledge of how normal development and everyday functioning are genetically controlled – a goal whose attainment is clearly still far in the future. Nevertheless, we are beginning to discern some of the principles or rules which govern the development of aneuploid phenotypes and to appreciate some of the mechanisms whereby genetic imbalance may affect development and function. It has been to the elucidation of these principles and mechanisms and to the consideration of appropriate models – both conceptual and real – that this volume has been devoted. In closing, therefore, I shall pull together these principles, mechanisms, and models for one last look at what has been accomplished and where we appear to be heading.

Principles

What do I mean by the principles or rules governing the development of aneuploid phenotypes? I am not referring to mechanistic rules which allow us to translate imbalance of a specific gene or chromosome into a set of inevitable phenotypic consequences. While I believe that such rules must ultimately exist, in general terms if not in precise detail, we are still very far from knowing them. The principles of which I speak are much more primitive ones, which merely anticipate in broad strokes what the mechanistic rules may ultimately permit us to do. However, they do provide the basis for believing that the latter do exist and that it is of value to search for specific mechanisms to explain how aneuploidy produces its effects.

In Part II of this volume, I reviewed a large amount of clinical data on a variety of aneuploid states to determine what general principles might be inferred. Although this review was greatly complicated by the fact that aneuploid phenotypes are often couched in vague and subjective terms, it was encouraging to find that attempts are being made to place phenotype description on a more objective and quantifiable basis. However, even this improvement in description does not alter the fact that the phenotypic features at which we are looking are unlikely, in general, to represent the end result of a single developmental aberration resulting from imbalance of a single locus. Rather, they are more likely to be, as Weiss (1973) wrote, "the last scene of a long play of interactions" affecting complex morphogenetic processes. However, it is precisely because of this developmental complexity that it is so gratifying to find that general principles of aneuploid phenotypes do exist. Stated briefly, these principles are the following:

1. Aneuploid phenotypes, while frequently overlapping, are differentiable from one another.
2. Although there may be a significant degree of variability in the expression of the phenotype of a given type of chromosome imbalance, the phenotype, in terms of overall pattern, is nonetheless a specific one.
3. Individual phenotypic features can often be assigned or mapped to specific regions of the genome.
4. The features of segmental aneuploidies can often be added together to generate the phenotypes of combined aneuploidies.

The evidence for and the exceptions to these principles have already been discussed in detail, and they are therefore being presented here in a relatively unqualified form to emphasize the fact that specific relationships between particular aneuploid states and the phenotypes associated with them do exist. While these relationships may be clouded to some extent by a variety of forms of phenotypic noise, their specificity is still not obscured. However, a further comment about the third and fourth principles, those pertaining to additivity and the ability of features to be mapped, is required. Statement of these principles is not intended to negate the fact that the combination of segmental aneuploidies may, through an interaction of the effects produced by each, result either in the expression of new phenotypic features not characterisitic of either alone or in the masking of certain features characteristic of one or both. The existence of such interactive effects, especially those involved in the development of new phenotypic features, indicates that particular abnormalities need not – indeed, should not – always be attributed to imbalance of only a single locus. Nevertheless, it is not unlikely that it will often turn out to be the case that only a single locus is involved.

Much comment has been devoted in the literature to what I have termed "phenotypic noise," the failure of aneuploid phenotypes to be wholly reproducible at all times. In general, the issue of noise or of phenotypic variability has been raised by others to introduce some element of doubt about the specificity of the relationship between genomic imbalance and the resulting phenotype. I think that this emphasis has been misplaced. Clearly, variability exists, as does the overlapping of phenotypes. The latter is readily explicable in terms of the interrelatedness of developmental processes. Very few, if any, developmental events can be visualized as being controlled by only a single genetic locus. Therefore, it is not at all surprising that two quite unrelated forms of genetic imbalance may produce the same phenotypic feature or group of features. [In fact, Opitz (1985) used this type of reasoning, extended to nongenetic as well as genetic etiologies, in defining the concept of developmental fields.] The probability that similar features result from different causes is further increased by the likelihood that many conceivable and perhaps unique phenotypic effects would be incompatible with viability and are therefore not represented among the sets of features that are discernible in aneuploid states. Those phenotypic features that are recognized represent, to a very large extent, relatively minor developmental aberrations, and it is for these, rather than for the more major defects, that most of the phenotypic overlaps appear to occur.

The variability of phenotypes, while quite a different matter in terms of possible causes, is also not unexpected. Three types of causes for variability were suggested earlier (genetic, stochastic, and environmental), and it is likely that all three of these do indeed contribute to phenotypic variability. It is worth recalling once again the lack of complete concordance for a variety of congenital malformations in identical twins. Thus, even holding all genetic factors constant, to a greater extent than would be possible for any other two individuals with the same genetic imbalance or abnormality, is no guarantee that all aspects of the phenotype will be identical. Therefore, in thinking about the significance of the fact that variability of phenotypic expression does exist, one must avoid falling into semantic and conceptual traps. That variability *does* exist in no way means that specificity does *not*. It means only that a variety of factors can, by mechanisms readily understandable in general terms, interfere with or alter the expression of certain aspects of an aneuploid phenotype without altering the highly specific relationship between the genomic imbalance and the overall phenotypic pattern. There is no need, therefore, to invoke any one of a variety of nonspecific mechanisms to explain variability of phenotypic expression and, incidentally, the overlap of phenotypic features. While such mechanisms could conceivably operate to affect some aspects of aneuploid phenotypes, any major role for them would have the force of obscuring those relationships, expressed in the first three principles, which clearly do exist. Therefore, nonspecific effects of aneuploidy, such as altered cell cycle, reduced developmental buffering, or generalized regulatory disturbance, if they have any role at all, cannot have a major one in specifying the aneuploid phenotype.

What implications do these principles have for the consideration of mechanisms regarding the pathogenesis of the phenotypes of aneuploid states? Overall, I consider the implications to be the following: It is legitimate to look for and to expect to find specific mechanisms to explain the relationships between particular phenotypic characteristics and the imbalance of individual loci or sets of loci. Contributions to the phenotype are likely to come from all parts of the unbalanced genome and not solely from just one or a very few loci. While some loci may have a greater phenotypic effect or representation than others, it is the cumulative effect of imbalance of many loci that determines the overall phenotype. As complicated as the relationships between genetic loci and phenotypic effects may be, and with whatever element of randomness they may be associated, these relationships should nonetheless exist and be discoverable. The generation of aneuploid phenotypes is not, therefore, a game of chance but is an exercise of the conventional rules of developmental genetics and developmental biology.

There are three additional and somewhat more specialized phenotypic principles or rules that are concerned not with issues of overlap and specificity but with quantitative effects on the phenotype. Again stated in brief terms, these principles are the following:

5. While true pairs of syndromes and antisyndromes (types and countertypes) do not appear to exist when homologous duplications and deletions (or trisomies and monosomies) are compared, a limited number of individual phenotypic features may represent real countercharacters.
6. The less severe a trisomic state is in terms of its overall phenotypic effects, the more severe the phenotype associated with the corresponding tetrasomy is likely to be.
7. Aneuploidy-induced lethality is a function of the amount of active genome that is unbalanced, although certain regions, probably few in number, may play a disproportionately large role.

The implication of the first of these principles (no. 5) is that, for most phenotypic features, it is unlikely that there is a readily discernible direct quantitative relationship between the genetic imbalance and the generation of the feature. This does not mean that the relationship between gene and developmental effect is any less specific, but only that it may not follow simple quantitative rules, possibly because the effect we are considering is several developmental steps removed from the action of the primary gene product. On the other hand, there are probably some situations in which such direct relationships do exist, implying that a single locus or set of loci do play the major role in affecting the development of the character in question. While the second of these principles (no. 6) has connotations similar to the first, it also suggests that there may exist quantitative threshold phenomena that may be involved in the pathogenesis of particular phenotypic effects. These effects may thus require a certain degree (in quantitative terms) of genetic imbalance before they become manifest, and a triplex state may be insufficient to bring about their development. Finally, the last principle (no. 7), which appears to hold true for mouse and *Drosophila* as well as man, indicates that in general the lethality of an aneuploidy state, like the overall phenotype, represents the cumulative effect of the imbalance of many loci and not merely the result of the imbalance of just one or a very few.

Mechanisms

The term "mechanisms" is used in the plural, and this is important. My overall sense of the clinical, theoretical, and experimental considerations applicable to an understanding of the effects of aneuploidy is that no single mechanism can explain how aneuploidy produces its deleterious consequences. There is no simple solution to the problem of aneuploidy – only a series of multifaceted individual solutions, each applicable to a particular aneuploid state.

The one given in the analysis of mechanisms – call it another principle or rule if you will – is that strictly proportional, quantitative gene dosage effects do exist. While it will never be possible to state that this rule is inviolable for all classes of gene products, it certainly seems clear that it holds quite well for at least one particular class, that of enzymes. While similar answers are likely to be forthcoming when other types of primary gene products, such as membrane and structural proteins and regulatory molecules, are examined (thereby making their examination perhaps of lesser interest to investigators), it is still necessary to validate this principle for them as well. Nevertheless, despite the absence of such general validation, I have taken the existence of proportional gene dosage effects as the starting premise for the detailed exploration of a variety of potential mechanisms in Part III of this volume. What has this consideration taught us?

For the most part, clinical or experimental precedents for potential mechanisms, if they exist at all, are derived from the study of inherited metabolic diseases and not of aneuploid states per se. It is necessary, therefore, to be quite careful to discriminate between the effects of changes in the concentration of gene products as opposed to the effects of either qualitative alterations or the total absence of the products. Keeping this in mind, it has been possible to make arguments that changes in the concentrations of enzymes, structural

Table 15.1. *Estimate of relative impact on development of phenotype of various mechanisms mediating the secondary effects of aneuploidy*

System affected	Impact on phenotype in	
	Trisomy	Monosomy
Cellular interactions (recognition, adhesion, communication)	+ +	+ +
Receptors	+ +	+ +
Growth factors and morphogens	+ +	+ +
Regulatory systems (specific)	+	+
Assembly of macromolecules	±	+ +
Metabolic pathways (enzymes)	±	+
Transport of nutrients and metabolites	±	+

proteins, transport system components, regulatory molecules, receptors, and cell surface constituents can each produce an effect on the development or functioning of an organism. However, not all types of mechanisms are likely to be of equal importance in all situations, and my best guess of their relative importance in generating the phenotypes associated with trisomies and monosomies is shown in Table 15.1.

While the most intensively studied because of their accessibility, changes in enzyme concentration probably have little impact on the development of phenotypic features, particularly in trisomic situations. Theoretical considerations suggest that increases in enzyme activity are, in general, likely to have only a minimal, if any effect. The effect of a 50% decrease in enzyme activity, as in a deletion or monosomy, is potentially greater, but the relatively innocuous nature of the numerous heterozygous states for human enzyme deficiencies indicates that only in rare exceptions is deficiency of an enzyme likely to have a major phenotypic impact. However, the cumulative effect of reductions in the activities of several enzymes could have subtle but nonetheless significant deleterious consequences for complex functions such as those involved in learning, cognition, and memory. Similar considerations also apply to changes in transport systems concerned with nutrients and metabolites.

While increases in the concentrations of structural protein constituents involved in the assembly of macromolecules are unlikely to be of great significance, the potential impact is more likely to be marked when reductions in concentration associated with deletions are present. In fact, it can easily be visualized that defects in the structural constituents of the cell could well play an important role in contributing to the very early time of death of monosomic embryos. This is a proposition which could be readily tested in monosomic mouse embryos.

In general, it is easier to come to grips with monosomies than trisomies, since lack of a product seems, at least in theory, to lead to more severe effects than does an excess in nearly any type of system under consideration. It is, therefore, not difficult to understand why monosomies as a group should do less well than the corresponding trisomies. Nevertheless, it is still necessary to explain the abnormalities associated with the trisomies, and certain systems – in particular, receptors, cell surface recognition and adhesion molecules, growth factors and morphogens, and specific regulatory molecules – are quite attractive as candidates for playing a major role in the pathogenesis of trisomic phenotypes. These systems seem especially constituted to permit the types of discontinuous or threshold effects whose existence was inferred above from principle no. 6, and very interesting precedents for some of them already exist in the literature on aneuploidy – in particular, the amplified cellular response to interferon and the enhanced adhesiveness of fetal cardiac and lung cells in trisomy 21. The principal precedents for a significant role for regulatory molecules are derived from the studies of somatic cell hybrids, which point to the existence of trans-acting regulatory molecules that can modulate the expression of differentiated functions. Because these effects, which are quite dosage dependent, are highly specific in nature, it would be expected that regulatory disturbances in aneuploid cells would be highly specific as well. It is not surprising, therefore, that the considerable experimental evidence now available, while not actually demonstrating the existence of such specific regulatory effects, does argue against the existence of generalized regulatory abnormalities in aneuploid cells.

Unfortunately, a consideration of the role of both constitutional and acquired aneuploidy in the etiology of malignancy does not contribute very much to our understanding of the other effects of chromosome imbalance, and, for the present, the reverse is also true. Most of the evidence that now exists regarding chromosome imbalance and neoplasia implicates mechanisms that do not involve simple gene-dosage effects. Rather, those conditions associated with deletions seem to make use of mechanisms closely akin to those which might be operative in autosomal recessive conditions in which both alleles are abnormal or absent. Similarly, the one well-studied trisomic state associated with malignancy, trisomy 15 in the mouse lymphomas and leukemias, appears to have an obligatory associated mutation in an oncogene or in a gene for a regulator of an oncogene. While it has been suggested that oncogenes and their regulators are involved in virtually all aspects of aneuploidy associated with malignancy, this still remains to be proven.

However, the findings relating to mechanisms by which constitutional deletions give rise to neoplasia do raise an interesting prospect with regard to a possible mechanism for variability in monosomy or deletion states (but not for trisomies or duplications). In the same way that mutation or deletion of a locus on the homologous chromosome may lead to malignancy in a clone of cells, similar changes in other constitutional monosomies or deletions could give rise to the random or sporadic occurrence of localized benign developmental lesions. A possible example could be the nevi associated with monosomy X.

In making judgements about the likely impact of the various mechanisms on the generation of aneuploid phenotypes, I have not distinguished between effects on morphogenesis and effects on function. Such a distinction would, to some extent, be an arbitrary one, since any perturbation of cellular growth and function during the period of morphogenesis could certainly affect tissue differentiation and morphogenesis. Nevertheless, certain types of mechanisms could be of particular importance during morphogenesis – in particular, those involving cellular interactions, growth factors and morphogens, and receptors (insofar as they were concerned with growth factors and morphogens). These mechanisms, as well as the others listed in Table 15.1 and discussed earlier in this volume, could all be involved in the postmorphogenetic phenotypic abnormalities associated with aneuploidy.

Although the mechanisms have been listed and discussed as though they were independent of one another, it must be emphasized that this is not necessarily the case. Thus, cellular recognition molecules may actually be multiple-subunit macromolecules, as might also receptors of various types and channels involved in ion transport. Conversely, abnormalities of many types could affect the function of a single tissue or part of the cell. For example, abnormalities in enzymes, receptors, cellular interactions, transport systems, and macromolecular assemblies could all affect the structure and/or function of cellular membranes, and an aberration in one component of the membrane could be reflected in the altered structure or function of another. Therefore, the identification of different mechanisms for the development of the phenotypic features of aneuploidy should be regarded as involving relative and not absolute distinctions, distinctions made for the sake of conceptualization and research.

Despite the plausibility of the several mechanisms that have been proposed, it must be admitted that it has not as yet been possible to forge a mechanistic link between any phenotypic feature of an aneuploid state and imbalance of a specific locus. The closest that we have come is in our understanding of the pathogenesis of the ovarian dysgenesis in monosomy X and possibly of the neoplastic conditions, such as retinoblastoma and Wilms tumor, associated with the constitutional chromosomal deletions. However, as has just been noted, the latter involves special considerations which may not be generally applicable to other aneuploid states.

While this failure to have made greater inroads into discovering the mechanisms underlying the development of aneuploid phenotypes might be regarded as discouraging, I really do not feel that it should be. The reasons for my optimism are several and relate mainly to the fact that the necessary information and techniques required for attacking the problem are really just beginning to become available. Critical to this endeavor has been the mapping of the human and mouse genomes, so that, for the first time, it is possible to think in terms of the specific genes that are unbalanced. The results of genomic mapping to date have been quite impressive, and considerable new power has been gained from the use of recombinant DNA techniques. Nevertheless, the process of mapping must still be regarded as being in an early phase. Only a small fraction of loci have been mapped, and even these, being largely involved with enzymes, are still but a skewed representation of the totality of loci operative during the life of the organism. From the point of view of understanding aneuploidy, what is still most conspicuously lacking is the mapping of loci involved in morphogenesis and in the organization and function of the central nervous system.

The ability to map loci involved in morphogenesis and brain function requires, in addition to the development of mapping techniques for dealing with such specialized functions, knowledge of what these functions actually are. Once again there is reason for optimism. Rapid strides are being made in both general developmental biology and in neurobiology. Therefore, although we do not as yet have the requisite knowledge to permit us to delineate the loci of interest, there is every promise that such knowledge will eventually be obtained. In fact, it is quite likely that the analysis of aneuploid material may assist in this endeavor and in the mapping of loci of importance in development and central nervous system function. While mapping by gene dosage effects is certainly not without its perils, it nevertheless still offers, in carefully controlled situations, an approach to determining which of the genes that are expressed at restricted times or in restricted tissues are coded for by which chromosomes. Being able to do such mapping will require, of course, knowledge of what the primary gene products are, at the level of either their RNA or protein forms.

Yet another reason for my optimism is that investigators interested in aneuploidy are now beginning to think in mechanistic terms. In the past, the problem has been regarded as being so massive and complex as to defy solution and therefore not worth the effort. I do not think that anyone yet believes that the solution, or better, the solutions, will be simple. Nevertheless, research on the effects of aneuploidy is increasingly turning from being primarily phenomenological to being concerned with considerations of mechanism. These considerations operate in two directions: from known genetic imbalances to their potential phenotypic consequences, and from observed phenotypic aberrations to their likely genetic causes. Examples of these approaches have been discussed in Part V. Ultimately, of course, these two approaches will converge, and when they do we shall truly be able to say that we understand how genetic imbalance produces aneuploid phenotypes. But, to reach this point it will constantly be necessary to raise the question of mechanisms each and every time the phenotypic consequences of any state of chromosome imbalance are considered.

Models

The fact that many of the consequences of aneuploidy in humans arise during the period of morphogenesis places a special stumbling block in the way of their investigation. Research on events occurring during gestation, especially early gestation, is both technically impractical and, at the present time and for the forseeable future, ethically and legally impossible. It is for this reason that our interests, and those of others, have turned to the development of models.

In a sense, there are numerous models for studying the effects of aneuploidy, and many of the precedents for the several mechanisms discussed in Part III can validly be regarded as models for aneuploidy. Thus, heterozygosity for an enzyme deficiency is a model for deletion of one enzyme locus; an increased concentration of an adhesive molecule in an in vitro aggregation experiment is a model for a duplication of the relevant gene; and a 2s × 1s somatic cell hybrid can be regarded, depending on the circumstances, as a model for either tetrasomy or monosomy. Models of this type are very helpful, and their usefulness will increase further as we begin to identify more loci that will be of particular interest. However, none of these types of models permits us to deal with more complex developmental or functional issues and to appreciate the consequences of aneuploidy at the level of the whole organism. This is what we hope that specific organismic models of aneuploidy will do for us.

Since ultimate concern is with man and with human disease, we require models that will duplicate the human condition – in developmental and functional terms – as closely as possible. No model can be an exact one, since no other organism duplicates the human with respect to all of his or her biological and genetic attributes. Nevertheless, models based on other mammals, which share numerous biological similarities, seem most appropriate. While much useful information will undoubtedly be obtained from systems such as *Drosophila*, plants, worms, and even yeast, it is only with mammalian models that a close biological approximation will be accomplished.

Part IV and a section of Part V of this volume have been heavily concerned with the development and utilization of mammalian models for aneuploidy, and both the particular virtues and the potential limitations of the mouse as a model system have already been discussed. Its greatest attractiveness, beyond ease of manipulation and genetic control, is that the processes of morphogenesis and probably of central nervous system function (in neurobiological if not psychological terms) are probably quite similar to those of man. And, despite considerable rearrangement of the mammalian genome, sizable regions still appear to be intact and structurally similar in both man and mouse. On the other hand, it may not always be easy or even possible to obtain post-natally viable animals with the desired region of imbalance for functional studies. And, even if one could, these animals will never be able to duplicate the higher central nervous system functions, such as cognition, which appear to be so vulnerable to the effects of aneuploidy in man, probably because of the great numbers of loci which are involved. It is difficult enough to define mental retardation in an aneuploid human. It is even more difficult to specify the proper functional homology in an aneuploid mouse.

Despite the latter difficulties, the aneuploid mouse as it now exists still offers a very attractive system for the study of the developmental consequences of aneuploidy, and potentially better model systems can already be envisioned. For example, as loci are mapped to a human region of interest and are then cloned into appropriate vectors by recombinant DNA techniques, it will become possible to develop mice carrying increased doses (in the genome) of one or a few, or even several of these genes in either their human or mouse forms. This will permit the genomic imbalance and hopefully the resulting phenotypic effects to be built up step by step. This could be viewed as a synthetic approach to unraveling the problem of the effects of aneuploidy – an approach which would be the fulfillment of Lederberg's (1966) earlier prediction. By contrast, it is now even possible to visualize the opposite situation, the use of similar recombinant DNA and mouse manipulation methods to develop animals in which the concentrations of specific gene products are titrated downward rather than up, thereby creating synthetic models for deletions. While the technology is still in its infancy, the likelihood of success appears to be quite high (Izant and Weintraub, 1985). And, if this approach does prove successful, perhaps it is not too much to hope that similar methods may ultimately make it possible to counteract the effects – by turning them off – of extra copies of genes in trisomic individuals.

Given our limited but still not insignificant successes to date and promise of much greater successes in the future, why, it might be asked, should we devote attention to studying the effects of aneuploidy? Three answers to this question were suggested in the first chapter, and may, in closing, be reiterated here. First, aneuploidy is an interesting problem in its own right – a biological phenomenon found in numerous and diverse species that is as deserving of understanding as any other. Second, attempting to understand the mechanisms of the effects of aneuploidy cannot help but increase our understanding of the normal processes of development and function and of the structure and regulation of the genome. One test for the validity of the explanations proposed for normal events will be their ability to predict and explain the consequences of genetic imbalance. And, finally, understanding the mechanisms of the deleterious effects of aneuploidy may, in particular situations, permit us to develop strategies to counteract these effects. Such strategies may not be applicable to morphogenetic aberrations that are beyond the point of intervention by the time they are recognized. However, they may ultimately offer hope for counteracting some of the functional deficits brought about by or associated with aneuploidy in man, including even the tumor progression associated with acquired aneuploidy. And, of these functional deficits, one – mental retardation – is probably of the greatest importance, touching as it does so close to what distinguishes humans from the other living creatures: the ability to think. Being able to remove this burden from so many affected individuals would certainly make all of the effort worthwhile.

Appendix

Standard karyotypes of man and mouse and human cytogenetic nomenclature

Descriptions of human chromosomal abnormalities are based on the International System for Human Cytogenetic Nomenclature (1978) and make use of the standard human chromosome banding pattern shown in Appendix Fig. 1. A similar convention is used for mouse chromosomal abnormalities, and the mouse chromosome banding pattern is shown in Appendix Fig. 2. These banding patterns are obtained according to staining methods described in the references to the figures.

The following examples illustrate the terminology used for the description of human karyotypes:

46,XX	diploid female complement with 46 chromosomes and an XX sex chromosome constitution
45,X	monosomy X
45,XY, − 21	male with monosomy 21
47,XX, + 13	female with trisomy 13
48,XX, + 13, + 21	female with trisomy 13 and trisomy 21 (double aneuploidy)
47,XXY	abnormal sex chromosome complement with two X and one Y chromosomes
47,XXY/47,XX, + 21	mosaicism or chimerism with two cell lines; one, 47,XXY, and the other, with trisomy 21 and two X chromosomes
69,XXX	triploidy with XXX sex chromosome constitution
92,XXYY	tetraploidy with XXYY sex chromosome constitution
dup(13q)	duplication (partial trisomy) of all or part of long arm (q) of chromosome 13
del(4p) or 4p −	deletion (partial monosomy) of all or part of short arm (p) of chromosome 4
del(8q22→q23) or del(8)(q22q23)	interstitial deletion of long arm of chromosome 8 between bands 22 and 24
dup(9pter→q12)	duplication of chromosome 9 in region between terminus of short arm (pter) and band 12 of long arm
r(13)	ring chromosome 13

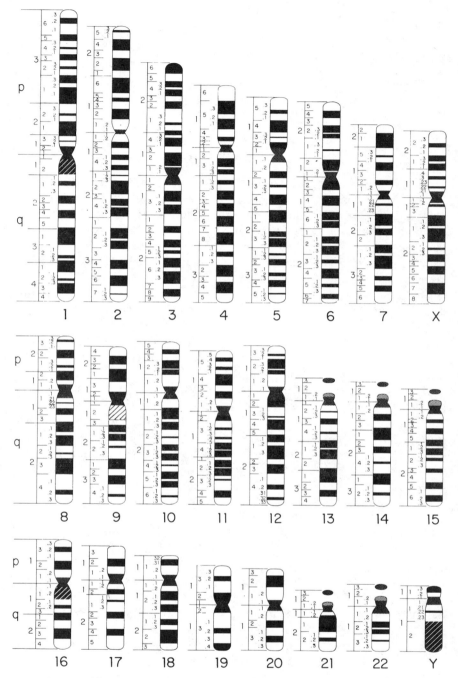

Appendix Fig. 1. Human karyotype at 550 band resolution. Reprinted from Yunis (1981) by permission.

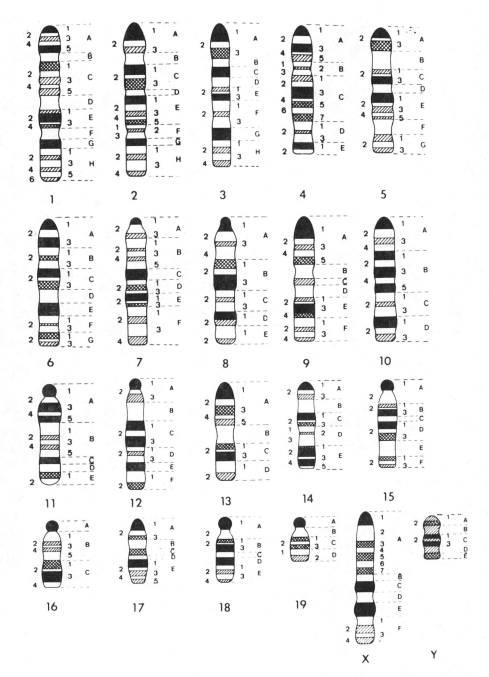

Appendix Fig. 2. Mouse idiogram. Reprinted from Green (1975) by permission.

inv(15)	inversion of segment of chromosome 15
idic(15)	isodicentric chromosome (dicentric isochromosome) 15
inv dup(15)	duplication of part of chromosome 15 with one of the segments in an inverted orientation
ins(1)(p32;q25q31)	insertion of segment q25→q31 of chromosome 1 into band p32 of chromosome 1
i(18p)	isochromosome of short arm of chromosome 18
t(4;22)(q35.5;q11.2)	translocation between chromosome 4 and 22, with breakpoints at band q35.5 of chromosome 4 and q11.2 of chromosome 22.
rcp(14;15)(q11.2;q13)	reciprocal balanced translocation between chromosomes 14 and 15, with breakpoints at band q11.2 of chromosome 14 and band q13 of chromosome 15

In descriptions of unbanded chromosomes, group designations are sometimes used: A, chromosomes 1–3; B, chromosomes 4, 5; C, chromosomes 6–12; D, chromosomes 13–15; E, chromosomes 16–18; F, chromosomes 19, 20; G, chromosomes 21, 22.

References

Aaronson, S.A. (1970). Susceptibility of human cell strains to transformation by simian virus 40 and simian virus 40 deoxyribonucleic acid. *J. Virol.* 6, 470–5.

Abe, T., Morita, M., Kawai, K., Misawa, S., Takino, T., Hashimoto, H., and Nakagome, Y. (1979). Partial tetrasomy 9(9pter → 9q2101) due to an extra iso-dicentric chromosome. *Ann. Génét.* 20, 111–14.

Abeliovich, D., Yagupsky, P., and Bashan, N. (1982). 3:1 meiotic disjunction in a mother with a balanced translocation, 46,XX,t(5,14) (p15;q13) resulting in tertiary trisomy and tertiary monosomy offspring. *Am. J. Med. Genet.* 12, 83–9.

Abraham, I., and Doane, W.W. (1978). Genetic regulation of tissue-specific expression of *amylase* structural genes in *Drosophila melanogaster*. *Proc. Natl. Acad. Sci. USA* 75, 4446–50.

Abrams, W.B., and Solomon, H.M. (1969). The human platelet as a pharmacologic model for the adrenergic neuron. *Clin. Pharmacol. Ther.* 10, 702–9.

Adamson, E.D., and Grover, A. (1983). The production and maintenance of a functioning epithelial layer from embryonal carcinoma cells. In *Teratocarcinoma Stem Cells*, ed. L.M. Silver, G.M. Martin, and S. Strickland, pp. 69–81. Cold Spring Harbor Laboratory, Cold Spring Harbor, New York.

Adinolfi, M., Gardner, B., and Martin, W. (1967). Observations on the levels of γG, γA, and γM globulins, anti-A and anti-B agglutinins, and antibodies to *Escherichia coli* in Down's anomaly. *J. Clin. Pathol.* 20, 860–4.

Agarwal, S.S., Blumberg, B.S., Gerstley, B.J.S., London, W.T., Sutnick, A.I., and Loeb, L.L. (1970). DNA polymerase activity as an index of lymphocyte stimulation: studies in Down's syndrome. *J. Clin. Invest.* 49, 161–9.

Aguet, M., and Blanchard, B. (1981). High affinity binding of ^{125}I-labeled mouse interferon to a specific cell surface receptor. II. Analysis of binding properties. *Virology* 115, 249–61.

Ahlström, U. (1974). Chromosomes of primary carcinomas induced by 7, 12-dimethylbenz(α)anthracene in the rat. *Hereditas* 78, 235–44.

Aitken, D.A., and Ferguson-Smith, M.A. (1978). Gene dosage evidence for the regional assignment of the GOT$_s$ structural gene locus to 10q2→10q25, *Cytogenet. Cell Genet.* 22, 468–71.

Aitken, D.A., and Ferguson-Smith, M.A. (1979). Intrachromosomal assignment of the structural gene for GALT to the short arm of chromosome 9 by gene dosage studies. *Cytogenet. Cell Genet.* 25, 131.

Alberman, E.D., and Creasy, M.R. (1977). Frequency of chromosomal abnormalities in stillbirths and perinatal deaths. *J. Med. Genet.* 14, 313–15.

Aldenhoff, P., Wegner, R.-D., and Sperling, K. (1980). Different sensitivity of diploid and trisomic cells from patients with Down syndrome mosaic after treatment with the trifunctional alkylating agent Trenimon. *Hum. Genet.* 56, 123–5.

Alfi, O.S., Donnell, G.N., Allerdice, P.W., and Derencsenyi, A. (1976). The 9p − syndrome. *Ann. Génét.* 19, 11–16.

Alimena, G., Billström, R., Casalone, R., Gallo, E., Mitelman, F., and Pasquali,

P. (1985). Cytogenetic pattern in leukemic cells of patients with constitutional chromosome anomalies. *Cancer Genet. Cytogenet.* 16, 207–18.

Alitalo, K., Winqvist, R., Lin, C.C., de la Chapelle, A., Schwab, M. and Bishop, J.M. (1984). Aberrant expression of an amplified c-*myb* oncogene in two cell lines from a colon carcinoma. *Proc. Natl. Acad. Sci. USA* 81, 4534–8.

Allerdice, P.W., Eales, B., Onyett, H., Sprague, W., Henderson, K., Lefeuvre, P.A., and Pal, G. (1983). Duplication 9q34 syndrome. *Am. J. Hum. Genet.* 35, 1005–19.

Allison, A.C., and Blumberg, B.S. (1958). Dominance and recessivity in medical genetics. *Am. J. Med.* 25, 933–41.

Al-Saadi, A., and Beierwaltes, W.H. (1967). Sequential cytogenetic changes in the evolution of transplanted thyroid tumors to metastatic carcinoma in the Fischer rat. *Cancer Res.* 27, 1831–42.

Alter, B.P., Goff, S.C., Hillman, D.G., Deisseroth, A.B., and Forget, B.G. (1977). Production of mouse globin in heterokaryons of mouse erythroleukaemia cells and human fibroblasts. *J. Cell Sci.* 26, 347–57.

Ambros, V., and Horvitz, H.R. (1984). Heterochronic mutants of the nematode *Caenorhabditis elegans. Science* 226, 409–16.

Ammann, A.J., and Hong, R. (1971). Selective IgA deficiency: presentation of 30 cases and a review of the literature. *Medicine* 50, 223–36.

Andrle, M., Erlach, A., and Rett, A. (1981). Partielle Monosomie and Trisomie 5p als Folge einer balanciertën väterlichen Translokation t(3,5). *Wien. Klin. Wochnschr.* 95, 16–19.

Andrle, M., Fiedler, W., Rett, A., Ambros, P., and Schweizer, D. (1979). A case of trisomy 22 in *Pongo pygmaeus. Cytogenet. Cell Genet.* 24, 1–6.

Annerén, G., and Björkstén, B. (1984). Low superoxide levels in blood phagocytic cells in Down's syndrome. *Acta Paediatr. Scand.* 73, 345–8.

Annerén, G., and Gustavson, K.-H. (1984). Partial trisomy 3q(3q25→qter) syndrome in two siblings. *Acta Paediatr. Scand.* 73, 281–4.

Annerén, G., Edqvist, L.-E., Gebre-Medhin, and Gustavson, K.-H. (1984). Glutathione peroxidase activity in erythrocytes in Down's syndrome. Abnormal variation in relation to age and sex through childhood and adolescence. Submitted for publication.

Antonarakis, S.E., Valle, D., Moser, H.W., Moser, A., Qualman, S.J., and Zinkham, W.H. (1984). Phenotypic variability in siblings with Farber disease. *J. Pediatr.* 104, 406–9.

Antonarakis, S.E., Kittur, S.D., Metaxotou, C., Bartsocas, C., Kitsiou, S., Watkins, P.C., Patel, A.S., Warren, A.C., Gusella, J.F., Groner, Y., Chakravarti, A., Meyers, D.A., and Kazazian H.H. Jr. (1985). Linkage map of chromosome 21q and the association of a DNA haplotype with a propensity to nondisjunction and trisomy 21. *Ann. N.Y. Acad. Sci.* 450, 95–107.

Aoki, N., Moroi, M., Sakata, Y., and Yoshida, N. (1978). Abnormal plasminogen. A hereditary molecular abnormality found in a patient with recurrent thrombosis. *J. Clin. Invest.* 61, 1186–95.

Ariëns, E.J., Beld, A.J., Rodrigues de Miranda, J.F., and Simonis, A.M. (1979). In *The Receptors. A Comprehensive Treatise. Vol. 1: General Principles and Procedures*, ed. R.D. O'Brien, pp. 33–91. Plenum, New York.

Armendares, S., and Salamanca-Gómez, F. (1978). Partial 2p trisomy (p21→pter) in two siblings of a family with a 2p-;15q+ translocation. *Clin. Genet.* 13, 17–24.

Aronson, M.M., Nichols, W.W., Mulivor, R.A., Greene, A.E., and Coriell, L.L. (1983). Chromosome maps of cell lines with specific monosomic or trisomic portions of the genome in the NIGMS Human Genetic Mutant Cell Repository. *Cytogenet. Cell Genet.* 36, 652–8.

Ashley, T. (1984). A re-examination of the case for homology between the X and Y chromosomes of mouse and man. *Hum. Genet.* 67, 372–7.

Athanasiou, K., and Bartsocas, C.S. (1982). The implications of S-phase exchanges for the mechanisms of radiosensitivity in trisomy 21. *Am. J. Med. Genet.* 12, 141–6.

Atkin, J.F., and Patil, S. (1983). Duplication of the distal segment of 14q. *Am. J. Med. Genet.* 16, 357–66.

Audit, I., Deparis, P., Flavin, M., and Rosa, R. (1976). Erythrocyte enzyme activities in diploid and triploid salamanders (*Pleurodeles waltlii*) of both sexes. *Biochem. Genet.* 14, 759–69.

Augustinsson, K.-B., and Olssen, B. (1961). The genetic control of arylesterase in the pig. *Hereditas* 47, 1–22.

Ayala, F.J., and McDonald, J.F. (1980). Continuous variation: possible role of regulatory genes. *Genetica* 52/53, 1–15.

Aziz, A.A., Blair, J.A., Leeming, R.J., and Sylvester, P.E. (1982). Tetrahydrobiopterin metabolism in Down's syndrome and in non-Down's syndrome mental retardation. *J. Ment. Defic. Res.* 26, 67–71.

Aziz, M.A. (1981). Possible "atavistic" structure in human aneuploids. *Am. J. Phys. Anthropol.* 54, 347–53.

Babior, B.M. (1978). Oxygen-dependent microbial killing by phagocytes. *N. Engl. J. Med.* 298, 659–68.

Babu, V.R., Van Dyke, D.L., and Jackson, C.E. (1984). Chromosome 20 deletions in human multiple endocrine neoplasia types 2A and 2B: a doubleblind study. *Proc. Natl. Acad. Sci. USA* 81, 2525–8.

Bachmann, K., Goin, O.B., and Goin, C.J. (1972). Nuclear DNA amounts in vertebrates. In *Evolution of Genetic Systems*, ed. H.H. Smith (*Brookhaven Symp. Biol.* 23), pp. 419–47, Gordon and Breach, New York.

Back, E., Stier, R., Böhm, N., Adlung, A., and Hameister, H. (1980). Partial monosomy 22pter → q11 in a newborn with the clinical features of trisomy 13 syndrome. *Ann. Génét.* 23, 244–8.

Baeteman, M.A., Baret, A., Courtière, A., Rebuffel, P., and Mattei, J.F. (1983). Immunoreactive Cu-SOD and Mn-SOD in lymphocyte subpopulations from normal and trisomy 21 subjects according to age. *Life Sci.* 32, 895–902.

Baghdassarian, A., Bayard, F., Borgaonkar, D.S., Arnold, E.A., Solez, K., and Migeon, C.J. (1975). Testicular function in XYY men. *Johns Hopkins Med. J.* 136, 15–24.

Baglioni, C. (1979). Interferon induced enzymatic activities and their role in the antiviral state. *Cell* 17, 255–64.

Baikie, A.G., Loder, P.B., de Grouchy, G.C., and Pitt, D.B. (1965). Phosphohexokinase activity of erythrocytes in mongolism: another possible marker for chromosome 21. *Lancet* 1, 412–14.

Balaban, G., Gilbert, F., Nichols, W., Meadows, A.T., and Shields, J. (1982). Abnormalities of chromosome #13 in retinoblastomas from individuals with normal constitutional karyotypes. *Cancer Genet. Cytogenet.* 6, 213–21.

Balazs, I., Purvello, M., Rochi, M., Rinaldi, A., and Siniscalco, M. (1982). Is the gene for steroid sulfatase X-linked in man? An appraisal of data from humans, mice and their hybrids. *Cytogenet. Cell Genet.* 32, 251–2.

Balestrazzi, P., Giovannelli, G., Rubini, L.L., and Dallapiccola, B. (1979). Partial trisomy 16q resulting from maternal translocation. *Hum. Genet.* 49, 229–35.

Balestrazzi, P., Croci, G., Frasi, C., Franchi, F., and Giovannelli (1983). Tetrasomy 9p confirmed by GALT. *J. Med. Genet.* 20, 396–9.

Ball, M.J., and Nuttall, K. (1980). Neurofibrillary tangles, granulovacuolar degeneration, and neuron loss in Down syndrome: quantitative comparison with Alzheimer dementia. *Ann. Neurol.* 7, 462–5.

Ball, M.J., and Nuttall, K. (1981). Topography of neurofibrillary tangles and granulovacuoles in hippocampi of patients with Down's syndrome: quantitative comparison with normal aging and Alzheimer's disease. *Neuropathol. Appl. Neurobiol.* 7, 13–20.

Banerjee, A., Jung, O., and Huang, C.C. (1977). Response of hematopoietic cell lines derived from patients with Down's syndrome and from normal individuals to mitomycin C and caffeine. *J. Natl. Cancer Inst.* 59, 37–9.

Banik, N.L., Davison, A.N., Palo, J., and Savolainen, H. (1975). Biochemical studies on myelin isolated from the brains of patients with Down's syndrome. *Brain* 98, 213–8.

Barak, Y., Mogilner, B.M., Karov, Y., Nir, E., Schlesinger, M., and Levin, S. (1982). Transient acute leukaemia in a newborn with Down's syndrome. Prediction of its reversibility by bone marrow cultures. *Acta Paediatr. Scand.* 71, 699–701.

Baranov, V.S. (1983a). Chromosomal control of early embryonic development in mice. I. Experiments on embryos with autosomal monosomy. *Genetica* 61, 165–77.

Baranov, V.S. (1983b). Chromosomal control of early embryonic development in mice. II. Experiments on embryos with structural aberrations of autosomes 7, 9, 14 and 17. *Genet. Res.* 41, 227–39.

Baranov, V.S., and Dyban, A.P. (1972). Disturbance of embryogenesis in trisomy of autosomes arising in offspring of mice with Robertsonian translocation (centric fusion) of T1Iem. *Arch. Anat. Leningrad* 113, 67–77.

Baranov, V.S., Dyban, A.P., and Chebotar, N.A. (1980). Peculiarities of preimplantation mouse development in mice with monosomy for autosome 17. *Ontogenez* 11, 148–59.

Baranov, V.S., Vaisman, B.L., and Udalova, L.D. (1982). Some morphological peculiarities and state of rRNA synthesis in mouse embryos with trisomy for chromosome 19. *Ontogenez* 13, 46–54.

Barden, H.S. (1983). Growth and development of selected hard tissues in Down syndrome. A review. *Hum. Biol.* 55, 539–76.

Baret, A., Baeteman, M.A., Mattei, J.F., Michel, P., Broussolle, B., and Giraud, F. (1981). Immunoreactive Cu SOD and Mn SOD in the circulating blood cells from normal and trisomy 21 subjects. *Biochem. Biophys. Res. Commun.* 98, 1035–43.

Barkin, R.M., Weston, W.L., Humbert, J.R., and Sunada, K. (1980). Phagocytic function in Down syndrome – II. Bactericidal activity and phagocytosis. *J. Ment. Defic. Res.* 24, 251–6.

Barlow, P.W. (1972). Differential cell division in human X chromosome mosaics. *Humangenetik* 14, 122–7.

Barlow, P. (1973). The influence of inactive chromosomes on human development. Anomalous sex chromosome complements and the phenotype. *Humangenetik* 17: 105–36.

Barr, M. Jr. (1984). Aberrant growth patterns of chromosomally abnormal midtrimester fetuses. *Proc. Greenwood Genet. Center* 3, 121–2.

Barsh, G.S., David, K.E., and Byers, P.H. (1982). Type I osteogenesis imperfecta:

a nonfunctional allele for pro α1(I) chains of type I procollagen. *Proc. Natl. Acad. Sci. USA* 79, 3838–42.

Bartels, H., and Kruse, K. (1968). Enzymbestimmungen in Erythrocyten bei Kindern mit Down-Syndrome. *Humangenetik* 5, 305–9.

Bartley, J.A., and Epstein, C.J. (1980). Gene dosage effect for glycinamide ribonucleotide synthetase in human fibroblasts trisomic for chromosome 21. *Biochem. Biophys. Res. Commun.* 93, 1286–9.

Bartoshesky, L., Lewis, M.B., and Pashayan, H.M. (1978). Developmental abnormalities associated with long arm deletion of chromosome no. 6. *Clin. Genet.* 13, 68–71.

Bass, H.N., Sparkes, R.S., and Miller, A.A. (1979). Features of trisomy 18 and 18p− syndromes in an infant with 46,XY,i(18q). *Clin. Genet.* 16, 163–8.

Batista, D.A.S., Vianna-Morgante, A.M., and Richiere-Costa, A. (1983). Tetrasomy 18p: tentative delineation of a syndrome. *J. Med. Genet.* 10, 144–7.

Batshaw, M.L., Roan, Y., Jung, A.L., Rosenberg, L.A., and Brusilow, S.W. (1980). Central dysfunction in asymptomatic carriers of ornithine transcarbamylase deficiency. *N. Engl. J. Med.* 302, 482–5.

Baxter, J.D., and Funder, J.W. (1979). Hormone receptors. *N. Engl. J. Med.* 301, 1149–61.

Baxter, J.D., and MacLeod, K. (1980). Molecular basis for hormone action. In *Metabolic Control and Disease*, ed. P.K. Bondy and L.E. Rosenberg, pp. 103–60. Saunders, Philadelphia.

Bayer, S.M., and McCoy, E.E. (1974). A comparison of the serotonin and ATP content in platelets from subjects with Down's syndrome. *Biochem Med.* 9, 225–32.

Bazelon, M., Paine, R.S., Cowie, V.A., Hunt, P., Houck, J.C., and Mahanand, D. (1967). Reversal of hypotonia in infants with Down's syndrome by adminstration of 5-hydroxytryptophan. *Lancet* 1, 1130–3.

Bazelon, M., Barnet, A., Lodge, A., and Shelburne, S.A. Jr. (1968). The effect of high doses of 5-hydroxytryptophan on a patient with trisomy 21. *Brain Res.* 11, 397–411.

Beatty, R.A. (1957). *Parthenogenesis and Polyploidy in Mammalian Development.* Cambridge University Press, Cambridge.

Beatty, R.A. (1978). The origin of human triploidy: an integration of qualitative and quantitative evidence. *Ann. Hum. Genet.* 41, 299–314.

Beçak, W., and Pueyo, M.T. (1970). Gene regulation in the polyploid amphibian *Odontophrynus americanus. Exp. Cell Res.* 63, 448–51.

Becker, K.L., Burke, E.C., and Albert, A. (1963). Double autosomal trisomy (D trisomy plus mongolism). *Mayo Clinic Staff Meetings* 38, 242–8.

Becker, M.A., Losman, N., Itkin, P., and Simkin, P.A. (1982). Gout with superactive phosphoribosylpyrophosphate synthetase due to increased enzyme catalytic rate. *J. Lab. Clin. Med.* 99, 495–511.

Bedin, M., Weil, D., Fournier, T., Cedard, L., and Frezal, J. (1981). Biochemical evidence for the non-inactivation of the steroid sulfatase locus in human placenta and fibroblasts. *Hum. Genet.* 59, 256–8.

Beechey, C.V., and Speed, R. (1981). Personal communication. *Mouse News Letter* No. 64, 56.

Beechey, C.V., Kirk, M., and Searle, A.G. (1980). A reciprocal translocation induced in an oocyte and affecting fertility in male mice. *Cytogenet. Cell Genet.* 27, 129–46.

Beemer, F.A., de France, H.F., Rosina-Angelista, I.J.M., Gerards, L.J., Cats, B.P., and Guyt, R. (1984). Familial partial monosomy 5p and trisomy 5q; three cases due to paternal pericentric inversion 5(p151q333). *Clin. Genet.* 26, 209–15.

Benda, C.E., and Strassmann, G.S. (1965). The thymus in mongolism. *J. Ment. Defic. Res.* 9, 109–17.

Bendel, R., Baldinger, S., Millard, C., and Arthur, D.C. (1982). Two successive partial trisomies for opposite halves of chromosome 22 in a mother with a balanced translocation. *J. Med. Genet.* 19, 313.

Benedict, W.F., Banerjee, A., Mark, C., and Murphree, A.L. (1983*a*). Nonrandom chromosomal changes in untreated retinoblastomas. *Cancer Genet. Cytogenet.* 10, 311–33.

Benedict, W.F., Murphree, A.L., Banerjee, A., Spina, C.A., Sparkes, M.C., and Sparkes, R.S. (1983*b*). Patient with 13 chromosome deletions: evidence that the retinoblastoma gene is a recessive cancer gene. *Science* 219, 973–5.

Benirschke, K., Bogart, M.H., McClure, H.M., and Nelson-Rees, W.A. (1974). Fluorescence of the trisomic chimpanzee chromosomes. *J. Med. Prim.* 3, 311–14.

Bennett, F.C., Sells, C.J., and Brand, C. (1979). Influences on measured intelligence in Down's syndrome. *Am. J. Dis. Child.* 133, 700–3.

Bennhold, H., Peters, H., and Roth, E. (1954). Ueber Einem Fall von kompletter Analbuminaemie ohne wesentliche klinische Krankheitszeichen. *Verh. Dtsch. Ges. Inn. Med.* 60, 630–4.

Benson, P.F., Linacre, B., and Taylor, A.I. (1968). Erythrocyte ATP: D-fructose–6-phosphate 1-phosphotransferase (phosphofructokinase) activity in children with normal/G trisomic mosaic Down's syndrome and in normal and Down's syndrome controls. *Nature, Lond.* 220, 1235–6.

Berberich, M.S., Carey, J.C., Lawce, H.J., and Hall, B.D. (1978). Duplication (partial trisomy) of the distal long arm of chromosome 17: a new clinically recognizable chromosome disorder. *Birth Defects: Orig. Art. Ser.* 14(6C), 287–95.

Berg, J.M., Brandon, M.W.G., and Kirman, B.H. (1959). Atropine in mongolism. *Lancet* 2, 441–2.

Berger, R., Bernheim, A., Sasportes, M., Hauptmann, G., Hors, J., Legrand, L., and Fellous, M. (1979). Regional mapping of the HLA locus on the short arm of chromosome 6. *Clin. Genet.* 15, 245–51.

Berman, J., Hultén, M., and Lindsten, J. (1967). Blood-serotonin in Down's syndrome. *Lancet* 1, 730.

Bernard, R., Sitruk, S., Bernard, D., and Mattei, J.-F. (1976). Etude de l'immunité humorale et de la tranformation blastique des lymphocytes chez le trisomique 21. *Arch. Fr. Pediatr.* 33, 37–53.

Bernstein, E.G., Russell, L.B., and Cain, C.S. (1978). Effect of gene dosage on expression of mitochondrial malic enzyme activity in the mouse. *Nature, Lond.* 271, 748–50.

Bersu, E.T. (1984). Morphological development of the fetal trisomy 19 mouse. *Teratology* 29, 117–29.

Bersu, E.T., and Mieden, G.D. (1980). Ontogeny of the retarded phenotype in the trisomy 19 mouse. *Teratology* 21, 28A.

Bersu, E.T., Crandall, C., and White, B.J. (1982). Growth characteristics of the murine trisomy 19 thymus. *Teratology* 26, 85–94.

Bersu, E.T., Crandall, C., and White, B.J. (1983). Erythropoiesis in the fetal trisomy 19 mouse. I. Characterization of erythrocyte populations in peripheral blood. *Teratology* 27, 271–6.

Bertolotti, R., and Weiss, M.C. (1974). Expression of differentiated functions in hepatoma cell hybrids. V. Reexpression of aldolase B *in vitro* and *in vivo*. *Differentiation* 2, 5–17.

Biederman, B., and Bowen, P. (1977). Sister chromatid exchanges in Down's syndrome. *Mammalian Chromosome Newsletter 18*, 10.

Bilheimer, D.W., Ho, Y.K., Brown, M.S., Anderson, R.G.W., and Goldstein, J.L. (1978). Genetics of the low density lipoprotein receptor. Diminished receptor activity in lymphocytes from heterozygotes with familial hypercholesterolemia. *J. Clin. Invest.* 61, 678–96.

Birchler, J.A. (1979). A study of enzyme activities in a dosage series of the long arm of chromosome one in maize. *Genetics* 92, 1211–29.

Birchler, J.A. (1981). The genetic basis of dosage compensation of alcohol dehydrogenase–1 in maize. *Genetics* 97, 625–37.

Birchler, J.A. (1983a). Allozymes in gene dosage studies. In *Isozymes in Plant Genetics and Breeding, Part A*, ed. S.D. Tanksley and T.J. Orton, pp. 85–108. Elsevier, Amsterdam.

Birchler, J.A. (1983b). Genetic basis of autosomal dosage compensation in Drosophila. *Genetics* 104, s8.

Birchler, J.A., and Newton, K.J. (1981). Modulation of protein levels in chromosomal dosage series of maize: the biochemical basis of aneuploid syndromes. *Genetics* 99, 247–66.

Bischoff, W.L. (1976). Genetic control of soluble NAD-dependent sorbitol dehydrogenase in *Drosophila melanogaster*. *Biochem. Genet.* 14, 1019–39.

Björkstén, B., Marklund, S., and Hägglöf, B. (1984). Enzymes of leukocyte oxidative metabolism in Down's syndrome. *Acta Paediatr. Scand.* 73, 97–101.

Björkstén, B. Bäck, O., Gustavson, K.H., Hallmans, G., Hägglöf, B., and Tärnvik, A. (1980). Zinc and immune function in Down syndrome. *Acta Paediatr. Scand.* 69, 183–7.

Blackburn, E.H. (1985). Artificial chromosomes in yeast. *Trends Genet.* 1, 8–12.

Blackburn, W.R., Miller, W.P., Superneau, D.W., Cooley, N.R. Jr., Zellweger, H., and Wertelecki, W. (1982). Comparative studies of infants with mosaic and complete triploidy: an analysis of 55 cases. *Birth Defects: Orig. Art. Ser.* 18(3B), 251–74.

Blank, C.E., Colver, D.C.B., Potter, A.M., McHugh, J., and Lorber, J. (1975). Physical and mental defect of chromosomal origin in four individuals of the same family. Trisomy for the short arm of 9. *Clin. Genet.* 7, 261–73.

Blau, H.M., Chiu, C.-P., and Webster, C. (1983). Cytoplasmic activation of human nuclear genes in stable heterokaryons. *Cell* 32, 1171–80.

Blazak, W.F., and Fechheimer, N.S. (1981). Gonosome-autosome translocations in fowl: the development of chromosomally unbalanced embryos sired by singly and doubly heterozygous cockerels. *Genet. Res.* 37, 161–71.

Bloom, S.E. (1972). Chromosome abnormalities in chicken (*Gallus domesticus*) embryos: types, frequencies and phenotypic effects. *Chromosoma* 37, 309–26.

Bloomer, J.R. (1980). Characterization of deficient heme synthase activity in protoporphyria with cultured skin fibroblasts. *J. Clin. Invest.* 65, 321–8.

Blue, M.E., Molliver, M.E., Gearhart, J.D., and Coyle, J.T. (1984). Abnormalities of cortical development in murine trisomy 16, an animal model for Down's syndrome. *Soc. Neurosci. Abstr.* 10, 953.

Boer, P. de (1973). Fertile tertiary trisomy in the mouse (*Mus musculus*). *Cytogenet. Cell Genet.* 12, 435–42.

Boer, P. de, and Branje, H.E.B. (1979). Association of the extra chromosome of tertiary trisomic male mice with the chromosomes during first meiotic prophase, and its significances for impairment of spermatogenesis. *Chromosoma* 73, 369–79.

Boer, P. de, and Groen, A. (1974). Fertility and meiotic behavior of male *T70H* tertiary trisomics of the mouse (*Mus musculus*). A case of preferential telomeric meiotic pairing in a mammal. *Cytogenet. Cell Genet.* 13, 489–510.

Boer, P. de, and Wauben-Penris (1984). Personal communication. *Mouse News Letter* No. 70, 105.

Boer, P. de, van der Hoeven, F.A., and Chardan, J.A.P. (1976). The production, morphology, karyotypes and transport of spermatozoa from tertiary trisomic mice and the consequences of egg fertilization. *J. Reprod. Fert.* 48, 249–56.

Boggs, S.S., Patrene, K.D., Downer, W.R., Schwartz, G.N., and Saxe, D.F. (1983). Trisomy of chromosome 6.15 is not necessary for proliferation of AKR[Rb6.15]1Ald lymphoma cells. *Cancer Genet. Cytogenet.* 9, 151–66.

Boman, H., Hermodson, M., Hammond, C.A., and Motulsky, A.G. (1976). Analbuminemia in an American Indian girl. *Clin. Genet.* 9, 513–26.

Bomsel-Helmreich, O. (1965). Heteroploidy and embryonic death. In *Preimplantation Stages of Pregnancy*, ed. G.E.W. Wolstenholme and M. O'Connor, pp. 246–67. Churchill, London.

Bond, D.J., and Chandley, A.C. (1983). *Aneuploidy*. Oxford University Press, Oxford.

Bonkowsky, H.L., Bloomer, J.R., Ebert, P.S., and Mahoney, M.J. (1975). Heme synthetase deficiency in human protoporphyria: demonstration of the defect in liver and cultured skin fibroblasts. *J. Clin. Invest.* 56, 1139–48.

Boué, A., Boué, J., Cure, S., Deluchat, C., and Perrandin, N. (1975). In vitro cultivation of cells from aneuploid human embryos. Initiation of cell lines and longevity of the culture. *In Vitro* 11, 409–13.

Boué, J., Boué, A., and Lazar, P. (1975). Retrospective and prospective epidemiological studies of 1500 karyotyped spontaneous human abortions. *Teratology* 12, 11–26.

Boué, J.G., Boué, A., and Spira, A. (1974). Evolution de dosages hormonaux urinaires dans les grossesses avec anomalies chromosomiques léthales ou viables. In *L'exploration hormonale dans la surveillance biologique de la grossesse humaine*, ed. R. Scholler, pp. 167–89. SEPE, Paris.

Boué, J., Philippe, E., Giroud, A., and Boué, A. (1976). Phenotypic expression of lethal chromosome anomalies in human abortuses. *Teratology* 14, 3–20.

Boué, J., Deluchat, C.C., Nicolas, H., and Boué, A. (1981). Prenatal losses of trisomy 21. In *Trisomy 21. An International Symposium*, ed. G.R. Burgio, M. Fracarro, L. Tiepolo, and U. Wolf, pp. 183–93, Springer-Verlag, Berlin.

Boullin, D.J. (1975). The use of the human blood platelets as a model for the function and pharmacology of central serotoninergic systems in health and disease. *Agents Actions* 5, 494–5.

Bourgeois, S., and Newby, R.F. (1979). Correlation between glucocorticoid receptor and cytolytic response of murine lymphoid cell lines. *Cancer Res.* 39, 4749–51.

Boveri, T.H. (1914). *Zur Frage der Entstehung maligner Tumoren*. Gustav Fischer, Jena.

Bowen, P., Fitzgerald, P.H., Gardner, R.J.M., Biederman, B., and Veale, A.M.O. (1983). Duplication 8q syndrome due to familial chromosome ins(10;8) (q21;q212q22). *Am. J. Med. Genet.* 14, 635–46.

Boyse, E.A., and Cantor, C. (1978). Immunogenetic aspects of biologic communication: a hypothesis of evolution by program duplication. *Birth Defects: Orig. Art. Ser.* 14 (2), 249–83.

Boyse, E.A., Stockert, E., and Old, L.J. (1968). Isoantigens of the *H–2* and *Tla* loci of the mouse. Interactions affecting their representation on thymocytes. *J. Exp. Med.* 128, 85–95.

Bradley, C.M., Cox, D.M., Patterson, D., and Robinson, A. (1982). Gene dosage effect for phosphoribosylglycineamide synthetase (GARS) in a patient with Down syndrome and non-Robertsonian t(21;21) translocation. *Pediatr. Res.* 16, 190A.

Braunger, R., Kling, H., Krone, W., Schmid, M., and Olert, J. (1977). Gene dosage effect for fumarate hydratase (FH; E.C. 4.2.1.2) in partial trisomy 1. *Hum. Genet.* 38, 65–77.

Bravo, R., Schafer, R., Willecke, K., MacDonald-Bravo, H., Fey, S.J., and Celis, J.E. (1982). More than one-third of the discernible mouse polypeptides are not expressed in a Chinese hamster-mouse embryo fibroblast hybrid that retains all mouse chromosomes. *Proc. Natl. Acad. Sci. USA* 79, 2281–5.

Brenner, D.A., and Bloomer, J.R. (1980). The enzymatic defect in variegate porphyria. Studies with human cultured skin fibroblasts. *N. Engl. J. Med.* 302, 765–9.

Brent, R.L., and Jensh, R.P. (1967). Intra-uterine growth retardation. *Adv. Teratol.* 2, 139–227.

Bricarelli, F.D., Magnani, M., Arslanian, A., Camera, G., Coviello, D.A., DiPietro, P., and Dallipiccola, B. (1981). Expression of GALT in two unrelated 9p – patients. Evidence for assignment of the GALT locus to the 9p21 band. *Hum. Genet.* 59, 112–14.

Bridges, C.B. (1921). Genetic and cytologic proof of non-disjunction of the fourth chromosome of Drosophila melanogaster. *Proc. Natl. Acad. Sci. USA* 7, 186–92.

Brinster, R.L., Ritchie, K.A., Hammer, R.E., O'Brien, R.L., Arp, B., and Storb, U. (1983). Expression of a microinjected immunoglobulin gene in the spleen of transgenic mice. *Nature, Lond.* 306, 332–6.

Brinster, R.L., Chen, H.Y., Trumbauer, M.E., Yagle, M.K., and Palmiter, R.D. (1985). Factors affecting the efficiency of introducing foreign DNA into mice by microinjecting eggs. *Proc. Natl. Acad. Sci. USA* 82, 4438–42.

Britten, R.J., and Davidson, E.H. (1969). Gene regulation for higher cells: a theory. *Science* 165, 349–57.

Brodeur, G.M., Dahl, G.V., Williams, D.L., Tipton, R.E., and Kalwinsky, D.K. (1980). Transient leukemoid reaction and trisomy 21 mosaicism in a phenotypically normal newborn. *Blood* 55, 691–3.

Brodeur, G.M., Green, A.A., Hayes, F.A., Williams, K.J., and Tsiafis, H.A. (1981). Cytogenetic features of human neuroblastomas and cell lines. *Cancer Res.* 41, 4678–86.

Brodeur, G.M., Seeger, R.C., Schwab, M., Varmus, H.E., and Bishop, J.M. (1984). Amplification of N-*myc* in untreated human neuroblastomas correlated with advanced disease state. *Science* 224, 1121–4.

Brook, C.G.D., Gasser, T., Werder, E.A., Prader, A., and Vanderschueren-Lodewykx, M.A. (1977). Height correlations between parents and mature offspring in normal subjects and in subjects with Turner's and Klinefelter's and other syndromes. *Ann. Hum. Biol.* 4, 17–22.

Brook, J.D. (1983). X-chromosome segregation, maternal age and aneuploidy in the *XO* mouse. *Genet. Res.* 41, 85–95.

Brooksbank, B.W.L., and Balazs, R. (1984). Superoxide dismutase, glutathione peroxidase and lipoperoxidation in Down's syndrome fetal brain. *Dev. Brain Res.* 16, 37–44.

Brown, D.D. (1981). Gene expression in eukaryotes. *Science* 211, 667–74.

Brown, D.D., and David, I.B. (1968). Specific gene amplification in oocytes. *Science* 160, 272–80.

Brown, J.E., and Weiss, M.C. (1975). Activation of production of mouse liver enzymes in rat hepatoma-mouse lymphoid cell hybrids. *Cell* 6, 481–94.

Brown, W.T., Devine, E.A., Nolin, S.L., Houck, G.E. Jr., and Jenkins, E.C. (1985). Localization of chromosome 21 probes by *in situ* hyridization. *Ann. N.Y. Acad. Sci.* 450, 69–83.

Bruce, M.E., and Fraser, H. (1975). Amyloid plaques in the brains of mice infected with scrapie: morphological variation and staining properties. *Neuropathol. Appl. Neurobiol.* 1, 189–202.

Buckle, V.J., Edwards, J.H., Evans, E.P., Jonasson, J.A., Lyon, M.F., Peters, J., Searle, A.G., and Wedd, N.S. (1984). Chromosome maps of man and mouse II. *Clin. Genet.* 26, 1–11.

Buckton, K.E., Whalley, L.J., Lee, M., and Christie, J.E. (1983). Chromosome changes in Alzheimer's presenile dementia. *J. Med. Genet.* 20, 46–51.

Bühler, E.M., and Malik, N. (1984). The tricho-rhino-phalangeal syndrome(s): chromosome 8 long arm deletion: is there a shortest region of overlap between reported cases? TRPI and TRPII syndromes: are they separate entities? *Am. J. Med. Genet.* 19, 113–19.

Bühler, E.M., Bühler, U.K., and Christen, R. (1983). Terminal or interstitial deletion in chromosome 8 long arm in Langer-Giedion syndrome (TRPII syndrome)? *Hum. Genet.* 64, 163–6.

Bulfield, G., and Kacser, H. (1974). Histidinaemia in mouse and man. *Arch. Dis. Child.* 49, 545–52.

Bulfield, G., Andrews, K., Kacser, H. and Stewart, J.A. (1974). Personal communication. *Mouse News Letter* No. 50, 35.

Burger, P.C., and Vogel, F.S. (1973). The development of the pathologic changes of Alzheimer's disease and senile dementia in patients with Down's syndrome. *Am. J. Pathol.* 73, 457–76.

Burgio, G.R., Ugazio, A.G., Nespoli, L., Marcioni, A.E., Bottelli, A.M., and Pasquali, F. (1975). Derangements of immunoglobulin levels, phytohemagglutinin responsiveness and T and B cell markers in Down's syndrome at different ages. *Eur. J. Immunol.* 5, 600–3.

Burgio, G.R., Fraccaro, M., Tiepolo, L., and Wolf, U. (eds.) (1981). *Trisomy 21. An International Symposium*, Springer-Verlag, Berlin.

Burgio, G.R., Ugazio, A., Nespoli, L., and Maccario, R. (1983). Down syndrome: a model of immunodeficiency. *Birth Defects: Orig. Art. Ser.* 19(3), 325–7.

Burgoyne, P.S. (1978). The role of the sex chromosomes in mammalian germ cell differentiation. *Ann. Biol. Anim. Biochem. Biophys.* 18(2B), 317.

Burgoyne, P.S. (1979). Evidence for an association between univalent Y chromosomes and spermatocyte loss in XYY mice and men. *Cytogenet. Cell Genet.* 23, 84–9.

Burgoyne, P.S. (1981). The genetics of sex development. In *Basic Reproductive Medicine, Vol. 1: Basics and Development of Reproduction*, ed. D. Hamilton and F. Naftolin, pp. 1–31. MIT Press, Cambridge.

Burgoyne, P.S. (1982). Genetic homology and crossing over in the X and Y chromosomes of mammals. *Hum. Genet.* 61, 85–90.

Burgoyne, P.S., and Baker, T.G. (1981a). The XO ovary – development and function. In *Development and Function of Reproductive Organs*, ed. A.G. Byskov and H. Peters, pp. 122–8. Excerpta Medica, Amsterdam, Oxford, and Princeton.

Burgoyne, P.S., and Baker, T.G. (1981*b*). Oocyte depletion in XO mice and their XX sibs from 12 to 200 days *post partum. J. Reprod. Fert.* 61, 207–12.

Burgoyne, P.S., and Biggers, J.D. (1976). The consequences of X-dosage deficiency in the germ line: impaired development *in vitro* of preimplantation embryos from XO mice. *Dev. Biol.* 51, 109–117.

Burgoyne, P.S., Evans, E.P., and Holland, K. (1983). XO monosomy is associated with reduced birth weight and lowered weight gain in the mouse. *J. Reprod. Fert.* 68, 381–5.

Burgoyne, P.S., Tam, P.P.L., and Evans, E.P. (1983). Retarded development of XO conceptus during early pregnancy in the mouse. *J. Reprod. Fert.* 68, 387–93.

Burrill, P., Bernardini, I., Kleinman, H.K., and Kretchmer, N. (1981). Effect of serum, fibronectin, and laminin on adhesion of rabbit intestinal epithelial cells in culture. *J. Supramol. Struct. Cell. Biochem.* 16, 385–92.

Bus, J.S., Aust, S.D., and Gibson, J.E. (1976). Paraquat toxicity: proposed mechanism of action involving lipid peroxidation. *Environ. Health Perspect.* 16, 139–46.

Butler, M.G., and Palmer, C.G. (1983). Paternal origin of chromosome 15 deletion in Prader-Willi syndrome. *Am. J. Hum. Genet.* 35, 128A.

Buul, P.P.W. van, and Boer, P. de (1982). Induction by X-rays of chromosomal aberrations in somatic and germ cells of mice with different karyotypes. *Mutation Res.* 92, 229–41.

Buyse, M.L., and Korf, B.R. (1983). "Killian syndrome," Pallister mosaic syndrome, or mosaic tetrasomy 12p? – an analysis. *J. Clin. Dysmorphol.* 1, 2–5.

Byers, P.H., Bonadio, J.F., Steinman, B., Barsh, G.S., Holbrook, K.A., Greenberg, C., Rowe, D.W., and Gelinas, R. (1983). Molecular heterogeneity in perinatal lethal osteogenesis (OI type II). *Am. J. Hum. Genet.* 35, 39A.

Cacheiro, N.L.A., and Generoso, W.M. (1975). Personal communication. *Mouse News Letter* No. 53, 52.

Cairns, J. (1981). The origin of human cancers. *Nature, Lond.* 289, 353–7.

Camerino, G., Koenig, M., Moisan, J.P., Morle, F., Jaye, M., de la Salle, H., Weil, D., Hellkuhl, B., Grzeschik, K.H., and Mandel, J.L. (1984). Regional mapping of coagulation factor IX gene and of several unique DNA sequences on the human X chromosome. *Cytogenet. Cell Genet.* 37, 431–2.

Cantrell, D.A., and Smith, K.A. (1984). The interleukin-2 T-cell system: a new cell growth model. *Science* 224, 1312–6.

Cantu, J.M., Hernandez, A., Vaca, G., Plascencia, L., Martinez-Basalo, C., Ibarra, B., and Rivera, H. (1981). Trisomy 22q1→qter: "aneusomie de recombination" of a pericentric inversion. *Ann. Génét.* 24, 37–40.

Caplan, A.I., and Ordahl, C.P. (1978). Irreversible gene repression model for control of development. *Science* 201, 120–30.

Caren, L.D., Bellavance, R., and Grumet, F.C. (1982). Demonstration of gene dosage effects on antigens in the Duffy, Ss, and Rh systems using an enzyme-linked immunosorbent assay. *Transfusion* 22, 475–8.

Carey, J.C. (1980). Spectrum of the DiGeorge "syndrome." *J. Pediatr.* 96, 955.

Carlin, M.E., and Neadle, M.M. (1978). Cri-du-chat syndrome – a correlation between abnormalities and size of deletion. *Birth Defects: Orig. Art. Ser.* 14(6C), 428–9.

Carlin, M.E., and Norman, C. (1978). Partial trisomy 12p associated with 4p deletion due to paternal t(12p-;4p+) translocation. *Birth Defects: Orig. Art. Ser.* 14(6C), 399–406.

Carlson, E.A., Letson, R.D., Ramsay, N.K.C., and Desnick, R.J. (1979). Factors for improved genetic counseling for retinoblastoma based on a survey of 55 families. *Am. J. Ophthalmol.* 87, 449–59.

Carlson, P.S. (1972). Locating genetic loci with aneuploids. *Mol. Gen. Genet.* 114, 273–80.

Carothers, A.D. (1983). Letter to the editor: gene dosage effects in trisomy; comment on a recent article by B.L. Shapiro. *Am. J. Med. Genet.* 16, 635–7.

Carr, D.H., Haggar, R.A., and Hart, A.G. (1968). Germ cells in the ovaries of XO female infants. *Am. J. Clin. Pathol.* 49, 521–6.

Casarett, G.W. (1964). Similarities and contrasts between radiation and time pathology. *Adv. Gerontol. Res.* 1, 109–63.

Cassidy, S.B., Thuline, H.C., and Holm, V.A. (1984). Deletion of chromosome 15 (q11–13) in a Prader-Labhart-Willi syndrome clinic population. *Am. J. Med. Genet.* 17, 485–95.

Castel, Y., Rivière, D., Boucly, J.-Y., and Toudic, L. (1976). Trisomie 15q partielle par translocation maternelle t(7;15)(q35;q14). *Ann. Génét.* 19, 75–9.

Cattanach, B.M. (1961). XXY mice. *Genet. Res.* 2, 156–60.

Cattanach, B.M. (1962). XO mice. *Genet. Res.* 3, 487–90.

Cattanach, B.M. (1964). Autosomal trisomy in the mouse. *Cytogenetics* 3, 159–66.

Cattanach, B.M. (1967). A test of distributive pairing between two specific nonhomologous chromosomes in the mouse. *Cytogenetics* 6, 67–77.

Cattanach, B.M. (1974). Genetic disorders of sex determination in mice and other mammals. In *Birth Defects. Proceedings of the Fourth International Conference*, ed. A. Motulsky and W. Lenz, pp. 129–41. Excerpta Medica, Amsterdam.

Cattanach, B.M., and Kirk, M. (1985). Differential activity of maternally and paternally derived chromosome regions in mice. *Nature, Lond.* 315, 496–8.

Cattanach, B.M., Pollard, C.E., and Hawkes, S.G. (1971). Sex-reversed mice: XX and XO males. *Cytogenetics* 10, 318–37.

Cattanach, B.M., Evans, E.P., Burtenshaw, M.D., and Barlow, J. (1982). Male, female, and intersex development in mice of identical chromosome constitution. *Nature, Lond.* 300, 445–6.

Cavenee, W.K., Dryja, T.P., Phillips, R.A., Benedict, W.F., Godbout, R., Gallie, B.L., Murphree, A.L., Strong, L.C., and White, R.L. (1983). Expression of recessive alleles by chromosomal mechanisms in retinoblastoma. *Nature, Lond.* 305, 779–84.

Centerwall, S.A., and Centerwall, W.R. (1960). A study of children with mongolism reared in the home compared to those reared away from home. *Pediatrics* 25, 678–85.

Centerwall, W.R., and Beatty-DeSana, J.W. (1975). The trisomy 9p syndrome. *Pediatrics* 56, 748–55.

Centerwall, W., and Francke, U. (1977). Familial trisomy 20p. Five cases and two cariers in three generations. A review. *Ann. Génét.* 20, 77–83.

Chadefaux, B., Allard, D., Rethoré, M.O., Raoul, O., Poissonnier, M., Gilgenkrantz, S., Cheruy, C., and Jérôme, H. (1984). Assignment of human phosphoribosylglycinamide synthetase locus to region 21q221. *Hum. Genet.* 66, 190–2.

Chadefaux, B., Rethoré, M.O., Raoul, O., Ceballos, I., Poissonnier, M., Gilgenkranz, S., and Allard, D. (1985). Cystathionine beta synthase: gene dosage effect in trisomy 21. *Biochem. Biophys. Res. Commun.* 128, 40–4.

Chaganti, R.S.K. (1983). Significance of chromosome change to hematopoietic neoplasms. *Blood* 62, 515–24.

Chan, F.P.H., Ball, J.K., and Sergovich, F.R. (1979). Trisomy #15 in murine thymoma induced by chemical carcinogens, X-irradiation, and an endogenous murine leukemia virus. *J. Natl. Cancer Inst.* 62, 605–10.

Chan, F.P.H., Ens, B., and Frei, J.V. (1981). Cytogenetics of murine thymic lymphomas induced by N-methyl-N-nitrosourea: difference in incidence of trisomy 15 in thymic lymphomas induced in neonatal and adult mice. *Cancer Genet. Cytogenet.* 4, 337–44.

Chan, M.M., Kano, K., Dorman, B., Ruddle, F., and Milgrom, F. (1979). Human cell surface antigens coded by genes on chromosome 21. *Immunogenetics* 8, 265–75.

Chance, P.F., and Gartler, S.M. (1983). Evidence for a dosage effect at the X-linked steroid sulfatase locus in human tissues. *Am. J. Hum. Genet.* 35, 234–40.

Chandley, A.C., and Fletcher, J.M. (1980). Meiosis in *Sxr* male mice. I. Does a Y-autosome rearrangement exist in sex-reversed (*Sxr*) mice. *Chromosoma* 81, 9–17.

Chang, P.L., Georgiadis, N., Joubert, G.I., and Davidson, R.G. (1983). Gene dosage effects in human diploid and tetraploid fibroblasts. *Exp. Cell Res.* 145, 277–84.

Chaudhari, N., and Hahn, W.E. (1983). Genetic expression in the developing brain. *Science* 220, 924–8.

Chaum, E., Ellsworth, R.M., Abramson, D.H., Haik, B.G., Kitchin, F.D., and Chaganti, R.S.K. (1984). Cytogenetic analysis of retinoblastoma: evidence for multifocal origin and in vivo gene amplification. *Cytogenet. Cell Genet.* 38, 82–91.

Chen, H.P., Walz, D.V., and Carroll, J.A. (1977). Immunoglobulinopathy in Down's syndrome. *Ohio St. Med. J.* 73, 27–9.

Chen, Y.-T., Worthy, T.E., and Krooth, R.S. (1978). Evidence for a striking increase in acetylcholinesterase activity in cultured human fibroblasts which are trisomic for chromosome two. *Somat. Cell Genet.* 4, 265–98.

Chernoff, G.F. (1977). The fetal alcohol syndrome in mice: an animal model. *Teratology* 15, 223–30.

Chi, T.P.L., and Krovetz, L.J. (1975). The pulmonary vascular bed in children with Down syndrome. *J. Pediatr.* 86, 533–8.

Chu, M.-L., Williams, C.J., Pepe, G., Hirsch, J.L., Prockop, D.J., and Ramirez, F. (1983). Internal deletion in a collagen gene in a perinatal lethal form of osteogenesis imperfecta. *Nature, Lond.* 304, 78–80.

Chudina, A.P., Malyutina, T.C., and Pogosyane, E.E. (1966). Sravneniye radiochustvitelnosti kromosom v kultviruyemikh leukocitakh perifericheskoy v norme i pri sindrome Downa. *Genetika* 4, 51–63.

Church, J.A., Koch, R., Donnell, G.N., Crowley, C., Fishler, K., and Nye, C. (1983). Abnormal T lymphocyte subsets and mitogen responses in patients with Down's syndrome. *Clin. Res.* 31, 117A.

Ciferri, O., Sora, S., and Tiboni, O. (1969). Effect of gene dosage on tryptophan synthetase activity in *Saccharomyces cerevisiae*. *Genetics* 61, 567–76.

Citoler, P., and Aechter, J. (1979). Histology of testis in XXY-fetuses. In *Prenatal Diagnosis*, ed. J.-D. Murken, S. Stengel-Rutkowski, and E. Schwinger, pp. 336–7. Enke, Stuttgart.

Clark, C.E., Cowell, H.R., Telfer, M.A., and Casey, P.A. (1980). Trisomy 6q25 → 6qter in two sisters, resulting from maternal 6;11 translocation. *Am. J. Med. Genet.* 5, 171–8.

Claussen, C.-P., Gropp, A., and Herbst, E.W. (1981). Disturbed haemoglobin-biosynthesis of haemopoietic cells with mouse trisomy 16 in radiation chimeras. *Abstracts of the 6th European Anatomical Congress.* Hamburg, September 28-October 2.

Cline, M.J., and Golde, D.W. (1979). Controlling the production of blood cells. *Blood* 53, 157–65.

Coetzer, T., and Zail, S.S. (1981). Tryptic digestion of spectrin in variants of hereditary elliptocytosis. *J. Clin. Invest.* 67, 1241–8.

Cohn, R., Dickerman, L.H., Johnson, W.E., and Zinn, A.B. (1984). A provisional phenotype/karyotype correlation map for duplications of chromosome 18. *Am. J. Hum. Genet.* 36, 90S.

Coleclough, C., Perry, R.P., Karjalainen, K., and Weigert, M. (1981). Aberrant rearrangements contribute significantly to the allelic exclusion of immunoglobulin gene expression. *Nature, Lond.* 290, 372–8.

Collins, M.K.L., Goodfellow, P.N., Spurr, N.K., Solomon, E., Tanigawa, G., Tonegawa, S., and Owen, M.J. (1985). The human T-cell receptor α-chain gene maps to chromosome 14. *Nature, Lond.* 314, 273–4.

Collins, S.J., Kubonishi, I., Miyoshi, I., and Groudine, M.T. (1984). Altered transcription of the c-*abl* oncogene in K–562 and other chronic myelogenous leukemia cells. *Science* 225, 72–4.

Colon, E.J. (1972). The structure of the cerebral cortex in Down's syndrome. A quantitative analysis. *Neuropädiatrie* 3, 362–76.

Comings, D.E. (1973). A general theory of carcinogenesis. *Proc. Natl. Acad. Sci. USA* 70, 3324–8.

Conley, M.E., Beckwith, J.B., Mancer, J.F.K., and Tenckhoff, L. (1979). The spectrum of the DiGeorge syndrome. *J. Pediatr.* 94, 883–90.

Connolly, B.A., Goodman, H.O., and Swanton, C.H. (1979). Evidence for inheritance of a renal beta-amino acid transport system and its localization to chromosome 21. *Am. J. Hum. Genet.* 31, 43A.

Conway, M.M., and Layzer, R.B. (1970). Blood cell phosphofructokinase in Down's syndrome. *Humangenetik* 9, 185–90.

Cooney, T.P., and Thurlbeck, W.M. (1982). Pulmonary hypoplasia in Down's syndrome. *N. Engl. J. Med.* 307, 1170–3.

Cooper, D.W., McAllan, B.M., Donald, J.A., Dawson, G., Dobrovic, A., and Marshall Greves, J.A. (1984). Steroid sulfatase is not detected on the X chromosome of Australian marsupials. *Cytogenet. Cell Genet.* 37, 439.

Cori, C.F., Gluecksohn-Waelsch, S., Shaw, P.A., and Robinson, C. (1983). Correction of a genetically caused enzyme defect by somatic cell hybridization. *Proc. Natl. Acad. Sci. USA* 80, 6611–14.

Cory, S., Bernard, O., Gerondakis, S., Webb, E., Mitchell, J., Corcoran, L.M., and Adams, J.W. (1983). Interchromosomal translocation of the c-*myc* oncogene in B-lymphoid tumors. Structural and functional consequences. *Progress in Immunology V*, ed. Y. Yamamura and T. Tada, pp. 149–58. Academic Press, Tokyo.

Costantini, F., and Lacy, E. (1981). Introduction of a rabbit β-globin gene into mouse germ line. *Nature, Lond.* 294, 92–4.

Costello, C., and Webber, A. (1976). White cell function in Down's syndrome. *Clin. Genet.* 9, 603–5.

Cottoni-Durand, M. (1979). Contribution à l'étude de la carte factorielle du chromosome 21. Medical dissertation, Nancy.

Countryman, P.J., Heddle, J.A., and Crawford, E. (1977). The repair of X-ray-induced chromosomal damage in trisomy 21 and normal diploid lymphocytes. *Cancer Res.* 37, 52–8.

Cousineau, A.J., Higgins, J.V., Hackel, E., Waterman, D.F., Toriello, H., Carlile, P.A., and Cook, P.J.L. (1981). Cytogenetic recognition of chromosomal duplication [dup(1)(p31.4→p22.1)] and the detection of three different alleles at the PGM_1 locus. *Ann. Hum. Genet.* 45, 337–40.

Cowan, W.M. (1978) Aspects of neural development. In *International Review of Physiology. Neurophysiology III, Vol. 17*, ed. R. Porter, pp. 149–91. University Park Press, Baltimore.

Cowell, J.K. (1980). Consistent chromosome abnormalities associated with mouse bladder epithelial cell lines transformed in vitro. *J. Natl. Cancer Inst.* 65, 955–61.

Cox, D.R., and Epstein, C.J. (1985). Comparative gene mapping of human chromosome 21 and mouse chromosome 16. *Ann. N.Y. Acad. Sci.* 450, 169–77.

Cox, D.R., Epstein, L.B., and Epstein, C.J. (1980). Genes coding for sensitivity to interferon (*IfRec*) and soluble superoxide dismutase (*SOD–1*) are linked in mouse and man and map to mouse chromosome 16. *Proc. Natl. Acad. Sci. USA* 77, 2168–72.

Cox, D.R., Goldblatt, D., and Epstein, C.J. (1981). Chromosomal assignment of mouse *PRGS*: further evidence for homology between mouse chromosome 16 and human chromosome 21. *Am. J. Hum. Genet.* 33, 145A.

Cox, D.R., Tucker, G.C., and Epstein, C.J. (1980). Mouse trisomy 16 as an animal model of human trisomy 21. *Am. J. Hum. Genet.* 32, 66A.

Cox, D.R., Kawashima, H., Vora, S., and Epstein, C.J. (1983). Regional mapping of *SOD–1*, *PRGS*, and *PFK-L* on human chromosome 21: implications for the role of these genes in the pathogenesis of Down syndrome. *Am. J. Hum. Genet.* 35, 188A.

Cox, D.R., Smith, S.A., Epstein, L.B., and Epstein, C.J. (1984). Mouse trisomy 16 as an animal model of human trisomy 21 (Down syndrome): formation of viable trisomy 16 ↔ diploid mouse chimeras. *Dev. Biol.* 101, 416–24.

Cox, D.W., Teshima, I.E., Markovic, V.D., and Juberg, R. (1983). Localization of the α_1-antitrypsin (PI) locus by dosage studies in individuals with unbalanced translocations. *Am. J. Hum. Genet.* 35, 40A.

Crapper-McLachlan, D.R., and De Boni, U. (1982). Models for the study of pathological neural aging. In *Neural Aging and its Implications in Human Neurological Pathology*, ed. R.D. Terry, C.L. Bolis, and G. Toffano, pp. 61–71. Raven Press, New York.

Crews, S., Barth, R., Hood, L., Prehn, J., and Calame, K. (1982). Mouse c-*myc* oncogene is located on chromosome 15 and translocated to chromosome 12 in plasmacytomas. *Science* 218, 1319–21.

Croce, C.M., Thierfelder, W., Erikson, J., Nishikura, K., Finan, J., Lenoir, G.M., and Nowell, P. (1983). Transcriptional activation of an unrearranged and untranslocated c-*myc* oncogene by translocation of a C_λ locus in Burkitt lymphoma cells. *Proc. Natl. Acad. Sci. USA* 80, 6922–6.

Crome, L., Cowie, V., and Slater, E. (1966). A statistical note on cerebellar and brain stem weight in mongolism. *J. Ment. Defic. Res.* 10, 69–72.

Cronmiller, C., and Mintz, B. (1978). Karotypic normalcy and quasi-normalcy of developmentally totipotent mouse teratocarcinoma cells. *Dev. Biol.* 67, 465–77.

Crossen, P.E., and Morgan, W.F. (1980). Sensitivity of Down's syndrome lymphocytes to mitomycin C and X-irradiation measured by sister chromatid exchange frequency. *Cancer Genet. Cytogenet.* 2, 281–5.

Crosti, N., Serra, A., Rigo, A., and Viglino, P. (1976). Dosage effect of *SOD-A* gene in 21-trisomic cells. *Hum. Genet.* 31, 197–202.

Crosti, N., Rigo, A., Stevanato, R., Bajer, J., Neri, G., Bova, R., and Serra, A. (1981). Lack of position effect on the activity of SOD $^{Cu/Zn}$ gene in subjects with 21/D and 21/G Robertsonian translocations. *Hum. Genet.* 57, 203–4.

Crosti, N., Bajer, J., Serra, A., Rigo, A., Scarpa, M., and Viglino, P. (1983). MnSOD and CuZnSOD genes regulation in 21 trisomic human fibroblasts. *Clin. Genet.* 24, 307–8.

Cummins, H., and Midlo, C. (1961). *Finger Prints, Palms and Soles. An Introduction to Dermatolglyphics.* Dover, New York.

Cummings, M.R., All, J., and Baro, D. (1981). Altered cell growth in Down syndrome fibroblasts cultured *in vitro. Am. J. Hum. Genet.* 33, 145A.

Cunningham, G.C., Day, R.W., Berman, J.L., and Hsia, D.Y.-Y. (1969). Phenylalanine tolerance tests in families with phenylketonuria and hyperphenylalaninemia. *Am. J. Dis. Child.* 117, 626–35.

Cuoco, C., Gimelli, G., Pasquali, F., Poloni, L., Zuffardi, O., Alicata, P., Battaglino, G., Bernardi, F., Cerme, R., Cotellessa, M., Ghidoni, A., and Motta, S. (1982). Duplication of the short arm of chromosome 9. Analysis of five cases. *Hum. Genet.* 61, 3–7.

Cupples, C.G., and Tan, Y.H. (1977). Effect of human interferon preparations on lymphoblastogenesis in Down's syndrome. *Nature, Lond.* 267, 165–7.

Curry, C.J.R., Loughman, W.D., Francke, U., Hall, B.D., Golbus, M.S., Derstine, J., and Epstein, C.J. (1979). Partial trisomy for the distal long arm of chromosome 5 (region q34→qter). A new clinically recognizable syndrome. *Clin. Genet.* 15, 454–61.

Dallapiccola, B., Magnani, M., Dacha, M., and Giorgi, P.L. (1979). Confirmation of regional assignment of nucleoside phosphorylase to band 14q13 by gene dosage studies. *Hum. Genet.* 50, 341–3.

Dallapiccola, B., Brinchi, V., Magnani, M., and Dacha, M. (1980*a*). Identification of the origin of a 22p+ chromosome by triplex dosage effect of LDHB, GAPHD, TPI, and ENO2. *Ann. Génét.* 23, 111–13.

Dallapiccola, B., Lungarotti, M.S., Falorni, A., Magnani, M., and Dacha, M. (1980*b*). Evidence for the assignment of GUK1 gene locus to 1q32→q43 segment from gene dosage effect. *Ann. Génét.* 23, 83–5.

Dallapiccola, B., Lungarotti, M.S., Magnani, M., and Dacha, M. (1981). Evidence of gene dosage effect for HK1 in the red cells of a patient with trisomy 10pter→p13. *Ann. Génét.* 24, 45–7.

Dalton, T.P., Edwards, J.H., Evans, E.P., Lyon, M.F., Parkinson, S.P., Peters, J., and Searle, A.G. (1981). Chromosome maps of man and mouse. *Clin. Genet.* 20, 407–15.

Danesino, C., D'Azzo, A., Maraschio, P., and Fraccaro, M. (1978). The gene for human peptidase A is on band 18q23 and shows triplex and uniplex dosage effect. *Clin. Genet.* 14, 287.

Danesino, C., Gimelli, G., Cuoco, C., and Ciccone, M.O. (1981). Triplex gene dosage effect for β-glucuronidase and possible assignment to band q22 in a partial duplication 7q. *Hum. Genet.* 56, 371–3.

Daniel, A. (1979). Structural differences in reciprocal translocations. Potential for a model of risk in Rcp. *Hum. Genet.* 51, 171–82.

Darlington, G.J., Bernhard, H.P., and Ruddle, F.H. (1974). Human serum albumin phenotype activation in mouse hepatoma-human leukocyte cell hybrids. *Science* 185, 859–62.

Darlington, G.J., Rankin, J.K., and Schlanger, G. (1982). Expression of human hepatic genes in somatic cell hybrids. *Somat. Cell Genet.* 8, 403–12.

Darnell, J.E., Jr. (1982). Variety in the level of gene control in eukaryotic cells. *Nature, Lond.* 297, 365–71.

Davidson, E.H., and Britten, R.J. (1973). Organization, transcription and regulation in the animal genome. *Q. Rev. Biol.* 48, 565–613.

Davidson, E.H., and Britten, R.J. (1979). Regulation of gene expression: possible role of repetitive sequences. *Science* 204, 1052–9.

Davidson, E.H., and Posakony, J.W. (1982). Repetitive sequence transcripts in development. *Nature, Lond.* 297, 633–5.

Davidson, R.L. (1972). Regulation of melanin synthesis in mammalian cells: effect of gene dosage on the expression of differentiation. *Proc. Natl. Acad. Sci. USA* 69, 951–5.

Davidson, R.L. (1974). Gene expression in somatic cell hybrids. *Annu. Rev. Genet.* 8, 195–218.

Davidson, R.L,. and Benda, P. (1970). Regulation of specific functions of glial cells in somatic hybrids, II. Control of inducibility of glycerol–3-phosphate dehydrogenase. *Proc. Natl. Acad. Sci. USA* 67, 1870–7.

Davies, K.E., Harper, K., Bonthron, D., Krumlauf, R., Polkey, A., Pembrey, M.E., and Williamson, R. (1984). Use of a chromosome 21 cloned DNA probe for the analysis of non-disjunction in Down syndrome. *Hum. Genet.* 66, 54–6.

Davis, F.M., and Adelberg, E.A. (1973). Use of somatic cell hybrids for analysis of the differentiated state. *Bacteriol. Rev.* 37, 197–214.

Davis, J.R., Rogers, B.B., Hagaman, R.A., Thies, C.A., and Veomett, I.C. (1985). Balanced reciprocal translocations: risk factors for aneuploid segregant viability. *Clin. Genet.* 27, 1–19.

Davisson, M.T. (1981). Standard Q-banded karyotype. In *Genetic Variants and Strains of the Laboratory Mouse*, ed. M.C. Green, pp. 318–19. Gustav Fischer Verlag, Stuttgart and New York.

Davisson, M.T., Wright, J.E., and Atherton, L.M. (1972). Centric fusion and trisomy for the LDH-B locus in brook trout, *Salvelinus fontinalis*. *Science* 178, 992–4.

Debrot, S., and Epstein, C.J. (1985). Tetrasomy 16 in the mouse: a more severe condition than the corresponding trisomy. *J. Embryol. Exp. Morphol.*, in press.

De Clercq, E., Edy, V.G., and Cassiman, J.-J. (1975). Nonantiviral activities of interferon are not controlled by chromosome 21. *Nature, Lond.* 256, 132–4.

De Clercq, E., Edy, V.G., and Cassiman, J.-J. (1976). Chromosome 21 does not code for an interferon receptor. *Nature, Lond.* 264, 249–51.

Deisseroth, A., Barker, J., Anderson, W.F., and Nieuhuis, A. (1975). Hemoglobin synthesis in somatic cell hybrids: coexpression of mouse with human or Chinese hamster globin genes in interspecific somatic cell hybrids of mouse erythroleukemia cells. *Proc. Natl. Acad. Sci. USA* 72, 2682–6.

Dekaban, A.S., Thron, R., and Steusing, J.K. (1966). Chromosomal aberrations in irradiated blood and blood cultures of normal subjects and selected patients with chromosomal abnormality. *Radiat. Res.* 27, 50–63.

de la Chapelle, A., and Berger, R. (1984). Report of the committee on chromosome rearrangements in neoplasia and on fragile sites. *Cytogenet. Cell Genet.* 37, 274–311.

de la Chapelle, A., Vuopio, P., and Icén, A. (1976). Trisomy 8 in the bone marrow associated with high red cell glutathione reductase activity. *Blood* 47, 815–26.

de la Chapelle, A., Herva, R., Koivisto, M., and Aula, P. (1981). A deletion in chromosome 22 can cause DiGeorge syndrome. *Hum. Genet.* 57, 253–6.

de la Cruz, F.F., and Gerald, P.S. (eds.) (1981). *Trisomy 21 (Down Syndrome). Research Perspectives.* University Park Press, Baltimore.

Delahunty, D., Kling, A., and McCormack, M.K. (1982). Oxygen metabolism in Alzheimer's disease (AD). *Am. J. Hum. Genet.* 34, 49A.

del Villano, B.C., and Tischfield, J.A. (1979). A radioimmune assay for human cupro-zinc superoxide dismutase and its application to erythrocytes. *J. Immunol. Methods* 29, 253–62.

De Maggio, A.E., and Lambrukos, J. (1974). Polyploidy and gene dosage effects in peroxidase activity in ferns. *Biochem. Genet.* 12, 429–40.

Dembić, Z., Bannwarth, W., Taylor, B.A., and Steinmetz, M. (1985). The gene encoding the T-cell receptor α-chain locus maps close to the *Np–2* locus on mouse chromosome 14. *Nature, Lond.* 314, 271–3.

Denegri, J.F., Rogers, P.C.J., Chan, K.W., Sadoway, J., and Thomas, J.W. (1981). In vitro cell growth in neonates with Down's syndrome and transient myeloproliferative disorder. *Blood* 58, 675–7.

Despoisses, S., Noel, L., Choiset, A., Portnoi, M.-F., Turleau, C., Quack, B., Taillemite, J.-L., Grouchy, J. de, and Junien, C. (1984). Regional mapping of FH to band 1q42.1 by gene dosage studies. *Cytogenet. Cell Genet.* 37, 450–1.

Deuel, T.F., Huang, J.S., Huang, S.S., Stroobant, P., and Waterfield, M.D. (1983). Expression of a platelet-derived growth factor-like protein in simian sarcoma virus transformed cells. *Science* 221, 1348–50.

D'Eustachio, P. and Ruddle, F.H. (1983). Somatic cell genetics and gene families. *Science* 220, 919–24.

de Verneuil, H., Phung, N., Nordmann, Y., Allard, D., Leprince, F., Jerome, H., Aurius, A., and Rethoré, M.O. (1982). Assignment of human urophorphyrinogen I synthase locus to region 11qter by gene dosage effect. *Hum. Genet.* 60, 212–13.

Devine, E.A., Nolin, S.L., Houck, G.E. Jr., Jenkins, E.C., Miller, D.L., and Brown, W.T. (1984). Isolation and regional localization by in situ hybridization of a unique gene segment to chromosome 21. *Biochem. Biophys. Res. Commun.* 121, 380–5.

Devlin, R.H., Holm, D.G., and Grigliatti, T.A. (1982). Autosomal dosage compensation in *Drosophila melanogaster* strains trisomic for the left arm of chromosome 2. *Proc. Natl. Acad. Sci. USA* 79, 1200–4.

Dewey, M.J., Martin, D.W. Jr., Martin, G.R., and Mintz, B. (1977). Mosaic mice with teratocarcinoma-derived mutant cells deficient in hypoxanthine phosphoribosyltransferase. *Proc. Natl. Acad. Sci. USA* 74, 5564–8.

Diamond, A., Devine, J.M., and Cooper, G.M. (1984). Nucleotide sequence of a human *Blym* transforming gene activated in a Burkitt's lymphoma. *Science* 225, 516–9.

Dickinson, A.G., Fraser, H., and Bruce, M. (1979). Animal models for dementias. In *Alzheimer's Disease. Early Recognition of Potentially Reversible Deficits,* ed. A.I.M. Glen and L.J. Whalley, pp. 42–5. Churchill Livingstone, Edinburgh.

Dofuku, R., Utakoji, T., and Matsuzawa, A. (1979). Trisomy of chromosome no. 13 in spontaneous mammary tumors of GR, C3H, and noninbred Swiss mice. *J. Natl. Cancer Inst.* 63, 651–6.

Dofuku, R., Biedler, J.L., Spingler, B.A., and Old, L.J. (1975). Trisomy of chromosome 15 in spontaneous leukemia of AKR mice. *Proc. Natl. Acad. Sci. USA* 72, 1515–17.

Donnell, G.N., Bergren, W.R., and Ng, W.G. (1967). Galactosemia. *Biochem. Med.* 1, 29–53.

Downward, J., Yarden, Y., Mayes, E., Scrace, G., Totty, N., Stockwell, P., Ullrich, A., Schlessinger, J., and Waterfield, M.D. (1984). Close similarity of epidermal growth factor receptor and v-*erb*-B oncogene protein sequences. *Nature, Lond.* 307, 521–7.

Dozy, A.M., Kan, Y.W., Embury, S.H., Mentzer, W.C., Wang, W.C., Lubin, B., Davis, J.R. Jr., and Koenig, H.M. (1979). α-Globin gene organization in blacks precludes the severe form of α-thalassaemia. *Nature, Lond.* 280, 605–7.

Dracopoli, N.C., Rettig, W.J., Albino, A.P., Esposito, D., Archidiacono, N., Rocchi, M., Siniscalco, M., and Old, L.J. (1985). Genes controlling gp25/30 cell-surface molecules map to chromosomes X and Y and escape X-inactivation. *Am. J. Hum. Genet.* 37, 199–207.

Dryja, T.P., Bruns, G.A.P., Gallie, B., Petersen, R., Green, W., Rapaport, J.M., Albert, D.M., and Gerald, P.S. (1983). Low incidence of deletion of the esterase D locus in retinoblastoma patients. *Hum. Genet.* 64, 151–5.

Dryja, T.P., Cavenee, W., White, R., Rapaport, J.M., Petersen, R., Albert, D.M., and Bruns, G.A.P. (1984). Homozygosity of chromosome 13 in retinoblastoma. *N. Engl. J. Med.* 310, 550–3.

Duesberg, P.H. (1983). Retroviral transforming genes in normal cells? *Nature, Lond.* 304, 219–26.

Dumble, L., and Morris, P.J. (1975). Improved HL-A typing of cadaveric renal donors. I. Quantitation of HL-A gene dosage. *Tissue Antigens* 5, 103–7.

Dunn, H.G. (1968). The Prader-Labhart-Willi syndrome: review of the literature and report of nine cases. *Acta Paediatr. Scand.* [*Suppl.*] *186*, pp. 1–38.

Duse, M., Brugo, M.A., Martini, A., Tassi, C., Ferrario, C., and Ugazio, A.G. (1980). Immunodeficiency in Down's syndrome: low levels of serum thymic factor in trisomic children. *Thymus* 2, 127–31.

Dutrillaux, B. (1979). Chromosomal evolution in primates: tentative phylogeny from *Microcebus murinus* (Prosimian) to man. *Hum. Genet.* 48, 251–314.

Dyban, A.P. (1982). Restoration of hemopoiesis in lethally irradiated mice by hemopoietic stem cells trisomic for chromosome 19. *Doklady Akad. Nauk. USSR* 263, 454–6.

Dyban, A.P., and Baranov, D.S. (1978). *Cytogenetics of Mammalian Development*. Nauka, Moscow.

Dyggve, H., and Clausen, J. (1970). The serum immunoglobulin level in Down's syndrome. *Devel. Med. Child Neurol.* 12, 193–7.

Edelman, G.M. (1983). Cell adhesion molecules. *Science* 219, 450–7.

Edelman, G.M. (1984). Cell adhesion and morphogenesis: the regulator hypothesis. *Proc. Natl. Acad. Sci. USA* 81, 1460–4.

Edwards, J.H. (1978). Skin age in Down's syndrome: a note on the findings of Murdoch and Evans. *J. Ment. Defic. Res.* 22, 223.

Edwards, J.H., Yunckeu, C., Rushton, D.I., Richards, S., and Mittwoch, U. (1967). Three cases of triploidy in man. *Cytogenetics* 6, 81–104.

Edwards, R.G. (1983). Chromosomal abnormalities in human embryos. *Nature, Lond.* 303, 283.

Efinski, D., Duma, H., Apostolovski, B., Sofijanov, N., Ristevski, B., and Darkovski, S. (1974). Klinefelter's and Down's syndrome in an adolescent with abnormal EEG. *Clin. Genet.* 5, 81–5.

Eglitis, M.A., and Wiley, L.M. (1981). Tetraploidy and early development: effects on developmental timing and embryonic metabolism. *J. Embryol. Exp. Morphol.* 66, 91–108.

Eicher, E.M. (1967). The genetic extent of the insertion involved in the flecked translocation in the mouse. *Genetics* 55, 203–12.

Eicher, E. (1970). X-autosome translocations in the mouse: total inactivation versus partial inactivation of the X chromosome. *Adv. Genet.* 15, 175–259.

Eicher, E. (1973). Translocation trisomic mice: production by female but not male translocation carriers. *Science* 180, 31.

Eicher, E.M. and Coleman, D.L. (1977). Influence of gene duplication and X-inactivation on mouse mitochondrial malic enzyme activity and electrophoretic patterns. *Genetics* 85, 647–58.

Elder, F.F.B., Ferguson, J.W., and Lockhart, L.H. (1984). Identical twins with deletion 16q syndrome: evidence that 16q12.2-q13 is the critical band region. *Hum. Genet.* 67, 233–6.

Elder, F.F.B., Nichols, M.M., Hood, O.J., and Harrison, W.R., III. (1985). Unbalanced translocation (15;17)(q133.3) with apparent Prader-Willi syndrome but without Miller-Dieker syndrome. *Am. J. Med. Genet.* 20, 519–24.

Elder, G.H., Evans, J.O., Thomas, N., Cox, R., Brodie, M.J., Moore, M.R., Goldberg, A., and Nicholson, D.C. (1976). The primary enzyme defect in hereditary coproporphyria. *Lancet* 2, 1217–19.

Elder, G.H., Sheppard, D.M., Tovey, J.A., and Urquhart, A.J. (1983). Immunoreactive uroporphyrinogen decarboxylase in porphyria cutanea tarda. *Lancet* 1, 1301–4.

Ellingson, R.J., Eisen, J.D., and Ottersberg, G. (1973). Clinical electroencephalographic observation on institutionalized mongoloids confirmed by karyotype. *Electroenceph. Clin. Neurophysiol.* 34, 193–6.

Ellis, W.G., McCulloch, J.R., and Corley, C.L. (1974). Presenile dementia in Down's syndrome. Ultrastructural identity with Alzheimer's disease. *Neurology* 24, 101–6.

Elsas, L.J., Hillman, R.E., Patterson, J.H., and Rosenberg, L.E. (1970). Renal and intestinal hexose transport in familial glucose-galactose malabsorption. *J. Clin. Invest.* 49, 576–85.

Elul, R., Hanley, J., and Simmons, J.Q., III. (1975). Non-Gaussian behavior of the EEG in Down's syndrome suggests decreased neuronal connections. *Acta Neurol. Scand.* 51, 21–8.

Endo, S., Takagi, N., and Sasaki, M. (1982). The late-replicating X chromosome in digynous mouse triploid embryos. *Dev. Genet.* 3, 165–76.

Engel, E. (1980). A new genetic concept: uniparental disomy and its potential effect, isodisomy. *Am. J. Med. Genet.* 6, 137–43.

Engel, E., and Forbes, A.P. (1965). Cytogenetic and clinical findings in 48 patients with congenitally defective or absent ovaries. *Medicine* 44, 135–64.

Englesberg, E., and Wilcox, G. (1974). Regulation: positive control. *Annu. Rev. Genet.* 8, 219–42.

Enns, L., McCoy, E.E., and Sneddon, J.M. (1983). 5-Hydroxytryptamine (5-HT) transport in Down's syndrome blood platelets: effect of K$^+$loading. *Biochem. Pharmacol.* 32, 1793–5.

Ephrussi, B. (1972). *Hybridization of Somatic Cells.* Princeton University Press, Princeton.

Epstein, C.J. (1964). Structural and control gene defects in hereditary diseases in man. *Lancet* 2, 1066.

Epstein, C.J. (1967). Cell size, nuclear content, and the development of polyploidy in the mammalian liver. *Proc. Natl. Acad. Sci. USA* 57, 327–34.

Epstein, C.J. (1969). Mammalian oocytes: X chromosome activity. *Science* 163, 1078–9.

Epstein, C.J. (1972a). Expression of the mammalian X chromosome before and after fertilization. *Science* 175, 1467–8.

Epstein, C.J. (1972b). Inactivation of the X chromosome. *N. Engl. J. Med.* 286, 318–9.

Epstein, C.J. (1977). Inferring from modes of inheritance to the mechanisms of genetic dosage. In *Pathogenesis of Human Muscular Dystrophies*, ed. L.P. Rowland, pp. 9–22. Excerpta Medica, Amsterdam and Oxford.

Epstein, C.J. (1978). Developmental mechanisms and abnormalities: toward a developmental genetics of man. In *Birth Defects*, ed. J.W. Littlefield and J. de Grouchy, pp. 387–95. Excerpta Medica, Amsterdam and Oxford.

Epstein, C.J. (1981a). Animal models for human trisomy. In *Trisomy 21 (Down Syndrome). Research Perspectives*, ed. F.F. de la Cruz and P.S. Gerald, pp. 263–73. University Park Press, Baltimore.

Epstein, C.J. (1981b). The effects of chromosomal aneuploidy on early development: experimental approaches. In *Fertilization and Embryonic Development In Vitro*, ed. L. Mastrioni, J.D. Biggers, and W. Sadler, pp. 273–81. Plenum, New York.

Epstein, C.J. (1981c). Inactivation of the X chromosome. In *The Biology of Normal Human Growth*, ed. M. Ritzén, A. Aperia, K. Hall, A. Larsson, A. Zetterberg, and R. Zetterström, pp. 79–90. Raven Press, New York.

Epstein, C.J. (1983a). The X chromosome in development. In *Cytogenetics of the Mammalian X Chromosome, Part A: Basic Mechanisms of X Chromosome Behavior*, ed. A.A. Sandberg, pp. 51–65. A.R. Liss, New York.

Epstein, C.J. (1983b). Cellular consequences of the state of X-chromosome activity. In *Cytogenetics of the Mammalian X Chromosome, Part A: Basic Mechanisms of X Chromosome Behavior*, ed. A.A. Sandberg, pp. 341–53. A.R. Liss, New York.

Epstein, C.J. (1983c). Down's syndrome and Alzheimer's disease: implications and approaches. In *Banbury Report 15: Biological Aspects of Alzheimer's Disease*, ed. R. Katzman, pp. 169–82. Cold Spring Harbor Laboratory, Cold Spring Harbor, New York.

Epstein, C.J. (1984). Early embryonic development: normal and abnormal. In *The Impact of Protein Chemistry on the Biomedical Sciences*, ed. A.N. Schechter, J. Dean, and R.F. Goldberger, pp. 331–48. Academic Press, New York.

Epstein, C.J. (1985a). The mouse trisomies: experimental systems for the study of aneuploidy. In *Issues and Reviews in Teratology, Vol. 3*, ed. H. Kalter, pp. 171–217. Plenum, New York.

Epstein, C.J. (1985b). Mouse monosomies and trisomies as experimental systems for studying mammalian aneuploidy. *Trends Genet.* 1, 129–34.

Epstein, C.J., and Epstein, L.B. (1983). Genetic control of the response to interferon in man and mouse. *Lymphokines* 8, 277–301.

Epstein, C.J., and Gatens, E.A. (1967). Nuclear ploidy in mammalian parenchymal liver cells. *Nature, Lond.* 214, 1050–51.

Epstein, C.J., and Travis, B. (1979). Preimplantation lethality of monosomy for mouse chromosome 19. *Nature, Lond.* 280, 144–5.

Epstein, C.J., Cox, D.R., and Epstein, L.B. (1985). Mouse trisomy 16: an animal model of human trisomy 21 (Down syndrome). *Ann. N.Y. Acad. Sci.* 450, 157–68.

Epstein, C.J., Goldberger, R.F., and Anfinsen, C.B. (1963). The genetic control of tertiary protein structure: studies with model systems. *Cold Spring Harbor Symp. Quant. Biol.* 28, 439–49.

Epstein, C.J., Smith, S., and Cox, D.R. (1984). Production and properties of mouse trisomy 15 ↔ in chimeras. *Dev. Genet.* 4, 159–65.

Epstein, C.J., Martin, G.M., Schultz, A.L., and Motulsky, A.G. (1966). Werner's syndrome. A review of its symptomatology, natural history, pathologic features, genetics and relationship to the natural aging process. *Medicine* 45, 177–221.

Epstein, C.J., Tucker, G., Travis, B., and Gropp, A. (1977). Gene dosage for isocitrate dehydrogenase in mouse embryos trisomic for chromosome 1. *Nature, Lond.* 267, 615–16.

Epstein, C.J., Smith, S., Travis, B., and Tucker, G. (1978). Both X-chromosomes function before visible X-chromosome inactivation in female mouse embryos. *Nature, Lond.* 274, 500–3.

Epstein, C.J., Epstein, L.B., Cox, D.R., and Weil, J. (1981). Functional implications of gene dosage effects in trisomy 21. In *Trisomy 21. An International Symposium*, ed. G.R. Burgio, M. Fraccaro, L. Tiepolo, and U. Wolf, pp. 155–72, Springer-Verlag, Berlin.

Epstein, C.J., McManus, N.H., Epstein L.B., Branca, A.A., D'Alessandro, S.B., and Baglioni, C. (1982*a*). Direct evidence that the gene product of the human chromosome 21 locus, *IFRC*, is the interferon-α receptor. *Biochem. Biophys. Res. Commun.* 107, 1060–66.

Epstein, C.J., Smith, S.A., Zamora, T., Sawicki, J.A., Magnuson, T.R., and Cox, D.R. (1982*b*). Production of viable adult trisomy 17 ↔ diploid mouse chimeras. *Proc. Natl. Acad. Sci. USA* 79, 4376–80.

Epstein, C.J., Cox, D.R., Epstein, L.B., Smith, S.A., Yee, D., and Hofmeister, B. (1983). Immunological defects in mice with trisomy 16. *Abstracts of the 5th International Congress of Immunology*, Kyoto, Japan, August 21–27.

Epstein, C.J., Cox, D.R., Epstein, L.B., and Magnuson, T.R. (1984). Animal models for human chromosome disorders. In *Research Perspectives in Cytogenetics*, ed. R.S. Sparkes and F.F. de la Cruz, pp. 75–95. University Park Press, Baltimore.

Epstein, C.J., Hofmeister, B.G., Yee, D., Smith, S.A., Philip, R., Cox, D.R., and Epstein, L.B. (1985). Stem cell deficiencies and thymic abnormalities in fetal mouse trisomy 16. *J. Exp. Med.* 162, 695–712.

Epstein, L.B. (1977). The effects of interferons on the immune response in vitro and in vivo. In *Interferons and Their Actions*, ed. W. Stewart II, pp. 92–132. CRC Press, Cleveland.

Epstein, L.B. (1979). The comparative biology of immune and classical interferons. In *Biology of the Lymphokines*, ed. S. Cohen, E. Pick, and J.J. Oppenheim, pp. 443–514. Academic Press, New York.

Epstein, L.B. (1984). The special significance of interferon-gamma. In *Interferons*, ed. N.B. Finter, *Vol. II. Interferons and the Immune System*, ed. J. Vilcek and E. DeMaeyer, pp. 185–220. Elsevier, Amsterdam.

Epstein, L.B., and Epstein, C.J. (1976). Localization of the gene *AVG* for the antiviral expression of immune and classical interferon to the distal part of the long arm of chromosome 21. *J. Infect. Dis.* [*Suppl.*] 133, A56-A62.

Epstein, L.B., and Epstein, C.J. (1980). T-lymphocyte function and sensitivity to interferon in trisomy 21. *Cell. Immunol.* 51, 303–18.

Epstein, L.B., Cox, D.R., and Epstein, C.J. (1980). Assignment of the genes for sensitivity to interferon (*IfRec*) and soluble superoxide dismutase (*SOD–1*) to mouse chromosome 16. *Ann. N.Y. Acad. Sci.* 350, 171–3.

Epstein, L.B., Lee, S.H.S., and Epstein, C.J. (1980). Enhanced sensitivity of tri-somy 21 monocytes to the maturation-inhibiting effect of interferon. *Cell. Immunol.* 50, 191–4.

Epstein, L.B., Weil, J., Lucas, D.O., Cox, D.R., and Epstein, C.J. (1981). The biology and properties of interferon-gamma: an overview, studies of production of T lymphocyte subsets, and analysis of peptide synthesis and antiviral effects in trisomy 21 and diploid human fibroblasts. In *The Biology of the Interferon System*, ed. E. DeMaeyer, G. Galasso, and H. Schellekens, pp. 247–56. Elsevier/North Holland, Amsterdam.

Erickson, R.P. (1979). Biology of penetrance and variable expressivity. *Birth Defects: Orig. Art. Ser.* 15(5B), 3–12.

Erickson, R.P., and Goodfellow, P.N. (1984). Sharing outside of pairing. *Nature, Lond.* 311, 106–7.

Evans, D.I.K., and Steward, J.K. (1972). Down's syndrome and leukaemia. *Lancet* 2, 1322.

Evans, E.P., and Phillips, R.J.S. (1975). Inversion heterozygosity and the origin of XO daughters of *Bpa/* + female mice. *Nature, Lond.* 256, 40–1.

Evans, E.P., Beechey, C.V., and Burtenshaw, M.D. (1978). Meiosis and fertility in XYY mice. *Cytogenet. Cell Genet.* 20, 249–63.

Evans, E.P., Brown, B.B., and Burtenshaw, M.D. (1980). Personal communication. *Mouse News Letter* No. 63, 30.

Evans, E.P., Burtenshaw, M.D., and Cattanach, B.M. (1982). Meiotic crossing-over between the X and Y chromosomes of male mice carrying the sex-reversing *(Sxr)* factor. *Nature, Lond.* 300, 443–5.

Evans, E.P., Ford, C.E., and Searle, A.G. (1969). A 39,X/41,XYY mosaic mouse. *Cytogenetics* 8, 87–96.

Evans, H.J. (1983). Chromosomes and cancer: from molecules to man – an overview. In *Chromosomes and Cancer. From Molecules to Man*, ed. J.D. Rowley and J.E. Ultman, pp. 333–51. Academic Press, New York.

Evans, M.J., and Kaufman, M.H. (1981). Establishment in culture of pluripotential cells from mouse embryos. *Nature, Lond.* 292, 154–6.

Eydoux, P., Junien, C., Despoisse, S., Chassevent, J., Bibring, C., and Gregori, C. (1981). Gene dosage effect for GALT in 9p trisomy and in 9p tetrasomy with an improved technique for GALT determination. *Hum. Genet.* 57, 142–4.

Fabris, N., Amadio, L., Licastro, F., Mocchegiani, E., Zannotti, M., and Franceschi, C. (1984). Thymic hormone deficiency in normal ageing and Down's syndrome: is there a primary failure of the thymus? *Lancet* 1, 983–6.

Fallon, J.F., and Caplan, A.I. (eds.) (1982). *Limb Development and Regeneration, Part A*. A.R. Liss, New York.

Farber, R.A. (1973). Gene dosage and the expression of electrophoretic patterns in heteroploid mouse cell lines. *Genetics* 74, 521–31.

Fearon, E.R., Vogelstein, B., and Feinberg, A.P. (1984). Somatic deletion and duplication of genes on chromosome 11 in Wilm's tumors. *Nature, Lond.* 309, 176–8.

Feaster, W.W., Kwok, L.W., and Epstein, C.J. (1977). Dosage effects for superoxide dismutase-1 in nucleated cells aneuploid for chromosome 21. *Am. J. Hum. Genet.* 29, 563–70.

Fekete, G., Kulcsár, G., Dán, P., Nász, I., Schuler, D., and Dobos, M. (1982). Immunological and virological investigations in Down's syndrome. *Eur. J. Pediatr.* 138, 59–62.

Felsher, B.F., Carpio, N.M., Engleking, D.W., and Nunn, A.T. (1982). Decreased hepatic uroporphyrinogen decarboxylase activity in porphyria cutanea tarda. *N. Engl. J. Med.* 306, 766–9.

Ferguson-Smith, M.A. (1965). Karyotype-phenotype correlation in gonadal dysgenesis and their bearing on the pathogenesis of malformations. *J. Med. Genet.* 2, 142–55.

Ferguson-Smith, M.A., and Aitken, D.A. (1982). The contribution of chromosome aberrations to the precision of human gene mapping. *Cytogenet. Cell Genet.* 32, 24–42.

Ferguson-Smith, M.A., Newman, B.F., Ellis, P.M., Thomson, D.M.G., and Riley, I.D. (1973). Assignment by deletion of human red cell acid phosphatase gene locus to the short arm of chromosme 2. *Nature New Biol.* 243, 271–4.

Ferguson-Smith, M.A., Aitken, D.A., Turleau, C., and Grouchy, J. de (1976). Localization of the human *ABO:Np–1:Ak–1* linkage group by regional assignment of *Ak–1* to 9q34. *Hum. Genet.* 34, 35–43.

Fialkow, P.J., Reddy, A.L., and Bryant, J.I. (1980). Clonal origin and trisomy of chromosome 15 in murine B-cell malignancies. *Int. J. Cancer* 26, 603–8.

Fialkow, P.J., Lisker, R., Giblett, E.R., and Zavala, C. (1970). Xg locus: failure to detect inactivation in females with chronic myelocytic leukaemia. *Nature, Lond.* 266, 367–8.

Filmus, J., Pollack, M.N., Cailleau, R., and Buick, R.N. (1985). MDA–468, a human breast cancer cell line with a high number of epidermal growth facter (EGF) receptors, has an amplified EGF receptor gene and is growth inhibited by EGF. *Biochem. Biophys. Res. Commun.* 128, 898–905.

Fineman, R.M., Ablow, R.C., Breg, W.R., Wing, S.D., Rose, J.S., Rothman, S.L.G., and Warpinski, J. (1979). Complete and partial trisomy of different segments of chromosome 8: case reports and review. *Clin. Genet.* 16, 390–8.

Finnell, R.H., and Chernoff, G.F. (1984). Variable patterns of malformation in the mouse fetal hydantoin syndrome. *Am. J. Med. Genet.* 19, 463–71.

First International Workshop on Chromosomes in Leukemia (1978). Chromosomes in Ph¹-positive chronic granulocytic leukemia. Chromosomes in acute non-lymphocytic leukemia. *Br. J. Haematol.* 39, 305–16.

Fisch, R.O., Doeden, D., Lansky, L.L., and Anderson, J.A. (1969). Maternal phenylketonuria. Detrimental effects on embryogenesis and fetal development. *Am. J. Dis. Child.* 118, 847–58.

Fitch, N. (1978). Partial trisomy 6. *Clin. Genet.* 14, 181–5.

Fitzgerald, P.H., Donald, R.A., and McCormick, P. (1984). Reduced fertility in women with X chromosome abnormality. *Clin. Genet.* 25, 301–9.

Flint, H.J., Tateson, R.W., Barthelmess, I.B., Porteous, D.J., Donachie, W.D., and Kacser, H. (1981). Control of the flux in the arginine pathway of *Neurospora crassa*. Modulation of enzyme activity and concentration. *Biochem. J.* 200, 231–46.

Fobes, J.F. (1980). Trisomic analysis of isozymic loci in tomato species: segregation and dosage effects. *Biochem. Genet.* 18, 401–21.

Fogel, S., and Welch, J.W. (1982). Tandem gene amplification mediates copper resistance in yeast. *Proc. Natl. Acad. Sci. USA* 79, 5342–6.

Folsom, R.C., Widen, J.E., and Wilson, W.R. (1983). Auditory brain-stem responses in infants with Down's syndrome. *Arch. Otolaryngol.* 109, 607–10.

Ford, C.E. (1981). Nondisjunction. In *Trisomy 21. An International Symposium*, ed. G.R. Burgio, M. Fraccaro, L. Tiepolo, and U. Wolf, pp. 103–43. Springer-Verlag, Berlin.

Ford, C.E., Jones, K.W., Miller, O.J., Mittwoch, U., Penrose, L.S., Ridler, M., and Shapiro, A. (1959*a*). The chromosomes in a patient showing both mongolism and the Klinefelter syndrome. *Lancet* 1, 709–10.

Ford, C.E., Jones, K.W., Polani, P.E., de Almeida, J.C., and Briggs, J.H. (1959*b*). A sex-chromosome anomaly in a case of gonadal dysgenesis (Turner's syndrome). *Lancet* 1, 711–3.

Ford, R.C., and Berman, J.L. (1977). Phenylalanine metabolism and intellectual functioning among carriers of phenylketonuria and hyperphenylalaninaemia. *Lancet* 1, 767–71.

Forejt, J., Capková, J., and Gregorová, S. (1980). *T(16;17)43H* translocation as a tool in analysis of the proximal part of chromosome 17 (including *T-t* gene complex) of the mouse. *Genet. Res.* 35, 165–77.

Fougère, C., and Weiss, M.C. (1978). Phenotypic exclusion in mouse melanoma-rat hepatoma hybrid cells: pigment and albumin production are not reexpressed simultaneously. *Cell* 15, 843–54.

Fougère, C., Ruiz, F., and Ephrussi, B. (1972). Gene dosage dependence of pigment synthesis in melanoma × fibroblast hybrids. *Proc. Natl. Acad. Sci. USA* 69, 330–4.

Fourth International Workshop on Chromosomes in Leukemia (1984). A prospective study of acute nonlymphocytic leukemia. *Cancer Genet. Cytogenet.* 11, 249–360.

Fowler, C.J., Wiberg, A., Gustavson, K.-H., and Winblad, B. (1981). Platelet monoamine oxidase activity in Down's syndrome. *Clin. Genet.* 19, 307–11.

Fowler, I., and Hollingsworth, D.R. (1973). Response to stimulation *in vitro* of lymphocytes from patients with Down's syndrome. *Proc. Soc. Exp. Biol. Med.* 144, 475–7.

Fraccaro, M., Zuffardi, O., Bühler, E., Schinzel, A., Simoni, G., Witkowski, R., Bonifaci, E., Caufin, D., Cignacco, G., Delendi, N., Gargantini, L., Losanowa, T., Marca, L., Ullrich, E., and Vigi, V. (1983). Deficiency, transposition, and duplication of one 15q region may be alternatively associated with Prader-Willi (or a similar) syndrome. Analysis of seven cases after varying ascertainment. *Hum. Genet.* 64, 388–94.

Fragoso, R., Hernandez, A., Plascencia, M.L., Nazara, Z., Martinez, Y., Martinez, R., and Cantú, J.M. (1982). 49,XXXXX syndrome. *Ann. Génét.* 25, 145–8.

Franceschi, C., Licastro, F., Paolucci, P., Masi, M., Cavicchi, S., and Zannotti, M. (1978). T and B lymphocyte subpopulations in Down's syndrome. A study of non-institutionalized subjects. *J. Ment. Defic. Res.* 22, 179–91.

Franceschi, C., Licastro, F., Chiricolo, M., Bonetti, F., Zannotti, M., Fabris, N., Mocchegiani, E., Fantini, M.P., Paolucci, P., and Masi, M. (1981). Deficiency of autologous mixed lymphocyte reactions and serum thymic factor level in Down's syndrome. *J. Immunol.* 126, 2161–4.

Francke, U. (1977). Abnormalities of chromosome 11 and 20. In *New Chromosomal Syndromes*, ed. J.J. Yunis, pp. 245–72. Academic Press, New York.

Francke, U. (1978). Clinical syndromes associated with partial duplications of chromosomes 2 and 3: dup(2p), dup(2q), dup(3p), dup(3q). *Birth Defects: Orig. Art. Ser.* 14(6C), 191–217.

Francke, U., and Taggart, R.T. (1979). Assignment of the gene for cytoplasmic superoxide dismutase (SOD–1) to a region of the chromosome 16 and of Hprt to a region of the X chromosome in the mouse. *Proc. Natl. Acad. Sci. USA* 76, 5230–3.

Francke, U., Holmes, L.B., Atkins, L., and Riccardi, V.M. (1979). Aniridia-Wilm's tumor association: evidence for specific deletion of 11p13. *Cytogenet. Cell Genet.* 24, 185–92.

Francke, U., de Martinville, B., D'Eustachio, P., and Ruddle, F.H. (1982). Comparative gene mapping: murine lambda light chains are located in region cen → B5 of mouse chromosome 16 not homologous to human chromosome 21. *Cytogenet. Cell Genet.* 33, 267–71.

Frankhauser, G. (1945). The effect of changes in chromosome number on amphibian development. *Q. Rev. Biol.* 20, 20–78.

Frants, R.R., Ericksson, A.W., Jongbloet, P.H., and Hammers, A.J. (1975). Superoxide dismutase in Down syndrome. *Lancet* 2, 42–3.

Fraser, F.C. (1976). The multifactorial/threshold concept – uses and misuses. *Teratology* 14, 267–80.

Fraser, F.C., and Sadovnick, A.D. (1976). Correlation of IQ in subjects with Down's syndrome and their parents and sibs. *J. Ment. Defic. Res.* 20, 179–82.

Fraser, J., and Mitchell, A. (1876). Kalmuc idiocy: report of a case with autopsy; with notes on sixty-two cases. *J. Ment. Sci.* 22, 169–79.

Frecker, M., Dallaire, L., Young, S.R., Chen, G.C.C., and Simpson, N.E. (1978). Confirmation of regional assignment of nucleoside phosphorylase (NP) on chromosome 14 by gene dosage studies. *Hum. Genet.* 45, 167–73.

Freeman, B.A., Young, S.L., and Crapo, J.D. (1983). Liposome-mediated augmentation of superoxide dismutase in endothelial cells prevents oxygen injury. *J. Biol. Chem.* 258, 12534–42.

Freeman, B.A., Turrens, J.F., Mirza, Z., Crapo, J.D., and Young, S.L. (1985). Modulation of oxidant lung injury by using liposome-entrapped superoxide dismutase and catalase. *Fed. Proc.* 44, 2591–5.

Frels, W.I., and Chapman, V.M. (1979). Paternal X chromosome expression in extraembryonic membranes of XO mice. *J. Exp. Zool.* 210, 553–60.

Frels, W.J., Bluestone, J.A., Hodes, R.J., Capecchi, M.R., and Singer, D.S. (1985). Expression of a microinjected porcine class I major histocompatibility complex gene in transgenic mice. *Science* 228, 577–80.

Frias, S., and Carnevale, A. (1983). Cell cycle in normal individuals and in patients with Down, cri-du-chat and Turner syndromes. *Ann. Génét.* 26, 60–2.

Fridovich, I. (1975). Superoxide dismutases. *Annu. Rev. Biochem.* 44, 147–59.

Fried, K., Tieder, M., Beer, S., Rosenblatt, M., and Krespin, M.I. (1977). Mental retardation with 45 chromosomes, 45,XX, − 5, − 14, + der(5), t(5;14)(p15;q13) not due to familial reciprocal translocation. *J. Med. Genet.* 14, 68–72.

Frischer, H., Chu, L.K., Ahmad, T., Justice, P., and Smith, G.F. (1981). Superoxide dismutase and glutathione peroxidase abnormalities in erythrocytes and lymphoid cells in Down syndrome. In *The Red Cell: Fifth Ann Arbor Conference*, ed. G.J. Brewer, pp. 269–83. A.L. Liss, New York.

Fryns, J.R., Jaeken, J., and van den Berghe, H. (1979). Parital trisomy 22q with elevated arylsulfatase-A activity. *Ann. Génét.* 22, 168–70.

Fryns, J.P., Kleczkowska, A., and van den Berghe, H. (1983). The X chromosome and sexual development: clinical aspects. In *Cytogenetics of the Mammalian X Chromosome, Part B: X Chromosome Anomalies and their Clinical Manifestations*, ed. A.A. Sandberg, pp. 115–26. A.R. Liss, New York.

Fryns, J.P., Petit, P., and van den Berghe, H. (1981). The various phenotypes in Xp deletion. Observations in eleven patients. *Hum. Genet.* 57, 385–7.

Fryns, J.P., Vinken, L., Geutjens, J., Marien, J., Deroover, J., and van den Berghe, H. (1980). Triploid-diploid mosaicism in a deeply mentally retarded adult. *Ann. Génét.* 23, 232–4.

Fryns, J.P., Petit, P., Vinken, J., Geutjens, J., Marien, J., and van den Berghe, H.

(1982). Mosaic tetrasomy 21 in severe mental handicap. *Eur. J. Pediatr.* 139, 87–9.

Fryns, J.P., Heremans, G., Marien, J., and van den Berghe, H. (1983*a*). Langer-Giedion syndrome and deletion of the long arm of chromosome 8. Confirmation of the critical segment to 8q23. *Hum. Genet.* 64, 194–5.

Fryns, J.P., Kleczkowska, A., Petit, P., and van den Bergh, H. (1983*b*). X-chromosome polysomy in the female: personal experience and review of the literature. *Clin. Genet.* 23, 341–9.

Fryns, J.P., Petit, P., Jaeken, J., Eggermont, E., Kleczkowska, A., and van den Berghe, H. (1983*c*). Partial distal 6p trisomy: a clinical entity. *Ann. Génét.* 26, 50–2.

Fujii, J.T., and Martin, G.R. (1983). Developmental potential of teratocarcinoma stem cells *in utero* following aggregation with cleavage-stage mouse embryos. *J. Embryol. Exp. Morphol.* 74, 79–96.

Fujimoto, A., Brown, D.L., Shinno, N.W., and Wilson, M.G. (1983). Nonimmune fetal hydrops and Down syndrome. *Am. J. Med. Genet.* 14, 533–7.

Fuks, A., and Guttmann, R.D. (1983). Genetic analysis of reactivities of allospecific monoclonal antibodies to rat histocompatability antigens. *Transplantation* 36, 568–71.

Fukushima, Y., Kuroki, Y., and Izawa, T. (1983). Two cases of the Langer-Giedion syndrome with the same interstitial deletion of the long arm of chromosome 8: 46,XY or XX,del(8)(q23.3q24.13). *Hum. Genet.* 64, 90–3.

Funa, K., Annerén, G., Alm, G.V., and Björkstén, B. (1984). Abnormal interferon production and NK cell responses to interferon in children with Down's syndrome. *Clin. Exp. Immunol.* 56, 493–500.

Fundele, R., Bücher, T., Gropp., A. and Winking, H. (1981). Enzyme patterns in trisomy 19 of the mouse. *Dev. Genet.* 2, 291–303.

Funderburk, S.J., Sparkes, R.S., and Sparkes, M.C. (1983). Trisomy 20p due to a paternal reciprocal translocation. *Ann. Génét.* 26, 94–7.

Gagnon, J., Katyk-Longten, N., de Groot, J.A., and Barbeau, A. (1961). Double trisomie autosomique à 48 chromosomes (21 + 18). *Union Med. Can.* 90, 1220–6.

Gahl, W.A., Bashan, N., Tietze, F., Bernardini, I., and Schulman, J.D. (1982). Cystine transport is defective in isolated leukocyte lysosomes from patients with cystinosis. *Science* 217, 1263–5.

Gahl, W.A., Bashan, N., Tietze, F., and Schulman, J.D. (1984). Lysosomal cystine counter-transport in heterozygotes for cystinosis. *Am. J. Hum. Genet.* 36, 277–82.

Gahmberg, C.G., Andersson, L.C., Ruutu, P., Timonen, T.T.T., Hänninen, A., Vuopio, P., and de la Chapelle, A. (1979). Decrease of the major high molecular weight surface glycoproteins of human granulocytes in monosomy 7 associated with defective chemotaxis. *Blood* 54, 401–6.

Gale, R.E., Clegg, J.B., and Huehns, E.R. (1979). Human embryonic haemoglobins Gower 1 and Gower 2. *Nature, Lond.* 280, 162–4.

Gallatin, W.M., Weissman, I.L., and Butcher, E.C. (1983). A cell-surface molecule involved in organ-specific homing of lymphocytes. *Nature, Lond.* 304, 30–4.

Garber, P., Sinet, P.M., Jerome, H., and Lejeune, J. (1979). Copper/zinc superoxide dismutase activity in trisomy 21 by translocation. *Lancet* 2, 914–15.

Garcia-Cruz, D., Vaca, G., Ibarra, B., Sánchez-Corona, J., Ocampo-Campus, R., Peregrina, S., Moller, M., Rivera, H., Rivas, F., González-Angulo, A., and

Cantú, J.M. (1982). Tetrasomy 9p: clinical aspects and enzymatic gene dosage expression. *Ann. Génét.* 25, 237–42.

Gardner, R.J.M., Rudd, N.L., Stevens, L.J., and Worton, R.G. (1978). Autosomal imbalance with a near-normal phenotype: the small effect of trisomy for the short arm of chomosome 18. *Birth Defects: Orig. Art. Ser.* 14(6C), 359–63.

Gardner, R.L. (1968). Mouse chimeras obtained by the injection of cells into the blastocyst. *Nature, Lond.* 220, 596–7.

Gardner, R.L., and Lyon, M.F. (1971). X chromosome inactivation studied by injection of a single cell into the mouse blastocyst. *Nature, Lond.* 231, 385–6.

Gartler, S.M., and Rivest, M. (1983). Evidence for X-linkage of steriod sulfatase in the mouse: steroid sulfatase levels in oocytes of XX and XO mice. *Genetics* 103, 137–41.

Gartler, S.M., and Sparkes, R.S. (1963). The Lyon-Beutler hypothesis and isochromosome X patients with the Turner syndrome. *Lancet* 2, 411.

Gartler, S., Andina, R., and Gant, N. (1975). Ontogeny of X-chromosome inactivation in the female germ line. *Exp. Cell Res.* 91, 454–7.

Gartler, S.M., Rivest, M., and Cole, R.E. (1980). Cytological evidence for an inactive X chromosome in murine oogonia. *Cytogenet. Cell Genet.* 28, 203–7.

Gartler, S.M., Liskay, R.M., Campbell, R.K., Sparkes, R., and Gant, N. (1972). Evidence for two functional X chromosomes in human oocytes. *Cell Differ.* 1, 215–8.

Garver, K.L., Cioca, A.M., and Turack, N.A. (1976). Partial monosomy or trisomy resulting from crossing over with a rearranged chromosome 1. *Clin. Genet.* 10, 319–24.

Gaudin, D. (1973). Some thoughts on a possible relationship between known gene dosage effects and neoplastic transformation. *J. Theor. Biol.* 41, 191–200.

Gaze, R.M., French, V., Snow, M., and Summerbell (eds.) (1981). *Growth and the Development of Pattern. J. Embryol. Exp. Morphol.* 65 (Supplement).

Gearhart, J., Oster-Granite, M.L., and Hatzidimitriou, G. (1985). Placental development in the trisomy 16 mouse. *Pediatr. Res.* 19, 325A.

Gearhart, J., Singer, H., Tiemeyer, M., and Coyle, J. (1983). Neurochemical studies of the trisomy 16 mouse. *Am. J. Hum. Genet.* 35, 164A.

Gelehrter, T.D. (1976). Enzyme induction. *N. Engl. J. Med.* 294, 522–6, 589–95, 646–51.

George, D.L., and Francke, U. (1976a). Gene dose effect: regional mapping of human glutathione reductase on chromosome 8. *Cytogenet. Cell Genet.* 17, 282–6.

George, D.L., and Francke, U. (1976b). Gene dosage effect: regional mapping of human nucleoside phosphorylase on chromosome 14. *Science* 194, 851–2.

German, J. (ed.) (1983). *Chromosome Mutation and Neoplasia.* A.R. Liss, New York.

Gershwin, M.E., Crinella, F.M., Castles, J.J., and Trent, J.K.T. (1977). Immunologic characteristics of Down's syndrome. *J. Ment. Defic. Res.* 21, 237–49.

Gibson, J., Ellis, P.M., and Forsyth, J.S. (1982). Interstitial deletion of chromosome 7: a case report and review of the literature. *Clin. Genet.* 22, 256–65.

Gilbert, E.F., and Opitz, J.M. (1982). Developmental and other pathological changes in syndromes caused by chromosome abnormalities. *Perspect. Pediatr. Pathol.* 7, 1–63.

Gilbert, F. (1983). Chromosomes, genes, and cancer: a classification of chromosome abnormalities in cancer. *J. Natl. Cancer Inst.* 71, 1107–14.

Gilgenkrantz, S., Defeche, C., Stehlin, S., and Gregoire, M.J. (1981). Proximal trisomy 13. A family with balanced reciprocal translocations t(8;13) in seven members and Robertsonian translocation t(13;14) in three members. *Hum. Genet.* 58, 436–40.

Gilgenkrantz, S., Vigneron, C., Gregoire, M.J., Pernot, C., and Raspiller, A. (1982). Association of del(11)(p15.1p12), aniridia, catalase deficiency, and cardiomyopathy. *Am. J. Med. Genet.* 13, 39–49.

Gilles, L., Ferradini, C., Foos, J., and Pucheault, J. (1976). The estimation of red cell superoxide dismutase activity by pulse radiolysis in normal and trisomic 21 subjects. *FEBS Lett.* 69, 55–8.

Gilmore, D.H., Boyd, E., McClure, J.P., Batstone, P., and Connor, J.M. (1984). Inv dup(15) with mental retardation but few dysmorphic features. *J. Med. Genet.* 21, 221–3.

Gladstone, P., Fueresz, L., and Pious, D. (1982). Gene dosage and gene expression in the *HLA* region: evidence from deletion variants. *Proc. Natl. Acad. Sci. USA* 79, 1235–9.

Glasgow, A.M., Kraegel, J.H., and Schulman, J.D. (1978). Studies of the cause and treatment of hyperammonemia in females with ornithine transcarbamylase deficiency. *Pediatrics* 62, 30–7.

Glenister, T.W. (1956). Determination of sex in early human embryos. *Nature, Lond.* 177, 1135–6.

Glenner, G.G., and Wong, C.W. (1984). Alzheimer's disease and Down's syndrome: sharing of a unique cerebrovascular amyloid fibril protein. *Biochem. Biophys. Res. Commun.* 122, 1131–5.

Gliddon, J.B., Busk, J., and Galbraith, G.C. (1975). Visual evoked responses as a function of light intensity in Down's syndrome and non-retarded subjects. *Psychophysiology* 12, 416–22.

Gluecksohn-Waelsch, S. (1979). Genetic control of morphogenetic and biochemical differentiation: lethal albino deletions in the mouse. *Cell* 16, 225–37.

Godbout, R., Dryja, T.P., Squire, J., Gallie, B.L., and Phillips, R.A. (1983). Somatic inactivation of genes on chromosome 13 is a common event in retinoblastoma. *Nature, Lond.* 304, 451–3.

Golbus, M.S. (1981). Chromosome aberrations and mammalian reproduction. In *Fertilization and Embryonic Development In Vitro*, ed. L. Mastroianni, Jr., and J.D. Biggers, pp. 257–72. Plenum, New York.

Golbus, M.S., Bachman, R., Wiltse, S., and Hall, B.D. (1976). Tetraploidy in a liveborn infant. *J. Med. Genet.* 13, 329–32.

Goldberger, R.F. (1974). Autogenous regulation of gene expression. *Science* 183, 810–16.

Golden, W., and Pashayan, H.M. (1976). The effect of parental education on the eventual mental development of noninstitutionalized children with Down syndrome. *J. Pediatr.* 89, 603–5.

Goldfeld, A.E., Rubin, C.S., Siegel, T.W., Shaw, P.A., Schiffer, S.G., and Gluecksohn-Waelsch, S. (1981). Genetic control of insulin receptors. *Proc. Natl. Acad, Sci. USA* 78, 6359–61.

Goldfeld, A.E., Firestone, G.L., Shaw, P.A., and Gluecksohn-Waelsch, S. (1983). Recessive lethal deletion in mouse chromosome 7 affects glucocorticoid receptor binding activities. *Proc. Natl. Acad, Sci. USA* 80, 1431–4.

Goldman, B., Polani, P.E., Daker, M.G., and Angell, R.R. (1982). Clinical and cytogenetic aspects of X-chromosome deletions. *Clin. Genet.* 21, 36–52.

Goldstein, J.L., and Brown, M.S. (1975). Familial hypercholesterolemia. A genetic regulatory defect in cholesterol metabolism. *Am. J. Med.* 58, 147–50.

Goldstein, J.L., Kita, T., and Brown, M.S. (1983). Defective lipoprotein receptors and atherosclerosis. Lessons from an animal counterpart of familial hypercholesterolemia. *N. Engl. J. Med.* 309, 288–96.

Goldstein, J.L., Sabhani, M.K., Faust, J.R., and Brown, M.S. (1976). Heterozygous familial hypercholesterolemia: failure of normal allele to compensate for mutant allele at a regulated genetic locus. *Cell* 9, 195–203.

Goodfellow, P.N., and Tippett, P. (1981). A human quantitative polymorphism related to Xg blood groups. *Nature, Lond.* 289, 404–5.

Goodfellow, P., Banting, G., Sheer, D., Ropers, H.H., Caine, A., Ferguson-Smith, M.A., Povey, S., and Voss, R. (1983). Genetic evidence that a Y-linked gene in man is homologous to a gene on the X chromosome. *Nature, Lond.* 302, 346–9.

Goodfellow, P., Pym, B., Mohandas, T., and Shapiro, L.J. (1984). The cell surface antigen locus, *MIC2X*, escapes X-inactivation. *Am. J. Hum. Genet.* 36, 777–82.

Goossens, M., Dozy, A.M., Embury, S.H., Zachariades, Z., Hadjiminas, M.G., Stamatoyannopoulos, G., and Kan, Y.W. (1980). Triplicated α-globin loci in humans. *Proc. Natl. Acad. Sci. USA* 77, 518–21.

Gopalakrishnan, T.V. (1984). DNA-mediated restoration of phenylalanine hydroxylase gene expression in enzyme-deficient derivatives of enzyme-constitutive mouse cell hybrids. *Somat. Cell Mol. Genet.* 10, 3–16.

Gopalakrishnan, T.V., and Anderson, W.F. (1979). Epigenetic activation of phenylalanine hydroxylase in mouse erythroleukemia cells by the cytoplast of rat hepatoma cells. *Proc. Natl. Acad. Sci. USA* 76, 3932–6.

Gopalakrishnan, T.V., and Littlefield, J.W. (1983). RNA from rat hepatoma cells can activate the phenylalanine hydroxylase gene of mouse erythroleukemia cells. *Somat. Cell Genet.* 9, 121–31.

Gopalakrishnan, T.V., Thompson, E.R., and Anderson, W.F. (1977). Extinction of hemoglobin inducibility in Friend erythroleukemia cells by fusion with cytoplasm of enucleated mouse neuroblastoma or fibroblast cells. *Proc. Natl. Acad. Sci. USA* 74, 1642–6.

Gordon, J.W. (1983). Transgenic mice: a new and powerful experimental tool in mammalian developmental genetics. *Dev. Genet.* 4, 1–20.

Gordon, J.W., and Ruddle, F.H. (1981). Integration and stable germ line transmission of genes injected into mouse pronuclei. *Science* 214, 1244–6.

Gordon, M.C., Sinha, S.K., and Carlson, S.D. (1971). Antibody responses to influenza vaccine in patients with Down's syndrome. *Am. J. Ment. Defic.* 75, 391–9.

Gorlin, J.G., Červenka, J., Bloom, B.A., and Langer, L.O., Jr. (1982). No chromosome deletion found in prometaphase banding in two cases of Langer-Giedion syndrome. *Am. J. Med. Genet.* 13, 345–7.

Gosden, C.M., Wright, M.O., Paterson, W.G., and Grant, K.A. (1976). Clinical details, cytogenetic studies and cellular physiology of a 69,XXX fetus, with comments on the biological effect of triploidy in man. *J. Med. Genet.* 13, 371–80.

Gottlieb, D.I., and Glaser, L. (1980). Cellular recognition during neural development. *Annu. Rev. Neurosci.* 3, 303–18.

Graham, G.J., Hall, T.J., and Cummings, M.R. (1984). Isolation of repetitive DNA sequences from human chromosome 21. *Am. J. Hum. Genet.* 36, 25–35.

Graham, J.B., Barrow, E.S., Reisner, H.M., and Edgell, C.-J.S. (1983). The genetics of blood coagulation. *Adv. Hum. Genet.* 13, 1–81.

Graham, J.M., Jr., Hoehn, H., Lin, M.S., and Smith, D.W. (1981). Diploid-triploid mixoploidy: clinical and cytogenetic aspects. *Pediatrics* 68, 23–8.

Gray, S.W., and Skandalakis, J.E. (1972). *Embryology for Surgeons.* W.B. Saunders, Philadelphia.

Green, H., and Thomas, J. (1978). Pattern formation by cultured human epidermal cells: development of curved ridges resembling dermatoglyphs. *Science* 200, 1385–8.

Green, M.C. (1975). The laboratory mouse, *Mus musculus*. In *Handbook of Genetics, Vol. 4*, ed. R.C. King, pp. 203–41. Plenum Press, New York.

Green, M.C. (1981). Catalog of mutant genes and polymorphic loci. In *Genetic Variants and Strains of the Laboratory Mouse*, ed. M.C. Green, pp. 8–278. Gustav Fischer Verlag, Stuttgart and New York.

Green, T.R., Schaefer, R.E., and Makler, M.T. (1980). Orientation of the NADPH-dependent superoxide generating oxidoreductase on the outer membrane of human PMN's. *Biochem. Biophys. Res. Commun.* 94, 262–9.

Greenberg, F., Crowder, W.E., Paschall, V., Colon-Linares, J., Lubianski, B., and Ledbetter, D.H. (1984). Familial DiGeorge syndrome and associated partial monosomy of chromosome 22. *Hum. Genet.* 65, 317–19.

Greene, E.L., Shenker, I.R., and Karelitz, S. (1968). Serum protein fractions in patients with Down's syndrome (mongolism). The influence of age. *Am. J. Dis. Child.* 115, 599–602.

Gregorová, S., Baranov, V.S., and Forejt, J. (1981). Partial trisomy (including T-t gene complex) of the chromosome 17 of the mouse. The effect on male fertility and the transmission to progeny. *Folia Biol. (Praha)* 27, 171–7.

Gregory, L., Williams, R., and Thompson, E. (1972). Leukocyte function in Down's syndrome and acute leukaemia. *Lancet* 1, 1359–61.

Grell, E. (1962). The dose effect of *ma.l+* and *ry+* on the xanthine dehydrogenase activity in *Drosophila melanogaster*. *Z. Vererb.* 93, 371–7.

Griffin, R.F., and Elsas, L.J. (1975). Classic phenylketonuria: diagnosis through heterozygote detection. *J. Pediatr.* 86, 512–17.

Griffiths, A.W., and Sylvester, P.E. (1967). Mongols and non-mongols compared in their response to active tetanus immunisation. *J. Ment. Defic. Res.* 11, 263–6.

Griffiths, A.W., Sylvester, P.E., and Baylis, E.M. (1969). Serum globulins and infection in mongolism. *J. Clin. Pathol.* 22, 76–8.

Gripenberg, U., Hongell, K., Knuutila, S., Kähkönen, M., and Leisti, J. (1980). A chromosome survey of 1062 mentally retarded patients. Evaluation of a long-term study at the Rinnekoti Institution, Finland. *Hereditas* 92, 223–8.

Grohé, G., and Gropp, A. (1980). Personal communication. *Mouse News Letter* No. 63, 22.

Groner, Y., Lieman-Hurwitz, J., Dafni, N., Sherman, L., Levanon, D., Bernstein, Y., Danciger, E., and Elroy-Stein, O. (1985). Molecular structure and expression of the gene locus on chromosome 21 encoding the Cu/Zn superoxide dismutase and its relevance to Down syndrome. *Ann. N.Y. Acad. Sci.* 450, 133–156.

Gropp, A. (1978). Relevance of phases of development for expression of abnormality. Perspectives drawn from experimentally induced chromosome aberrations. In *Abnormal Fetal Growth: Biological Bases and Consequences*, ed. F. Naftolin, pp. 85–100. Dahlem Konferenzen, Berlin.

Gropp, A. (1981). Clinical and experimental pathology of fetal wastage. In *Human Reproduction. Proceedings of III World Congress*, ed. K. Semm and L. Mettler, pp. 208–16. Excerpta Medica, Amsterdam.

Gropp, A. (1982). Value of an animal model for trisomy. *Virchows Arch. Pathol. Anat.* 395, 117–31.

Gropp, A. and Grohé, G. (1981). Strain background dependance of expression of chromosome triplication in the mouse embryo. *Hereditas* 94, 7–8.

Gropp, A., and Winking, H. (1981). Robertsonian translocations: cytology, meiosis, segregation patterns and biological consequences of heterozygosity. *Zool. Soc. London Symp.* 47, 141–81.

Gropp, A., Giers, D., and Kolbus, U. (1974). Trisomy in the fetal backcross progeny of male and female metacentric heterozygotes of the mouse. I. *Cytogenet. Cell Genet.* 13, 511–35.

Gropp, A., Kolbus, U., and Giers, D. (1975). Systematic approach to the study of trisomy in the mouse. II. *Cytogenet. Cell Genet.* 14, 42–62.

Gropp, A., Putz, B., and Zimmerman, U. (1978). Autosomal monosomy and trisomy causing developmental failure. *Curr. Topics Pathol.* 62, 177–92.

Gropp, A., Winking, H., Herbst, E.W., and Claussen, C.P. (1983). Murine trisomy: developmental profiles of the embryo, and isolation of trisomic cellular systems. *J. Exp. Zool.* 228, 253–69.

Gropp, D., Gropp, A., and Winking, H. (1981). Personal communication. *Mouse News Letter* No. 64, 70.

Gropp, D., Winking, H., and Gropp, A. (1981). Personal communication. *Mouse News Letter* No. 65, 32.

Grosschedl, R., Weaver, D., Baltimore, D., and Costantini, F. (1984). Introduction of a μ immunoglobulin gene into the mouse germ line: specific expression in lymphoid cells and synthesis of functional antibody. *Cell* 38, 647–58.

Grosse, K.-P., and Schwanitz, G. (1977). Double autosomal trisomy: case report. (48,XX, + 18, + 21) and review of the literature. *J. Ment. Defic. Res.* 21, 299–308.

Grouchy, J. de (1965). Chromosome 18: a topological approach. *J. Pediatr.* 66, 414–31.

Grouchy, J. de, and Turleau, C. (1974). Tentative localization of a Hageman (factor XII) locus on 7q, probably the 7q35 band. *Humangenetik* 24, 197–200.

Grouchy, J. de, and Turleau, C. (1977). *Clinical Atlas of Human Chromosomes*. Wiley, New York.

Grouchy, J. de, and Turleau, C. (1984). *Clinical Atlas of Human Chromosomes, 2nd edit.* Wiley, New York.

Grumbach, M.M., and Conte, F.A. (1981). Disorders of sex differentiation. In *Textbook of Endocrinology, 6th edit.*, ed. R.H. Williams, pp. 423–514. Saunders, Philadelphia.

Guanti, G. (1981). The aetiology of the cat eye syndrome reconsidered. *J. Med. Genet.* 18, 108–18.

Guialis, A., Beatty, B.G., Ingles, C.J., and Crevar, M.M. (1977). Regulation of RNA polymerase II activity in α-amantin resistant CHO hybrid cells. *Cell* 10, 53–60.

Guidi, G., Schiavon, R., Biasioli, A., and Perona, G. (1984). The enzyme glutathione peroxidase in arachidonic metabolism of human platelets. *J. Lab. Clin. Med.* 104, 574–82.

Guillemin, C. (1980a). Meiosis in four trisomic and one double trisomic males of the newt *Pleurodeles waltlii*. *Chromosoma* 77, 145–55.

Guillemin, C. (1980b). Effects phénotypiques de six trisomies et de deux doubles trisomies chez *Pleurodeles waltlii* Michahelles (Amphibian, Urodèle). *Ann. Génét.* 23, 5–11.

Gulliya, K., and Dowben, R.M. (1983). Abnormalities of in vitro immune responsiveness in patients with Down's syndrome. *J. Clin. Hematol. Oncol.* 13, 9–18.

Gullotta, F., Rehder, H., and Gropp, A. (1981). Descriptive neuropathology of chromosomal disorders in man. *Hum. Genet.* 57, 337–44.

Gupta, S., Fikrig, S.M., Mariano, E., and Quazi, G. (1983). Monoclonal antibody defined T cell subsets and autologous mixed lymphocyte reaction in Down's syndrome. *Clin. Exp. Immunol.* 53, 25–30.

Gusella, J.F., Tanzi, R.E., Watkins, P.C., Gibbons, K.T., Hobbs, W.J., Faryniarz, A.G., Healey, S.T., and Anderson, M.A. (1985). Genetic linkage map for chromosome 21. *Ann. N.Y. Acad. Sci.* 450, 25–31.

Gustavson, K.-H., Ivemark, B.C., Zettergvist, P., and Böök, J.A. (1962). Postmortem diagnosis of a new double-trisomy associated with cardiovascular and other anomalies. *Acta Paediatr.* 51, 686–97.

Haag, M.M., and Soukup, S.W. (1984). Association of chromosome 4 abnormalities with ethylnitrosourea-induced neuro-oncogenesis in the rat. *Cancer Res.* 44, 784–90.

Haan, E.A., Danks, D.M., Grimes, A., and Hoogenraad, N.J. (1982). Carrier detection in ornithine transcarbamylase deficiency. *J. Inher. Metab. Dis.* 5, 37–40.

Haas, M., Altman, A., Rothenberg, E., Bogart, M.H., and Jones, O.W. (1984). Mechanism of T-cell lymphomagenesis: transformation of growth-factor-independent T-lymphoblastoma cells to growth-factor-independent T-lymphoma cells. *Proc. Natl. Acad. Sci. USA* 81, 1742–6.

Habedank, M., and Rodewald, A. (1982). Moderate Down's syndrome in three siblings having partial trisomy 21q22.2 → qter and therefore no SOD–1 excess. *Hum. Genet.* 60, 74–77.

Haddad, H.M., and Wilkins, L. (1959). Congenital anomalies associated with gonadal aplasia. *Pediatrics* 23, 885–902.

Hagemeijer, A., Smit, E.M.E., Govers, F., and de Both, N.J. (1982). Trisomy 15 and other nonrandom chromosome changes in Rauscher murine leukemia virus-induced leukemia cell lines. *J. Natl. Cancer Inst.* 69, 945–51.

Halal, F., Quenneville, G., Laurin, S., and Loulou, G. (1983). Clinical and genetic aspects of antithrombin III deficiency. *Am. J. Med. Genet.* 14, 737–50.

Hall, B. (1964). *Mongolism in Newborns: A Clinical and Cytogenetic Study.* Berlingska Boktryckeriet, Lund.

Hall, B. (1965). Delayed ontogenesis in human trisomy syndromes. *Hereditas* 52, 335–44.

Hall, B.D. (1985). Mosaic tetrasomy 21 is mosaic tetrasomy 12p some of the time. *Clin. Genet.* 27, 284–5.

Hall, B.D., Langer, L.O., Giedion, A., Smith, D.W., Cohen, M.M. Jr., Beals, R.K., and Brandner, M. (1974). Langer-Giedion syndrome. *Birth Defects: Orig. Art. Ser.* 10(12), 147–64.

Hall, J.C., and Kankel, D.R. (1976). Genetics of acetylcholinesterase in *Drosophila melanogaster*. *Genetics* 83, 517–35.

Hamers, A.J.H., Klep-de Pater, J.M., and de France, H.F. (1983). Partial duplication in a ring chromosome 22. A mentally retarded boy with increased levels of arylsulfatase A and α-galactosidase. *Ann. Génét.* 26, 34–7.

Hamerton, J.L. (1971). *Human Cytogenetics, Vol. II: Clinical Cytogenetics.* Academic Press, New York.

Hamerton, J.L. (1976). Human population cytogenetics: dilemmas and problems. *Am. J. Hum. Genet.* 28, 107–22.

Hammer, R.E., Brinster, R.L., Rosenfeld, M.G., Evans, R.M., and Mayo, K.E. (1985). Expression of human growth hormone-releasing factor in transgenic mice results in increased somatic growth. *Nature, Lond.* 315, 413–6.

Han, T., Ozer, H., Sadamori, N., Emrich, L., Gomez, G.A., Henderson, E.S., Bloom, M.L., and Sandberg, A.A. (1984). Prognostic importance of cytogenetic abnormalities in patients with chronic lymphocytic leukemia. *N. Engl. J. Med.* 310, 288–92.

Handzel, Z.T., Dolfin, Z., Levin, S., Altman, V., Hahn, T., Trainin, N., and Gadot, N. (1979). Effect of thymic humoral factor on cellular immune functions of normal children and of pediatric patients with ataxia telangiectasia and Down's syndrome. *Pediatr. Res.* 13, 803–6.

Hansteen, I.L., Schirmer, L., and Hestetun, S. (1978). Trisomy 12p syndrome. Evaluation of a family with a t(12;21)(p12.1;p11) translocation with unbalanced offspring. *Clin. Genet.* 13, 339–49.

Harnden, D.S., Miller, O.J., and Penrose, L.S. (1960). The Klinefelter-mongolism type of double aneuploidy. *Ann. Hum. Genet.* 24, 165–9.

Harper, M.J., Fosten, M., and Monk, M. (1982). Preferential paternal X inactivation in extraembryonic tissues in early mouse embryos. *J. Embryol. Exp. Morphol.* 67, 127–35.

Harris, A.K., Stopak, D., and Warner, P. (1984). Generation of spacially periodic patterns by a mechanical instability: a mechanical alternative to the Turing model. *J. Embryol. Exp. Morphol.* 80, 1–20.

Harris, H. (1980). *The Principles of Human Biochemical Genetics, 3rd, rev. edit.* Elsevier/North-Holland, Amsterdam.

Harris, M.J., Poland, B.J., and Dill, F.J. (1981). Triploidy in 40 human spontaneous abortuses: assessment of phenotype in embryos. *Obstet. Gynecol.* 57, 600–6.

Harris, W.S., and Goodman, R.M. (1968). Hyper-reactivity to atropine in Down's syndrome. *N. Engl. J. Med.* 279, 407–10.

Harrison, J.T., Anisoivicz, A., Gadi, I.K., Raffeld, M., and Sager, R. (1983). Aza-cytidine-induced tumorigenesis of CHEF/18 cells: correlated DNA methylation and chromosome changes. *Proc. Natl. Acad. Sci. USA* 80, 6606–10.

Hart, G.E., and Langston, P.J. (1977). Chromosomal location and evolution of isozyme structural genes in hexaploid wheat. *Heredity* 39, 263–77.

Hasegawa, D.K., Tyler, B.J., and Edson, J.R. (1982). Thrombotic disease in three families with inherited plasminogen deficiency. *Blood* 60, 213a.

Hasegawa, T., Hara, M., Ando, M., Osawa, M., Fukuyama, Y., Takahashi, M., and Yamada, K. (1984). Cytogenetic studies of familial Prader-Willi syndrome. *Hum. Genet.* 65, 325–30.

Hassold, T., and Sandison, A. (1983). The effect of chromosome constitution on growth in culture of human spontaneous abortions. *Hum. Genet.* 63, 166–70.

Hassold, T.J., Matsuyama, A., Newlands, J.M., Matsuura, J.S., Jacobs, P.A., Manuel, B., and Tsuei, J. (1978). A cytogenetic study of spontaneous abortions in Hawaii. *Ann. Hum. Genet.* 41, 443–54.

Hauschka, T.S. (1961). The chromosomes in ontogeny and oncogeny. *Cancer Res.* 21, 957–74.

Hawkey, C.J., and Smithies, A. (1976). The Prader-Willi syndrome with a 15/15 translocation. *J. Med. Genet.* 13, 152–6.

Hayakawa, H., Matsui, C., Higurashi, M., and Kobayoshi, N. (1968). Hyperblastic response to dilute P.H.A. in Down's syndrome. *Lancet* 1, 95–6.

Hayday, A.C., Gillies, S.D., Saito, H., Wood, C., Wiman, K., Hayward, W.S., and Tonegawa, S. (1984). Activation of a translocated c-*myc* gene by an enhancer in the immunoglobulin heavy-chain locus. *Nature, Lond.* 307, 334–40.

Heaton, D.C., Fitzgerald, P.H., Fraser, G.J., and Abbott, G.D. (1981). Transient leukemoid proliferation of the cytogenetically unbalanced +21 cell line of a constitutional mosaic boy. *Blood* 57, 883–7.

Hecht, F. (1982). Leukaemia and chromosome 21. *Lancet* 1, 286–7.

Hecht, F., Jones, R.T., and Koler, R.D. (1967). Newborn infants with Hb Portland 1, an indicator of α-chain deficiency. *Ann. Hum. Genet.* 31, 215–18.

Hecht, F., Nievaard, J.E., Duncanson, N., Miller, J.R., Higgins, J.V., Kimberling, W.J., Walker, F.A., Smith, G.S., Thuline, H.C., and Tischler, B. (1969). Double aneuploidy: the frequency of XXY in males with Down's syndrome. *Am. J. Hum. Genet.* 21, 352–9.

Hedman, K.D., and Boyer, C.D. (1982). Gene dosage at the *amylose-extender* locus of maize: effects on the levels of starch branching enzymes. *Biochem. Genet.* 20, 483–92.

Heidemann, A., Schmalenberger, B., and Zankl, H. (1983). Sister chromatid exchange and lymphocyte proliferation in a Down syndrome mosaic. *Clin. Genet.* 23, 139–42.

Heisterkamp, N., Stephenson, J.R., Groffen, J., Hanson, P.F., de Klein, A., Bartram, C.R., and Grosveld, G. (1983). Localization of the c-*abl* oncogene adjacent to a translocation break point in chronic myelocytic leukaemia. *Nature, Lond.* 306, 239–42.

Henderson, A.S., Warburton, D., and Atwood, K.C. (1972). Location of ribosomal DNA in the human chromosome complement. *Proc. Natl. Acad. Sci. USA* 69, 3394–8.

Herbst, E.W., and Gropp, A. (1982). Relevance of trisomy 15 and other chromosome abnormalities in spontaneous AKR leukemia of mice with and without Robertsonian rearrangement. *J. Cancer Res. Clin. Oncol.* 104, 207–18.

Herbst, E.W., Gropp, A., and Tietgen, C. (1981). Chromosome rearrangements involved in the origin of trisomy 15 in spontaneous leukemia of AKR mice. *Int. J. Cancer* 28, 805–10.

Herbst, E.W., Pluznik, D.H., Gropp, A., and Uthgenaant, H. (1981). Trisomic hemopoietic stem cells of fetal origin restore hemopoiesis in lethally irradiated mice. *Science* 211, 1175–7.

Herbst, E.W., Gropp, A., Nielsén, K., Hoppe, H., Freymann, M., and Pluznik, D.H. (1982a). Reduced ability of mouse trisomy 16 stem cells to restore hemopoiesis in lethally irradiated animals. In *Experimental Hematology Today*, ed. S.J. Baum, G.D. Ledney, and S. Thierfelder, pp. 119–26. S. Karger, Basel.

Herbst, E.W., Gropp, A., Stehle, W., Fohlmeister, I., Nielsén K., Claussen, C.-P., and Pluznik, D. (1982b). Rauscher-Leukämie der Maus bei Trisomie des Chromosome 19. *Verh. Dtsch. Ges. Pathol.* 66, 122–7.

Hernandez, A., Corona-Rivera, E., Plascencia, L., Nazara, Z., Ibarra, B., and Cantú, J.M. (1979a). De novo partial trisomy of chromosome 18 (pter→q11:). Some observations on the phenotype mapping of chromosome 18 imbalances. *Ann. Génét.* 22, 165–7.

Hernandez, A., Rivera, H., Jiminez-Sainz, M., Fragoso, R., Zazara, Z., and Cantú, J.M. (1979b). Type and countretype signs in monosomy and trisomy 9p. *Ann. Génét.* 22, 155–7.

Heston, L.L., and Mastri, A.R. (1977). The genetics of Alzheimer's disease. Association with hematologic malignancy and Down's syndrome. *Arch. Gen. Psychiatry* 34, 976–81.

Heston, L.L., Mastri, A.R., Anderson, V.E., and White, J. (1981). Dementia of the Alzheimer type. Clinical genetics, natural history, and associated conditions. *Arch. Gen. Psychiatry* 38, 1085–90.

Hethcote, H.W., and Knudson, A.G., Jr. (1978). Model for the incidence of embryonal cancers: application to retinoblastoma. *Proc. Natl. Acad. Sci. USA* 75, 2453–7.

Higginson, G., Weaver, D.D., Magenis, R.E., Prescott, G.H., Haag, C., and Hepburn, D.J. (1976). Interstitial deletion of the long arm of chromosome no.7 (7q −) in an infant with multiple anomalies. *Clin. Genet.* 10, 307–12.

Higgs, D.R., Old, J.M., Pressley, L., Clegg, J.B., and Weatherall, D.J. (1980). A novel α-globin gene arrangement in man. *Nature, Lond.* 284, 632–5.

Higurashi, M., and Conen, P.E. (1972). *In vitro* chromosomal radiosensitivity in patients and in carriers with abnormal non-Down's syndrome karyotypes. *Pediatr. Res.* 6, 514–20.

Higurashi, M., Tamura, T., and Nakatake, T. (1973). Cytogenetic observations in cultured lymphocytes from patients with Down's syndrome and measles. *Pediatr. Res.* 7, 582–7.

Higurashi, M., Tada, A., Miyahara, S., and Hirayama, M. (1976). Chromosome damage in Down's syndrome induced by chickenpox infection. *Pediatr. Res.* 10, 189–92.

Higurashi, M., Iijima, K., Ishikawa, N., Hoshina, H., and Watanabe, N. (1979). Incidence of major chromosome aberrations in 12,319 newborn infants in Tokyo. *Hum. Genet.* 46, 163–72.

Hittner, H.M., Riccardi, V.M., and Francke, U. (1979). Aniridia caused by a heritable chromosome 11 deletion. *Ophthalmology* 86, 1173–83.

Hochman, B. (1976). The fourth chromosome of *Drosophila Melanogaster*. In *The Genetics and Biology of Drosophila, Vol. 1b*, ed. M. Ashburner and E. Novitski, pp. 903–28. Academic Press, London.

Hodes, M.E., Cole, J., Palmer, C.G., and Reed T. (1978). Clinical experience with trisomies 18 and 13. *J. Med. Genet.* 15, 49–60.

Hoehn, H. (1975). Functional implications of differential chromosome banding. *Am. J. Hum. Genet.* 27, 676–86.

Hoehn, H., Simpson, M., Bryant, E.M., Rabinovitch, P.S., Salk, D., and Martin, G.M. (1980). Effects of chromosome constitution on growth and longevity of human skin fibroblast cultures. *Am. J. Med. Genet.* 7, 141–54.

Hoffee, P.A., Hunt, S.W. III, Chiang, J., Labant, M.C., Clarke, M., and Jargiello, P. (1983). Evidence for a trans-dominant regulator of purine nucleoside phosphorylase expression in rat hepatoma cells. *Somat. Cell Genet.* 9, 249–67.

Hoffman, S., and Edelman, G.M. (1983). Kinetics of homophilic binding by embryonic and adult forms of the neural cell adhesion molecule. *Proc. Natl. Acad. Sci. USA* 80, 5762–66.

Holley, R.W., Böhlen, P., Fava, R., Baldwin, J.H., Kleeman, G., and Armour, R. (1980). Purification of kidney epithelial cell growth inhibitors. *Proc. Natl. Acad. Sci. USA* 77, 5989–92.

Hollinger, F.B., Goyal, R.K., Hersh, T., Powell, H.C., Schulman, R.J., and Melnick, J.L. (1972). Immune response to hepatitis virus type B in Down's syndrome and other mentally retarded patients. *Am. J. Epidemiol.* 95, 356–62.

Holm, V.A. (1981). The diagnosis of Prader-Willi syndrome. In *Prader-Willi Syndrome*, ed. V.A. Holm, S.J. Sulzberger, and P.L. Pipes, pp. 27–53. University Park Press, Baltimore.

Honey, N.K., and Shows, T.B. (1983). The tumor phenotype and the human gene map. *Cancer Genet. Cytogenet.* 10, 287–310.

Hongell, K. (1981). Proliferation of trisomic and euploid mouse cells *in vitro*. *Clin. Genet.* 19, 510.

Hongell, K., and Gropp, A. (1982). Trisomy 13 in the mouse. *Teratology* 26, 95–104.

Hongell, K., Herbst, E., and Gropp, A. (1980). Competitive proliferation of trisomic and euploid haemopoietic cells from foetal mice in radiation chimeras. *Clin. Genet.* 17, 72.

Hongell, K., Herbst, E.W., and Gropp, A. (1984). Transplantation of mixed trisomic and normal fetal hemopoietic stem cells of the mouse to irradiated hosts. Submitted for publication.

Hoo, J.J., Koch, M., Ziemsen, B., Foerster, W., and Nishigaki, I. (1982). Confirmation of regional assignment of gene for human esterase-D to chromosome band 13q14. *Hum. Genet.* 60, 276–7.

Hook, E.B. (1981). Prevalence of chromosome abnormalities during human gestation and implication for studies of environmental mutagens. *Lancet* 2, 169–72.

Hook, E.B. (1982). Epidemiology of Down syndrome. In *Down syndrome. Advances in Biomedicine and the Behavioral Sciences*, ed. S.M. Pueschel and J.E. Rynders, pp. 11–88. Ware Press, Cambridge, Massachusetts.

Hook, E.B., and Warburton, D. (1983). The distribution of chromosomal genotypes associated with Turner's syndrome: livebirth prevalence rates and evidence for diminished fetal mortality and severity in genotypes associated with structural X abnormalities or mosaicism. *Hum. Genet.* 64, 24–7.

Hösli, P., and Vogt, E. (1979). High alkaline phosphatase activity in isoproterenol stimulated fibroblast cultures from patients with numerically unbalanced chromosomal aberrations. *Clin. Genet.* 15, 487–94.

Hsia, D.Y.-Y., Nadler, H.L., and Shih, L. (1968). Biochemical changes in chromosomal abnormalities. *Ann. N.Y. Acad. Sci.* 171, 526–36.

Hsia, D.Y.-Y., Justice, P., Smith, G.F., and Dowben, R.M. (1971). Down's syndrome. A critical review of the biochemical and immunological data. *Am. J. Dis. Child.* 121, 153–61.

Hsu, T.C. (1961). Chromosomal evolution in cell populations. *Int. Rev. Cytol.* 12, 69–161.

Huang, C.C., Banerjee, A., Tan, J.C., and Hou, Y. (1977). Comparison of radiosensitivity between human hematopoietic cell lines derived from patients with Down's syndrome and from normal persons. *J. Natl. Cancer Inst.* 59, 33–6.

Huebner, K., Palumbo, A.P., Isobe, M., Kozak, C.A., Monaco, S., Rovera, G., Croce, C.M., and Curtis, P.J. (1985). The α-spectrin gene is on chromosome 1 in mouse and man. *Proc. Natl. Acad. Sci. USA* 82, 3790–3.

Huehns, E.R., Hecht, F., Keil, J.V., and Motulsky, A.G. (1964). Developmental hemoglobin anomalies in a chromosomal triplication: D_1 trisomy syndrome. *Proc. Natl. Acad. Sci. USA* 51, 89–97.

Hulten, M. (1984). Somatic gene mutation and breast carcinoma. *Nature, Lond.* 310, 103–4.

Human Gene Mapping 7: Seventh International Workshop on Human Gene Mapping (1984). *Cytogenet. Cell Genet.* 37, nos. 1–4.

Hummel, K.P. (1958). The inheritance and expression of disorganization, an unusual mutation in the mouse. *J. Exp. Zool.* 137, 389–423.

Hummel, K.P. (1959). Developmental anomalies in mice resulting from the action of the gene, disorganization, a semi-dominant lethal. *Pediatrics* 23, 212–21.

Hunter, A.G.W., Clifford, B., Speevak, M., and MacMurray, S.B. (1982). Mosaic tetrasomy 21 in a liveborn male infant. *Clin. Genet.* 21, 228–32.

Hunter, T. (1984). The proteins of oncogenes. *Sci. Amer.* 251, 70–9.

Hyman, R., Cunningham, K., and Stallings, V. (1981). Effect of gene dosage on cell-surface expression of Thy–1 antigen in somatic cell hybrids between Thy–1⁻ Abelson-leukemia-virus induced lymphomas and Thy–1⁺ mouse lymphomas. *Immunogenetics* 12, 381–95.

Iba, H., Takeya, T., Cross, F.R., Hanafusa, T., and Hanafusa, H. (1984). Rous sarcoma virus variants that carry the cellular *src* gene instead of the viral *src* gene cannot transform chicken embryo fibroblasts. *Proc. Natl. Acad. Sci. USA* 81, 4424–8.

Igarashi, M., Takahashi, M., Alford, B.R., and Johnson, P.E. (1977). Inner ear morphology in Down's syndrome. *Acta Otolaryngol.* 83, 175–81.

Iijima, K., Morimoto, K., Koizumi, A., Higurashi, M., and Hirayama, M. (1984). Bleomycin-induced chromosomal aberrations and sister chromatid exchanges in Down lymphocyte cultures. *Hum. Genet.* 66, 57–61.

Iles, S.A., and Evans, E.P. (1977). Karyotype analysis of teratocarcinomas and embryoid bodies of C3H mice. *J. Embryol. Exp. Morphol.* 38, 77–92.

International System for Human Cytogenetic Nomenclature (1978). *Cytogenet. Cell Genet.* 21(6), 313–404.

Izant, J.G., and Weintraub, H. (1985). Constitutive and conditional suppression of exogenous and endogenous genes by anti-sense RNA. *Science* 229, 345–52.

Jabs, E.W., Stamberg, J., and Leonard, C.O. (1982). Tetrasomy 21 in an infant with Down syndrome and congenital leukemia. *Am. J. Med. Genet.* 12, 91–5.

Jackson, C.E., Block, M.A., Greenwald, K.A., and Tashjian, A., Jr. (1979). The two-mutational-event theory in medullary thyroid cancer. *Am. J. Hum. Genet.* 31, 704–10.

Jacob, F., and Monod, J. (1961). Genetic regulatory mechanisms in the synthesis of proteins. *J. Mol. Biol.* 3, 318–56.

Jacobs, P.A., Matsuura, J.S., Mayer, M., and Newlands, I.M. (1978). A cytogenetic survey of an institution for the mentally retarded: I. Chromosome abnormalities. *Clin. Genet.* 13, 37–60.

Jacobs, P.A., Matsuyama, A.M., Buchanan, I.M., and Wilson, C. (1979). Late replicating X chromosomes in human triploidy. *Am. J. Hum. Genet.* 31, 446–57.

Jacobs, P.A., Szulman, A.E., Funkhouser, J., Matsuura, J.S., and Wilson, C.C. (1982). Human triploidy: relationship between parental origin of the additional haploid complement and the development of partial hydatidiform mole. *Ann. Hum. Genet.* 46, 223–31.

Jacobs, P.F., Burdash, N.M., Manos. J.P., and Duncan, R.C. (1978). Immunologic parameters in Down's syndrome. *Ann. Clin. Lab. Sci.* 8, 17–22.

Jacobsen, P., and Mikkelson, M. (1968). Chromosome 18 abnormalities in a family with a translocation t(18p–,21p+). *J. Ment. Defic. Res.* 12, 144–61.

Jenkins, E.C., Duncan, C.J., Wright, C.E., Giordano, F.M., Wilbur, L., Wisniewski, K., Sklower, S.L., French, J.H., Jones, C., and Brown, W.T. (1983). Atypical Down syndrome and partial trisomy 21. *Clin. Genet.* 24, 97–102.

Jensen, P.K.A., Junien, C., Despoisse, S., Bernsen, A., Thelle, T., Friedrich, U., and de la Chapelle, A. (1982). Inverted tandem duplication of the short arm of chromosome 8: a non-random de novo structural aberration in man. Localization of the gene for glutathione reductase in subband 8p21.1. *Ann. Génét.* 25, 207–11.

Jérôme, H. (1968). Les anomalies du métabolisme du tryptophan dans la trisomie 21. *Union Med. Can.* 97, 929–35.

Jeziorowska, A., Jakubowski, L., Armatys, A., and Kaluzewski, B. (1982). Copper/zinc superoxide dismutase (SOD–1) activity in regular trisomy 21, trisomy 21 by translocation and mosaic trisomy 21. *Clin. Genet.* 22, 160–4.

Johnson, M.P., Ramsay, N., Cervenka, J., and Wang, N. (1982). Retinoblastoma and its association with a deletion in chromosome #13: a survey using high-resolution chromosome techniques. *Cancer Genet. Cytogenet.* 6, 29–37.

Johnston, K., Golabi, M., Schonberg, S., Jourgensen, D., Anderson, R., and Douglas, R. (1985). Bilateral ectrodactyly of feet and tibial aplasia associated with an unusual chromosome abnormality: 47XX + isodic(14)(q121). *Am. J. Hum. Genet.* 37, A61.

Jonas, A.J., Greene, A.A., Smith, M.L., and Schneider, J.A. (1982). Cystine accumulation and loss in normal, heterozygous, and cystinotic fibroblasts. *Proc. Natl. Acad. Sci. USA* 79, 4442–5.

Jones, M.B. (1979). Years of life lost through Down's syndrome. *J. Med. Genet.* 16, 379–83.

Joyner, A.L., Lebo, R.V., Kan, Y.W., Tjian, R., Cox, D.R., and Martin, G.R. (1985). Comparative chromosome mapping of a conserved homoeo box region in mouse and man. *Nature, Lond.* 314, 173–5.

Juberg, R.C., and Mowrey, P.N. (1984). Interstitial del(13q) associated with blindness and mental retardation. *Am. J. Med. Genet.* 17, 609–13.

Juberg, R.C., Christopher, C.R., Alvira, M.M., and Gilbert, E.E. (1984). Clinicopathologic conference: dup(10q), del(12p) in one abnormal, dizygotic twin infant of a t(10;12)(q22.1;p13.3) mother. *Am. J. Med. Genet.* 18, 201–13.

Jüdes, U., Winking, H., and Gropp, A. (1980). Personal communication. *Mouse News Letter* No. 63, 23.

Junien, C., Huerre, C., and Rethoré, M.-O. (1983). Direct gene dosage determination in patients with unbalanced chromosomal aberrations using cloned DNA sequences. Application to the regional assignment of the gene for α2(I) procollagen (COLIA2). *Am. J. Hum. Genet.* 35, 584–91.

Junien, C., Huerre, C., and Rethoré, M.O. (1984). Regional assignment of the alpha–2(I) collagen gene to band 7q21 by direct gene dosage determination. *Cytogenet. Cell Genet.* 37, 502.

Junien, C., Rubinson, H., Dreyfus, J.C., Meienhofer, M.C., Ravise, N., Boué, J., and Boué, A. (1976). Gene dosage effect in human triploid fibroblasts. *Hum. Genet.* 33, 61–6.

Junien, C., Choury, D., Suerinck, E., Noel, B., and Kaplan, J.C. (1979*a*). The gene for human LDHB is in region 12(p12.1→p12.3) and shows single and triple dosage effects. *Cytogenet. Cell Genet.* 25, 169.

Junien, C., Kaplan, J.C., Serville, F., and Lenoir, G. (1979*b*). Triplex gene dosage effect of TPI and G3PD in a human lymphoblastoid cell line with partial trisomy 12p15 and 18p. *Hum. Genet.* 49, 221–3.

Junien, C., Kaplan, J.C., Bernheim, A., and Berger, R. (1979*c*). Regional assignment of red cell acid phosphatase locus to band 2p25. *Hum. Genet.* 48, 17–21.

Junien, C., Turleau, C., Bugnon, C., Bresson, C., Roche, R., Grouchy, J. de (1979*d*). Localisation de TGO$_s$ en 10q24q262 et suggestion de localisation de HK–1 en 10q23. *Ann. Génét.* 22, 50–2.

Junien, C., Grouchy, J. de, Turleau, C., and Serville, F. (1980*a*). Confirmation of the regional assignment of peptidase A (PEPA) to 18q23 by gene dosage studies. *Ann. Génét.* 23, 89–90.

Junien, C., Rubinson-Skala, H., Dreyfus, J.C., Ravise, N., Boué, J., Boué, A., and Kaplan, J.-C. (1980*b*). PK3: a new chromosome enzyme marker for gene dosage studies in chromosome 15 imbalance. *Hum. Genet.* 54, 191–6.

Junien, C., Turleau, C., Grouchy, J. de, Saïd, R., Rethoré, M.-O., Tenconi, R., and

Dufier, J.L. (1980c). Regional assignment of catalase (CAT) gene to band 11p13. Association with the aniridia-Wilms' tumor-gonadoblastoma (WAGR) complex. *Ann. Génét.* 23, 165–8.

Junien, C., Kaplan, J.C., Raoul, O., Rethoré, M.-O., Turleau, C., and Grouchy, J. de (1980d). Effet de dosage sesquialtère de la nucléoside phosphorylase érythrocytaire et leukocytaire dans deux cas de trisomie partielle 14q. *Ann. Génét.* 23, 86–8.

Junien, C., Despoisse, S., Turleau, C., Grouchy, J. de, Bucher, T., and Fundele, R. (1982a). Assignment of phosphoglycerate mutase (PGAMA) to human chromosome 10. Regional mapping of GOT1 and PGAMA to subbands 10q26.1 (or q25.3). *Ann. Génét.* 25, 25–7.

Junien, C., Despoisse, S., Turleau, C., Nicolas, H., Picard, F., Le Marec, B., Kaplan, J.-C., and Grouchy, J. de (1982b). Retinoblastoma, deletion 13q14, and esterase D: application of gene dosage effect to prenatal diagnosis. *Cancer Genet. Cytogenet.* 6, 281–7.

Kaback, M.M., and Bernstein, L.H. (1970). Biologic studies of trisomic cells growing *in vitro*. *Ann. N.Y. Acad. Sci.* 171, 526–36.

Kacser, H., and Burns, J.A. (1973). The control of flux. *Symp. Soc. Exp. Biol.* 27, 65–104.

Kacser, H., and Burns, J.A. (1981). The molecular basis of dominance. *Genetics* 97, 639–66.

Kaffe, S., Hsu, L.Y.F., and Hirschhorn, K. (1983). Structural abnormalities of the human X chromosome and their clinical features. In *Cytogenetics of the Mammalian X Chromosome, Part B: X Chromosome Anomalies and their Clinical Manifestations*, ed. A.A. Sandberg, pp. 341–58. A.R. Liss, New York.

Kahn, C.R., Bertolotti, R., Ninio, M., and Weiss, M.C. (1981). Short-lived cytoplasmic regulators of gene expression in cell cybrids. *Nature, Lond.* 290, 717–20.

Kaina, B., Waller, H., Waller, M., and Rieger, R. (1977). The action of *N*-methyl-*N*-nitrosourea on non-established human cell lines in vitro. I. Cell cycle inhibition and aberration induction in diploid and Down's fibroblasts. *Mutation Res.* 43, 387–400.

Kaiser, P., Harprecht, W., Steuermagel, P., and Daume, E. (1984). Long arm deletions of the X chromosome and their symptoms: a new case (bp q24) and a short review of the literature. *Clin. Genet.* 26, 433–9.

Kajii, T., Ohama, K., and Mikamo, K. (1978). Anatomic and chromosomal anomalies in 944 induced abortuses. *Hum. Genet.* 43, 247–58.

Kamoun, P., Cathelineau, L., Lafourcade, G., Pham Dinh, D., Jerome, H., and Bomsel-Helmreich, O. (1978). X-chromosome inactivation and enzymatic activities in diploid and triploid fibroblasts. *Exp. Cell Res.* 114, 357–63.

Kano-Tanaka, K., and Tanaka, T. (1982). Specific chromosome changes associated with viral transformation of rat glial cells. *Int. J. Cancer* 30, 495–501.

Karp, L.E. (1983). New hope for the retarded? *Am. J. Med. Genet.* 16, 1–5.

Karttunen, R., Nurmi, T., Ilonen, J., and Surcel, H.-M. (1984). Cell-mediated immunodeficiency in Down's syndrome: normal IL–2 production but inverted ratio of T cell subsets. *Clin. Exp. Immunol.* 55, 257–63.

Katzman, R. (ed.) (1983). *Banbury Report 15. Biological Aspects of Alzheimer's Disease*. Cold Spring Harbor Laboratory, Cold Spring Harbor, New York.

Kaufman, M.H., Guc-Cubrilo, M., and Lyon, M.F. (1978). X chromosome inactivation in diploid parthenogenetic mouse embryos. *Nature, Lond.* 271, 547–9.

Kedziora, J., Hubner, H., Kenski, M., Jeske, J., and Leyko, W. (1972). Efficiency of the glycolytic pathway in erythrocytes of children with Down's syndrome. *Pediatr. Res.* 6, 10–17.

Keele, D.K., Richards, C., Brown, J., and Marshall, J. (1969). Catecholamine metabolism in Down's syndrome. *Am. J. Ment. Defic.* 74, 125–9.

Keitges, E., Rivest, M., Siniscalco, M., and Gartler, S.M. (1985). X-linkage of steroid sulphatase in the mouse is evidence for a functional Y-linked allele. *Nature, Lond.* 315, 226–7.

Kelley, R.I., Zackai, E.H., Emanuel, B.S., Kistenmacher, M., Greenberg, F., and Punnett, H.H. (1982). The association of the DiGeorge anomalad with partial monosomy of chromosome 22. *J. Pediatr.* 101, 197–200.

Kemphues, K.J., Raff, E.C., Raff, R.A., and Kaufman, T.C. (1980). Mutation in a testis-specific β-tubulin in *Drosophila*: analysis of its effects on meiosis and map location of the gene. *Cell* 21, 445–51.

Kennedy, J.F., Freeman, M.G., and Benirschke, K. (1977). Ovarian dysgenesis and chromosome abnormalities. *Obstet. Gynecol.* 50, 13–20.

Keppy, D.O., and Denell, R.E. (1979). A mutational analysis of the triplo-lethal region of *Drosophila melanogaster*. *Genetics* 91, 421–41.

Ketupanya, A., Crandall, B.F., Blanchard, K., and Rogers, D.W. (1984). Mosaic hexasomy 21. *J. Med. Genet.* 21, 228–30.

Khodr, G.S., Cadena, G., Le, K.L., and Kagan-Hallet, K.S. (1982). Duplication (5p13→pter): prenatal diagnosis and review of the literature. *Am. J. Med. Genet.* 12, 43–9.

Khush, G.S. (1973). *Cytogenetics of Aneuploids*. Academic Press, New York.

Khush, G.S., Singh, R.J., Sur, S.C., and Librojo, A.L. (1984). Primary trisomics of rice: origin, morphology, cytology and use in linkage mapping. *Genetics* 107, 141–63.

Kielty, C.M., Povey, S., and Hopkinson, D.A. (1982a). Regulation of expression of liver-specific enzymes. II. Activation and chromosomal localization of soluble glutamate-pyruvate transaminase. *Ann. Hum. Genet.* 46, 135–43.

Kielty, C.M., Povey, S., and Hopkinson, D.A. (1982b). Regulation of expression of liver-speciic enzymes. III. Further analysis of a series of rat hepatoma × human somatic cell hybrids. *Ann. Hum. Genet.* 46, 307–27.

Killary, A.M., and Fournier, R.E.K. (1984). A genetic analysis of extinction: *trans*-dominant loci regulate expression of liver-specific traits in hepatoma hybrid cells. *Cell* 38, 523–34.

Kim, H.J., Hsu, L.Y.F., Goldsmith, L.C., Strauss, L., and Hirschhorn, K. (1977). Familial translocation with partial trisomy of 13 and 22: evidence that specific regions of chromosomes 13 and 22 are responsible for the phenotype of each trisomy. *J. Med. Genet.* 14, 114–19.

Kirk, K.M. (1984). The genesis and effects of unbalanced genomes resulting from reciprocal translocations in the mouse. Ph.D. Thesis. University of Reading.

Kitani, N., Abo, W., and Kokubun, Y. (1977). Leukocyte dysfunction and T-system immunodeficiency in Down's syndrome. *Jekeikai Med. J.* 24, 91–9.

Kittur, S.D., Antonarakis, S.E., Tanzi, R.E., Meyers, D.A., Chakravarti, A., Groner, Y., Phillips, J.A., Watkins, P.C., Gusella, J.F., and Kazazian, H.H. Jr. (1985). A linkage map of three anonymous DNA fragments and SOD-1 on chromosome 21. *EMBO J.* 4, 2257–60.

Klein, G. (1979). Lymphoma development in mice and humans: diversity of initiation is followed by convergent cytogenetic evolution. *Proc. Natl. Acad. Sci. USA* 76, 2442–6.

Klein, G. (1981). The role of gene dosage and genetic transpositions in carcinogenesis. *Nature, Lond.* 294, 313–8.

Klein, G. (1983). Specific chromosomal translocations and the genesis of B-cell-derived tumors in mice and men. *Cell* 32, 311–5.

Klein, G., Ohno, S., Rosenberg, N., Wiener, F., Spira, J., and Baltimore, D. (1980). Cytogenetic studies on Abelson-virus-induced mouse leukemias. *Int. J. Cancer* 25, 805–11.

Klep-de Pater, J.M., Bijlsma, J.B., de France, H.F., Leschot, N.J., Duijndam-van den Berge, M., and van Hemel, J.O. (1979). Partial trisomy 10q. A recognizable syndrome. *Hum. Genet.* 46, 29–40.

Klose, J. (1982). Genetic variability of soluble proteins studied by two-dimensional electrophoresis on different inbred mouse strains and on different mouse organs. *J. Mol. Evol.* 18, 315–28.

Klose, J., and Putz, B. (1983). Analysis of two-dimensional protein patterns from mouse embryos with different trisomies. *Proc. Natl. Acad. Sci. USA* 80, 3753–7.

Klose, J., Zeindl, E., and Sperling, K. (1982). Analysis of protein patterns in two-dimensional gels of cultured human cells with trisomy 21. *Clin. Chem.* 28, 987–92.

Knudson, A.G., Jr. (1971). Mutation and cancer: statistical study of retinoblastoma. *Proc. Natl. Acad. Sci. USA* 68, 820–3.

Knudson, A.G., Jr., and Strong, L.C. (1972a). Mutation and cancer: a model for Wilm's tumor of the kidney. *J. Natl. Cancer Inst.* 48, 313–24.

Knudson, A.G., Jr., and Strong, L.C. (1972b). Mutation and cancer: neuroblastoma and pheochromocytoma. *Am. J. Hum. Genet.* 24, 514–32.

Knudson, A.G., Jr., Hethcote, H.W., and Brown, B.W. (1975). Mutation and childhood cancer: a probabilistic model for the incidence of retinoblastoma. *Proc. Natl. Acad. Sci. USA* 72, 5116–20.

Knudson, A.G., Jr., Meadows, A.T., Nichols, W.W., and Hill, R. (1976). Chromosomal deletion and retinoblastoma. *N. Eng. J. Med.* 295, 1120–3.

Knuutila, S., Harkki, A., Ellimäki, K., and Salunen, R. (1979). Decreased sister chromatid exchange in Down's syndrome after measles vaccination. *Hereditas* 90, 147–9.

Kolata, G. (1984). Globin gene studies create a puzzle. *Science* 223, 470–1.

Kono, T., and Barham, F.W. (1971). The relationship between the insulin-binding capacity of fat cells and the cellular response to insulin. Studies with intact and trypsin-treated fat cells. *J. Biol. Chem.* 246, 6210–16.

Korenberg, J.R., Therman, E., and Denniston, C. (1978). Hot spots and functional organization of human chromosomes. *Hum. Genet.* 43, 13–22.

Koufos, A., Hansen, M.F., Lampkin, B.C., Workman, M.L., Copeland, N.G., Jenkins, N.A., and Cavenee, W.K. (1984). Loss of alleles at loci on human chromosome 11 during genesis of Wilms' tumour. *Nature, Lond.* 309, 170–2.

Kousseff, B.G. (1982). The cytogenetic controversy in the Prader-Labhart-Willi syndrome. *Am. J. Med. Genet.* 13, 431–9.

Kozak, C.A., Sears, J.F., and Hoggan, M.D. (1983). Genetic mapping of the mouse proto-oncogene c-*sis* to chromosome 15. *Science* 221, 867–9.

Kozak, L.P., and Quinn, P.J. (1975). Evidence for dosage compensation of an X-linked gene in the 6-day embryo of the mouse. *Dev. Biol.* 45, 65–73.

Krane, S.M. (1978). Renal glycosuria. In *The Metabolic Basis of Inherited Disease*, 4th edit., ed. J.B. Stanbury, J.B. Wyngaarden, and D.S. Fredrickson, pp. 1607–17. McGraw-Hill, New York.

Kratzer, P.G., and Gartler, S.M. (1978). HGPRT activity changes in preimplantation mouse embryos. *Nature, Lond.* 274, 503–4.

Kretschmer, R.R., López-Osuna, M., De La Rosa, L., and Armendares, S. (1974). Leukocyte function in Down's syndrome. Quantitative NBT reduction and bactericidal capacity. *Clin. Immunol. Immunopathol.* 2, 449–55.

Krone, W., and Wolf, U. (1972). Chromosomes and protein variation. In *The Biochemical Genetics of Man*, ed. D.J.H. Brock and O. Mayo, pp. 71–127. Academic Press, London.

Krone, W., and Wolf, U. (1977). Chromosome variation and gene action. *Hereditas* 86, 31–6.

Krone, W., and Wolf, U. (1978). Chromosomes and protein variation. In *The Biochemical Genetics of Man, 2nd edit.*, ed. D.J.H. Brock and O. Mayo, pp. 93–154. Academic Press, London.

Krooth, R.S. (1964). Properties of diploid cell strains developed from patients with an inherited abnormality of uridine biosynthesis. *Cold Spring Harbor Symp. Quant. Biol.* 29, 189–212.

Kučerová, M. (1967). Comparison of radiation effects *in vitro* upon chromosomes of human subjects. *Acta Radiol. (Ther. Phys. Biol.)* 6, 441–8.

Kučerová, M., and Polikova, Z. (1978). In vitro comparison of normal and trisomic cell sensitivity to physical and chemical mutagens. In *Mutagen-induced Chromosome Damage in Man*, ed. H.J. Evans and D.C. Lloyd, pp. 185–90. University Press, Edinburgh.

Kuhn, E.M. (1976). Localization of Q-banding of mitotic chiasmata in cases of Bloom's syndrome. *Chromosoma* 57, 1–10.

Kuliev, A.M., Kukharenko, V.I., Grinberg, K.N., Terskikh, V.V., Tamarkina, A.D., Bogomazov, E.A., Redkin, P.S., and Vasileysky, S.S. (1974). Investigation of a cell strain with trisomy 14 from a spontaneously aborted human fetus. *Humangenetik* 21, 1–12.

Kuliev, S.M., Kukharenko, V.I., Grinberg, K.N., Mikhailov, A.T., and Tamarkina, A.D. (1975). Human triploid cell strain. Phenotype on cellular level. *Humangenetik* 30, 127–34.

Kunze, J., Stephan, E., and Tolksdorf, M. (1972). Ring-chromosome 18. Ein 18p−/ 18q− Deletionssyndrome. *Humangenetik* 15, 289–318.

Kurnit, D.M. (1979). Down syndrome: gene dosage at the transcriptional level in skin fibroblasts. *Proc. Natl. Acad. Sci. USA* 76, 2372–5.

Kurnit, D.M., Aldridge, J.F., Matsuoka, R., and Matthyse, S. (1985). Increased adhesiveness of trisomy 21 cells and atrioventricular malformations in Down syndrome: a stochastic model. *Am. J. Med. Genet.* 20, 385–99.

Kushnick, T., Rao, K.W., and Lamb, A.N. (1984). Familial 5p− syndrome. *Clin. Genet.* 26, 472–6.

Kwee, M.L., Barth, P.G., Arwert, F., and Madan, K. (1984). Mosaic tetrasomy 21 in a male child. *Clin. Genet.* 26, 150–5.

Kyritsis, A.P., Tsokos, M., Triche, T.J., and Chader, G.J. (1984). Retinoblastoma – origin from a primitive neuroectodermal cell? *Nature, Lond.* 307, 471–3.

Lacy, E., Roberts, S., Evans, E.P., Burtenshaw, M.D., and Costantini, F.D. (1983). A foreign β-globin gene in transgenic mice: integration at abnormal chromosomal positions and expression in inappropriate tissues. *Cell* 34, 343–58.

Lambert, B., Hansson, K., Bui, T.H., Funes-Cravioto, F., Lindsten, J., Holmberg, M., and Strausmanis, R. (1976). DNA repair and frequency of X-ray and U.V.-light induced chromosome aberrations in leukocytes from patients with Down's syndrome. *Ann. Hum. Genet.* 39, 293–303.

Lamon, J.M., Frykholm, B.C., Bennett, M., and Tschudy, D.P. (1978). Prevention of acute porphyric attacks by intravenous haematin. *Lancet* 2, 492–4.

Land, H., Parada, L.F., and Weinberg, R.A. (1983). Cellular oncogenes and multi-step carcinogenesis. *Science* 222, 771–8.

Landing, B.H., and Shankle, W.. (1982). Reduced number of skeletal muscle fiber nuclei in Down syndrome: speculation on a "shut off" role of chromosome 21 in control of DNA and nuclear replication rates, possibly via determination of cell surface area per nucleus. *Birth Defects: Orig. Art. Ser.* 18 (3B), 81–7.

Landing, B.H., Dixon, L.G., and Wells, T.R. (1974). Studies on isolated human skeletal muscle fibers, including a proposed pattern of nuclear distribution and a concept of nuclear territories. *Hum. Pathol.* 5, 441–61.

Lang, R., Nakagawa, S., and Nitowsky, H.M. (1983). Elevated superoxide dismutase (SOD) activity in erythrocyte lysates from 21 trisomy fetuses. *Pediatr. Res.* 17, 214A.

Langenbeck, U., Blum, E., Wilkert-Walter, C., and Hansmann, I. (1984). Developmental pathogenesis of chromosomal disorders: report on two newly recognized signs of Down syndrome. *Am. J. Med. Genet.* 18, 223–30.

Lanman, J.T., Sklarin, B.S., Cooper, H.L., and Hirschhorn, K. (1960). Klinefelter's syndrome in a ten-month-old mongolian idiot. Report of a case with chromosome analysis. *N. Engl. J. Med.* 263, 887–90.

Larizza, D., Abbati, G., Lorini, R., Salvatoni, A., and Severi, F. (1982). The Turner phenotype and the different types of human X chromosome. *Hum. Genet.* 62, 93.

Larson, L.M., Bruce, A.W., Saumur, J.H., and Wasdahl, W.A. (1982*a*). Further evidence by gene dosage for the regional assignment of erythrocyte acid phosphatase (ACP1) and malate dehydrogenase (MDH1) loci on chromosome 2p. *Clin. Genet.* 22, 220–5.

Larson, L.M., Wasdahl, W.A., Saumur, J.H., Coleman, M.L., Hall, J.G., Dolan, C.R., and Schutta, C.J. (1982*b*). Familial reciprocal translocation, t(2;10) (p24;q26), resulting in duplication 2p and deletion 10q. *Clin. Genet.* 21, 187–95.

Lau, C.C., Gadi, I.K., Kalvonjian, S., Anisowicz, A., and Sager, R. (1985). Plasmid-induced "hit-and-run" tumorigenesis in Chinese hamster embryo fibroblast (CHEF) cells. *Proc. Natl. Acad. Sci. USA* 82, 2839–43.

Lawler, S.D., and Sanger, R. (1970). Xg blood groups and clonal-origin theory of chronic myeloid leukaemia. *Lancet* 1, 584–5.

Layzer, R.B., and Epstein, C.J. (1972). Phosphofructokinase and chromosome 21. *Am. J. Hum. Genet.* 24, 533–43.

Ledbetter, D.H., Mascarello, J.T., Riccardi, V.M., Harper, V.D., Airhart, S.D., and Strobel, R.J. (1982). Chromosome 15 abnormalities and the Prader-Willi syndrome: a follow-up report of 40 cases. *Am. J. Hum. Genet.* 34, 278–85.

Leder, P., Battey, J., Lenoir, G., Moulding, C., Murphy, W., Potter, H., Stewart, T., and Taub, R. (1983). Translocations among antibody genes in human cancer. *Science* 222, 765–71.

Lederberg, J. (1966). Experimental genetics and human evolution. *Bull. Atom. Scientists* 22(8), 4–11.

Lee, W.-H., Murphree, A.L., and Benedict, W.F. (1984). Expression and amplification of the N-*myc* gene in primary retinoblastoma. *Nature, Lond.* 309, 458–60.

Legraverend, C., Kärenlampi, S.O., Bigelow, S.W., Lalley, P.A., Kozak, C.A., Womack, J.E., and Nebert, D.W. (1984). Aryl hydrocarbon hydroxylase induction by benzo[α]anthracene: regulatory gene localized to the distal portion of mouse chromosome 17. *Genetics* 107, 447–61.

Leibovitz, A., and Yannet, H. (1941). The production of humoral antibodies by the mongolian. *Am. J. Ment. Defic.* 46, 304–9.

Leibson, H.J., Gefter, M., Zlotnik, A., Marrack, P., and Kappler, J.W. (1984). Role of γ-interferon in antibody-producing responses. *Nature, Lond.* 309, 799–801.

Leisti, J., Kaback, M.M., and Rimoin, D.L. (1974). Cri-du-chat and trisomy 13 syndromes in an infant with an unbalanced chromosomal translocation. *Birth Defects: Orig. Art. Ser.* 11(5), 317–19.

Lejeune, J. (1966). Type et contretypes. In *Journées Parisiennes de Pédiatrie*, pp. 73–85, Flammarion édit., Paris.

Lejeune, J. (1977). On the mechanism of mental deficiency in chromosomal diseases. *Hereditas* 86, 9–14.

Lejeune, J. (1979). Investigations biochemiques et trisomie 21. *Ann. Génét.* 22, 67–75.

Lejeune, J., Bourdais, M., and Prieur, M. (1976). Sensibilité pharmacologique de l'iris des enfants trisomiques 21. *Thérapie* 31, 447–54.

Lejeune, J., Berger, R., Rethoré, M.-O., Archambault, L., Jérôme, H., Thieffry, S., Aicardi, J., Broyer, M., Lafourcade, J., Cruveiller, J., and Turpin, R. (1964a). Monosomie partielle pour un petit acrocentrique. *C.R. Séances Acad. Sci.* 259, 4187–90.

Lejeune, J., Lafourcade, J., Berger, R., and Turpin, R. (1964b). Ségrégation familiale d'une translocation 5~13 déterminant une monosomie et une trisomie partielles du bras court du chromosome 5: maladie du "cri du chat" et sa reciproque. *C.R. Séances Acad. Sci.* 258, 5767–70.

Lejeune, J., Lafourcade, J., Berger, R., Cruveiller, J., Rethoré, M.-O., Dutrillaux, B., Abonyi, D., and Jérôme, H. (1968). Le phénotype [Dr], étude de trois cas de chromosomes D en anneau. *Ann. Génét.* 11, 79–87.

Lenke, R.R., and Levy, H.L. (1980). Maternal phenylketonuria and hyperphenylalaninemia. An international survey of the outcome of untreated and treated pregnancies. *N. Engl. J. Med.* 303, 1202–8.

Leonard, J.C., and Merz, T. (1983). The influence of cell cycle kinetics on the radiosensitivity of Down's syndrome lymphocytes. *Mutation Res.* 109, 111–21.

Leschot, N.J., De Nef, J.J., Geraedts, J.P.M., Becker-Bloemkolk, M.J., Talma, A., Bijlsma, and Verjaal, M. (1979). Five familial cases with a trisomy 16p syndrome due to translocation. *Clin. Genet.* 16, 205–14.

Leschot, N.J., Slater, R.M., Joenje, H., Becker-Bloemkolk, M.J., and de Nef, J.J. (1981). SOD-A and chromosome 21. Conflicting findings in a familial translocation (9p24;21q214). *Hum. Genet.* 57, 220–3.

Levan, G., and Mitelman, F. (1970). G-banding in Rous rat sarcomas during serial transfer: significant chromosome aberrations and incidence of abnormal mitoses. *Hereditas* 84, 1–14.

Levan, G., Ahlström, U., and Mitelman, F. (1974). The specificity of chromosome A2 involvement in DMBA-induced rat sarcoma, *Hereditas* 77, 263–80.

Levin, S., Nir, E., and Mogilner, B.M. (1975). T system immune-deficiency in Down's syndrome. *Pediatrics* 56, 123–6.

Levin, S., Schlesinger, M., Handzel, Z., Hahn, T., Altman, Y., Czernobilsky, B., and Boss, J. (1979). Thymic deficiency in Down's syndrome. *Pediatrics* 63, 80–7.

Levitt, D., and Cooper, M.D. (1981). Immunoregulatory defects in a family with selective IgA deficiency. *J. Pediatr.* 98, 52–8.

Lewandowski, R.C. Jr., and Yunis, J.J. (1977). Phenotypic mapping in man. In *New Chromosomal Syndromes*, ed. J.J. Yunis, pp. 369–94. Academic Press, New York.

Lewin, B. (1980). *Gene Expression. Vol. 2: Eucaryotic Chromsomes. 2nd edit.* Wiley, New York.

Lewis, A.J. (1964). The pathology of 18-trisomy. *J. Pediatr.* 65, 92–101.

Lewis, D., and Vakeria, D. (1977). Resistance to *p*-fluorophenylalanine in diploid/ haploid dikaryons: dominance modifier gene explained as a controller of hybrid multimer formation. *Genet. Res.* 30, 31–43.

Lewis, D.S., Thompson, M., Hudson, E., Liberman, M.M., and Samson, D. (1983). Down's syndrome and acute megalokaryoblastic leukemia. Case report and review of the literature. *Acta Haematol.* 70, 236–42.

Lewis, J., Slack, J.M.W., and Wolpert, L. (1977). Thresholds in development. *J. Theor. Biol.* 65, 579–90.

Lewis, W.H. (ed.) (1980). *Polyploidy. Biological Relevance.* Plenum, New York.

Lewkonia, R.M., Lin, C.C., and Haslam, R.H.A. (1980). Selective IgA deficiency with 18q + and 18q − karyotypic anomalies. *J. Med. Genet.* 17, 453–6.

Lewontin, R.C., Rose, S., and Kamin, L.J. (1984). *Not in Our Genes. Biology, Ideology, and Human Nature.* Pantheon, New York.

Ley, T., DeSimone, J., Anagnou, N.P., Keller, G.H., Humphries, R.K., Turner, P.H., Young, N.S., Heller, P., and Nienhuis, A.W. (1982). 5-Azacytidine selectively increases γ-globin synthesis in a patient with β + thalassemia. *N. Engl. J. Med.* 307, 1469–75.

Libb, J.W., Myers, G.J., Graham, E., and Bell, B. (1983). Correlates of intelligence and adaptive behavior in Down's syndrome. *J. Ment. Defic. Res.* 27, 205–10.

Liberfarb, R.M., Atkins, L., and Holmes, L.B. (1980). A clinical syndrome associated with 5p duplication and 9p deletion. *Ann. Génét.* 23, 26–30.

Lieman-Hurwitz, J., Dafni, N., Lavie, V., and Groner, Y. (1982). Human cytoplasmic superoxide dismutase cDNA clone: a probe for studying the molecular biology of Down syndrome. *Proc. Natl. Acad. Sci. USA* 79, 2908–11.

Lin, H.-P., Menaka, H., Lim, K.-H., and Yong, H.-S. (1980). Congenital leukemoid reaction followed by total leukemia. A case with Down's syndrome. *Am. J. Dis. Child.* 134, 939–41.

Lin, P.M., Crawford, M.H., and Oronzi, M. (1979). Universals in dermatoglyphics. *Birth Defects: Orig. Art. Ser.* 15(6), 63–84.

Lindsley, D.L., Sandler, L., Baker, B., Carpenter, A., Denell, R.E., Hall, J.C., Jacobs, P.A., Miklos, G.L.G., Davis, B.K., Gethmann, R.C., Hardy, R.W., Hessler, A., Miller, S.M., Nozawa, H., Parry, D.M., and Gould-Somero, M. (1972). Segmental aneuploidy and the genetic gross structure of the Drosophila genome. *Genetics* 71, 157–84.

Lo, C.W. (1983). Transformation by iontophoretic microinjection of DNA: multiple integrations without tandem insertions. *Mol. Cell. Biol.* 3, 1803–14.

Loomis, W.F. (1983). Genetic analysis of cell adhesion. *Dev. Genet.* 4, 61–8.

Lopez, V., Ochs, H.D., Thuline, H.C., Davis, S.D., and Wedgwood, R.S. (1975). Defective antibody response to bacteriophage φX174 in Down syndrome. *J. Pediatr.* 86, 207–11.

Lorenzen, F., Pang, S., Neu, M.J., Dupont, B., Pollack, M., Chow, D.M., and Levine, L.S. (1979). Hormonal phenotype and HLA-genotype in families of patients with congenital adrenal hyperplasia (21-hydroxylase deficiency). *Pediatr. Res.* 13, 1356–60.

Lott, I.T., Chase, T.N., and Murphy, D.L. (1972). Down's syndrome: transport, storage, and metabolism of serotonin in blood platelets. *Pediatr. Res.* 6, 730–5.

Lott, I.T., Murphy, D.L., and Chase, T.N. (1972). Down's syndrome. Central monoamine turnover in patients with diminished platelet serotonin. *Neurology* 22, 967–72.

Lowry, O.H., and Passoneau, J.V. (1964). The relations between substrates and enzymes of glycolysis in brain. *J. Biol. Chem.* 139, 31–42.

Lowry, R.B., Lin, C.C., Heikkila, E.M., Hoo, J.J., and Chudley, A.E. (1983), Familial mental retardation in 3 generations due to 7q duplication or deletion. *Am. J. Hum. Genet.* 35, 141A.

Lu, T.-Y., and Markert, C.L. (1980). Manufacture of diploid/tetraploid chimeric mice. *Proc. Natl. Acad. Sci. USA* 77, 6012–16.

Lubiniecki, A.S., Blattner, W.A., and Fraumeni, J.F., Jr. (1977). Elevated expression of T-antigen in simian papovavirus 40-infected skin fibroblasts from individuals with cytogenetic defects. *Cancer Res.* 37, 1580–3.

Lubiniecki, A.S., Blattner, W.A., Crittenden, V., Gunnell, M., Tarr, G.C., and Fraumeni, J.F., Jr. (1978). T-antigen expression in human skin fibroblasts is not regulated by an endogenus interferon response to SV40 infection. *Arch. Virol.* 57, 34–54.

Lubiniecki, A.S., Blattner, W.A., Martin, G.R., Fialkow, P.J., Dosik, H., Eatherly, C., and Fraumeni, J.F., Jr. (1979). SV40 T-antigen expression in cultured fibroblasts from patients with Down syndrome and their parents. *Am. J. Hum. Genet.* 31, 469–77.

Lucchesi, J.C., and Rawls, J.M. Jr. (1973). Regulation of gene function: a comparison of enzyme activity levels in relation to gene dosage in diploids and triploids of *Drosophila melanogaster*. *Biochem. Genet.* 9, 41–51.

Ludlow, J.R., and Allen, L.M. (1979). The effect of early intervention and preschool stimulus on the development of the Down's syndrome child. *J. Ment. Defic. Res.* 23, 29–44.

Luk, K.-C., and Mark, K.-K. (1982). Gene dosage as a regulatory factor for gene expression. I. In λplac5-infected cells. *J. Gen. Virol.* 58, 297–304.

Lundin, L.-G. (1979). Evolutionary conservation of large chromosomal segments reflected in mammalian gene maps. *Clin. Genet.* 16, 72–81.

Lusis, A.J., and Paigen, K. (1975). Genetic determination of the α-galactosidase developmental program in mice. *Cell* 6, 371–8.

Lusis, A.J., Chapman, V.M., Wangenstein, R.W., and Paigen, K. (1983). *Trans*-acting temporal locus within the β-glucuronidase gene complex. *Proc. Natl. Acad. Sci. USA* 80, 4398–4402.

Lykkesfeldt, G., Lykkesfeldt, A.E. and Skakkebaek, N.E. (1984). Steroid sulfatase in man: a noninactivated X-locus with partial gene dosage compensation. *Hum. Genet.* 65, 355–7.

Lyon, M.F. (1962). Sex chromatin and gene action in the mammalian X-chromsome. *Am. J. Hum. Genet.* 14, 135–48.

Lyon, M.F. (1966). Lack of evidence that inactivation of the mouse X-chromosome is incomplete. *Genet. Res.* 8, 197–203.

Lyon, M.F. (1970). Genetic activity of sex chromosomes in somatic cells of mammals. *Philos. Trans. R. Soc. Lond. (Biol.)* 259, 41–52.

Lyon, M.F. (1971). Possible mechanisms of X chromosome inactivation. *Nature (New Biol.), Lond.* 232, 229–32.

Lyon, M.F., and Hawker, S.G. (1973). Reproductive lifespan in irradiated and unirradiated chromosomally *XO* mice. *Genet. Res.* 21, 184–94.

Lyon, M.F., and Meredith, R. (1966). Autosomal translocations causing male sterility and viable aneuploidy in the mouse. *Cytogenetics* 5, 335–54.

Lyon, M., Sayers, I.M., and Evans, E.P. (1978). Personal communication. *Mouse News Letter* No. 58, 44.

McBride, O.W., and Peterson, J.L. (1980). Chromosome-mediated gene transfer in mammalian cells. *Annu. Rev. Genet.* 14, 321–45.

McBride, O.W., Olsen, A.S., Aulakh, G.S., and Athwal, R.S. (1982). Measurement of transcribed human X-chromosomal DNA sequences transferred to rodent cells by chromosome-mediated gene transfer. *Mol. Cell Biol.* 2, 52–65.

McBurney, M.W. (1976). Clonal lines of teratocarcinoma cells *in vitro*: differentiation and cytogenetic characteristics. *J. Cell Physiol.* 89, 441–5.

McBurney, M.W., Jones-Villeneuve, E.M.V., Edwards, M.K.S., and Rudnicki, M. (1983). Controlled development of embryonic tissues in a differentiating embryonal carcinoma cell line. In *Teratocarcinoma Stem Cells*, ed. L.M. Silver, G.M. Martin, and S. Strickland, pp. 121–4. Cold Spring Harbor Laboratory, Cold Spring Harbor, New York.

McClure, H.M. (1972). Animal model: trisomy in a chimpanzee. *Am. J. Pathol.* 67, 413–16.

McClure, H.M., Belden, K.H., Pieper, W.A., and Jacobson, C.B. (1969). Autosomal trisomy in a chimpanzee: resemblance to Down's syndrome. *Science* 165, 1010–12.

McCoy, E.E., and Enns, L. (1978). Sodium transport, ouabain binding, and (Na$^+$/K$^+$)-ATPase activity in Down's syndrome platelets. *Pediatr. Res.* 12, 685–9.

McCoy, E.E., and Enns, L. (1980). Decreased serotonin uptake in Down's syndrome platelets is linked to abnormalities in Na$^+$ and K$^+$ transport. *Clin. Res.* 28, 100A.

McCoy, E.E., and Sneddon, J.M. (1984). Decreased calcium content and ^{45}Ca^{2+} uptake in Down's syndrome blood platelets. *Pediatr. Res.* 18, 914–6.

McCoy, E.E., Segal, D.J., Bayer, S.M., and Strynadka, K. (1974). Decreased ATPase and increased sodium content of platelets in Down's syndrome. *N. Engl. J. Med.* 291, 950–3.

McCoy, E.E., Strynadka, K., Pabst, H.F., and Crawford, J. (1982). Decreased polyamine content of concanavalin A stimulated lymphocytes in Down's syndrome subjects. *Pediatr. Res.* 16, 314–17.

McDaniel, R.G., and Ramage, R.T. (1970). Genetics of a primary trisomic series in barley: identification by protein electrophoresis. *Can. J. Genet. Cytol.* 12, 490–5.

McGeer, E.G., Norman, M., Boyes, B., O'Kusky, J., Suzuki, J., and McGeer, P.L. (1985). Acetylcholine and aromatic amine systems in postmortem brain of an infant with Down's syndrome. *Exp. Neurol.* 87, 557–70.

McGillivray, B.C., Dill, F.J., and Lowry, K.B. (1978). HLA investigation and phenotype correlation of two siblings with trisomy 6p12pter. *Am. J. Hum. Genet.* 30, 59A.

McKnight, G.S., Hammer, R.E., Kuenzel, E.A., and Brinster, R.L. (1983). Expression of the chicken transferrin gene in transgenic mice. *Cell* 34, 335–41.

McKusick, V.A., and Ruddle, F.H. (1977). The status of the gene map of the human chromosomes. *Science* 196, 390–405.

McLaren, A., and Burgoyne, P. (1984). Personal communication. *Mouse News Letter* No. 70, 91.

McLaren, A., and Monk, M. (1982). Fertile females produced by inactivation of an X chromosome of '*sex-reversed*' mice. *Nature, Lond.* 300, 446–8.

McSwigan, J.D., Hanson, D.R., Lubiniecki, A., Heston, L.L., and Sheppard, J.R. (1981). Down syndrome fibroblasts are hyperresponsive to β-adrenergic stimulation. *Proc. Natl. Acad. Sci. USA* 78, 7670–3.

Maccario, R., Ugazio, A.G., Nespoli, L., Alberini, C., Montagna, D., Porta, F., Bonetti, F., and Burgio, G.R. (1984). Lymphocyte subpopulations in Down's syndrome: high percentage of circulatory HNK–1⁺, Leu2a⁺ cells. *Clin. Exp. Immunol.* 57, 220–6.

Macera, M.J., and Bloom, S.E. (1981). Ultrastructural studies of the neucleoli in diploid and trisomic chickens. *J. Hered.* 72, 249–52.

Madsen, M. (1981). Synergism and gene dose effect in HLA-DR serology. *Tissue Antigens* 17, 269–76.

Magenis, R.E., Koler, R.D., Lovrien, E., Bigley, R.H., DuVal, M.C., and Overton, K.M. (1975). Gene dosage: evidence for assignment of erythrocyte acid phosphatase locus to chromosome 2. *Proc. Natl. Acad. Sci. USA* 72, 4526–30.

Magenis, E., Brown, M., Donlon, T., Bangs, D., Allen, L., Bigley, R., and Koler, R. (1982). X-inactivation dogma revisited: dosage effect of G6PD in an infant with inverted duplication of the X distal long arm, 46Xinvdup(X)(qter→q26). *Am. J. Hum. Genet.* 34, 134A.

Magnani, M., Stocchi, V., Piatti, E., Dachà, M., Dallipiccola, B., and Fornaini, G. (1983). Red blood cell glucose metabolism in trisomy 10p: possible role of hexokinase in the erythrocyte. *Blood* 61, 915–19.

Magnuson, T., Smith, S., and Epstein, C.J. (1982). The development of monosomy 19 mouse embryos. *J. Embryol. Exp. Morphol.* 69, 223–36.

Magnuson, T., Epstein, C.J., Silver, L.M., and Martin, G.R. (1982). Pluripotent embryonic stem cell lines can be derived from *t^{w5}/t^{w5}* blastocysts. *Nature, Lond.* 298, 750–3.

Magnuson, T., Debrot, S., Dimpfl, J., Zweig, A., Zamora, T., and Epstein, C.J. (1985). The early lethality of autosomal monosomy in the mouse. *J. Exp. Zool.*, in press.

Malamud, N. (1972). Neuropathology of organic brain syndromes associated with aging. In *Aging and the Brain*, ed. C.M. Gartz, pp. 63–87. Plenum Press, New York.

Malawista, S.E., and Weiss, M.C. (1974). Expression of differentiated function in hepatoma cell hybrids: high frequency of induction of mouse albumin production in rat hepatoma – mouse lymphoblast hybrids. *Proc. Natl. Acad. Sci. USA* 71, 927–31.

Malpuech, G., Kaplan, J.C., Rethoré, M.O., Junien, C., and Geneix, A. (1975). Une observation de délétion partielle du bras court de chromosome 12. Localisation du gène de la lactico deshydrogénase 3. *Lyon Medical* 233, 275–9.

Mangia, F., Abbo-Halbasch, G., and Epstein, C.J. (1975). X chromosome expression during oogenesis in the mouse. *Dev. Biol.* 45, 366–8.

Mangin, M., Ares, M. Jr., and Weiner, A.M. (1985). U1 small nuclear RNA genes subject to dosage compensation in mouse cells. *Science* 229, 272–5.

Mankinen, C.B., Holt, J.G., and Sears, J.W. (1976). Partial trisomy 15 in a young girl. *Clin. Genet.* 10, 27–32.

Maraschio, P. and Fraccaro, M. (1983). Phenotypic effects of X-chromosome deficiencies. In *Cytogenetics of the Mammalian X Chromosome, Part B: X Chromosome Anomalies and Their Clinical Manifestations*, ed. A.A. Sandberg, pp. 359–69. A.R. Liss, New York.

Maraschio, P., Zuffardi, O., Bernardi, F., Bozzola, M., DePaoli, C., Fonatsch, G., Flatz, S.D., Ghersini, L., Gimelli, G., Loi, M., Lorini, R., Peretti, D., Poloni, L., Tonetti, D., Vanni, R., and Zamboni, G. (1981). Preferential maternal derivation in inv dup(15). Analysis of eight new cases. *Hum. Genet.* 57, 345–50.

Maraschio, P., Danesino, C., Lo Curto, F., Zuffardi, O., Dalla Fior, T., and Pedrotti, D. (1984). A liveborn 69,XXX triploid. Origin, X chromosome activity and gene dosage. *Ann. Génét.* 27, 96–101.

Marcu, K.B., Harris, L.J., Stanton, L.W., Erickson, J., Watt, R., and Croce, C.M. (1983). Transcriptionally active c-*myc* oncogene is contained within NIARD, a DNA sequence associated with chromosome translocations in B-cell neoplasia. *Proc. Natl. Acad. Sci. USA* 80, 519–23.

Marimo, B., and Giannelli, F. (1975). Gene dosage effect in human trisomy 16. *Nature, Lond.* 256, 204–6.

Marin-Padilla, M. (1976). Pyramidal cell abnormalities in the motor cortex of a child with Down's syndrome. A Golgi study. *J. Comp. Neurol.* 167, 63–82.

Markannen, A., Somer, M., and Nordström, A.-M. (1984). Distal trisomy 14q syndrome; a case report. *Clin. Genet.* 26, 231–4.

Martin, G.M., and Hoehn, H. (1974). Genetics and human disease. *Hum. Pathol.* 5, 387–405.

Martin, G.R. (1975). Teratocarcinomas as a model system for the study of embryogenesis and neoplasia. *Cell* 5, 229–43.

Martin, G.R. (1980). Teratocarcinomas and mammalian embryogenesis. *Science* 209, 768–76.

Martin, G.R. (1981). Isolation of a pluripotent cell line of early mouse embryos cultured in medium conditioned by teratocarcinoma stem cells. *Proc. Natl. Acad. Sci. USA* 78, 7634–8.

Martin, G.R., Wiley, L.M., and Damjanov, I. (1977). The development of cystic embryonal bodies *in vitro* from clonal teratocarcinoma stem cells. *Dev. Biol.* 61, 230–44.

Martin, G.R., Epstein, C.J., Travis, B., Tucker, G., Yatziv, S., Martin, D.W. Jr., Clift, S., and Cohen, S. (1978). X-chromosome inactivation during differentiation of female teratocarcinoma stem cells *in vitro*. *Nature, Lond.* 271, 329–33.

Martin, R.H., Lin, C.C., Balkan, W., and Burns, K. (1982). Direct chromosomal analysis of human spermatozoa: preliminary results from 18 normal men. *Am. J. Hum. Genet.* 34, 459–68.

Marx, J. (1983). Specific expression of transferred genes. *Science* 222, 1001–2.

Masaki, M., Higurashi, M., Iijima, K., Ishikawa, N., Tanaka, F., Fujii, T., Kuroki, Y., Matsui, I., Iinuma, K., Matsuo, N., Takeshita, K., Hashimoto, S. (1981). Mortality and survival of Down syndrome in Japan. *Am. J. Hum. Genet.* 33, 629–39.

Masouredis, J.P., Dupuy, M.E., and Elliot, M. (1967). Relationship between $Rh_0(D)$ zygosity and red cell $Rh_0(D)$ antigen content in family members. *J. Clin. Invest.* 46, 681–94.

Matsunaga, E. (1979). Hereditary retinoblastoma: host resistance and age of onset. *J. Natl. Cancer Inst.* 63, 933–9.

Matsunaga, E. (1980). Hereditary retinoblastoma: host resistance and second primary tumors. *J. Natl. Cancer Inst.* 65, 47–51.

Matsunaga, E. (1981). Genetics of Wilms' tumor. *Hum. Genet.* 57, 231–46.

Matsunaga, E., and Tonomura, A. (1972). Parental age and birth weight in translocation Down's syndrome. *Ann. Hum. Genet.* 36, 209–19.

Matte, R., Saski, M., and Obara, Y. (1969). Blastic response of lymphocytes in mosaic Down's syndrome under serial PHA dilutions. *Jpn. J. Hum. Genet.* 14, 160–2.

Mattei, J.F., Mattei, M.G., and Giraud, F. (1983). Prader-Willi syndrome and chromsome 15. A clinical discussion of 20 cases. *Hum. Genet.* 64, 356–62.

Mattei, J.F., Mattei, M.C., Ardissone, J.P., and Giroud, F. (1979). Regional mapping of human glutathione reductase on chromosome 8. *Cytogenet. Cell Genet.* 25, 183.

Mattei, J.F., Mattei, M.G., Ardissone, J.P., Taramasco, H., and Giraud, F. (1980). Pericentric inversion, inv(9)(p22132), in the father of a child with a duplication-deletion of chromosome 9 and a gene dosage effect for adenylate kinase–1. *Clin. Genet.* 17, 129–36.

Mattei, J.F., Mattei, M.G., Balestrazzi, P., and Giraud, F. (1983). Familial pericentric inversion of chromosome 9, INV(9)(p22q32) with recurrent duplication-deletion. *Clin. Genet.* 24, 220–2.

Mattei, M.G., Souiah, N., and Mattei, J.F. (1984). Chromosome 15 anomalies and the Prader-Willi syndrome: cytogenetic analysis. *Hum. Genet.* 66, 313–34.

Mavilio, F., Giampaolo, A., Carè, A., Migliaccio, G., Calandrini, M., Russo, G., Pagliardi, G.L., Mastroberardino, G., Marinucci, M., and Peschle, C. (1983). Molecular mechanisms of human hemoglobin switching: selective undermethylation and expression of globin genes in embryonic, fetal, and adult erythroblasts. *Proc. Natl. Acad. Sci. USA* 80, 6907–11.

Mayes, J., Muneer, R., and Sifers, M. (1984). Superoxide dismutase activity and oxygen toxicity in Down syndrome fibroblasts. *Am. J. Hum. Genet.* 36, 15S.

Meinhardt, H. (1978). Space-dependent cell determination under the control of a morphogen gradient. *J. Theor. Biol.* 74, 307–21.

Mellman, W.J., Younkin, L.H., and Baker, D. (1970). Abnormal lymphocyte function in trisomy 21. *Ann. N.Y. Acad. Sci.* 171, 537–42.

Mellman, W.J., Oski, F.A., Tedesco, T.A., Machera-Coelho, A., and Harris, H. (1964). Leukocyte enzymes in Down's syndrome. *Lancet* 2, 674–5.

Mellman, W.J., Raab, S.D., Oski, F.A., and Tedesco, T.A. (1968). Abnormal leukokinetics in 21 trisomy. *Ann. N.Y. Acad. Sci.* 155, 1020–2.

Meneely, P.M., and Wood, W.B. (1984). An autosomal gene that affects X chromosome expression and sex determination in *Caenorhabditis elegans*. *Genetics* 106, 29–44.

Mével-Ninio, M. (1984). Immunofluorescence analysis of reexpression and activation: the origin of phenotypic diversity of rat hepatoma-mouse fibroblast hybrid colonies. *Differentiation* 24, 68–76.

Mével-Ninio, M., and Weiss, M.C. (1981). Immunofluorescence analysis of the time-course of extinction, reexpression, and activation of albumin production in rat hepatoma-mouse fibroblast heterokaryons and hybrids. *J. Cell Biol.* 90, 339–50.

Meyer, U.A., and Schmid, R. (1978). The porphyrias. In *The Metabolic Basis of Inherited Disease, 4th edit.*, ed. J.B. Stanbury, J.B. Wyngaarden, and D.S. Fredrickson, pp. 1166–1220. McGraw-Hill, New York.

Michels, V.V., Berseth, C.L., O'Brien, J.F., and Dewald, G. (1984). Duplication of part of chromosome 1q: clinical report and review of the literature. *Am. J. Med. Genet.* 18, 125–34.

Migeon, B.R., and Do, T.T. (1978). In search of nonrandom X inactivation: studies of the placenta from newborns heterozygous for glucose–6-phosphate dehydrogenase. In *Genetic Mosaics and Chimeras in Mammals*, ed. L.B. Russell, pp. 379–91. Plenum, New York.

Migeon, B.R., and Jelalian, K. (1977). Evidence for two active X chromosomes in germ cells of female before meiotic entry. *Nature, Lond.*, 269, 242–3.

Migeon, B.R., Shapiro, L.J., Norum, R.A., Mohandas, T., Axelman, J., and Dabova, R.L. (1982). Differential expression of steroid sulfatase locus on active and inactive X chromosome. *Nature, Lond.* 299, 838–40.

Migeon, B.R., Wolf, S.F., Axelman, J., Kaslow, D.C., and Schmidt, M. (1985). Incomplete X chromosome dosage compensation in chorionic villi of human placenta. *Proc. Natl. Acad. Sci. USA* 82, 3390–4.

Mikamo, K., Aguereif, M., Hazeghi, P., and Martin-DuPan, R. (1968). Chromatin-positive Klinefelter's syndrome. A quantitative analysis of spermatogonial deficiency at 3, 4, and 12 months of age. *Fertil. Steril.* 19, 731–9.

Miller, C.L., and Ruddle, F.H. (1978). Co-transfer of human X-linked markers into murine somatic cells via isolated metaphase chromosomes. *Proc. Natl. Acad. Sci. USA* 75, 3346–50.

Miller, J.C., Sherrill, J.G., and Hathaway, W.E. (1967). Thrombocythemia in the myeloproliferative disorders of Down's syndrome. *Pediatrics* 40, 847–50.

Miller, J.F., Williamson, E., Glue, J., Gordon, Y.B., Grudzinskas, J.G., and Sykes, A. (1980). Fetal loss after implantation. A prospective study. *Lancet* 2, 554–6.

Miller, M., and Cosgriff, J.M. (1983). Hematological abnormalities in newborn infants with Down syndrome. *Am. J. Med. Genet.* 16, 173–7.

Miller, M.E., Mellman, W.J., Oski, F.A., and Kohn, G. (1967). Immunoglobulins in Down's syndrome. *Lancet* 2, 257–8.

Miller, M., Kaufman, G., Reed, G., Bilenker, R., and Schinzel, A. (1979). Familial, balanced insertional translocation of chromosome 7 leading to offspring with deletion and duplication of the inserted segment, 7p15→7p21. *Am. J. Med. Genet.* 4, 323–32.

Miller, O.J., Drayna, D., and Goodfellow, P. (1984). Report of the committee on the genetic constitution of the X and Y chromosomes. *Cytogenet. Cell Genet.* 37, 176–204.

Miller, R.W. (1970). Neoplasia and Down's syndrome. *Ann. N.Y. Acad. Sci.* 171, 637–44.

Milunsky, A., Hackley, B.M., and Halsted, J.A. (1970). Plasma, erythrocyte and leukocyte zinc levels in Down's syndrome. *J. Ment. Defic. Res.* 14, 99–105.

Minna, J.D., Lalley, P.A., and Francke, U. (1976). Comparative mapping using somatic cell hybrids. *In Vitro* 12, 726–33.

Minna, J., Nelson, P., Peacock, J., Glazer, D., and Nirenberg, M. (1971). Genes for neuronal properties expressed in neuroblastoma × L cell hybrids. *Proc. Natl. Acad. Sci. USA* 68, 234–9.

Mintz, B. (1959). Continuity of the female germ cell line from embryo to adult. *Arch. Anat. Micros. Morphol. Exp.* 48(Suppl.), 155–72.

Mintz, B. (1971). Allophenic mice of multi-embryo origin. In *Methods in Mammalian Embryology*, ed. J.C. Daniel, Jr. pp. 186–214. Freeman, San Francisco.

Mintz, B., and Cronmiller, C. (1981). METT–1: a karyotypically normal in vitro line of developmentally totipotent mouse teratocarcinoma cells. *Somat. Cell Genet.* 7, 489–505.

Mintz, B., and Fleischman, R.A. (1981). Teratocarcinomas and other neoplasms as developmental defects in gene expression. *Adv. Cancer Res.* 34, 211–78.

Mintz, B., Illmensee, K., and Gearhart, J.D. (1975). Developmental and experimental potentialities of mouse teratocarcinoma cells from embryonal body cores. In *Teratocarcinomas and Differentiation*, ed. M. Sherman and D. Solter, pp. 59–82. Academic Press, New York.

Mir, G.H., and Cumming, G.R. (1971). Response to atropine in Down's syndrome. *Arch. Dis. Child.* 46, 61–65.

Miró, R., Templado, C., Ponsá, M., Serradell, J., Marina, S., and Egozcue, J. (1980). Balanced translocation (10;13) in a father ascertained through the study of meiosis in semen, and partial trisomy 10q in his son. Characterization of the region responsible for the partial 10q syndrome. *Hum. Genet.* 53, 179–82.

Mitelman, F. (1981). Tumor etiology and chromosome pattern – evidence from human and experimental neoplasms. In *Genes, Chromosomes, and Neoplasia*, ed. F.E. Arrighi, P.N. Rao, and E. Stubblefield, pp. 335–50. Raven, New York.

Mitelman, F. (1983). Catalogue of chromosome aberrations in cancer. *Cytogenet. Cell Genet.* 36, 1–515.

Mitelman, F. (1984). Restricted number of chromosomal regions implicated in aetiology of human cancer and leukaemia. *Nature, Lond.* 310, 325–7.

Mitelman, F., and Levan, G. (1981). Clustering of aberrations to specific chromosomes in human neoplasms. IV. A survey of 1,871 cases. *Hereditas* 95, 79–139.

Mitelman, F., Mark, J., and Levan, G. (1972). The chromosomes of six primary sarcomas induced in the Chinese hamster by 7,12-dimethylbenz(α)anthracene. *Hereditas* 72, 311–8.

Mittwoch, U. (1967). DNA synthesis in cells grown in tissue culture from patients with mongolism. In *Mongolism*. Ciba Foundation Study Group No. 25, ed. G.E.W. Wolstenholme and R. Porter, pp. 51–69. Churchill, London.

Mittwoch, U. (1971). Mongolism and sex: a common problem of cell proliferation? *J. Med. Genet.* 9, 92–5.

Mittwoch, U. (1973). *Genetics of Sex Differentiation*. Academic Press, New York.

Mittwoch, U. (1977). Sex chromatin and the biological effects of triploidy. *J. Med. Genet.* 14, 151.

Miyabara, S., Gropp, A., and Winking, H. (1982). Trisomy 16 in the mouse fetus associated with generalized edema, cardiovascular and urinary tract anomalies. *Teratology* 25, 369–80.

Mock, D.M., Perman, J.A., Thaler, M.M., and Morris, R.C. Jr. (1983). Chronic fructose intoxication after infancy in children with hereditary fructose intolerance. A cause of growth retardation. *N. Engl. J. Med.* 309, 764–70.

Moedjono, S.J., Crandall, B.F., and Sparkes, R.S. (1980). Tetrasomy 9p: confirmation by enzyme analysis. *J. Med. Genet.* 17, 227–30.

Moen, D.W., Werner, J.K., and Bersu, E.T. (1984). Analysis of gross anatomical variations in human triploidy. *Am. J. Med. Genet.* 18, 345–56.

Mohandas, T., Sparkes, R.S., and Shapiro, L.J. (1982). Genetic evidence for the inactivation of a human autosomal locus attached to an inactive X chromosome. *Am. J. Hum. Genet.* 34, 811–7.

Monk, M. (1981). A stem-line model for cellular and chromosomal differentiation in early mouse development. *Differentiation* 19, 71–6.

Monk, M., and Harper, M.I. (1978). X-chromosome activity in preimplantation mouse embryos from XX and XO mothers. *J. Embryol. Exp. Morphol.* 46, 53–64.

Monk, M., and Harper, M.I. (1979). Sequential X chromosome inactivation coupled with cellular differentiation in early mouse embryos. *Nature, Lond.* 281, 311–3.

Monk, M., and Kathuria, H. (1977). Dosage compensation for an X-linked gene in preimplantation mouse embryos. *Nature, Lond.* 270, 599–601.

Monk, M., and McLaren, A. (1981). X-chromosome activity in foetal germ cells of the mouse. *J. Embryol. Exp. Morphol.* 63, 75–84.

Moolgavkar, S.H., and Knudson, A.G., Jr. (1981). Mutation and cancer: a model for human carcinogenesis. *J. Natl. Cancer Inst.* 66, 1037–52.

Moore, C.M., Pfeiffer, R.A., Craig-Holmes, A.P., Scott, C.I. Jr., and Meisel-Stosiek, M. (1982). Partial trisomy 7p in two families resulting from different balanced translocations. *Clin. Genet.* 21, 112–21.

Moore, E.E., Jones, C., Kao, F.-T., and Oates, D.C. (1977). Synteny between glycinamide ribonucleotide synthetase and superoxide dismutase (soluble). *Am. J. Hum. Genet.* 29, 389–96.

Moorhead, P.S., and Heyman, A. (1983). Chromosome studies of patients with Alzheimer disease. *Am. J. Med. Genet.* 14, 545–56.

More, R., Amir, N., Meyer, S., Kopolovic, J., and Yarom, R. (1982). Platelet abnormalities in Down's syndrome. *Clin. Genet.* 22, 128–36.

Morimoto, K., Kaneko, T., Iijima, K., and Koizumi, A. (1984). Proliferative kinetics and chromosome damage in trisomy 21 lymphocyte cultures exposed to γ-rays and bleomycin. *Cancer Res.* 44, 1499–1504.

Morse, J.O. (1978). Alpha$_1$-antitrypsin deficiency. *N. Engl. J. Med.* 299, 1099–1105.

Morten, J.E.N., Harnden, D.G., and Taylor, A.M.R. (1981). Chromosome damage in G_0 X-irradiated lymphocytes from patients with hereditary retinoblastoma. *Cancer Res.* 41, 3635–8.

Mortimer, J.G., Chewings, W.E., and Gardner, R.J.M. (1980). A further report on a kindred with cases of 4p trisomy and monosomy. *Hum. Hered.* 30, 58–61.

Mortimer, J.G., Chewings, W., Miethke, P., and Smith, G.F. (1978). Trisomy 4p and deletion 4p − in a family having a translocation, t(4p − ;12p +). *Hum. Hered.* 28, 132–40.

Morton, C.C., Bieber, F.R., Mohanakumar, T., Nance, W.E., Redwine, F.O., and Brown, J.A. (1980). Codominant expression of major histocompatibility complex (MHC) in a case of partial trisomy 6p resulting from an insertion and inversion involving heterologous chromosomes. *Am. J. Hum. Genet.* 32, 81A.

Morton, C.C., Brown, J.A., Evans, G.A., Nance, W.E., Mohanakumar, T., and Kirsch, I.R. (1982). Detection and localization of an extra HLA locus in a karyotypically normal male by chromosomal *in situ* hybridization. *Am. J. Hum. Genet.* 34, 136A.

Morton, N.E., Lindsten, J., Iselius, L., and Yee, S. (1982). Data and theory for a revised chiasma map of man. *Hum. Genet.* 62, 266–70.

Moscona, A.A. (1980). In *Membranes, Receptors, and the Immune Response*, ed. E.P. Cohen and H. Köhler, pp. 171–88. A.R. Liss, New York.

Moses, M.J., Counce, S.J., and Paulson, D.F. (1975). Synaptonemal complex complement of man in spreads of spermatocytes, with details of the sex chromosome pair. *Science* 187, 363–5.

Moss, T.J., and Austin, G.E. (1980). Pre-atherosclerotic lesions in Down syndrome. *J. Ment. Defic. Res.* 24, 137–41.

Mottet, N.K., and Jensen, H. (1965). The anomalous embryonic development associated with trisomy 13–15. *Am. J. Clin. Pathol.* 43, 334–47.

Moustafa, L.A., and Brinster, R.L. (1972). Induced chimerism by transplanting embryonic cells into mouse blastocysts. *J. Exp. Zool.* 181, 193–202.

Msall, M., Oster-Granite, M.L., and Gearhart, J. (1985). Neural crest involvement in mouse trisomy 16 embryopathy. *Pediatr. Res.* 19, 328A.

Muchi, H., Ohira, M., Ise, T., Shimoyama, M., Minato, K., Saito, H., and Watanabe, S. (1980). *Rinsho Ketsueki* 21, 388–95.

Mücke, J., Trautmann, U., Sandig, K.-R., and Theile, H. (1982). The crucial band for phenotype of trisomy 18. *Hum. Genet.* 60, 205.

Mudd, S.H. and Levy, H.L. (1978). Disorders of transsulfuration. In *The Metabolic Basis of Human Disease, 4th edit.*, ed. J.B. Stanbury, J.B. Wyngaarden, and D.S. Fredrickson, pp. 458–503. McGraw-Hill, New York.

Mulcahy, M.T., Pemberton, P.J., and Sprague, P. (1979). Trisomy 3q: two clinically similar but cytogenetically different cases. *Ann. Génét.* 22, 217–20.

Mulcahy, M.T., Pemberton, P.J., Thompson, E., and Watson, M. (1982). Is there a monosomy 10qter syndrome? *Clin. Genet.* 21, 33–5.

Müller, C.R., Migl, B., Traupe, H., and Ropers, H.H. (1980). X-linked steroid sulfatase: evidence for different gene-dosage in males and females. *Hum. Genet.* 54, 197–9.

Müller-Hermelink, H.K. (1977). Thymusepithel in der Ontogenese bei primären Immunodefekten und Thymomen. *Verh. Dtsch. Ges. Pathol.* 61, 487.

Mulvihill, J.J., and Smith, D.W. (1969). The genesis of dermatoglyphics. *J. Pediatr.* 75, 579–89.

Murdoch, J.C., and Evans, J.H. (1978). An objective in vitro study of aging in the skin of patients with Down's syndrome. *J. Ment. Defic. Res.* 22, 131–5.

Murdoch, J.C., Rodger, J.C., Rao, S.S., Fletcher, C.D., and Dunnigan, M.G. (1977). Down's syndrome: an atheroma-free model? *Br. Med. J.* 2, 226–8.

Murphree, A.L., and Benedict, W.F. (1984). Retinoblastoma: clues to human oncogenesis. *Science* 223, 1028–33.

Murray, A.W., and Szostak, J.W. (1983). Construction of artificial chromosomes in yeast. *Nature, Lond.* 305, 189–93.

Nadeau, J.H., and Eicher, E.M. (1982). Conserved linkage of soluble aconitase and galactose–1-phosphate uridyl transferase in mouse and man: assignment of these genes to mouse chromosome 4. *Cytogenet. Cell Genet.* 34, 271–81.

Nadeau, J.H., and Taylor, B.A. (1984). Lengths of chromosomal segments conserved since divergence of man and mouse. *Proc. Natl. Acad. Sci. USA* 81, 814–18.

Nadler, H.L., Monteleone, P., and Hsia, D.Y.-Y. (1967). Enzyme studies during lymphocyte stimulation with phytohemagglutinin in Down's syndrome. *Life Sci.* 6, 2003–8.

Naeye, R.L. (1967). Prenatal organ and cellular growth with various chromosomal disorders. *Biol. Neonat.* 11, 248–60.

Nagase, S., Shimamane, K., and Shumiya, S. (1979). Albumin-deficient rat model. *Science* 205, 590–1.

Nair, M.P.N., and Schwartz, S.A. (1984). Association of decreased T-cell-mediated natural cytotoxicity and interferon production in Down's syndrome. *Clin. Immunol. Immunopathol.* 33, 412–24.

Najafzadeh, T.M., Littman, V.A., and Dumars, K.W. (1983). Familial t(4;13) with abnormal offspring in three generations. *Am. J. Med. Genet.* 16, 15–22.

Narahara, K., Kikkawa, K., Kimira, S., Kimoto, H., Ogata, M., Kasai, R., Hamawaki, M., and Matsuoka, K. (1984). Regional mapping of catalase and Wilms tumor-aniridia, genitourinary abnormalities, and mental retardation triad loci to the chromosome segment 11p1305 → p1306. *Hum. Genet.* 66, 181–5.

Nash, W.G., and O'Brien, S.J. (1982). Conserved regions of homologous G-banded chromosomes between orders in mammalian evolution: carnivores and primates. *Proc. Natl. Acad. Sci. USA* 79, 6631–5.

Nebert, D.W., and Jensen, N.M. (1979). The *Ah* locus: genetic regulation of the metabolism of carcinogens, drugs, and other environmental chemicals by cytochrome P–450 mediated monooxygenases. In *CRC Critical Reviews in Biochemistry, Vol. 6,* ed. G.D. Fasman, pp. 401–37. CRC Press, Cleveland.

Nebert, D.W., Atlas, S.A., Guenther, T.M., and Kouri, R.E. (1978). The *Ah* locus: genetic regulation of the enzymes which metabolize polycyclic hydrocarbons and the risk for cancer. In *Polycyclic Hydrocarbons and Cancer, Vol. 2: Molecules and Cell Biology,* ed. H.V. Gelboin and P.O.P. Ts'o, pp. 346–90. Academic Press, New York.

Nebert, D.W., Eisen, H.J., Negishi, M., Lang, M.A., and Hjelmeland, L.M. (1981). Genetic mechanisms controlling the induction of polysubstrate monooxygenase (P–450) activities. *Annu. Rev. Pharmacol. Toxicol.* 21, 431–62.

Nebert, D.W., Negishi, M., Lang, M.A., Hjelmeland, L.M., and Eisen, H.J.

(1982). The *Ah* locus, a multigene family necessary for survival in a chemically adverse environment: comparison with the immune system. *Adv. Genet.* 21, 1–52.

Nee, L., Polinsky, R.J., Eldridge, R., Weingartner, H., Smallberg, S., and Ebert, M. (1983). A family with histologically confirmed Alzheimer's disease. *Arch. Neurol.* 40, 203–8.

Neri, G. (1984). Some question on the significance of chromosome alterations in leukemias and lymphomas. *Am. J. Med. Genet.* 18, 471–81.

Nesbitt, M., and Francke, U. (1973). A system of nomenclature for band patterns of mouse chromosomes. *Chromosoma* 41, 145–58.

Nève, J., Sinet, P.M., Molle, L., and Nicole, A. (1983). Selenium, zinc and copper levels in Down's syndrome (trisomy 21): blood levels and relations with glutathione peroxidase and superoxide dismutase. *Clin. Chim. Acta* 133, 209–14.

Nichols, W.W., Miller, R.C., Hoffman, E., Albert, D., Weichselbaum, R.R., Nove, J., and Little, J.B. (1979). Interstitial deletion of chromosome 13 and associated congenital anomalies. *Hum. Genet.* 52, 169–73.

Nicolas, J.F., Avner, P., Gaillard, J., Guénet, J.L., Jakob, H., and Jacob, F. (1976). Cell lines derived from teratocarcinomas. *Cancer Res.* 36, 4224–31.

Niebuhr, E. (1974). Triploidy in man. Cytogenetical and clinical aspects. *Humangenetik* 21, 103–25.

Niebuhr, E. (1977). Partial trisomies and deletions of chromosome 13. In *New Chromosomal Syndromes*, ed. J.J. Yunis, pp. 273–99. Academic Press, New York.

Niebuhr, E. (1978). Cytologic observations in 35 individuals with a 5p− karyotype. *Hum. Genet.* 42, 143–56.

Nielsen G., and Scandalios, J.G. (1974). Chromosomal location by use of trisomics and new alleles of an endopeptidase in *Zea mays*. *Genetics* 77, 679–86.

Nielsén, K., Marcus, M., and Gropp, A. (1985). In vitro growth kinetics of mouse trisomy–12 and trisomy–19. *Hereditas* 102, 77–84.

Niemierko, A. (1981). Postimplantation development of CB-induced triploid mouse embryos. *J. Embryol. Exp. Morphol.* 66, 81–9.

Niikawa, N., Jinno, Y., Tomiyasu, T., Fukushima, Y., and Kudo, K. (1981). Ring chromosome 11 [46,XX,r(11)(p15q25)] associated with clinical features of the 11q syndrome. *Ann. Génét.* 24, 172–5.

Niikawa, N., Fukushima, Y., Taniguichi, N., Iizuka, S., and Kajii, T. (1982). Chromosome abnormalities involving 11p13 and low erythrocyte catalase activity. *Hum. Genet.* 60, 373–5.

Nilsen-Hamilton, M., and Holley, R.W. (1983). Rapid selective effects by a growth inhibitor and epidermal growth factor on the incorporation of [^{35}S]methionine into proteins secreted by African green monkey (BSC–1) cells. *Proc. Natl. Acad. Sci. USA* 80, 5636–40.

Nishikura, K., ar-Rushdi, A., Erikson, J., Watt, R., Rovera, G., and Croce, C.M. (1983). Differential expression of the normal and of the translocated human c-*myc* oncogenes in B cells. *Proc. Natl. Acad. Sci. USA* 80, 4822–6.

Noel, B., Quack, B., and Rethoré, M.O. (1976). Partial deletions and trisomies of chromosome 13; mapping of bands associated with particular malformations. *Clin. Genet.* 9, 593–602.

Nordenson, I., Beckman, G., and Beckman, L. (1976). The effect of superoxide dismutase and catalase on radiation-induced chromosome breaks. *Hereditas* 82, 125–6.

Nordenson, I., Adolfsson, R., Beckman, G., Bucht, G., and Winblad, B. (1980). Chromosomal abnormality in dementia of the Alzheimer type. *Lancet* 1, 481–2.

Nordmann, Y., Grandchamp, B., Phung, N., de Verneuil, H., Grelier, M., and

Noiré, J. (1977). Coproporphyrinogen-oxidase deficiency in hereditary coproporphyria. *Lancet* 1, 140.

Norum, R.A., van Dyke, D.L., and Weiss, L. (1979). Deletion mapping of esterase D (*ESD*) to chromosome 13q12.5→q21.2. *Cytogenet. Cell Genet.* 25, 192.

Nove, J., Little, J.B., Weichselbaum, R.R., Nichols, W.W., and Hoffman, E. (1979). Retinoblastoma, chromosome 13, and in vitro cellular radiosensitivity. *Cytogenet. Cell Genet.* 24, 176–84.

Nowell, P.C. (1976). The clonal evolution of tumor cell populations. *Science* 194, 23–8.

Nowell, P.C. (1982). Cytogenetics of preleukemia. *Cancer Genet. Cytogenet.* 5, 265–78.

Nowell, P.C., and Hungerford, D.A. (1960). A minute chromosome in human chronic granulocytic leukemia. *Science* 132, 1497.

Nurmi, T. (1982). *Disturbed Immune Functions Associated with Chromosome Abnormalities*. Academic Dissertation, University of Oulu, Oulu, Finland.

Nurmi, T., Huttunen, K., Lassila, O., Henttonen, M., Säkkinen, A., Linna, S.-L., and Tiilikainen, A. (1982*a*). Natural killer cell function in trisomy–21 (Down's syndrome). *Clin. Exp. Immunol.* 47, 735–41.

Nurmi, T., Leinonen, M., Häivä, V.-M., Tiilikainen, A., and Kouvalainen, K. (1982*b*). Antibody response to pneumococcal vaccine in patients with trisomy–21 (Down's syndrome). *Clin. Exp. Immunol.* 48, 485–90.

Nyberg, P., Carlsson, A., and Winblad, B. (1982). Brain monoamines in cases with Down's syndrome with and without dementia. *J. Neural Transm.* 55, 289–99.

O'Brien, D., Haake, M.W., and Braid, B. (1960). Atropine sensitivity and serotonin and mongolism. *Am. J. Dis. Child.* 100, 873–4.

O'Brien, R.L., Poon, P., Kline, E., and Parker, J.W. (1971). Susceptibility of chromosomes from patients with Down's syndrome to 7,12-dimethybenz(α)anthracene induced aberrations *in vitro*. *Int. J. Cancer* 8, 202–10.

O'Brien, S.J., and Gethmann, R.C. (1973). Segmental aneuploidy as a probe for structural genes in Drosophila: mitochondrial membrane proteins. *Genetics* 75, 155–67.

O'Brien, S.J., and MacIntyre, R.J. (1978). Genetics and biochemistry of enzymes and specific proteins of *Drosophila*. In *The Genetics and Biology of Drosophila, Vol. 2a*, ed. M. Ashburner and T.R.F. Wright, pp. 395–551. Academic Press, London and New York.

O'Brien, S.J., and Nash, W.G. (1982). Genetic mapping in mammals: chromosome map of domestic cat. *Science* 216, 257–65.

Odell, W., Wolfsen, A., Yoshimoto, Y., Weitzman, R., Fisher, D. and Hirose, F. (1977). Ectopic peptide synthesis: a universal concomitant of neoplasia. *Trans. Assoc. Am. Physicians* 90, 204–25.

O'Donnell, J.J., and Hall, B.D. (eds.) (1979). Penetrance and variability in malformation syndromes. *Birth Defects: Orig. Art. Ser.* 15(5B).

Ohara, P.T. (1972). Electron microscopical study of the brain in Down's syndrome. *Brain* 95, 681–4.

O'Hare, A.E., Grace, E., and Edmunds, A.T. (1984). Deletion of the long arm of chromosome 11 [46,XXdel(11)(q24.1→qter)]. *Clin. Genet.* 25, 373–7.

Ohno, H., Iizuka, S., Kondo, T., Yamamura, K., Sekiya, C., and Taniguchi, N. (1984). The levels of superoxide dismutase, catalase, and carbonic anhydrase in erythrocytes of patients with Down's syndrome. *Klin. Wochenschr.* 62, 287–8.

Ohno, S. (1963). Life history of female germ cells in mammals. In *Second Interna-*

tional Conference on Congenital Malformations, pp. 36–42. International Medical Congress, Ltd., New York.

Ohno, S. (1967). *Sex Chromosomes and Sex-linked Genes*. Springer-Verlag, Berlin, Heidelberg, and New York.

Ohno, S. (1971). Genetic implication of karyological instability of malignant somatic cells. *Physiol. Rev.* 51, 496–526.

Ohno, S. (1979). *Major Sex-Determining Genes*. Springer-Verlag, Berlin, Heidelberg, and New York.

Ohno, S., Kittrell, W.A., Christian, L.C., Stenius, C., and Witt, G.A. (1963). An adult triploid chicken (*Gallus domesticus*) with a left ovo-testis. *Cytogenetics* 2, 42–9.

Oliver, M.J., Huber, R.E., and Williamson, J.H. (1978). Genetic and biochemical aspects of trehalase from *Drosophila melanogaster*. *Biochem. Genet.* 16, 927–40.

O'Neill, H.C. (1980). Quantitative variation in H–2-antigen expression. II. Evidence for a dominance pattern in H–2K and H–2D expression in F_1 hybrid mice. *Immunogenetics* 11, 241–54.

O'Neill, H.C., and Blanden, R.V. (1979a). Quantitative differences in the expression of parentally-derived H–2 antigens in F_1 hybrid mice affect T-cell responses. *J. Exp. Med.* 149, 724–31.

O'Neill, H.C., and Blanden, R.V. (1979b). Variation in H–2 antigen expression in F_1 hybrid mice: analysis using monoclonal antibodies. *Aust. J. Exp. Biol. Med. Sci.* 57, 627–35.

Opitz, J.M. (1985). The developmental field concept. *Am. J. Med. Genet.* 21, 1–11.

Opitz, J.M., and Gilbert, E.F. (1982a). Pathogenetic analysis of congenital anomalies in humans. *Pathobiology Annual 1982* 12, 301–49.

Opitz, J.M., and Gilbert, E.F. (1982b). CNS anomalies and the midline as a "developmental field." *Am. J. Med. Genet.* 12, 443–55.

Orkin, S.H., Goldman, D.S., and Sallan, S.E. (1984). Development of homozygosity for chromosome 11p markers in Wilms' tumour. *Nature, Lond.* 309, 172–4.

Orkin, S.H., Old, J., Lazarus, H., Altay, C., Gurgey, A., Weatherall, D.J., and Nathan, D.G. (1979). The molecular basis of α-thalassemias: frequent occurrence of dysfunctional α loci among non-Asians with Hb H disease. *Cell* 17, 33–42.

Ornoy, A., Kohn, G., Ben Zur, Z., Weinstein, D., and Cohen, M.M. (1978). Triploidy in human abortions. *Teratology* 18, 315–20.

Orye, E., Verhaaren, H., van Egmond, H., and Devloo-Blancquaert, A. (1975). A new case of the trisomy 9p syndrome. Report of a patient with unusual chromosome findings (46,XX/47,XX, + i(9p)) and a peculiar congenital heart defect. *Clin. Genet.* 7, 134–43.

Osley, M.A., and Hereford, L.M. (1981). Yeast histone genes show dosage compensation. *Cell* 24, 377–84.

Øster, J., Mikkelsen, M., and Nielsen A. (1975). Mortality and life-table of Down's syndrome. *Acta Pediatr. Scand.* 64, 322–6.

Oster-Granite, M.L. (1983). Delayed basicranium and craniofacial development in murine trisomy 16. *Am. J. Hum. Genet.* 35, 164A.

Oster-Granite, M.L., and Hatzidimitriou, G. (1985). Development of the hippocampal formation in trisomy 16 mice. *Pediatr. Res.* 19, 328A.

Oster-Granite, M.L., Baker, C., and Ozand, P.T. (1983). Neuroanatomic, ocular and audiovestibular malformations in trisomy 16 mice. *Pediatr. Res.* 17, 300A.

Oster-Granite, M.L., Reed, W.D., Collins, R.M. Jr., and Ozand, P.T. (1983). The brain catecholamine system in mouse trisomy 16. *Pediatr. Res.* 17, 139A.

Ott, M.-O., Sperling, L., and Weiss, M.C. (1984). Albumin extinction without methylation of its gene. *Proc. Natl. Acad. Sci. USA* 81, 1738–41.

Ottolenghi, S., Lanyon, W.G., Paul, J., Williamson, R., Weatherall, D.J., Clegg, J.B., Pritchard, J., Pootrakul, S., and Boon, W.H. (1974). The severe form of α thalassaemia is caused by a haemoglobin gene deletion. *Nature, Lond.* 251, 389–92.

Owen, M.C., Brennan, S.O., Lewis, J.H., and Carrell, R.W. (1983). Mutation of antitrypsin to antithrombin. α₁-Antitrypsin Pittsburgh (358Met→Arg), a fatal bleeding disorder. *N. Engl. J. Med.* 309, 694–8.

Ozand, P.T., Hawkins, R.L., Collins, R.M., Jr., Reed, W.D., Baab, P.J., and Oster-Granite, M.L. (1984). Neurochemical changes in murine trisomy 16: delay in cholinergic and catecholaminergic systems. *J. Neurochem.* 43, 401–8.

Paasonen, M.K. (1968). Platelet 5-hydroxytryptamine as a model in pharmacology. *Ann. Med. Exp. Biol. Fenn.* 46, 416–22.

Page, D., DeMartinville, B., Barker, D., Wyman, A., White, R., Francke, U., and Botstein, D. (1982). Single-copy sequence hybridizes to polymorphic and homologous loci on human X and Y chromosomes. *Proc. Natl. Acad. Sci. USA* 79, 5352–6.

Paigen, K. (1979). Genetic factors in developmental regulation. In *Physiological Genetics*, ed. J.G. Scandalios, pp. 1–61. Academic Press, New York.

Paigen, K. (1980). Temporal genes and other developmental regulators in mammals. In *The Molecular Genetics of Development*, ed. T. Leighton and W.F. Loomis, pp. 419–70. Academic Press, New York.

Palmer, C.G., Provisor, A.J., Weaver, D.D., Hodes, M.E., and Heerema, N. (1983). Juvenile chronic granulocytic leukemia in a patient with trisomy 8, neurofibromatosis, and prolonged Epstein-Barr virus infection. *J. Pediatr.* 102, 888–92.

Palmiter, R.D., Chen, H.Y., and Brinster, R.L. (1982). Differential regulation of metallothionein-thymidine kinase fusion genes in transgenic mice and their offspring. *Cell* 29, 701–10.

Palmiter, R.D., Brinster, R.L., Hammer, R.E., Trumbauer, M.E., Rosenfeld, M.G., Birnberg, N.S., and Evans, R.M. (1982). Dramatic growth of mice that develop from eggs microinjected with metallothionein-growth hormone fusion genes. *Nature, Lond.* 300, 611–15.

Pan, S.F., Fatora, R., Sorg, R., Garver, K.L., and Steele, M.W. (1977). Meiotic consequences of an intrachromosomal insertion of chromosome no. 1: a family pedigree. *Clin. Genet.* 12, 303–13.

Pangalos, C., and Couturier, J. (1981). Partial trisomy 13 (q21.3→qter) resulting from a maternal translocation t(13;21). *Ann. Génét.* 24, 179–81.

Pant, S.S., Moser, H.W., and Krane, S.M. (1968). Hyperuricemia in Down's syndrome. *J. Clin. Endocrinol. Metab.* 28, 472–8.

Pantelakis, S.N., Karaklis, A.C., Alexiou, D., Vardus, E., and Valaes, T. (1970). Red cell enzymes in trisomy–21. *Am. J. Hum. Genet.* 22, 184–93.

Papaioannou, V.E., and West, J.D. (1981). Relationship between the parental origin of the X chromosomes, embryonic cell lineage and X chromosome expression in mice. *Genet. Res.* 37, 183–97.

Papaioannou, V.E., Gardner, R.L., McBurney, M.W., Babinet, C., and Evans, M.J. (1978). Participation of cultured teratocarcinoma cells in mouse embryogenesis. *J. Embryol. Exp. Morphol.* 44, 93–104.

Park, B.H. (1981). Impaired proliferative response to T-cells in Down's syndrome. *Fed. Proc.* 40, 1126.

Park, B.H., and Averion, R. (1982). Imbalance of T-cell subsets in Down's syndrome (DS) as analyzed by monoclonal antibodies and flow cytometry. *Fed. Proc.* 41, 801.

Park, W.W. (1957). The occurrence of sex chromatin in early human and macaque embryos. *J. Anat.* 91, 369–73.

Parsons, G.D., and Mackie, G.A. (1983). Expression of the gene for ribosomal protein S20: effects of gene dosage. *J. Bacteriol.* 154, 152–60.

Patau, K. (1962). Partial trisomy. In *Second International Conference on Congenital Malformations*, pp. 52–9. International Medical Congress Ltd., New York.

Pathak, S., Hsu, T.C., Trentin, J.J., Butel, J.S., and Panigrahy, B. (1981). Nonrandom chromosome abnormalities in transformed Syrian hamster cell lines. In *Genes, Chromosomes and Neoplasia*, ed. F.E. Arrighi, P.N. Rao, and E. Stubblefield, pp. 405–18. Raven, New York.

Paton, G.R., Silver, M.F., and Allison, A.C. (1974). Comparison of cell cycle time in normal and trisomic cells. *Humangenetik* 23, 173–82.

Patterson, D., Graw, S., and Jones, C. (1981). Demonstration, by somatic cell genetics, of coordinate regulation of genes for two enzymes of purine synthesis assigned to human chromosome 21. *Proc. Natl. Acad. Sci. USA* 78, 405–9.

Patterson, D., Jones, C., Scoggin, C., Miller, Y.E., and Graw, S. (1982). Somatic cell genetic approaches to Down's syndrome. *Ann. N.Y. Acad. Sci.* 396, 69–81.

Patterson, D., Henikoff, S., Sloan, J.S., Bleskan, J., Hards, R., and Keene, M. (1985a). Studies on the organization and structure of genes and enzymes of purine synthesis in animals and man. *Abstracts of V. International Symposium on Human Purine and Pyrimidine Metabolism.* San Diego, California. July 28-August 1.

Patterson, D., van Keuren, M., Drabkin, H., Watkins, P., Gusella, J., and Scoggin, C. (1985b). Molecular analysis of chromosome 21 using somatic cell hybrids. *Ann. N.Y. Acad. Sci.* 450, 109–20.

Pauli, R.M., Kirkpatrick, S.J., Meisner, L.F., Mijanovich, J.R., and Spritz, R.A. (1982). Neonatal death in cousins with trisomy 10q and monosomy 4p due to a familial translocation. *Clin. Genet.* 22, 340–7.

Pauli, R.M. (1983). Editorial comment: dominance and homozygosity in man. *Am. J. Med. Genet.* 16, 455–8.

Payne, F.E., and Schmickel, R.D. (1971). Susceptibility of trisomic and of triploid human fibroblasts to simian virus 40 (SV40). *Nature (New Biol.), Lond.* 230, 190.

Peacock, J., McMorris, F., and Nelson, P. (1973). Electrical excitability and chemosensitivity of mouse neuroblastoma × mouse or human fibroblast hybrids. *Exp. Cell Res.* 79, 199–212.

Pearson, G., Mann, J.D., Bensen, J., and Bull, R.W. (1979). Inversion duplication of chromosome 6 with trisomic codominant expression of HLA antigens. *Am. J. Hum. Genet.* 31, 29–34.

Pearson, N.J., Fried, H.M., and Warner, J.R. (1982). Yeast use translational control to compensate for extra copies of a ribosomal protein gene. *Cell* 29, 347–55.

Pearson, S.J., Tetri, P., George, D.L., and Francke, U. (1983). Activation of human α_1-antitrypsin gene in rat hepatoma × human fetal liver cell hybrids depends on the presence of human chromosome 14. *Somat. Cell Genet.* 9, 567–92.

Pedersen-Bjergaard, J., Haahr, S., Philip, P., Thomsen, M., Jensen, G., Ersbøll, T., and Nissen, N.I. (1980). Abolished production of interferon by leukocytes of patients with the acquired cytogenetic abnormalities 5q− or −5 in secondary and de novo acute non-lymphocytic leukemia. *Br. J. Haematol.* 46, 211–23.

Perlmutter, R.M., Klotz, J.L., Pravtcheva, D., Ruddle, F., and Hood, L. (1984). A novel 6:10 chromosomal translocation in the murine plasmocytoma NS–1. *Nature, Lond.* 307, 473–6.

Perry, R.P., Kelley, D.E., Coleclough, C., Seidman, J.G., Leder, P., Tonegawa, S., Matthyssens, G., and Weigert, M. (1980). Transcription of mouse ϰ chain genes: implications for allelic exclusion. *Proc. Natl. Acad. Sci. USA* 77, 1937–41.

Peschle, C., Migliaccio, G., Migliaccio, A.R., Covelli, A., Giuliani, A., Mavilio, F., and Mastroberardino, G. (1983). Hemoglobin switching in humans. In *Current Concepts in Erythropoiesis*, ed. C.D.R. Dunn, pp. 339–87. Wiley, Chichester.

Peterson, J.A., and Weiss, M.C. (1972), Expression of differentiated functions in hepatoma cell hybrids: induction of mouse albumin production in rat hepatoma – mouse fibroblast hybrids. *Proc. Natl. Acad. Sci. USA* 69, 571–5.

Petkau, A., and Chelack, W.S. (1976). Radioprotective effect of superoxide dismutase on model phospholipid membranes. *Biochim. Biophys. Acta*. 433, 445–56.

Petkau, A., Kelly, K., Chelack, W.S., Pleskach, S.D., Barefoot, C., and Meeker, B.E. (1975). Radioprotection of bone marrow stem cells by superoxide dismutase. *Biochem. Biophys. Res. Commun*. 67, 1167–74.

Petkau, A., Kelly, K., Chelack, W.S., and Barefoot, C. (1976). Protective effect of superoxide dismutase on erythrocytes of X-irradiated mice. *Biochem. Biophys. Res. Commun*. 70, 452–8.

Pettersen, J.C., and Bersu, E.T. (1982). A comparison of the anatomical variations found in trisomies 13, 18, and 21. In *Advances in the Study of Birth Defects*, ed. T.V.M. Persaud, pp. 161–79. University Park Press, Baltimore.

Pexieder, T., Miyabara, S., and Gropp, A. (1981). Congenital heart disease in experimental (fetal) mouse trisomies: incidence. In *Perspectives in Cardiovascular Research, Vol. 5: Mechanisms of Cardiac Morphogenesis and Teratogenesis*, ed. T. Pexieder, pp. 389–99. Raven Press, New York.

Pfeiffer, R.A. (1980a). Langer-Giedion syndrome and additional congenital malformations with interstitial deletion of the long arm of chromosome 8. 46,XY,del8(q13–22). *Clin. Genet*. 18, 142–6.

Pfeiffer, R.A. (1980b). Observations in a case of an X/Y translocation, t(X;Y)(p22;q11), in a mother and son. *Cytogenet. Cell Genet*. 26, 150–7.

Philip, R., Berger, A.C., McManus, N.H., Warner, N.H., Peacock, M.A., and Epstein, L.B. (1985a). Abnormalities of the in vitro cellular and humoral responses to bacterial and viral antigens with concomitant numerical alterations in lymphocyte subsets in Down syndrome (trisomy 21). *J. Immunol.*, in press.

Philip, R., Hebert, S.J., McManus, N.H., Warner, N.H., Peacock, M.A., and Epstein, L.B. (1985b). Natural killer cell function in Down syndrome (trisomy 21). Submitted for publication.

Philip, T., Lenoir, G., Rolland, M.O., Philip, I., Hamet, M., Lauras, B., and Fraisse, J. (1980). Regional assignment of the *ADA* locus on 20q13.2→qter by the gene dosage studies. *Cytogenet. Cell Genet*. 27, 187–9.

Phillips, R.J.S., Hawker, S.G., and Mosely, H.J. (1973). Bare-patches, a new sex-linked gene in the mouse, associated with a high production of *XO* females. I. A preliminary report of breeding experiments. *Genet. Res*. 22, 91–9.

Pihko, H., Therman, E., and Uchida, J.A. (1981). Partial 11q trisomy syndrome. *Hum. Genet*. 58, 129–34.

Piko, L., and Bomsel-Helmreich, O. (1960). Triploid rat embryos and other chromosomal deviants after colchicine treatment and polyspermy. *Nature, Lond*. 186, 737–9.

Pipkin, S.B., Chakrabartty, P.K., and Bremner, T.A. (1977). Location and regulation of *Drosophila* fumarase. *J. Hered*. 68, 245–52.

Pitt, D., Leversha, M., Sinfield, C., Campbell, P., Anderson, R., Bryan, D., and Rogers, J. (1981). Tetraploidy in a liveborn infant with spina bifida and other anomalies. *J. Med. Genet*. 18, 309–11.

Pittard, W.B. III., Sorensen, R.U., and Stallard, R. (1983). Lymphocyte proliferation in a 31-week premature neonate with 69,XXX chromosomal constitution. *Clin. Genet*. 24, 26–8.

Pluznik, D.H., Herbst, E.W., Lenz, R., Sellin, D., Hertogs, C.F., and Gropp, A. (1981). Controlled production of trisomic hematopoietic stem cells: an experimental tool in hematology and immunology. In *Experimental Hematology Today*, ed. S.J. Baum, G.D. Ledney, and A. Kahn, pp. 3–11. S. Karger, Basel.

Poissonnier, M., Saint-Paul, B., Dutrillaux, B., Chassaigne, X., Gruyer, P., and Blignières-Strouk, G. (1976). Trisomie 21 partielle 21q21(21q22.2). *Ann. Génét.* 79, 69–73.

Polani, P.E. (1982). Pairing of X and Y chromosomes, non-inactivation of X-linked genes, and the maleness factor. *Hum. Genet.* 60, 207–11.

Porter, I.H., and Paul, B. (1973). The cell cycle in patients with chromosomal anomalies. *Pediatr. Res.* 7, 314.

Potter, C.W., Potter, A.M., and Oxford, J.S. (1970). Comparison of transformation and T antigen induction in human cell lines. *J. Virol.* 5, 293–8.

Preston, R.J. (1981). X-ray-induced chromosome aberrations in Down lymphocytes: an explanation of their increased sensitivity. *Environ. Mutagen.* 3, 85–89.

Preus, M. (1980). The numerical versus intuitive approach to syndrome nosology. *Birth Defects: Orig. Art. Ser.* 16(5), 93–104.

Preus, M., and Aymé, S. (1983). Formal analysis of dysmorphism: objective methods of syndrome definition. *Clin. Genet.* 23, 1–16.

Preus, M., Schinzel, A., Aymé, S., and Kaijser, K. (1984). Trisomy 9(pter→q1 to q3): the phenotype as an objective aid to karyotypic interpretation. *Clin. Genet.* 26, 52–5.

Preus, M., Aymé, S., Kaplan, P., and Vekemans, M. (1985). A taxonomic approach to the del(4p) phenotype. *Am. J. Med. Genet.* 21, 337–45.

Price, W.H., Brunton, M., Buckton, K., and Jacobs, P.A. (1976). Chromosome survey of new patients admitted to the four maximum security hospitals in the United Kingdom. *Clin. Genet.* 9, 389–98.

Priest, J.H. (1960). Atropine response of the eyes in mongolism. *Am. J. Dis. Child.* 100, 869–72.

Priest, R.E., and Priest, J.H. (1969). Diploid and tetraploid clonal cells in culture: gene ploidy and synthesis of collagen. *Biochem. Genet.* 3, 371–82.

Primrose, D.A. (1983). Phenylketonuria with normal intelligence. *J. Ment. Defic. Res.* 27, 239–46.

Prişcu, R., and Sichitiu, S. (1975). Types of enzymatic overdosing in trisomy 21: erythrocytic dismutase-AJ and phosphoglucomutase. *Humangenetik* 29, 79–83.

Prockop, D.J., and Kivirikko, K.I. (1984). Heritable diseases of collagen. *N. Engl. J. Med.* 311, 376–86.

Prockop, D.J., Williams, C., de Wet, W.J., Sippola, M., Uitto, J., and Pihlajaniemi, T. (1983). Shortening and lengthening of pro α chains of type I procollagen in osteogenesis imperfecta. *Clin. Res.* 31, 533A.

Pueschel, S.M. (1982*a*). Biomedical aspects in Down syndrome: biochemistry. In *Down Syndrome. Advances in Biomedicine and the Behavioral Sciences*, ed. S.M. Pueschel and J.E. Rynders, pp. 249–69. Ware Press, Cambridge, Massachusetts.

Pueschel, S.M. (1982*b*). Biomedical aspects in Down syndrome: cardiology. In *Down Syndrome. Advances in Biomedicine and the Behavioral Sciences*, ed. S.M. Pueschel and J.E. Rynders, pp. 203–7. Ware Press, Cambridge, Massachusetts.

Pueschel, S.M. (1982*c*). Biomedical aspects in Down syndrome: phenotype. In *Down Syndrome. Advances in Biomedicine and the Behavioral Sciences*, ed. S.M. Pueschel and J.E. Rynders, pp. 169–190. Ware Press, Cambridge, Massachusetts.

Pueschel, S.M., and Rynders, J.E. (eds.) (1982). *Down Syndrome. Advances in Biomedicine and the Behavioral Sciences*. Ware Press, Cambridge, Massachusetts.

Pueschel, S.M., and Steinberg, L. (1980). *Down Syndrome: A Comprehensive Bibliography*. Garland STPM Press, New York and London.

Pueschel, S.M., Rothman, K.J., and Ogilby, J.D. (1976). Birth weight of children with Down's syndrome. *Am. J. Ment. Defic.* 80, 442–5.

Putz, B., and Morriss-Kay, G. (1981). Abnormal neural fold development in trisomy 12 and trisomy 14 mouse embryos. *J. Embryol. Exp. Morphol.* 66, 141–58.

Putz, B., Krause, G., Garde, T., and Gropp, A. (1980). A comparison between trisomy 12 and vitamin A induced exencephaly and associated malformations in the mouse embryo. *Virchows Arch. Pathol. Anat.* 368, 65–80.

Puukka, R., Puukka, M., Leppilampi, M., Linna, S.-L., and Kouvalainen, K. (1982). Erythrocyte adenosine deaminase, purine nucleoside phosphorylase and phosphoribosyltransferase activity in patients with Down's syndrome. *Clin. Chim. Acta.* 126, 275–81.

Pyeritz, R.E., Murphy, E.A., and McKusick, V.A. (1979). Clinical variability in the Marfan syndrome(s). *Birth Defects: Orig. Art. Ser.* 15(5B), 155–78.

Quesenberry, P., and Levitt, L. (1979). Hematopoietic stem cells. *N. Engl. J. Med.* 301, 755–60.

Quiroz, E., Orozco, A., and Salamanca, F. (1985). Diploid-tetraploid mosaicism in a malformed boy. *Clin. Genet.* 27, 183–6.

Rabbits, T.H., Forster, A., Hamlyn, P., and Baer, R. (1984). Effect of somatic mutation within translocated c-*myc* genes in Burkitt's lymphoma. *Nature, Lond.* 309, 592–7.

Rabin, M., Hart, C.P., Ferguson-Smith, A., McGinnis, W., Levine, M., and Ruddle, F.H. (1985). Two homoeo box loci mapped in evolutionarily related mouse and human chromosomes. *Nature, Lond.* 314, 175–8.

Radman, M., Jeggo, P., and Wagner, R. (1982). Chromosomal rearrangement and carcinogenesis. *Mutation Res.* 98, 249–64.

Rastan, S. (1982). Timing of X-chromosomal inactivation in postimplantation mouse embryos. *J. Embryol. Exp. Morphol.* 71, 11–24.

Rawls, J.M., Jr., and Lucchesi, J.C. (1974). Regulation of enzyme activities in *Drosophila*. I. The detection of regulatory loci by gene dosage responses. *Genet. Res.* 24, 59–72.

Reed, T. (1981). Dermatoglyphics in medicine – problems and use in suspected chromosome abnormalities. *Am. J. Med. Genet.* 8, 411–29.

Reed, W.D., Oster-Granite, M.L., Pischkoff, S.A., Baab, P., and Ozand, P.T. (1983). Beta adrenergic markers in cultured cells of human trisomy 21 and murine trisomy 16. *Pediatr. Res.* 17, 141A.

Rees, H. (1972). DNA in higher plants. In *Evolution of Genetic Systems*, ed. H.H. Smith (*Brookhaven Symp. Biol.* 23), pp. 419–47. Gordon and Breach, New York.

Reeve, A.E., Housiaux, P.J., Gardner, R.J.M., Chewings, W.E., Grindley, R.M., and Millow, L.J. (1984). Loss of a Harvey *ras* allele in sporadic Wilms tumour. *Nature, Lond.* 309, 174–6.

Rehder, H. (1981). Pathology of trisomy 21 – with particular reference to persistent common atrioventricular canal of the heart. In *Trisomy 21. An International Symposium*, ed. G.R. Burgio, M. Fraccaro, L. Tiepolo, and U. Wolf, pp. 57–73. Springer-Verlag, Berlin.

Rehder, H., and Friedrich, U. (1979). Partial trisomy 1q syndrome. *Clin. Genet.* 15, 534–40.

Reiser, K., Whitcomb, C., Robinson, K., and MacKenzie, M.R. (1976). T and B lymphocytes in patients with Down's syndrome. *Am. J. Ment. Defic.* 80, 613–19.

Reisman, L.E., Kasahara, S., Chung, C.-Y., Darnell, A., and Hall, B. (1966). Antimongolism. Studies in an infant with a partial monosomy of the 21 chromosome. *Lancet* 1, 394–7.

Rethoré, M.-O. (1977). Syndromes involving chromosomes 4, 9, and 12. In *New Chromosomal Syndromes*, ed. J.J. Yunis, pp. 119–83. Academic Press, New York.

Rethoré, M.-O. (1981). Structural variation of chromosome 21 and symptoms of Down's syndrome. In *Trisomy 21. An International Symposium*, ed. G.R. Burgio, M. Fraccaro, L. Tiepolo, and U. Wolf, pp. 173–82. Springer-Verlag, Berlin.

Rethoré, M.-O., Dutrillaux, B., Giovannelli, G., Forabosco, A., Dallapiccola, B., and Lejeune, J. (1974). La trisomie 4p. *Ann. Génét.* 17, 125–8.

Rethoré, M.-O., Kaplan, J.-C., Junien, C., Cruveiller, J., Dutrillaux, B., Aurias, A., Carpentier, S., Lafourcade, J., and Lejeune, J. (1975). Augmentation de l'activité de la LDH-B chez un garçon trisomique 12p par malségrégation d'une translocation maternelle t(12;14)(q12;p11). *Ann. Génét.* 18, 81–7.

Rethoré, M.-O., Junien, C., Malpuech, G., Baccichetti, C., Tenconi, R., Kaplan, J.-C., de Romeuf, J., and Lejeune, J. (1976). Localisation du gène de la glycéraldéhyde 3-phosphate déhydrogénase (G3PD) sur le segment distal du bras court du chromosome 12. *Ann. Génét.* 19, 140–2.

Rethoré, M.-O., Kaplan, J.-C., Junien, C., and Lejeune, J. (1977). 12pter→12p122 possible assignment of human triosephosphate isomerase. *Hum. Genet.* 36, 235–7.

Rethoré, M.-O., Junien, C., Aurias, A., Couturier, J., Dutrillaux, B., Kaplan, J.C., and Lejeune, J. (1980). Augmentation de la LDHA et trisomie 11p partielle. *Ann. Génét.* 23, 35–9.

Rethoré, M.-O., Lafourcade, J., Couturier, J., Harpey, J.P., Hamet, M., Engler, R., Alcindor, L.G., and Lejeune, J. (1982). Augmentation d'activité de l'adénine phosphoribosyl transférase, chez un enfant trisomique 16q22.2→16→ter per translocation t(16;21)(q22.2;q22.2)pat. *Ann. Génét.* 25, 36–42.

Rethoré, M.-O., Couturier, J., Villain, E., Hambourg, M., and Lejeune, J. (1984). Maladie du cri du chat et trisomie 8p par translocation paternelle t(5;8) (p1409;p12). *Ann. Génét.* 27, 118–21.

Rex, A.P., and Preus, M. (1982). A diagnostic index for Down syndrome. *J. Pediatr.* 100, 903–6.

Reynolds, J.F., Shires, M.A., Wyandt, H.E., and Kelly, T.E. (1983). Trisomy 4p in four relatives: variability and lack of distinctive features in phenotypic expression. *Clin. Genet.* 24, 365–74.

Riccardi, V.M., and Forgason, J. (1979). Chromosome 8 abnormalities as components of neoplastic and hematologic disorders. *Clin. Genet.* 15, 317–26.

Riccardi, V.M., Kleiner, B., and Lubs, M.-L. (1979). Neurofibromatosis: variable expression is not intrinsic to the mutant gene. *Birth Defects: Orig. Art. Ser.* 15(5B), 283–9.

Riccardi, V.M., Hittner, H.M., Francke, U., Pippin, S., Holmquist, G.P., Kretzer, F.L., and Perrell, R. (1979). Partial triplication and deletion of 13q: study of a family presenting with bilateral retinoblastomas. *Clin. Genet.* 15, 332–45.

Riccardi, V.M., Hittner, H.M., Francke, U., Yunis, J.J., Ledbetter, D., and Borges, W. (1980). The aniridia-Wilms tumor association: the critical role of chromosome band 11p13. *Cancer Genet. Cytogenet.* 2, 131–7.

Ricciuti, F.C., Gelehrter, T.D., and Rosenberg, L.E. (1976). X-chromosome inactivation in human liver: confirmation of X-linkage of ornithine transcarbamylase. *Am. J. Hum. Genet.* 28, 332–8.

Richards, B.W. (1969). Mosaic mongolism. *J. Ment. Defic. Res.* 13, 66–83.

Richards, B.W., and Enver, F. (1979). Blood pressure in Down's syndrome. *J. Ment. Defic. Res.* 23, 123–35.

Ridler, M.A.C., and McKeown, J.A. (1979). 11q aneuploidy: partial monosomy and trisomy in the children of a mother with a t(3;11)(p27;q23) translocation. *Hum. Genet.* 52, 101–6.

Rigas, D.A., Elsasser, P., and Hecht, F. (1970). Impaired *in vitro* response of circulating lymphocytes to phytohemagglutinin in Down's syndrome: dose- and time-response curves and relation to cellular immunity. *Int. Arch. Aller. Appl. Immunol.* 39, 587–608.

Ringertz, N., and Savage, R.E. (1976). *Cell Hybrids*. Academic Press, New York.

Rivas, F., Hernandez, A., Nazara, Z., Fragozo, R., Olivares, N., Rolón, A., and Cantú, J.M. (1979). On the deletion 4p16 Wolf-Hirschhorn syndrome. *Ann. Génét.* 22, 228–31.

Rivas, F., Rivera, H., Plascencia, M.L., Ibarra, B., and Cantú, J.M. (1984). The phenotype in partial 13q trisomies, apropos of a familial (13;15)(q22;q26) translocation. *Hum. Genet.* 67, 86–93.

Rivera, H., Turleau, C., Grouchy, J. de, Junien, C., Despoisse, S., and Zucker, J.-M. (1981). Retinoblastoma-del(13q14): report of two patients, one with a trisomic sib due to maternal insertion. Gene-dosage effect for esterase D. *Hum. Genet.* 59, 211–4.

Rivera, H., Rolón, A., Sánchez-Corona, J., and Cantú, J.M. (1985). *De novo* t(4;5)(q3100;q2200) with del(5)(q1500q2200). Tentative delineation of a 5q monosomy syndrome and assignment of the critical segment. *Clin. Genet.* 27, 105–9.

Robbins, K.C., Antoniades, H.N., Devare, S.G., Hunkapiller, M.W., and Aaronson, S.A. (1983). Structural and immunological similarities between simian sarcoma virus gene product(s) and human platelet-derived growth factor. *Nature, Lond.* 305, 605–8.

Roberts, M., and Ruddle, F.H. (1980). The Chinese hamster gene map. Assignment of four genes (*DTS, PGM2, GP6D, Enol*) to chromosome 2. *Exp. Cell Res.* 17, 47–54.

Robinson, A., Lubs, H.A., Nielsen, J., and Sorensen, K. (1979). Summary of clinical findings: profiles of children with 47,XXY,47,XXX and 47,XYY karyotypes. *Birth Defects: Orig. Art. Ser.* 15(1), 261–6.

Robinson, A., Bender, B., Borelli, J., Puck, M., and Salbenblatt, J. (1983). Sex chromosomal anomalies: prospective studies in children. *Behavioral Genet.* 13, 321–9.

Robison, L.L., Nesbit, M.E., Jr., Sather, H.N., Level, C., Shahidi, N., Kennedy, M.S., and Hammond, D. (1984). Down syndrome and acute leukemia in children: a 10-year retrospective survey from Children's Cancer Study Group. *J. Pediatr.* 105, 235–42.

Robson, E.B., and Meera Khan, P. (1982). Report of the committee on the genetic constitution of chromosomes 7, 8, and 9. *Cytogenet. Cell Genet.* 32, 144–52.

Roderick, T.H., and Davisson, M.T. (1981). Linkage map. In *Genetic Variants and Strains of the Laboratory Mouse*, ed. M.C. Green, pp. 279–82. Gustav Fischer Verlag, Stuttgart and New York.

Roderick, T.H., Lalley, P.A., Davisson, M.T., O'Brien, S.J., Womack, J.E., Créau-Goldberg, N., Echard, G., and Moore, K.L. (1984). Report of the committee on comparative mapping. *Cytogenet. Cell Genet.* 37, 312–39.

Rodewald, A., Zankl, M., Gley, E.-O., and Zang, K.D. (1980). Partial trisomie 5q: three different phenotypes. *Hum. Genet.* 55, 191–8.

Roehrdanz, R.L., and Lucchesi, J.C. (1981). An X chromosome locus in Drosophila melanogaster that enhances survival of the triplo-lethal genotype, *Dp-(Tpl). Dev. Genet.* 2, 147–58.

Rogers, J.F. (1984). Clinical delineation of proximal and distal partial 13q trisomy. *Clin. Genet.* 25, 221–9.

Rogers, P., Denegri, J.F., Thomas, J.W., Kalousek, D., Gillen, J., and Baker, M.A. (1978). Down's syndrome – leukemia vs pseudoleukemia. *Blood* 52 (Suppl. 1), 273.

Rogers, P.C., Kalousek, D.K., Denegri, J.F., Thomas, J.W., and Baker, M.A. (1983). Neonate with Down's syndrome and transient congenital leukemia. In vitro studies. *Am. J. Pediatr. Hematol. Oncol.* 5, 59–64.

Rohde, R.A. (1965). Congenital chromosomal syndromes. A model for pathogenesis. *Calif. Med.* 103, 249–53.

Rohde, R.A., and Berman, N. (1963). The Lyon hypothesis and further malformation postulates in the chromosomal syndromes. *Lancet* 2, 1169–70.

Rohde, R.A., Hodgman, J.E., and Cleland, R.S. (1964). Multiple congenital anomalies in the E_1-trisomy (group 16–18) syndrome. *Pediatrics* 33, 258–70.

Ropers. H.-H. and Wiberg, U. (1982). Evidence for X-linkage and noninactivation of steriod sulphatase locus in wood lemming. *Nature, Lond.* 296, 766–768.

Ropers, H.-H., Wolff, G., and Hitzeroth, H.W. (1978). Preferential X inactivation in human placenta membranes: is the paternal X inactive in early embryonic development of female mammals? *Hum. Genet.* 43, 265–73.

Ropers, H.-H., Migl, B., Zimmer, J., Fraccaro, M., Maraschio, P.P., and Westerveld, A. (1981). Activity of steroid sulfatase in fibroblasts with numerical and structural X chromosome aberrations. *Hum. Genet.* 57,354–6.

Ropper, A.H., and Williams, R.S. (1980). Relationship between plaques, tangles, and dementia in Down syndrome. *Neurology* 30, 639–44.

Rosenfeld, W., Verma, R.S., and Jhaveri, R.C. (1984). Cat-eye syndrome with unusual marker chromosome probably not chromosome 22. *Am. J. Med. Genet.* 18, 19–24.

Rosenthal, I.M., Bocian, M., and Krmpotic, E. (1972). Multiple anomalies including thymus aplasia associated with monosomy 22. *Pediatr. Res.* 6, 358.

Rosner, F., and Lee, S.L. (1972). Down's syndrome and acute leukemia: myeloblastic or lymphoblastic? *Am. J. Med.* 53, 203–18.

Rosner, F., Kozinn, P.J., and Jervis, G.A. (1973). Leukocyte function and serum immunoglobulins in Down's syndrome. *N.Y. State J. Med.* 73, 672–5.

Ross, J.D., Moloney, W.C., and Desforges, J.F. (1963). Ineffective regulation of granulopoiesis masquerading as congenital leukemia in a mongoloid child. *J. Pediatr.* 63, 1–10.

Ross, M.H., Galaburda, A.M., and Kemper, T.L. (1984). Down's syndrome: is there a decreased population of neurons? *Neurology* 34, 909–16.

Roth, M. (1982). Dementia in relation to aging in the central nervous system. *Aging* 18, 231–50.

Rowley, J.D. (1973). A new consistent chromosomal abnormality in chronic myelogenous leukemia identified by quinacrine fluorescence and Giemsa staining. *Nature, Lond.* 243, 290–3.

Rowley, J.D. (1975). Nonrandom chromosomal abnormalities in hematologic disorders of man. *Proc. Natl. Acad. Sci. USA* 72, 152–6.

Rowley, J.D. (1977). Mapping of human chromosomal regions related to neoplasia: Evidence from chromosomes 1 and 17. *Proc. Natl. Acad. Sci. USA* 74, 5729–33.

Rowley, J.D. (1981). Down syndrome and acute leukaemia: increased risk may be due to trisomy 21. *Lancet* 2, 1020–2.

Rowley, J.D. (1983a). Introduction to Part II. Chromosome pattern in animal and human tumors. In *Chromosomes and Cancer. From Molecules to Man*, ed. J.D. Rowley and J.E. Ultmann, pp. 57–60. Academic Press, New York.

Rowley, J.D. (1983*b*). Chromosome changes in leukemic cells as indicators of mutagenic exposure. In *Chromosomes and Cancer. From Molecules to Man*, ed. J.D. Rowley and J.E. Ultmann, pp. 139–59. Academic Press, New York.

Rowley, J.D. (1984). Biological implications of consistent chromosome rearrangements in leukemia and lymphoma. *Cancer Res.* 44, 3159–68.

Rowley, J.D., and Testa, J.R. (1982). Chromosome abnormalities in malignant hematologic diseases. *Adv. Cancer Res.* 36, 103–48.

Rowley, J.D., and Ultmann, J.E., ed. (1983). *Chromosomes and Cancer. From Molecules to Man*. Academic Press, New York.

Ruble, M.F., Bryant, E.M., Rabinovitch, P.S., and Hoehn, H. (1978). Growth kinetics and life spans of fibroblast cultures from patients with constitutional aneuploidy: effects of chromosome constitution. *J. Cell Biol.* 79, 393*a*.

Rudak, E., Jacobs, P.A., and Yanagimachi, R. (1978). Direct analysis of the chromosome constitution of human spermatozoa. *Nature, Lond.* 274, 911–13.

Rudd, N.L., and Teshima, I.E. (1983). The association of a lymphoreticular malignancy with an 11q deletion: a coincidence or a cancer susceptibility? *Hum. Genet.* 63, 323–6.

Rudd, N.L., Bain, H.W., Giblett, E., Chen, S.-H., and Worton, R.G. (1979). Partial trisomy 20 confirmed by gene dosage studies. *Am. J. Med. Genet.* 4, 357–64.

Ruppenthal, G.C., Caffery, S.A., Goodlin, B.L., Sackett, G.P., Vigufsson, N.V., and Peterson, V.G. (1983). Pigtailed macaques (*Macaca nemestrina*) with trisomy X manifest physical and mental retardation. *Am. J. Ment. Defic.* 87, 471–6.

Russell, L.B. (1976). Numerical sex-chromosome anomalies in mammals: their spontaneous occurrence and use in mutagenesis studies. In *Chemical Mutagens. Principles and Methods for their Detection, Vol. 4*, ed. A. Hollander, pp. 55–91. Plenum, New York and London.

Russell, L.B., and Chu, E.H.Y. (1961). An XXY male in the mouse. *Proc. Natl. Acad. Sci. USA* 47, 571–5.

Russell, W.L., Russell, L.B., and Grower, J.J. (1959). Exceptional inheritance of a sex-linked gene in the mouse explained on the basis that the *X/O* sex-chromosome constitution is female. *Proc. Natl. Acad. Sci. USA* 45, 554–60.

Rutishauser, U. (1984). Developmental biology of a neural crest adhesion molecule. *Nature, Lond.* 310, 549–54.

Rutten, F.J., Scheres, J.M.J.C., Hustinx, T.W.J., and ter Haar, B.G.A. (1974). A presumptive tetrasomy for the short arm of chromosome 9. *Humangenetik* 25, 163–70.

Ruutu, P., Ruutu, T., Vuopio, P., Kosunen, T.U., and de la Chapelle, A. (1977). Defective chemotaxis in monosomy 7. *Nature, Lond.* 265, 146–7.

Ryerse, J.S., and Nagel, B.A. (1984). Gap junction distribution in the *Drosophila* wing disc mutants *vg*, $l_{(2)}gd$, $l_{(3)}c43^{hs1}$, and $l_{(2)}gl^4$. *Dev. Biol.* 105, 396–403.

Sadler, J.R., and Novick, A. (1965). The properties of repressor and the kinetics of its action. *J. Mol. Biol.* 12, 305–27.

Sadovnick, A.D., and Baird, P.A. (1981). A cost-benefit analysis of prenatal detection of Down syndrome and neural tube defects in older mothers. *Am. J. Med. Genet.* 10, 367–78.

Sager, R. (1983). Genomic rearrangements and the origin of cancer. In *Chromosome Mutation and Neoplasia*, ed. J. German, pp. 333–46. A.R. Liss, New York.

St. Clair, D., and Blackwood, D. (1985). Premature senility in Down's syndrome. *Lancet* 2, 34.

Salin, M.L., and McCord, J.M. (1975). Free radicals and inflammation. Protection of phagocytosing leukocytes by superoxide dismutase. *J. Clin. Invest.* 56, 1319–23.

Sandberg, A.A. (1980). *The Chromosomes in Human Cancer and Leukemia.* Elsevier, New York.

Sandberg, A.A. (ed.) (1983*a*). *Cytogenetics of the Mammalian X Chromosome, Part A: Basic Mechanisms of X Chromosome Behavior.* A.R. Liss, New York.

Sandberg, A.A. (ed.) (1983*b*). *Cytogenetics of the Mammalian X Chromosome, Part B: X Chromosome Anomalies and Their Clinical Manifestations.* A.R. Liss, New York.

Sandberg, A.A. (1983*c*). A chromosomal hypothesis of oncogenesis. *Cancer Genet. Cytogenet.* 8, 277–85.

Sandison, A., Broadhead, D.M., and Bain, A.D. (1982). Elucidation of an unbalanced chromosome translocation by gene dosage studies. *Clin. Genet.* 22, 30–6.

Sandler, L., and Hecht, F. (1973). Genetic effects of aneuploidy. *Am. J. Hum. Genet.* 25, 332–9.

Sanger, R., Tippett, P., Gavin, J., Teesdale, P., and Daniels, G.L. (1977). Xg blood groups and sex chromosome abnormalities in people of northern European ancestry: an addendum. *J. Med. Genet.* 14, 210–13.

San Martin, V., Fernandez-Novoa, C., Hevia, A., Novales, A., Fornell, J., and Galera, H. (1981). Partial trisomy of chromosome 18 (pter→q11). A discussion of the identification of the critical segment. *Ann. Génét.* 24, 248–50.

Santesson, B., Böök, J.A., and Kjessler, B. (1973). The mortality of human XO embryos. *J. Reprod. Fertil.* 34, 51–5.

Sasaki, M.S., and Tonomura, A. (1969). Chromosomal radiosensitivity in Down's syndrome. *Jpn. J. Hum. Genet.* 14, 81–92.

Sassa, S., and Kappas, A. (1981). Genetic, metabolic, and biochemical aspects of the porphyrias. *Adv. Hum. Genet.* 11, 121–231.

Sassa, S., Zalar, G.L., Poh-Fitzpatrick, M.B., Anderson, K.E., and Kappas, A. (1982). Studies in porphyria. Functional evidence for a partial deficiency of ferrochelatase activity in mitogen-stimulated lymphocytes from patients with erythropoietic protoporphyria. *J. Clin. Invest.* 69, 809–15.

Sassaman, E.A. (1982). Biomedical aspects in Down syndrome: oncology. In *Down Syndrome. Advances in Biomedicine and the Behavioral Sciences.* ed. S.M. Pueschel and J.E. Rynders, pp. 237–41. Ware Press, Cambridge, Massachusetts.

Say, B., Barber, N., Bobrow, M., Jones, K., and Coldwell, J.G. (1976). Familial translocation (3p15p) with partial trisomy for the upper arm of chromosome 3 in two sibs. *J. Pediatr.* 88, 447–50.

Scarbrough, P.R., Hersh, J., Kukolich, M.K., Carroll, A.J., Finley, S.C., Hochberger, R., Wilkerson, S., Yen, F.F., and Althaus, B.W. (1984). Tetraploidy: a report of three live-born infants. *Am. J. Med. Genet.* 19, 29–37.

Scavarda, N.J., O'Tousa, J., and Pak, W.L. (1983). *Drosophila* locus with gene-dosage effects on rhodopsin. *Proc. Natl. Acad. Sci. USA* 80, 4441–5.

Schappert-Kimmijser, J., Hemmes, G.D., and Nijland, R. (1966). The heredity of retinoblastoma. *Ophthalmologica* 151, 197–213.

Schinzel, A. (1981*a*). Incomplete trisomy 22. II. Familial trisomy of the distal segment of chromosome 22q in two brothers from a mother with a translocation, t(6;22)(q27;q13). *Hum. Genet.* 56, 263–8.

Schinzel, A. (1981*b*). Incomplete trisomy 22. III. Mosaic trisomy 22 and the problem of full trisomy 22. *Hum. Genet.* 56, 269–73.

Schinzel, A. (1983). *Catalog of Unbalanced Chromosome Aberrations in Man.* de Gruyter, Berlin and New York.

Schinzel, A. (1984). Cyclopia and cebocephaly in two newborn infants with unbalanced segregation of a familial translocation rcp(1;7)(q32;q34). *Am. J. Med. Genet.* 18, 153–61.

Schinzel, A., and Tönz, O. (1979). Partial trisomy 7q and probable partial monosomy of 5p in the son of a mother with a reciprocal translocation between 5p and 7q. *Hum. Genet.* 53, 121–4.

Schinzel, A., Hayashi, K., and Schmid, W. (1976). Further delineation of the clinical picture of trisomy for the distal segment of chromosome 13. Report of three cases. *Hum. Genet.* 32, 1–12.

Schinzel, A., Schmid, W., and Mürset, G. (1974). Different forms of incomplete trisomy 13. Mosaicism and partial trisomy for the proximal and distal long arms. Report of three cases. *Humangenetik* 22, 287–98.

Schinzel, A., Schmid, W., Auf der Maur, P., Moser, H., Degenhardt, K.H., Geisler, M., and Grubisic, A. (1981*a*). Incomplete trisomy 22. I. Familial 11/22 translocation with 3:1 meiotic disjunction. Delineation of a common clinical picture and report of nine new cases from six families. *Hum. Genet.* 56, 249–62.

Schinzel, A., Schmid, W., Fraccaro, M., Tiepolo, L., Zuffardi, O., Opitz, J.M., Lindsten, J., Zetterqvist, P., Enell, H., Baccichetti, C., Tenconi, R., and Pagon, R.A. (1981*b*). The "cat eye syndrome": dicentric small marker chromosome probably derived from a no. 22 (tetrasomy 22pter→q11) associated with a characteristic phenotype. Report of 11 patients and delineation of the clinical picture. *Hum. Genet.* 57, 148–58.

Schlesselman, J.J. (1979). How does one assess the risk of abnormalities from human in vitro fertilization? *Am. J. Obstet. Gynecol.* 135, 135–48.

Schmickel, R.D., Silverman, E.M., Floyd, A.D., Payne, F.E., Pooley, J.M., and Beck, M.L. (1971). A live-born infant with 69 chromosomes. *J. Pediatr.* 79, 97–103.

Schmid, W. (1979). Trisomy for the distal third of the long arm of chromosome 19 in brother and sister. *Hum. Genet.* 46, 263–70.

Schmid, W., Müller, G., Schütz, G., and Gluecksohn-Waelsch, S. (1985). Deletions near the albino locus on chromosome 7 of the mouse affect the level of tyrosine aminotransferase mRNA. *Proc. Natl. Acad. Sci. USA* 82, 2866–9.

Schmutz, S.M., and Lin, C.C. (1983). Enzyme expression in tetraploid human fibroblasts. *Am. J. Hum. Genet.* 35, 53A.

Schnatterly, P., Bono, K.L., Robinow, M., Wyandt, H.E., Kardon, N., and Kelly, T.E. (1984). Distal 15q trisomy: phenotypic comparison of nine cases in an extended family. *Am. J. Hum. Genet.* 36, 444–51.

Schneider, E.L., and Epstein, C.J. (1972). Replication rate and lifespan of cultured fibroblasts in Down's syndrome. *Proc. Soc. Exp. Biol. Med.* 141, 1092–4.

Schneider, H.J., and Zellweger, H. (1968). Forme fruste of the Prader-Willi syndrome (HHHU) and balanced D/E translocation. *Helvet. Paediatr. Acta* 23, 128–36.

Schneider, J.A., Wong, V., Bradley, K., and Seegmiller, J.E. (1968). Biochemical comparisons of the adult and childhood forms of cystinosis. *N. Engl. J. Med.* 279, 1253–7.

Schochet, S.S. Jr., Lampert, P.W., and McCormick, W.F. (1973). Neurofibrillary tangles in patients with Down's syndrome: a light and electron microscopic study. *Acta Neuropathol.* 23, 342–6.

Scholl, T., Stein, Z., and Hansen, H. (1982). Leukemia and other cancers, anomalies and infections as causes of death in Down's syndrome in the United States during 1976. *Devel. Med. Child Neurol.* 24, 817–29.

Schroder, J.F., Suomalainen, H.A., Nikinmaa, B., Enns, C.A., Brown, J., Herzenberg, L.A., and Sussman, H.H. (1984). Chromosomal assignment of genes coding for human cell surface antigens with mouse/human lymphocyte hybrids. *Cytogenet. Cell Genet.* 37, 577–8.

Schuchman, E.H., Astrin, K.E., Aula, P., and Desnick, R.J. (1984). Regional assignment of the structural gene for human α-L-iduronidase. *Proc. Natl. Acad. Sci. USA* 81, 1169–73.

Schütten, H.J., Schütten, B.T., and Mikkelsen, M. (1978). Partial trisomy of chromosome 13. Case report and review of literature. *Ann. Génét.* 21, 95–9.

Schwab, M., Varmus, H.E., Bishop, J.M., Grzeschik, K.-H., Naylor, S.L., Sakaguchi, A.Y., Brodeur, G., and Trent, J. (1984). Chromosome localization in normal human cells and neuroblastomas of a gene related to c-*myc*. *Nature, Lond.* 308, 188–91.

Schwanitz, G., Schmid, P., Berthold, H.J., and Grosse, K.-P. (1978). Partial trisomy 13 with clinical signs of Pätau syndrome, resulting from a complex paternal rearrangement of chromosomes 6, 10, and 13. *Ann. Génét.* 21, 100–3.

Schwanitz, G., Zerres, K., Zang, K.D., and Schwinger, E. (1984). Partielle Trisomie 15q und partielle Monosomie 14q. *Therapiewoche* 34, 1348–54.

Schwartz, M., Duara, R., Haxby, J., Grady, C., White, B.J., Kessler, R.M., Kay, A.D., Cutler, N.R., and Rapoport, S.I. (1983). Down's syndrome in adults: brain metabolism. *Science* 221, 781–3.

Schwartz, S., Cohen, M.M., Panny, S., Beisel, J.H., and Vora, S. (1984). Duplication of chromosome 10p: confirmation of regional assignments of platelet-type phosphofructokinase and hexokinase I by gene-dosage studies. *Am. J. Hum. Genet.* 36, 750–9.

Schwartz, S., Beisel, J.H., Panny, S.R., and Cohen, M.M. (1985*a*). A complex rearrangement, including a deleted 8q, in a case of Langer-Giedion syndrome. *Clin. Genet.* 27, 175–82.

Schwartz, S., Max, S.R., Panny, S.R., and Cohen, M.M. (1985*b*). Deletions of proximal 15q and non-classical Prader-Willi syndrome phenotypes. *Am. J. Med. Genet.* 20, 255–63.

Schweber, M. (1985). A possible unitary genetic hypothesis for Alzheimer's disease and Down syndrome. *Ann. N.Y. Acad. Sci.* 450, 223–38.

Scoggin, C.H., Bleskan, J., Davidson, J.N., and Patterson, D. (1980). Gene expression of glycinamide ribonucleotide synthetase in Down syndrome. *Clin. Res.* 28, 31A.

Scoggin, C.H., Paul, S., Miller, Y.E., and Patterson, D. (1983). Two-dimensional electrophoresis of peptides from human-CHO cell hybrids containing human chromosome 21. *Somat. Cell Genet.* 9, 687–97.

Scola, P.S. (1982). Biomedical aspects in Down syndrome: musculoskeletal system. In *Down Syndrome. Advances in Biomedicine and the Behavioral Sciences*, ed. S.M. Pueschel and J.E. Rynders, pp. 213–9. Ware Press, Cambridge, Massachusetts.

Scott, B.S., Becker, L.E., and Petit, T.L. (1983). Neurobiology of Down's syndrome. *Prog. Neurobiol.* 21, 199–237.

Scott, B.S., Petit, T.L., Becker, L.E., and Edwards, B.A.V. (1982). Abnormal electric membrane properties of Down's syndrome DRG neurons in cell culture. *Dev. Brain Res.* 2, 257–70.

Scriver, C.R. (1983). Familial iminoglycinuria. In *The Metabolic Basis of Inherited Disease, 5th edit.*, ed. J.B. Stanbury, J.B. Wyngaarden, D.S. Frederickson, J.L. Goldstein, and M.S. Brown, pp. 1792–1803. McGraw-Hill, New York.

Searle, A. (1981). Chromosomal variants. In *Genetic Variants and Strains of the*

Laboratory Mouse, ed. M.C. Green, pp. 324–57. Gustav Fischer Verlag, Stuttgart and New York.

Searle, A.G., and Beechey, C.V. (1978). Complementation studies with mouse translocations. *Cytogenet. Cell Genet.* 20, 282–303.

Searle, A.G., Beechey, C.V., Evans, E.P., and Kirk, M. (1983). Two new X-autosome translocations in the mouse. *Cytogenet. Cell Genet.* 35, 279–92.

Second International Workshop on Chromosomes in Leukemia (1980). *Cancer Genet. Cytogenet.* 2, 89–113.

Segal, D.J., and McCoy, E.E. (1973). Studies on Down's syndrome in tissue culture. 1. Growth rates and protein contents of fibroblast cultures. *J. Cell Physiol.* 83, 85–90.

Segal, S., and Thier, S.O. (1983). Cystinuria. In *The Metabolic Basis of Inherited Disease, 5th edit.*, ed. J.B. Stanbury, J.B. Wyngaarden, D.S. Frederickson, J.S. Goldstein, and M.S. Brown, pp. 1774–91. McGraw-Hill, New York.

Seger, R., Buchinger, G., and Ströder, J. (1977). On the influence of age on immunity in Down's syndrome. *Eur. J. Pediatr.* 124, 77–87.

Seibel, N.L., Sommer, A., and Miser, J. (1984). Transient neonatal leukemoid reactions in mosaic trisomy 21. *J. Pediatr.* 104, 251–4.

Sellin, D., Schlizio, E., and Herbst, E.W. (1982). Immunokapazität trisomer Lymphozyten. Vergleichende Untersuchungen an Lymphozyten der Trisomie 12 und Trisomie 19 der Maus (Strahlenchimären) mit Stimulierung *in vitro* sowie Sensibilisierung *in vivo*. *Verh. Dtsch. Ges. Pathol.* 66, 261–4.

Serra, A., Arpaia, E., and Bova, R. (1978). Kinetics of 21-trisomic lymphocytes. I. In vitro response of 21-trisomic lymphocytes to PHA. *Hum. Genet.* 41, 157–67.

Serra, A., Bova, R., and Neri, G. (1981). Sensitivity of Down's syndrome lymphocyte DNA to the mutagenic action of busulfan. In *Trisomy 21. An International Symposium*, ed. G.R. Burgio, M. Fraccaro, L. Tiepolo, and U. Wolf, p. 254. Springer-Verlag, Berlin.

Serville, F., Junien, C., Kaplan, J.C., Gachet, M., Cadoux, J., and Broustet, A. (1978). Gene dosage effect for human triose phosphate isomerase and glyceraldehyde-3-phosphate dehydrogenase in partial trisomy 12p12 and trisomy 18p. *Hum. Genet.* 45, 63–9.

Shah, S.N. (1979). Fatty acid composition of lipids of human brain myelin and synaptosomes: changes in phenylketonuria and Down's syndrome. *Int. J. Biochem.* 10, 477–82.

Shani, M. (1985). Tissue-specific expression of rat myosin light-chain 2 gene in transgenic mice. *Nature, Lond.* 314, 283–6.

Shapiro, B.L. (1983*a*). Down syndrome – A disruption of homeostasis. *Am. J. Med. Genet.* 14, 241–69.

Shapiro, B.L. (1983*b*). Letter to the editor: Down syndrome. *Am. J. Med. Genet.* 16, 639.

Shapiro, I.M., Luka, J., Anderson-Anvret, M., and Klein, G. (1979). Relationship between Epstein-Barr virus genome number, amount of nuclear antigen, and early antigen inducibility in diploid and tetraploid lymphoma cells of related origin. *Intervirology* 12, 19–25.

Shapiro, L.J., and Mohandas, T. (1983). Noninactivation of X-chromosome loci in man. In *Cytogenetics of the Mammalian X Chromosome, Part A: Basic Mechanisms of X Chromosome Behavior*, ed. A.A. Sandberg, pp. 299–314. A.R.Liss, New York.

Shapiro, L.J., Mohandas, T., and Rotter, J.T. (1982). Dosage compensation at the steroid sulfatase (*STS*) locus. *Clin. Res.* 30, 119A.

Shapiro, L.R. (1983). Phenotypic expressions of numeric and structural X-chromosome abnormalities. In *Cytogenetics of the Mammalian X Chromosome, Part B: X Chromosome Anomalies and Their Clinical Manifestations*, ed. A.A. Sandberg, pp. 321–39, A.R. Liss, New York.

Shapiro, S.D., Hansen, K.L., and Littlefield, C.A. (1985). Non-mosaic partial tetrasomy and partial trisomy 9. *Am. J. Med. Genet.* 20, 271–6.

Share, J.B. (1976). Review of drug treatment for Down's syndrome persons. *Am. J. Ment. Defic.* 80, 388–93.

Shaw, P.A., and Gluecksohn-Waelsch, S. (1983). Epidermal growth factor and glucagon receptors in mice homozygous for a lethal chromosomal deletion. *Proc. Natl. Acad. Sci. USA* 80, 5379–82.

Sheppard, J.R., McSwigan, J.D., Wehner, J.M., White, J.G., Shows, T.B., Jakobs, K.H., and Schultz, G. (1982). The adrenergic responsiveness of Down syndrome cells. In *Membranes and Genetic Disease*, ed. J.R. Sheppard, V.E. Anderson, and J.W. Eaton, pp. 307–25. A.R. Liss, New York.

Sherman, L., Dafni, N., Lieman-Hurwitz, J., and Groner, Y. (1983). Nucleotide sequence and expression of human chromosome 21-encoded superoxide dismutase mRNA. *Proc. Natl. Acad. Sci. USA* 80, 5465–9.

Sherman, M.I., Paternoster, M.L., Eglitis, M.A., and McCue, P.A. (1983). Studies on the mechanism by which chemical inducers promote differentiation of embryonal carcinoma cells. In *Teratocarcinoma Stem Cells*, ed. L.M. Silver, G.M. Martin, and S. Strickland, pp. 83–95. Cold Spring Harbor Laboratory, Cold Spring Harbor, New York.

Shih, L.Y., Rosin, I., Suslak, L., Searle, B., and Desposito, F. (1982). Localization of the structural gene for galactose-1-phosphate uridyl transferase to band p13 of chromosome 9 by gene dosage studies. *Am. J. Hum. Genet.* 34, 62A.

Shih, V.E. (1978). Urea cycle disorders and other congenital hyperammonemic syndromes. In *The Metabolic Basis of Inherited Disease, 4th edit.*, ed. J.B. Stanbury, J.B. Wyngaarden, and D.S. Frederickson, pp. 362–86. McGraw-Hill, New York.

Shokeir, M.H.K. (1978). Complete trisomy 22. *Clin. Genet.* 14, 139–46.

Short, E.M., Conn, H.O., Snodgrass, P.J., Campbell, A.G.M., and Rosenberg, L.E. (1973). Evidence for X-linked dominant inheritance of ornithine transcarbamylase deficiency. *N. Engl. J. Med.* 288, 7–12.

Shulman, L.M., Kamarck, M.E., Slate, D.L., Ruddle, F.H., Branca, A.W., Baglioni, C., Maxwell, B.L., Gutterman, J., Anderson, P., Nagler, C., and Vilcek, J. (1984). Antibodies to chromosome 21 coded cell surface components block binding of human α interferon but not γ interferon to human cells. *Virology* 137, 422–7.

Shuttleworth, R.D., and O'Brien, J.R. (1981). Effect of age and operations on platelet serotonin (5HT) and platelet volume. *Thromb. Haemost.* 46, 198.

Sichitiu, S., Sinet, P.M., Lejeune, J., and Frézal, J. (1974). Surdosage de la forme dimérique de l'indophénoloxydase dans la trisomie 21, secondaire au surdosage génique. *Humangenetik* 23, 65–72.

Sidman, C.L., Marshall, J.D., Shultz, L.D., Gray, P.W., and Johnson, H.M. (1984). γ-Interferon is one of several direct B cell-maturing lymphokines. *Nature, Lond.* 309, 801–4.

Siegel, M. (1948a). Susceptibility of mongoloids to infection. I. Incidence of pneumonia, influenza A and *Shigella dysenteriae* (Sonne). *Am. J. Hyg.* 48, 53–62.

Siegel, M. (1948b). Susceptibility of mongoloids to infection. II. Antibody response to tetanus toxoid and typhoid vaccine. *Am. J. Hyg.* 48, 63–73.

Sills, J.A., Buckton, K.E., and Raeburn, J.A. (1976). Severe mental retardation in a boy with partial trisomy 10q and partial monosomy 2q. *J. Med. Genet.* 13, 507–10.

Silver, L.M., Martin, G.R., and Strickland, S. (eds.) (1983). *Cold Spring Harbor Conferences on Cell Proliferation, Vol. 10: Teratocarcinoma Stem Cells*. Cold Spring Harbor Laboratory, Cold Spring Harbor, New York.

Simpson, J.L. (1976). *Disorders of Sexual Differentiation*. Academic Press, New York.

Sinet, P.M. (1982). Metabolism of oxygen derivatives in Down's syndrome. *Ann. N.Y. Acad. Sci.* 396, 83–94.

Sinet, P.-M., Lejeune, J., and Jerome, H. (1979). Trisomy 21 (Down syndrome). Glutathione peroxidase, hexose monophosphate shunt and IQ. *Life Sci.* 24, 29–34.

Sinet, P.-M., Allard, D., Lejeune, J. and Jerome, H. (1974). Augmentation d'activité de la superoxyde dismutase érythrocytaire dans la trisomie pour le chromosome 21. *C.R. Séances Acad. Sci.* 278, 3267–70.

Sinet, P., Lavelle, F., Michelson, A.M., and Jerome, H. (1975). Superoxide dismutase activities of blood platelets in trisomy 21. *Biochem. Biophys. Res. Commun.* 67, 904–9.

Sinet, P.-M., Couturier, J., Dutrillaux, B., Poissonnier, M., Raoul, O., Rethoré, M.-O., Allard, D., Lejeune, J., and Jerome, H. (1976). Trisomie 21 et superoxyde dismutase–1 (IPO-A). Tentative de localisation sur la sous bande 21q22,1. *Exp. Cell Res.* 97, 47–55.

Sinet, P.-M., Bresson, J.-L., Couturier, J., Laurent, C., Prieur, M., Rethoré, M.-O., Taillemite, J.-L., Toudic, D., Jérome, H., and Lejeune, J. (1977). Localisation probable du gène de la glutathion reductase (EC 1.6.4.2.) sur la bande 8p21. *Ann. Génét.* 20, 13–17.

Sinex, F.M., and Merril, C.R. (ed.) (1982). Alzheimer's disease, Down's syndrome, and aging. *Ann. N.Y. Acad. Sci.* 396, 1–199.

Singer, H.S., Tiemeyer, M., Hedreen, J.C., Gearhart, J., and Coyle, J.T. (1984). Morphologic and neurochemical studies of embryonic brain development in murine trisomy 16. *Dev. Brain Res.* 15, 155–66.

Singh, L., and Jones, K.W. (1982). Sex reversal in the mouse (*Mus musculus*) is caused by a recurrent non-reciprocal crossover involving the X and an aberrant Y chromosome. *Cell* 28, 205–16.

Singh, R.P. (1970). Hygroma of the neck in XO abortuses. Report of two cases. *Am. J. Clin. Pathol.* 53, 104–7.

Singh, R.P., and Carr, D.H. (1966). The anatomy and histology of XO human embryos and fetuses. *Anat. Res.* 155, 369–75.

Skibsted, L., Westh, H., and Niebuhr, E. (1984). X long-arm deletions. A review of non-mosaic cases studied with banding techniques. *Hum. Genet.* 67, 1–5.

Sklower, S., Jenkins, E., Nolin, S., Warburton, D., Ane-Any Yaboa, K., Merkrebs, A., Schwartz, R., Wisniewski, K., Stimson, C., and Brown, T. (1982). Familial distal trisomy 14q. *Am. J. Hum. Genet.* 34, 145A.

Skovby, F., Krassikoff, N., and Francke, U. (1984). Assignment of the gene for cystathionine-β-synthase to human chromosome 21 in somatic cell hybrids. *Hum. Genet.* 65, 291–4.

Slack, J.M.W. (1983). *From Egg to Embryo. Determinative Events in Early Development*. Cambridge University Press, Cambridge.

Slamm, D.J., deKernion, J.B., Verma, I.M., and Cline, M.J. (1984). Expression of cellular oncogenes in human malignancies. *Science* 224, 256–62.

Slate, D.L., Shulman, L., Lawrence, J.B., Revel, M., and Ruddle, F.H. (1978). Presence of human chromosome 21 alone is sufficient for hybrid cell sensitivity to human interferon. *J. Virol.* 25, 319–25.

Slinde, S., and Hansteen I.L. (1982). Two chromosomal syndromes in the same family: monosomy and trisomy for part of the short arm of chromosome 10. *Eur. J. Pediatr.* 139, 153–7.

Smith, A., Broe, G.A., and Williamson, M. (1983). Chromosome aneuploidy in Alzheimer's disease. *Clin. Genet.* 24, 54–7.

Smith, A., Broe, G.A., and Williamson, M. (1984). Chromosome fragility in Alzheimer's disease. *Clin. Genet.* 25, 416–21.

Smith, C.A.B. (1979). Note on the forms of dermatoglyphic patterns. *Birth Defects: Orig. Art. Ser.* 15(6), 43–52.

Smith, D.W. (1977). Letter to the editor. *J. Pediatr.* 91, 346.

Smith, D.W. (1982). *Recognizable Patterns of Human Malformation. Genetic, Embryologic and Clinical Aspects. 3rd edit.* W.B. Saunders, Philadelphia.

Smith, G.F., and Berg, J.M. (1976). *Down's Anomaly.* Churchill Livingstone, Edinburgh.

Smith, G.F., Hoffman, W.L., Radunz, S.A., Singh, K., and Dowben, R.M. (1976). Studies of lymphocyte function in patients with Down's syndrome. *Excerpta Medica, International Congress Series No. 397. V. International Congress of Human Genetics,* p. 154.

Smith, G.F., Spiker, D., Peterson, C.P., Cicchetti, D., and Justice, P. (1984). Use of megadoses of vitamins and minerals in Down syndrome. *J. Pediatr.* 105, 228–34.

Smith, H.H., and Conklin, M.E. (1975). Effects of gene dosage on peroxidase isozymes in *Datura stramonium* trisomics. In *Isozymes III. Developmental Biology,* ed. C.L. Markert, pp. 603–18. Academic Press, New York.

Smith, J.C., and Wolpert, L. (1981). Pattern formation along the anteroposterior axis of the chick wing: the increase in width following a polarizing region graft and the effect of X-irradiation. *J. Embryol. Exp. Morphol.* 63, 127–44.

Snow, M.H.L. (1973). Tetraploid mouse embryos produced by cytochalasin B during cleavage. *Nature, Lond.* 244, 513–15.

Snow, M.H.L. (1975). Embryonic development of tetraploid mice during the second half of gestation. *J. Embryol. Exp. Morphol.* 34, 707–21.

Snow, M.H.L. (1976). The immediate postimplantation development of tetraploid mouse blastocysts. *J. Embryol. Exp. Morphol.* 35, 81–6.

Snow, M.H.L., Tam, P.P.L., and McLaren, A. (1981). On the control and regulation of size and morphogenesis in mammalian embryos. In *Levels of Genetic Control in Development,* ed. S. Subtelny and U.K. Abbott, pp. 201–17. A.R. Liss, New York.

Snyder, F.F., Hoo, J.J., Shearer, J.E., Heikkila, E.M., Rudd, N.L., and Lin, C.C. (1984). Gene dosage studies of inorganic pyrophosphatase (PP) and hexokinase (HK1) on human chromosome 10. *Cytogenet. Cell Genet.* 37, 588.

Solari, A.J. (1980). Synaptonemal complexes and associated structure in microspread human spermatocytes. *Chromosoma* 81, 315–37.

Solari, A.J., and Tres, L.L. (1970). The three-dimensional reconstruction of the XY synaptonemal pair in human spermatocytes. *J. Cell Biol.* 45, 43–53.

Somers, D.G., Pearson, M.L., and Ingles, C.J. (1975). Regulation of RNA polymerase II activity in a mutant rat myoblast cell line resistant to α-amanitin. *Nature, Lond.* 253, 372–4.

Somssich, J.E., Spira, J., Hameister, H., and Klein, G. (1982). Cytogenetic replication studies on murine T-cell leukemias with special consideration to chromosome 15. *Chromosoma* 86, 197–208.

Soos, M., Shade, M., Bell, H., Moxley, M., and Steel, C.M. (1981). Mapping by gene dosage, using aneuploid human lymphoid cell lines. *Ann. Hum. Genet.* 45,

169–79.

Sorbi, S., and Blass, J.P. (1983). Fibroblast phosphofructokinase in Alzheimer's disease and Down's syndrome. In *Banbury Report 15. Biological Aspects of Alzheimer's Disease*, ed. R. Katzman, pp. 297–307. Cold Spring Harbor Laboratory, Cold Spring Harbor, New York.

Sparkes, R.S., and Baughan, M.A. (1969). Blood cell enzymes in translocation Down's syndrome. *Am. J. Hum. Genet.* 21, 430–9.

Sparkes, R.S., and Weiss, M.C. (1973). Expression of differentiated functions in hepatoma cell hybrids: alanine aminotransferase. *Proc. Natl. Acad. Sci. USA* 70, 377–81.

Sparkes, R.S., Bass, H.N., and Sparkes, M.C. (1978). 10q(q23→qter) duplication: GOT_s, HK_1, and other gene markers. *Hum. Genet.* 42, 267–70.

Sparkes, R.S., Sparkes, M.C., and Crist, M. (1979). Expression of esterase D and other gene markers in trisomy 13. *Hum. Genet.* 52, 179–83.

Sparkes, R.S., Sparkes, M.C., Funderburk, S.J., and Moedjono, S. (1980a). Expression of GALT in 9p chromosome alterations: assignment of *GALT* locus to 9cen→9p22. *Ann. Hum. Genet.* 43, 343–7.

Sparkes, R.S., Sparkes, M.C., Wilson, M.G., Touner, J.W., Benedict, W., Murphree, A.L., and Yunis, J.J. (1980b). Regional assignment of genes for human esterase D and retinoblastoma to chromosome band 13q14. *Science* 208, 1042–4.

Sparkes, R.J., Murphree, A.L., Lingua, R.W., Sparkes, M.C., Field, L.L., Funderburk, S.J., and Benedict, W.F. (1983). Gene for hereditary retinoblastoma assigned to human chromosome 13 by linkage to esterase D. *Science* 219, 971–3.

Sperling, O., Boer, P., Brosh, S., Zoref, E., and de Vries, A. (1978). Overproduction disease in man due to enzyme feedback resistance mutation. Purine overproduction in gout due to excessive activity of mutant feedback-resistant phosphoribosylpyrophosphate synthetase. *Enzyme* 23, 1–9.

Spina, C.A., Smith, D., Korn, E., Fahey, J.L., and Grossman, H.J. (1981). Altered cellular immune functions in patients with Down's syndrome. *Am. J. Dis. Child.* 135, 251–5.

Spira, J., and Dofuku, R. (1980). Trisomy of chromosome 13 in spontaneous mammary tumors of GR, C3H, and noninbred Swiss mice (2 letters to editor). *J. Natl. Cancer Inst.* 65, 669–70.

Spira, J., Wiener, F., Ohno, S., and Klein, G. (1979). Is trisomy cause or consequence of murine T cell leukemia development? Studies on Robertsonian translocation mice. *Proc. Natl. Acad. Sci. USA* 76, 6619–21.

Spira, J., Babonits, M., Wiener, F., Ohno, S., Wirschubski, Z., Haran-Ghera, N., and Klein, G. (1980). Nonrandom chromosomal changes in Thy–1 positive and Thy–1 negative lymphomas induced by 7,12-dimethylbenzanthracene in SJL mice. *Cancer Res.* 40, 2609–16.

Spira, J., Wiener, F., Babonits, M., Gamble, J., Miller, J., and Klein, G. (1981). The role of chromosome 15 in murine leukemogenesis. I. Contrasting behavior of the tumor vs. normal parent-derived chromosomes no. 15 in somatic hybrids of varying tumorigenicity. *Int. J. Cancer* 28, 785–98.

Spritz, R.A., Emanuel, B.S., Chern, C.J., and Mellman, W.J. (1979). Gene dosage effect: intraband mapping of human soluble glutamic oxaloacetic transaminase. *Cytogenet. Cell Genet.* 23, 149–56.

Squire, J., Phillips, R.A., Boyce, S., Godbout, R., Rogers, B., and Gallie, B.L. (1984). Isochromosome 6p, a unique chromosomal abnormality in retinoblastoma:

verification by standard staining techniques, new densitometric methods, and somatic cell hybridization. *Hum. Genet.* 66, 46–53.

Squires, N., Aine, C., Buchwald, J., Norman, R., and Galbraith, G. (1980). Auditory brain stem response abnormalities in severely and profoundly retarded adults. *Electroenceph. Clin. Neurophysiol.* 50, 172–85.

Stacey, D.W., and Kung, H.-F. (1984). Transformation of NIH 3T3 cells by microinjection of Ha-*ras* p21 protein. *Nature, Lond.* 310, 508–11.

Stallings, R.L., and Siciliano, M.J. (1981). Confirmational, provisional, and/or regional assignment of 15 enzyme loci onto Chinese hamster autosomes 1, 2, and 7. *Somat. Cell Genet.* 7, 683–98.

Stamatoyannopoulos, G., and Nienhius, A.W. (eds.) (1983). *Globin Gene Expression and Hematopoietic Differentiation.* A.R. Liss, New York.

Stanbridge, E.J., Flandermeyer, R.R., Daniels, D.W., and Nelson-Rees, W.A. (1981). Specific chromosome loss associated with the expression of tumorigenicity in human cell hybrids. *Somat. Cell Genet.* 7, 699–712.

Stanbridge, E.J., Der, C.J., Doersen, C.-J., Nishimi, R.Y., Peehl, D.M., Weissman, B.E., and Wilkinson, J.E. (1982). Human cell hybrids: analysis of transformation and tumorigenicity. *Science* 215, 252–9.

Stanbury, J.B., Wyngaarden, J.B., and Frederickson, D.S. (eds.) (1978). *The Metabolic Basis of Inherited Disease, 4th edit.* McGraw-Hill, New York.

Stanbury, J.B., Wyngaarden, J.B., Frederickson, D.S., Goldstein, J.L., and Brown, M.S. (eds.) (1983). *The Metabolic Basis of Inherited Disease, 5th edit.* McGraw-Hill, New York.

Standen, G., Philip, M.A., and Fletcher, J. (1979). Reduced number of peripheral blood granulocytic progenitor cells in patients with Down's syndrome. *Br. J. Haematol.* 42, 417–23.

Stedman, D.J., and Eichorn, D.H. (1964). A comparison of the growth and development of institutionalized and home-reared mongoloids during infancy and early childhood. *Am. J. Ment. Defic.* 69, 391–401.

Steele, M.W. (1976). On the evolution of X-chromosome inactivation in mammals and the clinical consequences to man – a hypothesis. *Med. Hypotheses* 2, 195–8.

Steele, M.W., Pan, S., Mickell, J., and Senders, V. (1974). Trisomy for the distal half of the long arm of chromosome no. 18. *J. Pediatr.* 85, 827–9.

Steinbach, P., and Benz, R. (1983). Demonstration of gene dosage effects for AK$_3$ and GALT in fibroblasts from a fetus with 9p trisomy. *Hum. Genet.* 63, 290–1.

Steinherz, R., Tietze, F., Triche, T., Modesti, A., Gahl, W.A., and Schulman, J.D. (1982). Heterozygote detection in cystinosis, using leukocytes exposed to cystine dimethyl ester. *N. Engl. J. Med.* 306, 1468–70.

Stengel-Rutkowski, S., Murken, J.D., Frankenberger, R., Riechert, N., Spiess, H., Rodewald, A., and Stene, J. (1977). New chromosomal dysmorphic syndromes. 2. Trisomy 10p. *Eur. J. Pediatr.* 126, 109–25.

Stengel-Rutkowski, S., Warkotsch, A., Schimanek, P., and Stene, J. (1984). Familial Wolf's syndrome with a hidden 4p deletion by translocation of an 8p segment. Unbalanced inheritance from a maternal translocation (4;8)(p15.3;p22). Case report, review and risk estimates. *Clin. Genet.* 25, 500–21.

Stevenson, R.E., and Huntley, C.C. (1967). Congenital malformations in offspring of phenylketonuric mothers. *Pediatrics* 40, 33–45.

Stewart, B.R., and Merriam, J.R. (1975). Regulation of gene activity by dosage compensation at the chromosomal level in Drosophila. *Genetics* 79, 635–47.

Stewart, D.A., Netley, C.T., and Park, E. (1982). Summary of clinical findings with 47,XXY, 47,XYY, and 47,XXX karyotypes. *Birth Defects: Orig. Art. Ser.* 18(4), 1–5.

Stewart, G.D., Gallagher, J., Harris, P., Young, B.D., and Ferguson-Smith, M.A. (1984). Isolation and characterization of cloned DNA from a DNA bank enriched for human chromosome 21. *Cytogenet. Cell Genet.* 37, 268.

Stewart, J.M., Go, S., Ellis, E., and Robinson, A. (1970). Absent IgA and deletions of chromosome 18. *J. Med. Genet.* 7, 11–19.

Stewart, T.A., Wagner, E.F., and Mintz, B. (1982). Human β-globin gene sequences injected into mouse eggs, retained in adults, and transmitted to progeny. *Science* 217, 1046–8.

Stiehm, E.R., and Fudenberg, H.H. (1966). Serum levels of immune globulins in health and disease: a survey. *Pediatrics* 37, 715–27.

Stiller, C.A., and Kinnier Wilson, L.M. (1981). Down syndrome and leukaemia. *Lancet* 2, 1343.

Stoler, A., and Bouck, N. (1985). Identification of a single chromosome in the normal human genome essential for suppression of hamster cell transformation. *Proc. Natl. Acad. Sci. USA* 82, 570–4.

Stoll, C., Rethoré, M.-O., Laurent, C., and Lejeune, J. (1975). Le contretype de la maladie du cri du chat: la trisomie 5p. *Arch. Fr. Pediatr.* 32, 551–61.

Stolle, C.A., Myers, J.C., and Pyeritz, R.E. (1983). A structural defect in type III procollagen causes one form of Ehlers-Danlos type IV. *Am. J. Hum. Genet.* 35, 54A.

Storb, R., Thomas, E.D., Buckner, C.D., Cleft, R.A., Deeg, H.J., Fefer, A., Goodell, B.W., Sale, G.E., Sanders, J.E., Singer, J., Stewart, P., and Weiden, P.L. (1980). Marrow transplantation in thirty "untransfused" patients with severe aplastic anemia. *Ann. Int. Med.* 92, 30–6.

Strand, L.J., Felsher, B.F., Redeker, A.G., and Marver, H.S. (1970). Heme biosynthesis in intermittent acute porphyria: decreased hepatic conversion of porphobilinogen to porphyrins and increased delta aminolevulinic acid synthetase activity. *Proc. Natl. Acad. Sci. USA* 67, 1315–20.

Strand, L.J., Meyer, U.A., Felsher, B.F., Redeker, A.G., and Marver, H.S. (1972). Decreased red cell uroporphyrinogen I synthetase activity in intermittent acute porphyria. *J. Clin. Invest.* 51, 2530–6.

Strickland, S. (1981). Mouse teratocarcinoma cells: prospects for the study of embryogenesis and neoplasia. *Cell* 24, 277–8.

Strickland, S., Smith, K.K., and Marotti, K.R. (1980). Hormonal induction of differentiation in teratocarcinoma stem cells: generation of parietal endoderm by retinoic acid and dibutyryl cAMP. *Cell* 21, 347–55.

Strobel, R.J., Riccardi, V.M., Ledbetter, D.H., and Hittner, H. (1980). Duplication 11p11.3→14.1 due to meiotic crossing-over. *Am. J. Med. Genet.* 7, 15–20.

Strong, L.C., Riccardi, V.M., Ferrell, R.E. and Sparkes, R.S. (1981). Familial retinoblastoma and chromosome 13 deletion transmitted via an insertional translocation. *Science* 213, 1501–3.

Struwe, F. (1929). Histopathologische Untersuchungen über Entstehung und Wesen der senilen Plaques. *Z. Ges. Neurol. Psychiatry* 122, 291–307.

Sudharsan Raj, A., and Heddle, J.A. (1980). The effect of superoxide dismutase, catalase and L-cysteine on spontaneous and on mitomycin C induced chromosomal breakage in Fanconi's anemia and normal fibroblasts as measured by the micronucleus method. *Mutation Res.* 78, 59–66.

Suerinck, E., Suerinck, A., Kaplan, J.C., Meyer, J., Junien, C., Noel, B., and Rethoré, M.-O. (1978). Trisomy 12p par malségrégation d'une translocation paternelle t(12;22)(p11;p11). *Ann. Génét.* 21, 243–6.

Suetsuga, M., and Mehraein, P. (1980). Spine distribution along the apical dendrites of the pyramidal neurons in Down's syndrome. *Acta Neuropathol.* 50, 207–10.

Suh, H.W., Goforth, D.R., Cunningham, B.A., and Liang, G.H. (1977). Biochemical characterization of six trisomies of grain sorghum, *Sorghum bicolor* (L.) Moench. *Biochem. Genet.* 15, 611–19.

Sullivan, D.T., Kitos, R.J., and Sullivan, M.C. (1973). Developmental and genetic studies on kynurenine hydroxylase from *Drosophila melanogaster*. *Genetics* 75, 651–61.

Summerbell, D. (1983). The effect of local application of retinoic acid to the anterior margin of the developing chick limb. *J. Embryol. Exp. Morphol.* 78, 269–89.

Summitt, R.L. (1981). Chromosome specific segments that cause the phenotype of Down syndrome. In *Trisomy 21 (Down Syndrome). Research Perspectives*, ed. F.F. de la Cruz and P.S. Gerald, pp. 225–35. University Park Press, Baltimore.

Sutherland, G.R., Murch, A.R., Gardener, A.J., Carter, R.F., and Wiseman, C. (1976). Cytogenetic survey of a hospital for the mentally retarded. *Hum. Genet.* 34, 231–45.

Sutnick, A.I., London, W.T., and Blumberg, B.S. (1969). Effects of host and environment on immunoglobulins in Down's syndrome. *Arch. Int. Med.* 124, 722–5.

Swift, G.H., Hammer, R.E., MacDonald, R.J., and Brinster, R.L. (1984). Tissue-specific expression of the rat pancreatic elastase I gene in transgenic mice. *Cell* 38, 639–46.

Sylvester, P.E. (1974). Aortic and pulmonary valve fenestrations as aging indices in Down's syndrome. *J. Ment. Defic. Res.* 18, 367–76.

Sylvester, P.E. (1983). The hippocampus in Down's syndrome. *J. Ment. Defic. Res.* 27, 227–36.

Takagi, N. (1983). Cytogenetic aspects of X-chromosome inactivation in mouse embryogenesis. In *Cytogenetics of the Mammalian X Chromosome, Part A: Basic Mechanisms of X Chromosome Behavior*, ed. A.A. Sandberg, pp. 21–50. A.R. Liss, New York.

Takagi, N., and Sasaki, M. (1975). Preferential inactivation of the paternally derived X chromosome in the extraembryonic membranes of the mouse. *Nature, Lond.* 256, 640–2.

Takagi, N., and Sasaki, M. (1976). Digynic triploidy after superovulation in mice. *Nature, Lond.* 264, 278–81.

Takagi, N., Sugawara, O., and Sasaki, M. (1982). Regional and temporal changes in the pattern of X-chromosome replication during the early post-implanatation development of the female mouse. *Chromosoma* 85, 275–86.

Takashima, S., Becker, L.E., Armstrong, D.L., and Chan, F. (1981). Abnormal neuronal development in the visual cortex of the human fetus and infant with Down's syndrome: a quantitative and qualitative Golgi study. *Brain Res.* 225, 1–21.

Tam, C.F., and Walford, R.L. (1980). Alterations in cyclic nucleotides and cyclase-specific activities in T lymphocytes of aging normal humans and patients with Down's syndrome. *J. Immunol.* 125, 1665–70.

Tan, Y.H. (1976). Chromosome 21 and the cell growth inhibitory effect of human interferon preparations. *Nature, Lond.* 260, 141–3.

Tan, Y.H., Chou, E.L., and Lundh, N. (1975). Regulation of chromosome 21-directed antiviral gene(s) as a consequence of age. *Nature, Lond.* 257, 310–12.

Tan, Y.H., Tischfield, J., and Ruddle, F. (1973). The linkage of genes for the human interferon induced antiviral protein and indophenol oxidase-B traits to chromosome G–21. *J. Exp. Med.* 137, 317–30.

Tan, Y.H., Schneider, E.L., Tischfield, J., Epstein, C.J., and Ruddle, F.H. (1974). Human chromosome 21 dosage: effect on the expression of the interferon-induced antiviral state. *Science* 186, 61–3.

Tanaka, A., Nonoyama, M., and Glaser, R. (1977). Transcription of latent Epstein-Barr virus genomes in human epithelial/Burkitt hybrid cells. *Virology* 82, 63–8.

Tang, S.W., Berg, J., Bruni, J., and Davis, A. (1985). Decreased platelet [^{3}H] imipramine binding in Down's syndrome. *Soc. Neurosci. Abstr.* 11, 137.

Tangye, S.R. (1979). The EEG and incidence of epilepsy in Down's syndrome. *J. Ment. Defic. Res.* 23, 17–24.

Tarkowski, A.K., Witkowska, A., and Opas, J. (1977). Development of cytochalasin B-induced tetraploid and diploid/tetraploid mosaic mouse embryos. *J. Embryol. Exp. Morphol.* 41, 47–64.

Taub, R.A., Hollis, G.F., Hieter, P.A., Korsmeyer, S., Waldman, T.A., and Leder, P. (1983). Variable amplification of immunoglobulin λ light-chain genes in human popoulations. *Nature, Lond.* 304, 172–4.

Taylor, J.M., Dozy, A., Kan, Y.W., Varmus, H.E., Lie-Injo, L.E., Ganesan, J., and Todd, D. (1974). Genetic lesion in homozygous α thalassaemia (hydrops fetalis). *Nature, Lond.* 251, 392–3.

Taylor, K.M., Wolfinger, H.L., Brown, M.G., and Chadwick, D.L. (1975). Origin of a small metacentric chromosome: familial and cytogenetic evidence. *Clin. Genet.* 8, 364–9.

Taysi, K., Sparkes, R., O'Brien, T.J., and Dengler, D.R. (1982). Down's syndrome phenotype and autosomal gene inactivation in a child with presumed (X;21) de novo translocation. *J. Med. Genet.* 19, 144–8.

Taysi, K., Chao, W.T., Monaghan, N., and Monaco, M.P. (1983). Trisomy 6q22→6qter due to maternal 6;21 translocation. Case report and review of the literature. *Ann. Génét.* 26, 243–6.

Tchernia, G., Mohandas, N., and Shohet, S.B. (1981). Deficiency of skeletal membrane protein band 4.1 in homozygous hereditary elliptocytosis. Implications for erythrocyte membrane stability. *J. Clin. Invest.* 68, 454–60.

Tenconi, R., Baccichetti, Anglani, F., Pellegrino, P.-A., Kaplan, J.-C., and Junien, C. (1975). Partial deletion of the short arm of chromosome 12 (p11;p13). *Ann. Génét.* 18, 95–8.

Tenconi, R., Giorgi, P.L., Tarantino, E., and Formica, A. (1978). Trisomy 12p due to an adjacent 1 segregation of a maternal reciprocal translocation t(12;18)(p11;q23). *Ann. Génét.* 21, 229–33.

Ternaux, J.R., Mattei, J.F., Faudon, M., Barritt, M.C., Ardissone, J.P., and Giraud, F. (1979). Peripheral and central 5-hydroxytryptamine in trisomy 21. *Life Sci.* 25, 2017–22.

Teyssier, J.R., and Bajolle, F. (1980). Duplication-deficiency of chromosome 18, resulting from recombination of a paternal pericentric inversion, with a note for genetic counselling. *Hum. Genet.* 53, 195–200.

Teyssier, J.R., Bajolle, F., and Caron, J. (1981). Complete deletion of long arm of X chromosome in woman without Turner syndrome. *Lancet* 1, 1158–9.

Teyssier, J.R., Behar, C., Bajolle, F., and Potron, G. (1984). Selective involvement of cells carrying extra chromosome 21 in a child with acute non-lymphocytic leukaemia. *Lancet* 1, 290–1.

Thalhammer, O., Lubec, G., and Königshofer, H. (1979). Intracellular phenylalanine and tyrosine concentration in 19 heterozygotes for phenylketonuria (PKU) and 26 normals. Do the higher values in heterozygotes explain their lowered intellectual level? *Hum. Genet.* 49, 333–6.

Thalhammer, O., Havelec, L., Kroll, E., and Wehle, E. (1977). Intellectual level (IQ) in heterozygotes for phenylketonuria (PKU). Is the PKU gene also acting by means other than phenylalanine-blood level elevation? *Hum. Genet.* 38, 285–8.

Tharapel, S.A., Wilroy, R.S. Jr., and Lewandowski, R.C. (1984). Partial trisomies of chromosome 13. *Am. J. Hum. Genet.* 36, 77S.

Tharapel, A.T., Wilroy, R.S., Martens, P.R., Holbert, J.M., and Summitt, R.L. (1983). Diploid-triploid mosaicism: delineation of the syndrome. *Ann. Génét.* 26, 229–33.

Thase, M.E. (1982). Longevity and mortality in Down's syndrome. *J. Ment. Defic. Res.* 26, 177–92.

Therman, E. (1983). Mechanisms through which abnormal X-chromosome constitutions affect the phenotype. In *Cytogenetics of the Mammalian X Chromosome, Part B: X Chromosome Anomalies and their Clinical Manifestations*, ed. A.A. Sandberg, pp. 159–73. A.R. Liss, New York.

Therman, E. and Sarto, G.E. (1983). Inactivation center on the human X chromosome. In *Cytogenetics of the Mammalian X Chromosome, Part A: Basic Mechanisms of X Chromosome Behavior*, ed. A.A. Sandberg, pp. 315–25, A.R. Liss, New York.

Therman, E., Sarto, G.E., Distèche, C., and Denniston, C. (1976). A possible active segment on the inactive human X chromosome. *Chromosoma* 59, 137–45.

Therman, E., Denniston, C., Sarto, G.E., and Ulber, M. (1980). X chromosome constitution and the human female phenotype. *Hum. Genet.* 54, 133–43.

Third International Workshop on Chromosomes in Leukemia (1981). *Cancer Genet. Cytogenet.* 4, 95–142.

Tickle, C. (1980). The polarizing region and limb development. In *Development in Mammals, Vol. 4*, ed. M.H. Johnson, pp. 101–36. Elsevier/North Holland, Amsterdam.

Tickle, C. (1981). The number of polarizing region cells required to specify additional digits in the developing chick wing. *Nature, Lond.* 289, 295–8.

Tickle, C., Summerbell, D., and Wolpert, L. (1975). Positional signalling and specification of digits in chick limb morphogenesis. *Nature, Lond.* 254, 199–202.

Tickle, C., Shellswell, G., Crawley, A., and Wolpert, L. (1976). Positional signalling by mouse limb polarizing region in the chick wing bud. *Nature, Lond.* 259, 396–7.

Tickle, C., Alberts, B., Wolpert, L. and Lee, J. (1982). Local application of retinoic acid to the limb bud mimics the action of the polarizing region. *Nature, Lond.* 296, 564–6.

Timko, M.P., Vasconcelos, A.C., and Fairbrothers, D.E. (1980). Euploidy in *Ricinus*. I. Euploidy and gene dosage effects on cellular proteins. *Biochem. Genet.* 18, 171–83.

Tippett, P., Shaw, M.-A., Daniels, G.L., and Green, C.A. (1984). Family studies with anti-Xg^a and 12E7. *Cytogenet. Cell Genet.* 37, 595.

Todaro, G.J., and Martin, G.M. (1967). Increased susceptibility of Down's syndrome fibroblasts to transformation by SV40. *Proc. Soc. Exp. Biol. Med.* 124, 1232–6.

Tolksdorf, M., and Wiedemann, H.-R. (1981). Clinical aspects of Down's syndrome from infancy to adult life. In *Trisomy 21. An International Symposium*, ed. G.R. Burgio, M. Fraccaro, L. Tiepolo, and U. Wolf, pp. 3–31. Springer-Verlag, Berlin.

Tomaselli, M.B., John, K.M., and Lux, S.E. (1981). Elliptical erythrocyte membrane skeletons and heat-senstive spectrin in hereditary elliptocytosis. *Proc. Natl. Acad. Sci. USA* 78, 1911–15.

Tomkins, D.J., Hunter, A.G.W., Uchida, I., and Roberts, M.H. (1982). Two children with deletion of the long arm of chromosome 4 with breakpoint at band q33. *Clin. Genet.* 22, 348–55.

Tomkins, D.J., Gitelman, B.J., and Roberts, M.H. (1983). Confirmation of a de novo duplication, dup(10)(q24→q26), by *GOT1* gene dosage studies. *Hum. Genet.* 63, 369–73.

Tomkins, G.M., Gelehrter, T.D., Granner, D., Martin, D., Jr., Samuels, H.H., and Thompson, E.B. (1969). Control of specific gene expression in higher organisms. *Science* 166, 1474–80.

Toomey, K.E., Mohandas, T., Sparkes, R.S., Kaback, M.M., and Rimoin, D.L. (1978). Segregation of an insertional chromosome rearrangement in 3 generations. *J. Med. Genet.* 15, 382–7.

Toussi, T., Halal, F., Lesage, R., Delorme, F., and Bergeron, A. (1980). Renal hypodysplasia and unilateral ovarian agenesis in the penta-X syndrome. *Am. J. Med. Genet.* 6, 153–62.

Townes, P.L., White, M.R., Stiffler, S.J., and Goh, K. (1975). Double aneuploidy. Turner-Down syndrome. *Am. J. Dis. Child.* 129, 1062–5.

Tu, J.-B., and Zellweger, H. (1965). Blood-serotonin deficiency in Down's syndrome. *Lancet* 2, 715–17.

Tukey, R.H., Hannah, R.R., Negishi, M., Nebert, D.W., and Eisen, H.J. (1982). The *Ah* locus: correlation of intranuclear appearance of inducer-receptor complex with induction of cytochrome P_1-450 in RNA. *Cell* 31, 275–84.

Turleau, C., and Grouchy, J. de (1977). Trisomy 18qter and trisomy mapping of chromosome 18. *Clin. Genet.* 12, 361–71.

Turleau, C., Grouchy, J. de, and Klein, M. (1972). Phylogénie chromosomique de l'homme et des primates hominiens (*Pan troglodytes, Gorilla gorilla*, et *Pan pygmaeus*): essai de reconstitution du caryotype de l'ancêtre commun. *Ann. Génét.* 15, 225–40.

Turleau, C., Grouchy, J. de, Ponsot, G., and Bouygues, D. (1979). Monosomy 10qter. *Hum. Genet.* 47, 233–7.

Turleau, C., Chavin-Colin, F., Narbouton, R., Asensi, D., and Grouchy, J. de (1980). Trisomy 18q−. Trisomy mapping of chromosome 18 revisited. *Clin. Genet.* 18, 20–6.

Turleau, C., Chavin-Colin, F., Grouchy, J. de, Maroteaux, P., and Rivera, H. (1982). Langer-Giedion syndrome with and without del 8q. Assignment of critical segment to 8q23. *Hum. Genet.* 62, 183–7.

Turleau, C., Grouchy, J. de, Tournade, M.-F., Gagnadoux, M.-F., and Junien, C. (1984). Del 11p/aniridia complex. Report of three patients and review of 37 observations from the literature. *Clin. Genet.* 26, 356–62.

Uchigata, Y., Yamamoto, H., Kawamura, A., and Okamoto, H. (1982). Protection by superoxide dismutase, catalase, and poly(ADP-ribose) synthetase inhibitors against alloxan- and streptozotocin-induced islet DNA strand breaks and against inhibition of proinsulin synthesis. *J. Biol. Chem.* 257, 6084–8.

Ugazio, A.G. (1981). Down's syndrome: problems of immunodeficiency. In *Trisomy 21. An International Symposium*, ed. G.R. Burgio, M. Fraccaro, L. Tiepolo, and U. Wolf, pp. 33–9. Springer-Verlag, Berlin.

Ugazio, A.G., Lanzavecchia, A., Jayakar, S., Plebani, A., Duse, M., and Burgio, G.R. (1978). Immunodeficiency in Down's syndrome: titres of "natural" antibodies to *E. coli* and rabbit erythrocytes at different ages. *Acta Paediatr. Scand.* 67, 705–8.

van de Vooren, M.J., Planteydt, H.T., Hagemeijer, A., Peters-Slough, M.F., and Timmerman, M.J. (1984). Familial balanced insertion (5;10) and monosomy and trisomy (10) (q24.2→q25.3). *Clin. Genet.* 25, 52–8.

van der Hoeven, F.A., and Boer, P. de (1984). Personal communication. *Mouse News Letter* No. 70, 104.

Van Keuren, M.L., Merril, C.K., and Goldman, D. (1983). Proteins affected by chromosome 21 and ageing *in vitro.* In *Gene Expression in Normal and Transformed Cells*, ed. J.E. Celis and R. Bravo, pp. 349–78. Plenum, New York.

Veenema, H., Tasseron, E.W.K., and Geraedts, J.P.M. (1982). Mosaic tetraploidy in a male neonate. *Clin. Genet.* 22, 295–8.

Vianna-Morgante, A., Nozaki, M.J., Ortega, C.C., Coates, V., and Yamamura, Y. (1976). Partial monosomy and parital trisomy 18 in two offspring of carrier of pericentric inversion of chromosome 18. *J. Med. Genet.* 13, 366–70.

Vijayalaxmi, and Evans, H.J. (1982). Bleomycin-induced chromosomal aberrations in Down's syndrome lymphocytes. *Mutation Res.* 105, 107–13.

Vogel, F. (1973). Genotype and phenotype in human chromosome aberrations and in the minute mutants of *Drosophila melanogaster*. *Humangenetik* 19, 41–56.

Vogel, F. (1984). Clinical consequences of heterozygosity for autosomal-recessive diseases. *Clin. Genet.* 25, 381–415.

Vogel, F., and Motulsky, A.G. (1979). *Human Genetics. Problems and Approaches*. Springer-Verlag, Berlin.

Vogel, W., Trautman, T., Hörler, H., and Pentz, S. (1983). Cytogenetic and biochemical investigations on fibroblast cultures and clones with one and two active X chromosomes of a 69,XXY triploidy. *Hum. Genet.* 64, 246–8.

Vogel, W., Grompe, M., Storz, R., and Pentz, S. (1984). A comparative study on steroid sulfatase and arylsulfatase C in fibroblast clones from 45,X/47,XXX and 69,XXY. *Hum. Genet.* 66, 367–9.

Vora, S. (1982). Isozymes of phosphofructokinase. *Isozymes: Curr. Top. Biol. Med. Res.* 6, 119–67.

Vora, S., and Francke, U. (1981). Assignment of the human gene for liver-type–6-phosphofructokinase isozyme (*PFKL*) to chromosome 21 by using somatic cell hybrids and monoclonal anti-L antibody. *Proc. Natl. Acad. Sci. USA* 78, 3738–42.

Waddington, C.H. (1942). The canalization of development and the inheritance of acquired characters. *Nature, Lond.* 150, 563–5.

Wagner, T.E., Hoppe, P.C., Jollick, J.D., Scholl, D.R., Hodenka, R., and Gault, J.B. (1981). Microinjection of a rabbit β-globin gene into zygotes and subsequent expression in adult mice and their offspring. *Proc. Natl. Acad. Sci. USA* 78, 6376–80.

Wallace, M.E. (1976). A modifier mapped in the mouse. *Genetica* 46, 529.

Walford, R.L., Gossett, T.C., Naeim, F., Tam, C.F., Van Lancker, J.L., Barnett, E.V., Chia, D., Sparkes, R.S., Fahey, J.L., Spina, C., Gatti, R.A., Medici, M.A., Grossman, H., Hibrawi, H., and Motola, M. (1982). Immunological and biochemical studies of Down's syndrome as a model of accelerated aging. In *Immunological Aspects of Aging*, ed. D. Segre and L. Smith, pp. 479–532. Marcel Dekker, New York and Basel.

Walsh, K., and Koshland, D.E. Jr. (1985). Characterization of rate-controlling steps *in vivo* by use of an adjustable expression vector. *Proc. Natl. Acad. Sci. USA* 82, 3577–81.

Walzer, S., Gerald, P.S., Breau, G., O'Neill, D., and Diamond, L.K. (1966). Hematologic changes in the D_1 trisomy syndrome. *Pediatrics* 38, 419–29.

Wang, N., Soldat, L., Fau, S., Figenshau, S., Clayman, R., and Frayley, E. (1983). The consistent involvement of chromosome 3 and 6 aberrations in renal cell carcinoma. *Am. J. Hum. Genet.* 35, 73A.

Warburton, D. (1983). Parental age and X-chromosome aneuploidy. In *Cytogenetics of the Mammalian X Chromosome, Part B: X Chromosome Abnormalities and their Clinical Manifestations*, ed. A.A. Sandberg, pp. 23–33. A.R. Liss, New York.

Warburton, D., Stein, Z., Kline, S., and Susser, M. (1980). Chromosome abnormalities in spontaneous abortions: data from the New York City study. In *Human Embryonic and Fetal Death*, ed. I.H. Porter and E.B. Hook, pp. 261–8. Academic Press, New York.

Ward, B.E., Cook, R.H., Robinson, A., and Austin, J.H. (1979). Increased aneuploidy in Alzheimer disease. *Am. J. Med. Genet.* 3, 137–44.

Ward, J.C., Sharpe, C.R., Luthardt, F.W., Martens, P.R., and Palmer, C.G. (1983). Regional gene mapping of human β-glucuronidase region 7q11.23→7q21. *Am. J. Hum. Genet.* 35, 56A.

Ward, P., Packman, S., Loughman, W., Sparkes, M., Sparkes, R., McMahon, A., Gregory, T., and Ablin, A. (1984). Location of the retinoblastoma susceptibility gene(s) and the human esterase D locus. *J. Med. Genet.* 21, 92–5.

Wark, H.J., Overton, J.H., and Marian, P. (1983). The safety of atropine premedication in children with Down's syndrome. *Anesthesia* 38, 871–4.

Warkany, J. (1971). *Congenital Malformations. Notes and Comments*. Year Book, Chicago.

Warter, S., Ruch, J.-V., and Lehman, M. (1973). Karyotype with chromosomal abnormality with various inherited defects in the offspring (recombination aneusomy). *Humangenetik* 20, 355–9.

Washburn, L., and Eicher, E. (1975). Personal communication. *Mouse News Letter* No. 56, 43.

Watkins, P., Tanzi, R., Landes, G., Tricoli, J.V., Shows, T.B., and Gusella, J.H. (1984). DNA segments for human chromosome 21. *Cytogenet. Cell Genet.* 37, 281.

Watts, R.W.E., Perera, Y.S., Allsop, J., Newton, C., Platts-Mills, T.A.E., and Webster, A.D.B. (1979). Immunological and purine enzyme studies on hyperuricaemic and normouricaemic patients with Down's syndrome. *Clin. Exp. Immunol.* 36, 355–63.

Weatherall, D.J. (1978). The thalassemias. In *The Metabolic Basis of Inherited Disease, 4th edit.*, ed. J.B. Stanbury, J.B. Wyngaarden, and D.S. Frederickson, pp. 1508–23. McGraw-Hill, New York.

Weatherall, D.J., and Clegg, J.B. (1981). *The Thalassaemia Syndromes, 3rd edit.* Blackwell Scientific Publications, Oxford.

Weaver, D.D., Gartler, S.M., Boué, A., and Boué, I.G. (1975). Evidence for two active X chromosomes in a human XXY triploid. *Humangenetik* 28, 39–42.

Weber, F., Muller, H., and Sparkes, R. (1975). The 9p trisomy syndrome due to inherited translocation. *Birth Defects: Orig. Art. Ser.* 11(5), 201–4.

Webster, W.S., Walsh, D.A., McEwen, S.E., and Lipson, A.H. (1983). Some teratogenic properties of ethanol and acetaldehyde in C57Bl/6J mice: implications for the study of the fetal alcohol syndrome. *Teratology* 27, 231–43.

Weichselbaum, R.R., Nove, J., and Little J.B. (1980). X-ray sensitivity of fifty-three human diploid fibroblast cell strains from patients with characterized genetic disorders. *Cancer Res.* 40, 920–5.

Weil, J., and Epstein, C.J. (1979). The effect of trisomy 21 on the patterns of polypeptide synthesis in human fibroblasts. *Am. J. Hum. Genet.* 31, 478–88.

Weil, J., Epstein, L.B., and Epstein, C.J. (1980). Synthesis of interferon-induced polypeptides in normal and chromosome–21 aneuploid human fibroblasts: relationship to relative sensitivities in antiviral assays. *J. Interferon Res.* 1, 111–24.

Weil, J., Epstein, C.J., Epstein, L.B., van Blerkom, J., and Xuong, N.H. (1983a). Computer-assisted analysis demonstrates that polypeptides induced by natural and recombinant interferon-α are the same and that several have related primary structures. *Antiviral Res.* 3, 303–14.

Weil, J., Epstein, C.J., Epstein, L.B., Sedmak, J.J., Sabran, J., and Grossberg, S.E. (1983*b*). A unique set of polypeptides is induced by γ-interferon in addition to those induced in common with α- and β-interferons. *Nature, Lond.* 301, 437–9.

Weil, J., Tucker, G., Epstein, L.B., and Epstein, C.J. (1983*c*). Interferon induction of (2′–5′)oligoisoadenylate synthetase in diploid and trisomy 21 fibroblasts: relation to dosage of the interferon receptor gene (*IFRC*). *Hum. Genet.* 65, 108–11.

Weinberger, M.M., and Oleinick, A. (1970). Congenital marrow dysfunction in Down's syndrome. *J. Pediatr.* 77, 273–9.

Weisbrod, S. (1982). Active chromatin. *Nature, Lond.* 297, 289–95.

Weise, P., Koch, R., Shaw, K.N.F., and Rosenfeld, M.J. (1974). The use of 5-HTP in the treatment of Down's syndrome. *Pediatrics* 54, 165–8.

Weiss, G., Weick, R.F., Knobil, E., Wolman, S.R., and Gorstein, F. (1973). An X-O anomaly and ovarian dysgenesis in a rhesus monkey. *Folia Primat.* 19, 24–7.

Weiss, L. (1971). Additional evidence of gradual loss of germ cells in the pathogenesis of streak ovaries in Turner's syndrome. *J. Med. Genet.* 8, 540–4.

Weiss, M.C. (1977). The use of somatic cell hybridization to probe the mechanisms which maintain cell differentiation. In *Human Genetics*, ed. S. Armendares and R. Lisker, pp. 284–92. Excerpta Medica, Amsterdam.

Weiss, M.C. (1980). The analysis of cell differentiation by hybridization of somatic cells. *Results and Problems in Cell Differentiation* 11, 87–92.

Weiss, M.C., and Chaplain, M. (1971). Expression of differentiated functions in hepatoma cell hybrids: reappearance of tyrosine aminotransferase inducibility after the loss of chromosomes. *Proc. Natl. Acad. Sci. USA* 68, 3026–30.

Weiss, P.A. (1973). Differentiation and its three facets: facts, terms and meaning. *Differentiation* 1, 3–10.

Weiss, S.J., LoBuglio, A.F., and Kessler, H.B. (1980). Oxidative mechanisms of monocyte-mediated cytotoxicity. *Proc. Natl. Acad. Sci. USA* 77, 584–7.

Weissenbach, J., Geldwerth, D., Guellaeu, G., Fellous, M., and Bishop, C. (1984). Sequences homologous to the human Y chromosome detected in the female genome. *Cytogenet. Cell. Genet.* 37, 604.

Welshons, W.J., and Russell, L.B. (1959). The Y-chromosome as the bearer of male determining factors in the mouse. *Proc. Natl. Acad. Sci. USA* 45, 560–6.

Wenger, S.L., and Steele, M.W. (1981). Meiotic consequences of pericentric inversions of chromosome 13. *Am. J. Med. Genet.* 9, 275–83.

Wenstrup, R.J., Hunter, A., and Byers, P.M. (1983). Abnormality in pro α(2)I chain of type I collagen in a form of osteogenesis imperfecta (OI). *Am. J. Hum. Genet.* 35, 57A.

Wertelecki, W., Graham, J.M. Jr., and Sergovich, F.R. (1976). The clinical syndrome of triploidy. *Obstet. Gynecol.* 47, 69–76.

West, J.D., Frels, W.I., Chapman, V.M., and Papaioannou, V.E. (1977). Preferential expression of the maternally derived X chromosome in the mouse yolk sac. *Cell* 12, 873–82.

Westerveld, A., and Naylor, S. (1984). Report of the committee on the genetic constitution of chromosomes 18, 19, 20, 21, and 22. *Cytogenet. Cell Genet.* 37, 155–75.

Westerveld, A., Visser, R.P.L.S., Freeke, M.A., and Bootsma, D. (1972). Evidence for linkage of 3-phosphoglycerate kinase, hypoxanthine-guanine-phosphoribosyl transferase, and glucose 6-phosphate dehydrogenase loci in Chinese hamster cells studied by using a relationship between gene multiplicity and enzyme activity. *Biochem. Genet.* 7, 33–40.

Wetterberg, L., Gustavson, K.-H., Bäckström, M., Ross, S.B., and Frödén, U.

(1972). Low dopamine-β-hydroxylase activity in Down's syndrome. *Clin. Genet.* 3, 152–3.

Whalley, L.B., and Buckton, K.E. (1979). Genetic factors in Alzheimer's disease. In *Alzheimer's Disease. Early Recognition of Potentially Reversible Deficits*, ed. A.I.M. Glen and L.J. Whalley, pp. 36–41. Churchill Livingstone, Edinburgh.

White, A.G., da Costa, A.J., and Darg, C. (1973). Determination of HL-A2 gene dosage effect on lymphocytes by release of radioactive chromium. *Tissue Antigens* 3, 123–9.

White, B.J., Tjio, J.-H., Van de Water, L.C., and Crandall, C. (1972). Trisomy for the smallest autosome of the mouse and identification of the T1Wh translocation chromosome. *Cytogenetics* 11, 363–78.

White, B.J., Tjio, J.-H., Van de Water, L.C., and Crandall, C. (1974*a*). Trisomy 19 in the laboratory mouse. I. Frequency in different crosses at specific developmental stages and relationship of trisomy to cleft palate. *Cytogenet. Cell Genet.* 13, 217–31.

White, B.J., Tjio, J.-H., Van de Water, L.G., and Crandall, C. (1974*b*). Trisomy 19 in the laboratory mouse. II. Intra-uterine growth and histological studies of trisomics and their normal littermates. *Cytogenet. Cell Genet.* 13, 232–45.

White, B.J., Crandall, C., Goudsmit, J., Morrow, C.M., Alling, D.W., Gajdusek, D.C., and Tjio, J.-H. (1981). Cytogenetic studies of familial and sporadic Alzheimer disease. *Am. J. Med. Genet.* 10, 77–89.

Whittingham, S., Sharma, D.L.B., Pitt, D.B., and Mackay, I.R. (1977). Stress deficiency of the T-lymphocyte system exemplified by Down syndrome. *Lancet* 1, 163–6.

Wiener, F., Spira, J., Ohno, S., Haran-Ghera, N., and Klein, G. (1979). Nonrandom duplication of chromosome 15 in murine T-cell leukemias induced in mice heterozygous for translocation T(14;15)6. *Int. J. Cancer* 23, 504–7.

Wiener, F., Babonits, M., Spira, J., Bregula, U., Klein, G., Merwin, R.M., Asofsky, R., Lynes, M., and Haughton, G. (1981). Chromosome 15 trisomy in spontaneous and carcinogen-induced murine thymomas of B-cell origin. *Int. J. Cancer* 27, 51–8.

Wiener, F., Babonits, M., Spira, J., Klein, G., and Bazin, H. (1982*a*). Nonrandom chromosomal changes involving chromosomes 6 and 7 in spontaneous rat immunocytomas. *Int. J. Cancer* 29, 431–7.

Wiener, F., Spira, J., Babonits, M., and Klein, G. (1982*b*). Non-random duplication of chromosome 15 in T-cell leukemias induced in mice heterozygous for reciprocal and Robertsonian translocations. *Int. J. Cancer* 30, 479–87.

Wiener, F., Babonits, M., Bregula, U., Klein, G., Léonard, A., Wax, J.S., and Potter, M. (1984*a*). High resolution banding analysis of the involvement of strain BALB/c- and AKR-derived chromosomes no. 15 in plasmocytoma-specific translocations. *J. Exp. Med.* 159, 276–91.

Wiener, G., Ohno, S., Babonits, M., Sümergi, J., Wirschubsky, Z., Klein, G., Mushinski, J.F., and Potter, M. (1984*b*). Hemizygous interstitial deletion of chromosome 15 (band D) in three translocation-negative murine plasmocytomas. *Proc. Natl. Acad. Sci. USA* 81, 1159–63.

Wiktor-Jedrzejezak, W., Sharkis, S., Ahmed, A., and Sell, K.W. (1977). Theta-sensitive cell and erythropoiesis: identification of a defect in W/W^v anemic mice. *Science* 196, 313–5.

Willecke, K., and Schafer, R. (1984). Human oncogenes. *Hum. Genet.* 66, 132–42.

Williams, L.T., and Lefkowitz, R.J. (1978). *Receptor Binding Studies in Adrenergic Pharmacology*. Raven, New York.

Williamson, R.A., Donlan, M.A., Dolan, C.R., Thuline, H.C., Harrison, M.T., and Hall, J.G. (1981). Familial insertional translocation of a portion of 3q into 11q resulting in duplication and deletion of region 3q22.1→q24 in different offspring. *Am. J. Med. Genet.* 9, 105–11.

Willing, M.C., Niehuis, A.W., and Anderson, W.F. (1979). Selective activation of human β- but not γ-globin gene expression in human fibroblast × mouse erythroleukemia cell hybrids. *Nature, Lond.* 277, 534–8.

Wilroy, R.S., Summitt, R.L., Martens, P., and Gooch, W.M. (1977). Partial monosomy and partial trisomy for different segments of chromosome 13 in several individuals of the same family. *Ann. Génét.* 20, 237–42.

Wilson, G.N., Raj, A., and Baker, D. (1985). The phenotype and cytogenetic spectrum of partial trisomy 9. *Am. J. Med. Genet.* 20, 277–82.

Wilson, M.G., Schroeder, W.A., Graves, D.A., and Kach, V.D. (1967). Hemoglobin variations in D-trisomy syndrome. *N. Engl. J. Med.* 277, 953–8.

Wilson, M.G., Towner, J.W., Forsman, I., and Siris, E. (1979). Syndromes associated with deletion of the long arm of chromosome 18[del(18q)]. *Am. J. Med. Genet.* 3, 155–74.

Wilson, M.G., Towner, J.W., Coffin G.S., Ebbin, A.J., Siris, E., and Brager, P. (1981). Genetic and clinical studies in 13 patients with the Wolf-Hirschhorn syndrome [del(4p)]. *Hum. Genet.* 59, 297–307.

Wilson, W.G., Wyandt, H.E., and Shah, H. (1983). Interstitial deletion of 8q. Occurrence in a patient with multiple exostoses and unusual facies. *Am. J. Dis. Child.* 137, 444–8.

Winking, H., and Gropp, H. (1978). Developmental pattern of triplication of chromosome 17 and association of trisomy with T^{Hp}. *Hereditas* 89, 150.

Winsor, E.J.T., and Welch, J.P. (1983). Prader-Willi syndrome associated with inversion of chromosome 15. *Clin. Genet.* 24, 456–61.

Winter, J.H., Bennett, B., Watt, J.L., Brown, T., San Román, C., Schinzel, A., King, J., and Cook, P.S.L. (1982). Confirmation of linkage between antithrombin III and Duffy blood group and assignment of *AT3* to 1q22→q25. *Ann. Hum. Genet.* 46, 29–34.

Wirschubsky, Z., Wiener, F., Spira, J., Sümegi, J., and Klein, G. (1984). Triplication of one chromosome no. 15 with an altered c-*myc* containing EcoRI fragment and elimination of the normal homologue in a T-cell lymphoma line of AKR origin (TIKAUT). *Int. J. Cancer* 33, 477–81.

Wisniewski, K.E., Laure-Kamionowska, M., and Wisniewski, H.M. (1984). Evidence of arrest of neurogenesis and synaptogenesis in brains of patients with Down's syndrome. *N. Engl. J. Med.* 311, 1187–8.

Wisniewski, K.E., Laure-Kamionowska, M., and Wisniewski, H.M. (1985). Morphometric studies on the cortex during postnatal brain development in Down syndrome. Submitted for publication.

Wisniewski, K.E., Wisniewski, H.M., and Wen, G.Y. (1985). Occurrence of neuropathological changes and dementia of Alzheimer's disease in Down's syndrome. *Ann. Neurol.* 17, 278–82.

Wisniewski, K., Howe, J., Gwyn Williams, D., and Wisniewski, H.M. (1978). Precocious aging and dementia in patients with Down's syndrome. *Biol. Psychiatry* 13, 619–27.

Wisniewski, K., Dambska, M., Jenkins, E.C., Sklower, S., and Brown, W.T. (1983). Monosomy 21 syndrome: further delineation including clinical, neuropathological, cytogenetic and biochemical studies. *Clin. Genet.* 23, 102–10.

Wisniewski, K.E., Laure-Kamionowska, M., Connell, F., and Wisniewski,H.M.

(1985). Synaptogenesis in Down syndrome visual cortex during postnatal brain development. Submitted for publication.

Wisniewski, L., Chan, R., and Higgins, J.V. (1978). Partial trisomy 2q and familial translocation t(2;18)(q31;p11). *Hum. Genet.* 45, 25–8.

Wisniewski, L., Politis, G.D., and Higgins, J.V. (1978). Partial tetrasomy 9 in a liveborn infant. *Clin. Genet.* 14, 147–53.

Wisniewski, L., Hassold, T., Heffelfinger, J., and Higgins, J.V. (1979). Cytogenetic and clinical studies in five cases of inv dup(15). *Hum. Genet.* 50, 259–70.

Witschi, E. (1951). Embryogenesis of the adrenal and the reproductive glands. *Recent Prog. Horm. Res.* 6, 1–27.

Wolf, B., and Rosenberg, L.E. (1978). Heterozygote expression in propionyl coenzyme A carboxylase deficiency. Differences between major complementation groups. *J. Clin. Invest.* 62, 931–6.

Wolf, U., Fraccaro, M., Mayerova, A., Hecht, T., Maraschio, P., and Hameister, H. (1980). A gene controlling H-Y antigen on the X chromosome. Tentative assignment by deletion mapping to Xp223. *Hum. Genet.* 54, 149–54.

Wolfe, J., Erickson, R.P., Rigby, P.W., Jr., and Goodfellow, P.N. (1984). Regional localization of 3 Y-derived sequences on the human X and Y chromosomes. *Ann. Hum. Genet.* 48, 253–9.

Wolfe, L.C., John, K.M., Falcone, J.C., Byrne, A.M., and Lux, S.E. (1982). A genetic defect in the binding of protein 4.1 to spectrin in a kindred with hereditary spherocytosis. *N. Engl. J. Med.* 307, 1367–74.

Wolman, S.R. (1984). Cytogenetics and cancer. *Arch. Pathol. Lab. Med.* 108, 15–19.

Wolpert, L. (1978). Pattern formation and the development of the chick limb. *Birth Defects: Orig. Art. Ser.* 14(2), 547–59.

Wolpert, L., and Hornbruch, A. (1981). Positional signalling along the anteriorposterior axis of the chick wing. The effect of multiple polarizing region grafts. *J. Embryol. Exp. Morphol.* 63, 145–59.

Wright, T.C., Orkin, R.W., Destrempes, M., and Kurnit, D.M. (1984). Increased adhesiveness of Down syndrome fetal fibroblasts *in vitro*. *Proc. Natl. Acad. Sci. USA* 81, 2426–30.

Wróblewska, J. (1971). Developmental anomaly in the mouse associated with triploidy. *Cytogenetics* 10, 199–207.

Wurster-Hill, D.H., Benirschke, K., and Chapman, D.I. (1983). Abnormalities of the X chromosome in mammals. In *Cytogenetics of the Mammalian X Chromosome, Part B: X Chromosome Anomalies and their Clinical Manifestations*, ed. A.A. Sandberg, pp. 283–300. A.L. Liss, New York.

Wyngaarden, J.B., and Kelley, W.N. (1978). Gout. In *The Metabolic Basis of Inherited Disease, 4th edit.*, ed. J.B. Stanbury, J.B. Wyngaarden, and D.S. Frederickson, pp. 916–1010. McGraw-Hill, New York.

Wyss, D., DeLozier, C.D., Daniell, J., and Engel, E. (1982). Structural anomalies of the X chromosome: personal observation and review of nonmosaic cases. *Clin. Genet.* 21, 145–59.

Xue, Q.-M., Shen, D.-G., and Dong, W. (1984). ATPase activity of erythrocyte membrane in patients with trisomy 21 (Down's syndrome). *Clin. Genet.* 26, 429–32.

Yachie, A., Yokoi, T., Uwadana, N., Nagaoki, T., Miyawaki, T., and Taniguchi, N. (1984). Plasma zinc status and T-cell immune aberrations in institutionalized patients with Down's syndrome: a trial of restoration. Submitted for publication.

Yamada, K.M., Olden, K., and Hahn, L-.H.E. (1980). Cell surface protein and cell interactions. In *The Cell Surface: Mediator of Developmental Processes*, ed. S. Subtelny and N.K. Wessells, pp. 43–77. Academic Press, New York.

Yamamato, M., and Watanabe, G. (1979). Epidemiology of gross chromosomal anomalies at the early embryonic stages of pregnancy. In *Epidemiologic Methods for Detection of Teratogens. Contributions to Epidemiology and Biostatistics, Vol. 1*, ed. M.A. Klingberg and J.A.C. Weatherall, pp. 101–6. S. Karger, Basel.

Yamamoto, M., Ito, T., Watanabe, M., and Watanabe, G. (1981). Estimated prevalence of chromosome anomalies in the first trimester of the Japanese pregnant population. *Tohoku J. Exp. Med.* 135, 109–13.

Yates, C.M., Simpson, J., Maloney, A.J.F., Gordon, A., and Reid, A.H. (1980). Alzheimer-like cholinergic deficiency in Down syndrome. *Lancet* 2, 979.

Yates, C.M., Ritchie, I.M., Simpson, J., Maloney, A.J.F., and Gordon, A. (1981). Noradrenaline in Alzheimer-type dementia and Down syndrome. *Lancet* 2, 39–40.

Yatziv, S., Erickson, R.P., and Epstein, C.J. (1977). Mild and severe Hunter syndrome (MPS II) within the same sibship. *Clin. Genet.* 11, 319–26.

Yielding, K.L. (1967). Chromosome redundancy and gene expression: an explanation for trisomy abnormalities. *Nature, Lond.* 214, 613–14.

Yokoyama, M., Ball, C., Lou, K., and Alepa, F.P. (1967). Immunogenetic studies on mongolism. *Am. J. Ment. Defic.* 71, 597–601.

Yoshida-Noro, C., Suzuki, N., and Takeichi, M. (1984). Molecular nature of the calcium-dependent cell-cell adhesion system in mouse teratocarcinoma and embryonic cells studied with a monoclonal antibody. *Dev. Biol.* 101, 19–27.

Yoshimitsu, K., Hatano, S., Kobayashi, Y., Takeoka, Y., Hayashidani, M., Ueda, K., Nomura, K., Ohama, K., and Usui, T. (1983). A case of 21q− syndrome with half normal SOD–1 activity. *Hum. Genet.* 64, 200–2.

Yotti, L.P., Glover, T.W., Trosko, J.E., and Segal, D.J. (1980). Comparative study of X-ray and UV induced cytotoxicity, DNA repair, and mutagenesis in Down's syndrome and normal fibroblasts. *Pediatr. Res.* 14, 88–92.

Young, D. (1971). The susceptibility to SV40 virus transformation of fibroblasts obtained from patients with Down's syndrome. *Eur. J. Cancer* 7, 337–9.

Young, R.S., Reed, T., Hodes, M.E., and Palmer, C.G. (1982). The dermatoglyphic and clinical features of the 9p trisomy and partial 9p monosomy syndromes. *Hum. Genet.* 62, 31–9.

Young, R.S., Weaver, D.D., Kukolich, M.K., Heerema, N.A., Palmer, C.G., Kawira, E.L., and Bender, H.A. (1984). Terminal and interstitial deletions of the long arm of chromosome 7: a review with five new cases. *Am. J. Med. Genet.* 17, 437–50.

Yu, C.W., Chen, H., and Fowler, M. (1983). Specific terminal DNA replication sequence of X chromosomes in different tissues of a live-born triploid infant. *Am. J. Med. Genet.* 14, 501–11.

Yunis, E., Ramirez, E., and Uribe, J.G. (1980). Full trisomy 7 and Potter syndrome. *Hum. Genet.* 54, 13–18.

Yunis, J.J. (1965). Interphase deoxyribonucleic acid condensation, late deoxyribonucleic acid replication, and gene inactivation. *Nature, Lond.* 205, 311–12.

Yunis, J.J. (1981). Chromosomes and cancer: new nomenclature and future directions. *Hum. Pathol.* 12, 494–503.

Yunis, J.J. (1983). The chromosomal basis of human neoplasia. *Science* 221, 227–36.

Yunis, J.J., and Lewandowski, R.C. (1983). High resolution cytogenetics. *Birth Defects: Orig. Art. Ser.* 19(5), 11–37.

Zabel, B.V., and Baumann, W.A. (1982). Langer-Giedion syndrome with interstitial 8q deletion. *Am. J. Med. Genet.* 11, 353–8.

Zadeh, T.M., Funderburk, S.J., Carrel, R., and Dumars, K. (1981). 9p duplication confirmed by gene dosage effect. Report of two patients. *Ann. Génét.* 24, 242–4.

Zaletajev, D.V., and Marincheva, G.S. (1983). Langer-Giedion syndrome, in a child with complex structural aberration of chromosome 8. *Hum. Genet.* 63, 178–82.

Zang, K.D. (1982). Cytological and cytogenetical studies on human meningioma. *Cancer Genet. Cytogenet.* 6, 249–74.

Zavodny, P.J., Roginski, R.S., and Skoultchi, A.I. (1983). Regulated expression of human globin genes and flanking DNA in mouse erythroleukemia – human cell hybrids. In *Globin Gene Expression and Hematopoietic Differentiation*, ed. G. Stamatoyannopoulos and A.W. Nienhius, pp. 53–62. A.R. Liss, New York.

Zech, L., Gahrton, G., Hammarström, L., Juliusson, G., Mellstedt, H., Robert, K.H., and Smith, C.J.F. (1984). Inversion of chromosome 14 marks human T-cell chronic lymphocytic leukaemia. *Nature, Lond.* 308, 858–60.

Zuckerkandl, E. (1978). Multilocus enzymes, gene regulation and genetic sufficiency. *J. Mol. Evol.* 12, 57–89.

Index

abnormalities: correction in trisomic mouse fetuses, 232; mechanisms involved in aneuploidy-associated postmorphogenetic phenotypic, 376; receptors and genetic, 108–12

aborted fetuses, chromosomal abnormalities in spontaneously and therapeutically, 3, 5

abortion, aneuploidy and spontaneous, 5–6

acetylcholinesterase activity, 137–8·

activation in interspecific hybrid cells, 129, 131

aging, Alzheimer disease and trisomy 21 and, 309–11

ALA; *see* aminolevulinic acid (ALA)

alleles, suppression of second set of, 82

Alzheimer disease, 263, 300, 306–11

aminolevulinic acid (ALA), porphyrias and synthesis of, 96

AMP, *see* cyclic AMP

aneuploid cells, regulatory disturbances in, 135–40, 375

aneuploid embryos (human), death of, 3

aneuploidy: abnormalities from, 86; acquired, 351, 353, 355–65; altered cell-cycle hypothesis and, 175–7; autosomal (in mouse), 237–8, 241–6; causes of phenotypic changes associated with, 114; cell hybrid investigations and, 134–5; cell replication hypothesis and, 174–7; characters that remain unaltered in phenotype of double, 39; combined sex chromosome and autosomal, 23; defined, 9; dermoglyphic changes and, 170–2; development and, 181; direct and indirect secondary effects of, 86; economic burden of, 5; effects of double, 28, 38, 39, 50; effects of X-chromosomal, 324; eukaryotic regulation changes and, 113–14; frequency at conception, 3; full vs. segmental, 33–5, 38, 81–2; gene dosage effects and, 65, 66, 76, 259–73, 283; genesis of, 235; imbalances due to, 100; incidence at age 40, 5; interactive effects of combination

of segmental, 370; malignancy and acquired, 6; malignancy and constitutional, 346–7, 350; mechanisms responsible for abnormalities associated with, 76, 112, 373, 376; mechanistic basis for effects of, 6–8, 373; mechanism as focus of research on, 377; models for study of effects of, 378; mental retardation and, 5; morbidity and, 6; negative effects of, 143–5; nervous system and, 299–300; nonspecific effects of, 174–204, 278, 372; primary and secondary effects of, 85, 86; reasons for studying effects of, 379–80; receptor-mediated phenomena and, 103–4, 106–8, 374, 376; reductionist approach to understanding effects of, 7–8; regulatory systems and, 117, 143–5; relationships among mechanisms underlying phenotypic features of, 376; segmental, 23, 25, 28–9, 31–5, 38, 50, 81–2, 210, 220–1, 370; specificity of effects of, 39; syndromes from very small amounts of, 52; synergistic effect in double, 32–3; teratocarcinoma cells for studying, 233–4; *see also names of specific aneuploid conditions*

animal models, 157–8, 201, 207, 235–6, 379; for chromosome-mediated gene transfer, 248–9; nonhuman primates as, 236–7; with single extra copies of particular genes, 248; for study of autosomal aneuploidy, 237–8, 241–6; for study of X-chromosomal aneuploidy, 244–6; for trisomy 21, 311–13, 315–23; *see also names of specific animals*

Aniridia–Wilms tumor, 47

antienzymes: deficiencies of, 97; increase in, 98

antimongolism, 159

antisyndrome concept, 159–60, 161, 165, 167–70, 172–3

antithrombin III, excess of, 98

antitrypsin: abnormal increase in, 98; deficiency of, 97

477

DATE DUE